# Homogeneous Catalysis with Metal Phosphine Complexes

*Edited by*
## Louis H. Pignolet
University of Minnesota
Minneapolis, Minnesota

PLENUM PRESS • NEW YORK AND LONDON

Library of Congress Cataloging in Publication Data

Main entry under title:

Homogeneous catalysis with metal phosphine complexes.

(Modern inorganic chemistry)
Includes bibliographies and index.
1. Catalysis. 2. Phosphine. 3. Complex compounds. I. Pignolet, Louis H., 1943– .
II. Series.
QD505.H65   1983                    660.2′995                    83-17609
ISBN 0-306-41211-X

©1983 Plenum Press, New York
A Division of Plenum Publishing Corporation
233 Spring Street, New York, N.Y. 10013

All rights reserved

No part of this book may be reproduced, stored in a retrieval system, or transmitted in any form or by any means, electronic, mechanical, photocopying, microfilming, recording, or otherwise, without written permission from the Publisher

Printed in the United States of America

# Homogeneous Catalysis with Metal Phosphine Complexes

# MODERN INORGANIC CHEMISTRY

Series Editor: John P. Fackler, Jr.
*Texas A&M University*

---

METAL INTERACTIONS WITH BORON CLUSTERS
Edited by Russell N. Grimes

HOMOGENEOUS CATALYSIS
WITH METAL PHOSPHINE COMPLEXES
Edited by Louis H. Pignolet

THE JAHN–TELLER EFFECT AND
VIBRONIC INTERACTIONS IN MODERN CHEMISTRY
I. B. Bersuker

---

A Continuation Order Plan is available for this series. A continuation order will bring delivery of each new volume immediately upon publication. Volumes are billed only upon actual shipment. For further information please contact the publisher.

# *Contributors*

**Dr. Alan Balch** • Department of Chemistry, University of California, Davis California

**Dr. John M. Brown** • Dyson Perrins Laboratory, South Parks Road, Oxford, England

**Dr. Penny A. Chaloner** • Department of Chemistry, Rutgers University, Piscataway, New Jersey

**Dr. Joseph Chatt** • Department of Chemistry, School of Chemistry and Molecular Sciences, The University of Sussex, Brighton, Sussex, England

**Dr. Robert H. Crabtree** • Sterling Chemistry Laboratory, Yale University, New Haven, Connecticut

**Dr. Daniel H. Doughty** • Exploratory Chemistry, Div. 8315, Sandia National Laboratories, Livermore, California

**Dr. Jack W. Faller** • Department of Chemistry, Yale University, New Haven, Connecticut

**Dr. T. Adrian George** • Department of Chemistry, University of Nebraska-Lincoln, Nebraska

**Dr. Bálint Heil** • University of Chemical Industries, Institute of Organic Chemistry, Hungarian Academy of Sciences, Research Group for Petrochemistry, Veszprém, Shönherz Zoltán u. 12, Hungary

**Dr. Norman L. Holy** • Department of Chemistry, Western Kentucky University, Bowling Green, Kentucky

**Dr. James A. Ibers** • Department of Chemistry, Northwestern University, Evanston, Illinois

**Nancy L. Jones** • Department of Chemistry, Northwestern University, Evanston, Illinois

**Dr. László Markó** • University of Chemical Industries, Institute of Organic Chemistry, Hungarian Academy of Sciences, Research Group for Petrochemistry, Veszpŕem, Shõnherz Zoltán u. 12, Hungary

**Dr. Devon W. Meek** • Department of Chemistry, Ohio State University, Columbus, Ohio

**Dr. Louis H. Pignolet** • Department of Chemistry, University of Minnesota, Minneapolis, Minnesota

**Dr. Thomas B. Rauchfuss** • School of Chemistry, University of Illinois, Urbana-Champaign, Illinois

**Dr. D. Max Roundhill** • Department of Chemistry, Tulane University, New Orleans, Louisiana

**Dr. Alan R. Sanger** • Alberta Research Council, 11315-87 Avenue, Edmonton, Alberta, Canada

**Dr. Chad A. Tolman** • Central Research and Development Department, Experimental Station, E. I. du Pont de Nemours and Company, Wilmington, Delaware

**Dr. Szilárd Törös** • University of Chemical Industries, Institute of Organic Chemistry, Hungarian Academy of Sciences, Research Group for Petrochemistry, Veszpŕem, Schõnherz Zoltán u. 12, Hungary

# *Preface*

The field of transition metal catalysis has experienced incredible growth during the past decade. The reasons for this are obvious when one considers the world's energy problems and the need for new and less energy-demanding syntheses of important chemicals. Heterogeneous catalysis has played a major industrial role; however, such reactions are generally not selective and are exceedingly difficult to study. Homogeneous catalysis suffers from on-site engineering difficulties; however, such reactions usually provide the desired selectivity. For example, Monsanto's synthesis of optically-active amino acids employs a chiral homogeneous rhodium diphosphine catalyst.

Industrial uses of homogeneous catalyst systems are increasing. It is not by accident that many homogeneous catalysts contain tertiary phosphine ligands. These ligands possess the correct steric and electronic properties that are necessary for catalytic reactivity and selectivity. This point will be emphasized throughout the book. Thus the stage is set for a comprehensive treatment of the many ways in which phosphine catalyst systems can be designed, synthesized, and studied.

The book is intended to provide an up-to-date treatise by experts in the field of homogeneous catalysis with metal phosphine complexes. Reaction mechanisms are emphasized, and the fine tuning of catalyst reactivity and selectivity is thoroughly discussed. The techniques and methodology that are required to determine catalytic reaction pathways are comprehensively described, and numerous case studies are presented. The reader will gain insight into the subject's many facets ranging from the historical development of the field to the newest techniques in catalyst design and study.

# Contents

*Chapter 1*

**Historical Introduction**

*Joseph Chatt*

| | |
|---|---|
| 1. The Phosphines | 1 |
|    1.1. The Complexes | 3 |
|    1.2. Catalysis | 8 |
|    References | 10 |

*Chapter 2*

**Mechanistic Studies of Catalytic Reactions Using Spectroscopic and Kinetic Techniques**

*Chad A. Tolman and Jack W. Faller*

| | |
|---|---|
| 1. Introduction | 13 |
|    1.1. The Nature of Catalyst Systems | 14 |
|    1.2. Mechanisms, Rate Laws, and Dominant Species | 16 |
|    1.3. The Use of Isotopic Techniques | 22 |
|    1.4. The Implications of Product Stereochemistry on Mechanisms | 34 |
|    1.5. Distinguishing Radical Pathways | 43 |
| 2. Applications Illustrating the Methods | 49 |
|    2.1. Butene Isomerization by $NiL_4$ and $H_2SO_4$ | 49 |
|    2.2. Olefin Hydrogenation with Wilkinson's Catalyst | 53 |
|    2.3. The Nickel Catalyzed Cyclooligomerization of Butadiene | 64 |
|    2.4. Rhodium Catalyzed Olefin Hydroformylation | 81 |
|    References | 102 |

*Chapter 3*

**Structurally-Characterized Transition-Metal Phosphine Complexes of Relevance to Catalytic Reactions**

Nancy L. Jones and James A. Ibers

| | |
|---|---|
| 1. Introduction | 111 |
| 2. Alkene Hydrogenation | 112 |
| 3. Hydroformylation | 116 |
| 4. Olefin Oligomerization | 122 |
| 5. Metallacycles | 126 |
| 6. Summary | 129 |
| References | 132 |

*Chapter 4*

**Asymmetric Hydrogenation Reactions Using Chiral Diphosphine Complexes of Rhodium**

John M. Brown and Penny A. Chaloner

| | |
|---|---|
| 1. Introduction | 137 |
| 2. The Variation of Reaction Conditions | 139 |
|    2.1. Changing the Catalyst | 139 |
|    2.2. Changing the Substrate | 143 |
|    2.3. Changing the Solvent | 148 |
|    2.4. Changing the Pressure | 149 |
|    2.5. Summary of the Features of Effective Catalysts | 149 |
| 3. Studies of Reaction Mechanism | 149 |
|    3.1. Reaction Rates | 151 |
|    3.2. X-ray Structural Analysis of Catalysts and Reaction Intermediates | 151 |
|    3.3. Structure and Dynamics in Solution | 153 |
|    3.4. Conclusions | 159 |
| 4. The Origin of Stereoselectivity | 160 |
| References | 163 |

*Chapter 5*

**Binuclear, Phosphine-Bridged Complexes: Progress and Prospects**

Alan L. Balch

| | |
|---|---|
| 1. Introduction | 167 |
| 2. Structural Aspects | 170 |

| | |
|---|---|
| 3. Nuclear Magnetic Resonance Spectroscopic Probes | 183 |
| 4. Reactivity Patterns for Bridged, Binuclear Complexes | 188 |
|    4.1. Insertions of Small Molecules into Metal-Metal Bonds | 189 |
|    4.2. Introduction of a Bridging Ligand with Contraction of the Metal-Metal Separation | 193 |
|    4.3. Loss of Carbon Monoxide Accompanied by Carbon Monoxide Bridge Formation | 195 |
|    4.4. Two-Center, Two-Fragment Oxidative Addition | 197 |
|    4.5. Two-Center, Three-Fragment Oxidative Addition | 200 |
|    4.6. The Formation or Stabilization of Other Novel Bridging Ligands | 201 |
|    4.7. Capture of a Second Metal Induced by 2(Diphenylphosphino)pyridine | 204 |
| 5. Demonstrated Catalytic Activity of Binuclear Compounds | 206 |
| 6. Concluding Remarks | 208 |
|    References | 210 |

*Chapter 6*

### Hydrogenation and Hydroformylation Reactions Using Binuclear Diphosphine-Bridged Complexes of Rhodium

*Alan R. Sanger*

| | |
|---|---|
| 1. Introduction | 216 |
| 2. Hydrogenation of Alkenes or Alkynes | 217 |
|    2.1. Complexes of $Ph_2PCH_2PPh_2$ | 218 |
|    2.2. Complexes of Diphosphines with Longer Chain Lengths | 225 |
| 3. Hydroformylation of Alkenes | 227 |
| 4. Concluding Remarks | 233 |
|    References | 235 |

*Chapter 7*

### Functionalized Tertiary Phosphines and Related Ligands in Organometallic Coordination Chemistry and Catalysis

*Thomas B. Rauchfuss*

| | |
|---|---|
| 1. Introduction | 239 |
| 2. Ether Phosphines | 240 |
| 3. Aminophosphines | 244 |
| 4. Carbonylphosphines | 247 |
| 5. Alkenylphosphines | 251 |
| 6. Cyclopentadienylphosphines | 253 |

| 7. Concluding Remarks | 254 |
|---|---|
| References | 255 |

## Chapter 8

### Polydentate Ligands and Their Effects on Catalysis

*Devon W. Meek*

| 1. Introduction | 257 |
|---|---|
| 2. Advantages of Chelating Polydentate Ligands | 258 |
| 3. Methods for Syntheses of Polydentate Ligands | 261 |
|     3.1. Grignard and Alkali-Metal Reagents | 261 |
|     3.2. Other Synthetic Methods | 266 |
| 4. Polyphosphine Homogeneous Catalysts | 274 |
|     4.1. Hydrogenation Catalysis with Chelating Triphosphine Ligands | 275 |
|     4.2. Hydroformylation Catalysis with Chelating Diphosphine Ligands | 279 |
| 5. Solid-Supported Polyphosphine Catalysts | 285 |
|     5.1. Coordination of Multiple Phosphine Groups from Supported Monophosphines | 285 |
|     5.2. A Triphosphine Ligand Attached to Glass Beads | 287 |
|     5.3. A $NP_2$ Ligand Attached to Organic Compounds | 288 |
|     5.4. Chelating Diphosphines Attached to Organic Polymers | 288 |
| 6. Concluding Remarks | 289 |
| References | 293 |

## Chapter 9

### Cationic Rhodium and Iridium Complexes in Catalysis

*Robert H. Crabtree*

| 1. Introduction | 297 |
|---|---|
| 2. The Effects of Net Ionic Charge | 298 |
|     2.1. Effects of the Reactivity of Bound Ligands | 298 |
|     2.2. Effects on the Types of Ligand Bound | 299 |
|     2.3. Effects on Redox Properties | 301 |
|     2.4. Effects on Solubility and Catalyst Separation | 301 |
|     2.5. Counter Ion Effects | 302 |
|     2.6. Mechanistic Effects | 302 |
|     2.7. Conclusion | 303 |

3. Synthetic Methods . . . . . . . . . . . . . . . . . . . . . 303
   3.1. Halide Abstraction . . . . . . . . . . . . . . . . . . 303
   3.2. Protonation . . . . . . . . . . . . . . . . . . . . . . 304
   3.3. Electrophilic Attack at a Ligand . . . . . . . . . . 304
   3.4. Redox . . . . . . . . . . . . . . . . . . . . . . . . . 304
4. Cationic Rhodium and Iridium Complexes in Catalysis . . . 304
   4.1. Hydrogenation of C=C Groups . . . . . . . . . . . 304
   4.2. Hydrogenation of C=O Groups . . . . . . . . . . . 310
   4.3. Alkane Activation . . . . . . . . . . . . . . . . . . 310
   4.4. Hydroformylation . . . . . . . . . . . . . . . . . . 313
   4.5. Decarbonylation . . . . . . . . . . . . . . . . . . . 313
   4.6. Water Gas Shift (WGS) . . . . . . . . . . . . . . . 313
   4.7. Polymerization . . . . . . . . . . . . . . . . . . . 314
5. Concluding Remarks . . . . . . . . . . . . . . . . . . . . 314
   References . . . . . . . . . . . . . . . . . . . . . . . . . 314

## Chapter 10

### Hydrogenation Reactions of CO and CN Functions Using Rhodium Complexes

*Bálint Heil, László Markó, and Szilárd Törös*

1. Introduction . . . . . . . . . . . . . . . . . . . . . . . . . 317
2. Hydrogenation of Aldehydes . . . . . . . . . . . . . . . . 318
3. Hydrogenation of Ketones Not Containing Other Functional Groups . . . . . . . . . . . . . . . . . . . . . . . . . . . . 319
   3.1. Catalytic Activity and Stereoselectivity . . . . . . . 319
   3.2. Enantioselective Hydrogenation of Ketones . . . . . 324
4. Hydrogenation of Ketones Containing Other Functional Groups 329
   4.1. Selective Hydrogenation of Unsaturated Ketones . . . 329
   4.2. Hydrogenation of Keto Acids and Keto Esters . . . . 331
   4.3. Hydrogenation of Aminoketones and Other Biologically Active Derivatives . . . . . . . . . . . . . . . . . . 333
5. Homogeneous Hydrogenation of Carbon-Nitrogen Double Bonds . . . . . . . . . . . . . . . . . . . . . . . . . . . . 335
6. Abbreviations for Ligand Names . . . . . . . . . . . . . . 337
   References . . . . . . . . . . . . . . . . . . . . . . . . . 338

*Chapter 11*

**Decarbonylation Reactions Using Transition Metal Complexes**

Daniel H. Doughty and Louis H. Pignolet

1. Introduction . . . . . . . . . . . . . . . . . . . . . 343
2. Discussion of Decarbonylation Mechanism with $RhCl(PPh_3)_3$ . 347
    2.1. Stoichiometric Decarbonylation of Acid Chlorides . . . 347
    2.2. Stoichiometric Decarbonylation of Aldehydes . . . . . 352
    2.3. Catalytic Decarbonylation of Acid Chlorides and Aldehydes . . . . . . . . . . . . . . . . . . 355
3. Catalytic Decarbonylation of Aldehydes with Cationic Diphosphine Complexes of Rh(I) . . . . . . . . . . . . . . . 358
    3.1. Discussion of Mechanism with $[Rh(P-P)_2]^+$ Complexes . 362
4. Catalytic Decarbonylation of Aldehydes Using Cationic Diphosphine Complexes of Ir(I) . . . . . . . . . . . . . . 369
5. Additional Studies with Bis-Chelate Complexes of Rh(I) . . 371
6. Decarbonylation of Benzoylchloride with $[Rh(dppp)_2]^+$ . . . 372
    References . . . . . . . . . . . . . . . . . . . . . 372

*Chapter 12*

**Homogeneous Catalysis of Oxidation Reactions Using Phosphine Complexes**

D. Max Roundhill

1. Significance of Metal-Catalyzed Oxidation Reactions . . . . 377
    1.1. Mechanistic Features of Metal-Catalyzed Oxidations . . 378
2. Transition Metal Phosphine Oxygen Complexes . . . . . . 378
    2.1. Synthesis and Structure . . . . . . . . . . . . . 378
    2.2. Reactions with Electrophiles . . . . . . . . . . . 380
3. Oxidation of Alkenes . . . . . . . . . . . . . . . . . 383
4. Oxidation of Other Substrates . . . . . . . . . . . . . 385
    4.1. Isocyanides . . . . . . . . . . . . . . . . . . 385
    4.2. Carbon Monoxide . . . . . . . . . . . . . . . . 385
    4.3. Aldehydes and Ketones . . . . . . . . . . . . . 386
    4.4. Cumene . . . . . . . . . . . . . . . . . . . 387
    4.5. Tertiary Phosphines . . . . . . . . . . . . . . 387
5. CO-Oxidations . . . . . . . . . . . . . . . . . . . 389
    5.1. Alkenes and Tertiary Phosphines . . . . . . . . . 390
    5.2. Alkenes and Hydrogen . . . . . . . . . . . . . 392
    5.3. Isocyanides and Carbon Monoxide . . . . . . . . . 392
    5.4. Triphenylphosphine and Carbon Monoxide . . . . . . 393

| | |
|---|---:|
| 6. Oxygen Atom Transfer from Metal Phosphine Hydroperoxides and Superoxides | 393 |
|    6.1. Metal Hydroperoxides and Alkylperoxides | 394 |
|    6.2. Metal Peracyls | 396 |
| 7. Oxygen Atom Transfer from Coordinated Nitrite Ligands | 396 |
|    7.1. Transfer from Metal Nitrites to Carbon Monoxide and Triphenylphosphine | 396 |
|    7.2. Transfer from Metal Nitrites to Alkenes | 398 |
| References | 399 |

## Chapter 13

### *Catalysis of Nitrogen-Fixing Model Studies*

*T. Adrian George*

| | |
|---|---:|
| 1. Introduction | 405 |
|    1.1. Scope and Limitations | 406 |
| 2. Nitrogen-Fixing Reactions | 408 |
|    2.1. Titanium, Zirconium, and Hafnium | 409 |
|    2.2. Vanadium, Niobium, and Tantalum | 409 |
|    2.3. Chromium, Molybdenum, and Tungsten | 413 |
|    2.4. Manganese, Technetium, and Rhenium | 429 |
|    2.5. Iron, Ruthenium, and Osmium | 429 |
|    2.6. Cobalt, Rhodium, and Iridium | 430 |
|    2.7. Nickel, Palladium, and Platinum | 430 |
|    2.8. Copper | 430 |
| 3. Nitriding Reactions | 431 |
| 4. Dinitrogen Binding and Reactivity | 431 |
| 5. Future Prospects | 434 |
| 6. Concluding Remarks | 436 |
| References | 436 |

## Chapter 14

### *Polymer-Bound Phosphine Catalysts*

*Norman L. Holy*

| | |
|---|---:|
| 1. Introduction and Scope | 443 |
| 2. Preparation of Polymer-Bound Phosphine Catalysts | 444 |
|    2.1. Modification of Preformed Polymers | 444 |
|    2.2. Polymerization of Phosphine Monomers | 448 |
| 3. Physiochemical Characterization of Catalysts | 450 |

4. Influence of the Support . . . . . . . . . . . . . . . . . . . 454
   4.1. Changes in Selectivity . . . . . . . . . . . . . . . . . 455
   4.2. Asymmetric Induction . . . . . . . . . . . . . . . . 457
   4.3. Matrix Isolation . . . . . . . . . . . . . . . . . . . . 457
   4.4. Coordinative Unsaturation . . . . . . . . . . . . . 459
5. Reactions of Olefins and Dienes . . . . . . . . . . . . . 460
   5.1. Isomerization . . . . . . . . . . . . . . . . . . . . . 460
   5.2. Hydrogenation . . . . . . . . . . . . . . . . . . . . 461
   5.3. Dimerization, Oligomerization, and Polymerization . . . 466
   5.4. Addition . . . . . . . . . . . . . . . . . . . . . . . . 470
   5.5. Reactions with CO . . . . . . . . . . . . . . . . . . 472
   5.6. Metathesis . . . . . . . . . . . . . . . . . . . . . . . 476
6. Trends . . . . . . . . . . . . . . . . . . . . . . . . . . . 478
   6.1. Catalyst Characterization . . . . . . . . . . . . . . 478
   6.2. Reactions . . . . . . . . . . . . . . . . . . . . . . . 479
   References . . . . . . . . . . . . . . . . . . . . . . . . 480

*Index* . . . . . . . . . . . . . . . . . . . . . . . . . . . . . 485

# 1

# Historical Introduction

## Joseph Chatt

The phosphine metal complexes now finding application in homogeneous catalysis are those of the tertiary organic phosphines. They were discovered around the middle of the last century and their ability to combine with heavy metal salts noted almost immediately, but the application of their metal complexes to homogeneous catalysis came only after the lapse of about 100 years.

### 1. THE PHOSPHINES

Trimethylphosphine and poly(dimethylphosphane) $[P(CH_3)_2]_n$ were the first tertiary phosphines to be discovered. They were discovered in 1847 by Paul Thénard,[1] even before the aliphatic amines. He obtained them by passing methyl chloride over an impure calcium phosphide at 180–300°C. Triethylphosphine was prepared similarly from ethyl iodide and an impure sodium phosphide by Ferdinand Bérle in 1855.[2]

These highly reactive substances did not receive the attention they deserved until after the discovery of the aliphatic amines and their relation to phosphine excited the interest of Augustus William Hofmann. It was to him and his co-workers, especially Augustus Cahours (over the period 1857–1871) that we owe the early development of organic phosphorus chemistry.[3–5] Hofmann was a remarkable man who came from Justus von Leibig's laboratory at Giesen. He was remarkable not only for his meticulously accurate researches, as attested by Leibig himself in a footnote

---

*Dr. Joseph Chatt* • School of Chemistry and Molecular Sciences, University of Sussex, Brighton BN1 9QJ, United Kingdom.

to one of his early papers on indigo, but he also contributed substantially to the establishment of chemistry as a proper subject for academic study. Coming to London in 1845, he successfully launched, as its Director, the newly founded Royal College of Chemistry now part of the Imperial College of Science and Technology in the University of London. He was the fifth President of the Chemical Society of London, and on his return to Berlin in 1864 he played an important role in the founding of the "Deutsche Chemische Gesellschaft zu Berlin" of which he was elected its first President. The difficulties he overcame in preparing the methylphosphines are well illustrated by his account of the reaction of methyl iodide on sodium phosphide, his first improved method. "On the application of heat these substances act on one another with great energy, producing combustible and detonating compounds, so that the experiment is not without danger. Often the product of the operation is lost; and if the reaction takes place without explosion the separation of the constituents of the very complicated mixture which results can be effected only with the greatest difficulty."[4,5] He developed more controllable methods, as for example, the reaction of diethylzinc in ether with phosphorus trichloride and wrote of this: "The reaction between these two bodies is very violent and readily gives rise to dangerous explosions if the necessary precautions are neglected." His precautions were to keep the reaction vessel full of carbon dioxide and well cooled. Other safer methods, usually starting from phosphine or phosphonium iodide, were developed later, but none was particularly convenient.[6]

It says much for Hofmann's and Cahours's skill that they were able to purify and characterize such highly reactive materials. Certainly to my knowledge the preparation of trimethyl- and triethylphosphine defeated some inorganic and organic chemists even in the 1950s. It was this difficulty of access to the phosphines which delayed the extensive study of tertiary phosphine complexes for a whole century.

The early development of aromatic phosphine chemistry took place during the last quarter of the nineteenth century mainly in Michaelis's school.[7,8] He prepared dichloro(phenyl)phosphine by the passage of the mixed vapours of benzene and phosphorus trichloride through a red hot porcelain tube.[7] This was quickly followed by the preparation of phenylphosphine, dialkylphenylphosphines, and triphenylphosphine.[8-10] This latter phosphine was destined some 75 years later to become one of the most important phosphines in the development of homogeneous catalysis by phosphine metal complexes. Michaelis obtained it by the reaction of sodium with dichloro(phenyl)phosphine or phosphorus trichloride and bromobenzene in boiling ether.[8-10] The development of aromatic phosphorus chemistry rapidly overtook that of their intractable aliphatic analogs. It was well established by the 1920s whereas the aliphatic had to await the application of Grignard reagents and their reaction with dichloro(phenyl)-phosphine and phosphorus trichloride, first in the aromatic series[11] and then by Hibbert to prepare triethylphosphine in 1906.[12] Hibbert's method of work-up gave poor yields and was difficult, but it was the best method available even in the 1930s when the present interest in phosphine complex chemistry started to stir. The difficulty with triethylphosphine was possibly the reason that the more easily isolated tripropylphosphine and its higher homologues were not prepared before 1929.[13,14]

## 1.1. The Complexes

There are references to metal complexes in the early Hofmann literature but most of these are to organophosphonium halometallates, e.g., $[PH(CH_3)_3]ZnI_4$, which crystallize from halogen acid solutions on the addition of the appropriate metal salt.[3] Nevertheless true coordination compounds of platinum(II) were isolated almost immediately after the discovery of the tertiary aliphatic phosphines.[15] This doubtless arose because Hofmann used the then current technique of amine chemistry to identify his phosphines through determination of the platinum content of their sparingly soluble and easily purified hexachloroplatinates(VI). In fact, he obtained white $[PtCl_2(PMe_3)_2]$ and triethylphosphine complexes, but serious investigation started only in the 1870s[16] when Cahours and Gal[16] showed that on boiling an aqueous solution of platinum(IV) chloride with triethylphosphine they obtained the yellow $\alpha$-form of $[PtCl_2(PEt_3)_2]$ and the white $\beta$-form. They also prepared the salt $[Pt(PEt_3)_4]Cl_2$. Complexes of gold(I), copper(I), and palladium(II) were prepared by these and other workers about the same time, but they aroused little interest beyond the nature of the isomeric platinum(II) complexes. With the advent of Alfred Werner and the establishment of the planar configurations of the platinum(II) diammines, $[Pt(NH_3)_2Cl_2]$, the nature of the isomerism was ellucidated and, by color analogy with the diammines, the yellow $\alpha$-isomer was assigned the *cis*-configuration and the white $\beta$-isomer, the *trans*.[17] The few known analogous platinum(II) complexes, e.g., those of the tertiary arsines and organic sulphides, had their configurations similarly and incorrectly assigned. There the matter rested until the 1930s. In the meantime other complexes had been prepared, particularly the "mercurichlorides" $[\{HgCl_2(PR_3)\}_2]$ which served as readily purified crystalline derivatives for identification by melting point of the liquid phosphines.

The present phase in the development of the complex chemistry of the tertiary organic phosphines started in the early 1930s and was interrupted by the second World War. It is intimately connected with that of the corresponding arsines and sulphides and less so with the stibines, selenides, and tellurides—all being easier to prepare than the phosphines.

Early in this century, the organic chemistry of arsenic and sulfur underwent rapid development owing mainly to the discovery that organoarsenicals have valuable chemotherapeutic properties and, because of the 1914–1918 war, that some arsenicals and sulphides have sternutatory or vesciant properties of possible use in warfare. More chemists thus obtained knowledge of these potentially ligand substances which in the fullness of time would be used to render transition metal ions soluble in organic solvents. Naturally the complex chemistry of the more readily obtainable organic sulphides and organoarsines was developed first, and the scene was set for new development in the area of organophosphine complexes by 1930.

It was the application to coordination chemistry of the physicochemical methods finding their way into organic chemistry during the 1920s and 1930s which raised the need for organic soluble or easily fusible coordination compounds of high stability. Two notable studies then revived interest in trialkylphosphine and related complexes, which was to grow in strength rapidly after the second World War. These studies were attempts by F. G. Mann[18] to apply Sugden's parachor[19] to determine whether the coordinate bond, also known as a semipolar double bond, is single or double, and K. A. Jensen's determination of the dipole moments of complex compounds.[20]

Mann's study required low-melting compounds for the determination of their surface tensions in the liquid phase and he chose organo-phosphines, arsines and -sulphides as ligands with palladium(II) and mercury(II) halides for his study. He soon showed that the parachor was useless for his purpose, but discovered that the trialkylarsine $(AsR_3)$ complexes, e.g., $[PdCl_2(AsR_3)_2]$, lost some of their arsine on fusion *in vacuo* to give a new series of halogen bridged complexes, e.g., $[Pd_2Cl_4(AsR_3)_2]$. Then even the existence of halogen bridges was contentious. This was resolved by an X-ray structure paralleled by an extensive study of the chemistry of organophosphine and -arsine complexes in the late 1930s; it increased enormously the number of known complexes of the tertiary phosphines and arsines.[21,22]

K. A. Jensen's dipole moment study of complex compounds required complexes soluble in nonpolar solvents, and he also chose organo-sulphide, -arsine and -phosphine complexes.[20] He established that a high separation of charge occurs in asymmetric coordination compounds, such as *cis*-$[PtCl_2(PR_3)_2]$ ($PR_3$ = trialkylphosphine) with dipole moments around 11 Debye units, and he showed that all those platinum(II) complexes except the ammines, hitherto thought to have a *trans*-configuration, had a *cis* one and *visa versa*. He also demonstrated that tertiary phosphines would stabilize unusual oxidation states by his preparation of $[NiBr_3\{P(C_2H_5)_3\}_2]$.[23]

The first event of direct relevance to the application of phosphine complexes to catalysis occurred immediately after the second World War.

This was the appearance of publications by W. Reppe and co-workers in 1948[24,25] followed by Badische Anilin und Soda Fabrik patents.[26] They showed that various triphenylphosphine complexes of nickel, especially [Ni(CO)$_2$(PPh$_5$)$_2$](Ph = C$_6$H$_5$), were more effective than other nickel complex catalysts for the polymerization of olefinic and acetylenic substances and that others, especially [NiBr$_2$(PPh$_3$)$_2$], catalyzed the formation of acrylic acid esters from alcohols (ROH), acetylene, and carbon monoxide:

$$ROH + CO + C_2H_2 \rightarrow CH_2\!=\!CHCOOR$$

The mechanisms of these catalytic reactions were obscure, but the demonstration that phosphine complexes had potentially useful catalytic properties attracted the attention of the petrochemicals industry worldwide. Thus, the postwar studies of organic soluble transition metal complexes, especially carbonyl, olefin, acetylene, phosphine, and arsine complexes, occurred mainly in research laboratories financed by industry or governments and tended to concentrate on the first three of the above ligands in reaction with molecular hydrogen as showing greater industrial potential. Phosphines and arsines were involved only incidentally. Despite Reppe's discovery, the immediate postwar studies of phosphine and arsine complexes were directly derived from prewar studies and aimed at coordination chemistry for its own sake. This was true of my work in Imperial Chemical industries Limited, U.K., and in the study of arsine complexes by F. P. Dwyer in the University of New South Wales during the 1940s.[27] This latter work led directly to the establishment of an important school of arsine complex chemistry in University College, London, under one of Dwyers's students, R. S. Nyholm.[28] He had an enormous influence on the development of coordination chemistry, especially of phosphine and arsine complexes over the two decades from 1950. It was not so much through his research, important though it was, but more by his infectious enthusiasm, his ability to produce excellent "popular" reviews, and his interaction with chemists of all disciplines.

It was during this period from the second World War to around 1965 that the unique character of the phosphines and arsines as ligands was established. It was also during this period that the fundamentally important qualities of hydride and ligands attached through carbon to transition metals were discovered. Here it is not possible to do more than summarize these developments,† which are elaborated more fully where appropriate in other chapters.

There was first the ability of phosphine and arsine ligands to stabilize unusually high oxidation states of the later transition metals,[23] much

---

† For a bibliography relevant to phosphine chemistry over this period, see Reference 29.

exploited by Nyholm.[28] This stabilization is a consequence of their high ligand field strengths, a factor which renders them compatible with ligands such as carbon monoxide, hydride ion, saturated organic substances, and alkyl and aryl groups, all important in homogeneous catalysis. Thus, it was found that phosphines substituted readily into the binary metal carbonyls and hydridocarbonyls, that they stabilized as coligands, hydride, alkyl, and aryl complexes of transition metals, especially of Group VIII. Moreover phosphines, especially triphenylphosphine, also stabilized low-oxidation state complexes, e.g., [RhCl(PPh$_3$)$_3$] and Pt(PPh$_3$)$_n$ ($n$ = 3 or 4), a very important property in preventing the precipitation of the heavier Group VIII metals under the reducing conditions inherent in such important reactions as hydrocarbonylation and hydrogenation.

It was also shown that phosphines and arsines as ligands have fairly high *trans*-effects, so that although one, two, or rarely three monophosphine ligands might coordinate strongly to a metal ion or atom, three or more usually gave easily dissociable complexes thus freeing metal sites for the activation of reactant molecules. The inability of the metal to hold strongly a greater number of phosphine or arsine molecules is caused not only by their high *trans*-effects but also by their steric bulk. More recently in the late 1960s it was found that phosphines tend to stabilize dinitrogen complexes which are prepared from molecular nitrogen, often with hydride as coligand. This has led to attempts to produce phosphine complexes, especially from molybdenum or tungsten, to mediate catalytically the production of nitrogen hydrides or amines directly from molecular nitrogen. Truly catalytic systems have not yet been discovered but stoichiometric and cyclic systems with ammonia, hydrazine, or amines as products have been evolved.[30]

Other qualities of ligands pertinent to the important reactions of homogeneous catalysis by tertiary phospine complexes were established also during the period to 1965. These were:

(a) the exceptionally high *trans*-effects of hydride, alkyl, and aryl ligands; their high ligand field strengths; and their strong $\sigma$-electron donor abilities as notional anionic ligands, a consequence of the low electronegativities of hydrogen and carbon as compared with the more common ligand atoms;

(b) the oxidative addition and related reactions of alkyl and hydrogen halides; of aromatic and aliphatic hydrocarbon groups; and of the utmost importance, of dihydrogen by the insertion of electron-rich metal centers into C–X, H–X, C–H, and H–H bonds, respectively, e.g.:

$$[\text{PtMeI}(\text{PE}t_3)_2] + \text{MeI} \rightarrow [\text{PtMe}_2\text{I}_2(\text{PE}t_3)_2],$$

(c) the insertion of carbon monoxide into M–C bonds, where C is the carbon of an alkyl or aryl ligand, to produce acyl or aroyl complexes;
(d) the insertion of olefins into M–H bonds to produce metal alkyls;
(e) the insertion of olefins into metal alkyl bonds to produce higher alkyls and similar insertion of alkynes leading to polymerizations of various kinds; and
(f) the removal of notionally anionic carbon ligands from the metal as an uncharged species by addition of hydrogen across the M–C bond: $H_2 + M–C \rightarrow MH + HC$, or by its elimination as an olefin: $M–CH_2–CH_2R \rightarrow M–H + CH_2=CHR$.

These reactions provided the basis for the mechanistic understanding of such industrially important reactions as the hydrocarbonylation of olefins and the then recently discovered Ziegler–Natta polymerization. They held out promise of more useful catalyses and led to a flourishing activity in the area of organotransition metal chemistry world-wide during the 1960s and 1970s.

Another important development during the period to 1965 was the appearance of certain "key" phosphines as commercial products, especially triphenylphosphine, tri-$n$-butylphosphine, dichloro(phenyl)phosphine, and chlorodi(phenyl)phosphine, the latter two being percursors for the easily isolated dialkyl(phenyl)- and alkyldi(phenyl)-phosphines. Also during this time, the development for routine use of new physical methods considerably eased the investigation of phosphine complexes, first infrared spectroscopy and its later development to longer wavelengths where metal-halogen stretching frequencies occur, then the more useful $^1H$ nmr, and later $^{31}P$ nmr spectroscopies. These latter particularly allowed quick determinations of the configurations of phosphine complexes, replacing the time-consuming measurement of dipole moments, which had hitherto been the only reliable method, apart from the even more laborious X-ray crystal-structure determination. Incidently, the dipole moments gave much valuable information about the distribution of electric charge in complex compounds.[31]

For the study of the fundamental chemistry of transition metal complexes of phosphines and arsines and their organometallic derivatives, it was necessary to have stable complexes of known structures and configurations; hence, the studies up to 1945 involved mainly the strongly bonding trialkylphosphines, and arsine complexes stabilized by chelation as in the famous $o$-phenylenebis(dimethylarsine). For catalysis, it is necessary to create vacant sites in the coordination shell of the metal where the reactants can be activated or brought together by coordination for reaction. Triphenylphosphine and other aromatic phosphines with their greater steric bulk and more weakly bonding affinity for the metals were ideal for the launching of phosphine complex chemistry into its next phase of development, that of homogeneous catalysis.

## 1.2. Catalysis

The study of homogeneous catalysis by phosphine complexes started to flourish in the 1950s. It is so recent that here it is necessary only to indicate the sequence and time-scale of its development.† It was spurred by K. Ziegler's discovery that the products of the reaction of triethylaluminium with certain complexes of zirconium or titanium were excellent catalysts for the polymerization of ethene at ordinary temperatures and pressures.[33] Hitherto pressures of over 1000 atm had been used in the manufacture of polyethylene. Furthermore, G. Natta immediately extended Ziegler's process using triethylaluminium and titanium halides to produce isotactic, i.e., stereoregular, polypropylene, and other stereoregular polymers.[34] These discoveries quickly led to a new and very successful process for polyethylene and to a new commercially important product in isotactic polypropylene. It also led to a flurry of research activity into further uses of transition metal compounds as catalysts, many of which were rendered homogeneous by the use of triphenylphosphine as a nonreacting ligand to hold the metal in solution under reducing conditions.

The metals involved were mainly those of Group VIII, and the various reactions mediated homogeneously by phosphine complexes form the detailed subjects of the later chapters in this book. The most significant have involved alkenes and, to a lesser extent, alkynes under reducing conditions. Of these, hydrogenation has received most study and will be used to illustrate the general growth of awareness of the unique advantages of phosphine complexes in catalysis. Other important reactions mediated advantageously by phosphines as ligands are carbonylation in its various aspects including hydroformylation, the oligomerization of alkenes, and their isomerization by double-bond migration.

Despite M. Iguchi's discovery in 1939[35] that certain unstable rhodium complexes catalyzed the hydrogenation of organic substances, such as fumaric acid, and Reppe's discovery in the mid 1940s that phosphines enhanced the activity of some of his nickel-based polymerization and carbonylation catalysts,[24-26] it was not until 1965 that the next major breakthrough occurred. This was the discovery of a phosphine-activated homogeneous hydrogenation catalyst, $[RhCl(PPh_3)_3]$, known as Wilkinson's catalyst, which catalyzes the hydrogenation of alkenes at 25°C under one atmosphere pressure of dihydrogen.[36,37] Its remarkable activity immediately directed attention to other similar complexes including

---

† For a general bibliography leading into this phase of development see Reference 32.

[RuHCl(PPh$_3$)$_3$], formed by dihydrogen reduction of [RuCl$_2$(PPh$_3$)$_x$] ($x$ = 3 or 4), and one of the most active homogeneous catalysts for the hydrogenation of 1-alkenes.[38]

A very important factor to emerge from the study of these homogeneous systems was their selectivity as compared with heterogeneous systems. By the correct choice of conditions, they could be operated to hydrogenate alkynes in the presence of alkenes, or one alkyne selectively in the presence of another. The steric bulk of the phosphines usually means that 1-alkenes are hydrogenated some orders of magnitude faster than internal alkenes. Dienes can be reduced selectively to monoenes, as for example by *trans*-[Pt(SnCl$_3$)H(PPh$_3$)$_2$].[39] The complex [RuCl$_2$(PPh$_3$)$_3$] will even catalyze the reduction by hydrogen of nitro-compounds to amines without hydrogenating the aromatic ring or reducing other substituents, including CN.[40]

The selectivity of phosphine complexes in catalysis was further emphasized in 1968[41,42] when it was found that the hydrogenation of atropic acid, CH$_2$=C(Ph)COOH, catalyzed by [RhCl$_3$(P*MePhPr$^n$)$_3$] using (S)-(+)-methylphenyl-*n*-propylphosphine, gave asymmetric hydrogenation to MeC*H(Ph)COOH with a 15% enantiomeric excess. Shortly afterwards, it was shown that the asymmetric center in the catalyst need not be at phosphorus but could be advantageously situated in one of the organic groups of the phosphine.[43] Thus, a rhodium catalyst prepared from neomenthyldiphenylphosphine gave an enantiomeric excess of 61% of (s)-(+)-3-phenylbutanoic acid on hydrogenation of $\beta$-methylcinnamic acid. Such asymmetric phosphines have the advantage that they can be obtained from naturally occurring, optically-active precursors. The best known of such phosphines is the bidentate ligand (−)-2,3-*O*-isopropylidene-2,3-dihydroxy-1,4-bis(diphenyl-phosphino)butane, usually referred to as (−)-DIOP and prepared from L(+)-tartaric acid.[44] A rhodium catalyst prepared from (−)-DIOP catalyzes the hydrogenation of PhCH=C(NHCOOMe)-COOH to give (*R*)-PhCH$_2$C*H(NHCOOMe)COOH in 72% optical purity, and recently an enantioselectivity of 90% has been obtained in the reduction of an appropriate ketone to its alcohol using another chelate diphosphine rhodium catalyst.[45] Such asymmetrical syntheses are of obvious importance to the pharmaceutical industry.

The great strength of homogeneous catalysis and especially of that involving phosphine ligands is the opportunity it affords to tailor ligands so as to enhance the reactivity and selectivity of metal centered catalysts. Only slight changes in the ligand can effect considerable changes in selectivity. That the reaction occurs in homogeneous solutions also facilitates the study of the mechanistic steps involved in the catalytic cycle and, by change of phosphine or metal center, may allow the isolation of stable substances analogous to metal-containing intermediates of the catalytic cycle. The study of homogeneous catalysis has thus been very useful in elucidating the important reactions of heterogeneous catalysis, but the large-scale reactions of the petrochemicals industry are still catalyzed heterogeneously. Homogeneous catalysis has the great disadvantage that the catalyst remains in the product and must usually be separated from it. Ideally, one would prefer a heterogeneous catalyst with the reactivity and selectivity of the homogeneous. The present thrust in homogeneous catalysts is not only to increase their reactivity and by tailoring the ligands to attempt to promote selectivity in the desired reaction, but also to render them heterogeneous for convenience in use by attaching them to a surface, hopefully without loss of reactivity or selectivity.[46]

## REFERENCES

1. P. Thènard, *Jahresber.* **1847–1848,** 645–646; *Compt. rend.* **25,** 892–895 (1847).
2. F. Bérle, *Jahresber.* **1855,** 590–591; *J. Prakt. Chem.* **66,** 73–75 (1855); *Annalen* **97,** 334–355 (1856).
3. A. Cahours and A. W. Hofmann, *Ann.* **104,** 1–39 (1957).
4. A. W. Hofmann and A. Cahours, *Quart. J. Chem. Soc.* **11,** 56–78 (1859).
5. A. W. Hofmann, *Phil. Trans.* **150,** 409–448 (1860).
6. ———, *Ber.* **4,** 205–209 (1871).
7. A. Michaelis, *Ber.* **6,** 601–603 (1873); *Annalen* **181,** 265–363 (1876).
8. A. Michaelis and L. Gleichman, *Ber.* **15,** 801–804 (1882).
9. A. Michaelis and A. Reese, *Ber.* **15,** 1610 (1882).
10. A. Michaelis and H. V. Soden, *Annalen.* **229,** 295–340 (1885).
11. P. Pfeiffer, *Ber.* **37,** 4620–4623 (1904).
12. H. Hibbest, *Ber.* **39,** 160–162 (1906).
13. W. C. Davies and W. J. Jones, *J. Chem. Soc.* **1929,** 33–35.
14. W. C. Davies, P. L. Pearce, and W. J. Jones, *J. Chem. Soc.* **1929,** 1262–1268.
15. A. W. Hofmann, *Annalen* **103,** 357–358 (1857).
16. A. Cahours and H. Gal, *Compt. rend.* **70,** 897–903 (1870); *Jahresber.* **1870,** 808–814.
17. A. Werner, *Z. Anorg. Chem.* **3,** 265–330 (1893), p. 318.
18. F. G. Mann and D. Purdie, *J. Chem. Soc.* **1935,** 1549–1563.
19. S. Sugden, *J. Chem. Soc.* **1924,** 1177–1189.
20. K. A. Jensen, *Z. Anorg. Allg. Chem.* **229,** 225–251 (1936).
21. J. Chatt and F. G. Mann, *J. Chem. Soc.* **1938,** 1622–1634 and references therein.
22. F. G. Mann and D. Purdie, *J. Chem. Soc.* **1940,** 1235–1239 and earlier Parts in the series.
23. K. A. Jensen, *Z. Anorg. Allg. Chem.* **229,** 265–281 (1936).
24. See for example, W. Reppe and W. J. Schweckendiek, *Annalen* **560,** 104–116 (1948).
25. W. Reppe cited in J. W. Copenhaver and M. H. Bigelow, *Acetylene and Carbon Monoxide Chemistry* (Reinhold, New York, 1949).
26. See for example, W. Reppe and W. Schweckendiek, German Patent 871, 494 (1953); *Chem. Abstr.* **52,** 10175 (1958).
27. See for example, F. P. Dwyer and D. M. Stewart, *J. Proc. Roy. Soc. N.S. Wales* **83,** 177 (1949); [*Chem. Abstr.* **45,** 7462 (1951)] and references therein.

28. D. P. Craig, "Ronald Sydney Nyholm," *Biographical Memories of the Royal Society* **18**, 445–475 (1972).
29. G. Booth, *Adv. Inorg. Chem. Radiochem.* **6**, 1–69 (1964).
30. J. Chatt, J. R. Dilworth, and R. L. Richards, *Chem. Rev.* **78**, 589–625 (1978).
31. J. Chatt and G. J. Leigh, *Angew. Chem. Int. Ed. Engl.* **17**, 400–407 (1978).
32. C. Masters, *Homogeneous Transition-Metal Catalysis* (Chapman and Hall, London and New York, 1981).
33. K. Ziegler, E. Holzkamp, H. Breil, and H. Martin, *Angew. Chem.* **67**, 541–547 (1955).
34. G. Natta, *J. Polymer Sci.* **16**, 143–154 (1955).
35. M. Iguchi, *J. Chem. Soc. (Jpn.)* **60**, 1787–1792 (1939).
36. J. F. Young, J. A. Osborn, F. H. Jardine, and G. Wilkinson, *J. Chem. Soc., Chem. Comm.* **1965**, 131–132.
37. R. S. Coffey and Imperial Chemical Industries, British Patent no. 1 121 642 (1965).
38. D. Evans, J. A. Osborn, F. H. Jardine, and G. Wilkinson, *Nature (London)* **208**, 1203–1204 (1965).
39. E. N. Frankel, E. A. Emken, H. Itatani, and J. C. Bailar, Jr., *J. Org. Chem.* **32**, 1447–1452 (1967).
40. J. F. Knifton, *J. Org. Chem.* **40**, 519–521 (1975); **41**, 1200–1206 (1976).
41. W. S. Knowles and M. J. Sabacky, *J. Chem. Soc., Chem. Commun.* **1968**, 1445–1446.
42. Compare L. Horner, H. Siegel, and H. Büthe, *Angew. Chem. Int. Ed. Engl.* **7**, 942 (1968).
43. J. D. Morrison, R. E. Burnett, A. M. Aguiar, C. J. Morrow, and C. Phillips, *J. Am. Chem. Soc.* **93**, 1301–1303 (1971).
44. H. B. Kagan and T. P. Dang, *J. Am. Chem. Soc.* **94**, 6429–6433 (1972).
45. T. Hayashi, A. Katsumura, M. Konishi, and M. Kumada, *Tetrahedron. Lett.* **1979**, 425–428.
46. See for example, F. R. Hartley and P. N. Vezey, *Adv. Organomet. Chem.* **15**, 189–234 (1977).

# 2

# Mechanistic Studies of Catalytic Reactions Using Spectroscopic and Kinetic Techniques

## C. A. Tolman and J. W. Faller

## 1. INTRODUCTION

The rapid growth of the organometallic chemistry of the transition metals during the last 15–20 years owes much to the development of homogeneous catalyst systems which are capable of synthesizing organic molecules under mild conditions and occasionally with remarkable selectivities. Several have been commercialized and are now used on a large scale.[1] A few have received considerable detailed study—including spectroscopic identification of the species present in solution under reaction conditions, isolation of reactive intermediates in some cases, determination of the overall rate law and measurement of rate and equilibrium constants of several individual steps, and isotopic labeling studies—so that we have a reasonably clear picture of how they operate. It is these systems that form the focus of this chapter. For more general reviews the reader is referred to some recent

---

Dr. Chad A. Tolman • Central Research and Development Department, Experimental Station, E. I. du Pont de Nemours and Company, Wilmington, Delaware 19898. Dr. Jack W. Faller • Department of Chemistry, Yale University, New Haven, Connecticut 06511.

books,[1-3] which also discuss the special electronic properties of transition metals which are in part responsible for their catalytic behavior.†

## 1.1. The Nature of Catalyst Systems

Before discussing specific examples, we should define what we mean by the word "catalyst," and point out some of the important general features of homogeneous catalytic systems. By "catalyst" we mean *the transition metal compound added to a chemical reaction to accelerate its rate or to change the distribution of products formed*.‡ Generally, many product molecules are produced per metal atom added. We shall see that catalyst systems consist of many different metal-containing species and that, frequently, the compound originally added may change before a substrate first becomes coordinated; the original compound may not even be recoverable at the end of the reaction. This fact is recognized by some authors by introducing terms like "the catalyst precursor," "the real catalyst," or "the actual catalyst." For example, Collman and Hegedus[2a] define "the actual catalyst" as "the dominant complex present during the catalytic cycle". We shall see that the dominant complex present in an active catalyst system may depend on the substrates used, their concentrations, the concentration of phosphine ligands, and even the time elapsed after a reaction is initiated. The dominant complex or the "principal species" present may not even be involved in the catalytic cycle. If during catalysis we can spectroscopically observe a complex that is part of the catalytic cycle (one of what we shall term a "loop species"), it often occurs in observable concentrations because it precedes the slowest or rate-determining step and is the least active species in the cycle! This loop species in highest concentration might be identified as the "principal active species" or the "major active species" and may in reality be the least active. These difficulties in nomenclature can produce problems, such as requiring references to "Wilkinson's catalyst precursor" rather than "Wilkinson's catalyst." The term "catalyst" will inevitably mean different things to different people and this vagueness leads us to prefer the simple original empirical definition of "catalyst" given above.

The catalytic system may also require a cocatalyst. By "cocatalyst" we mean a compound added to the system to improve the effectiveness of the catalyst. These are usually nontransition-metal compounds and would

---

† Masters' book[3] has a particularly nice discussion of the important features of homogeneous metal catalysts.

‡ It is possible to distinguish reactions which are catalytic in the metal from stoichiometric ones if several product molecules are produced per metal atom charged. Some stoichiometric reactions can be made catalytic, e.g., by reoxidizing the Pd(0) produced in the Wacker process[1,2] by Cu(II), which is in turn regenerated by oxidation with air.

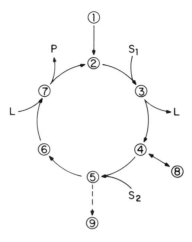

Figure 2.1. A schematic representation of a mechanism of a typical homogeneous catalytic system. $S_1$ and $S_2$ are substrates, and P product. $L$ represents a phosphorus ligand.

include, for example, Bronsted or Lewis acids or bases, or reducing or oxidizing agents.

Figure 2.1 shows a schematic representation of a typical homogeneous catalyst system. Each numbered circle represents a discrete metal complex. The heart of the system is the loop, consisting in this case of complexes *2–7*, which are connected in a cyclic sequence by elementary reaction steps. It is by repeated cycles around this loop that a single metal atom can convert many substrate molecules $S_1$ and $S_2$ into product molecules P. In addition to the loop species in Figure 2.1, there are other species in the system off the loop (compounds *1* and *8*) connected to it by reversible reactions and another (*9*) connected by an irreversible reaction (indicated by the dashed arrow) going from the loop. Note that any compound that is in the loop or can be connected to a loop species could be charged into the system as a catalyst. A compound like *9* is commonly referred to as a "dead catalyst" or a "catalyst deactivation product." Reactions which irreversibly bleed metal out of the loop are responsible for the finite life of all real catalyst systems. A useful measure of catalyst life is the number of catalytic cycles before deactivation occurs (the turnover number), which is given by the relative rate of cycling about the loop (product formation) to catalyst deactivation. A useful measure of instantaneous catalyst activity is the turnover rate, which is the number of product molecules produced, per metal atom in the system, per unit time. Depending on the equilibrium constants connecting complexes *1* through *8* to the loop, only a small fraction of the metal atoms may be present in the loop at any one time, as we shall see in the subsequent examples.

It should be noted that the solid arrows in the loop are drawn in one direction to indicate the major flow of the reactions; actually, the individual reaction steps can be, and often are, reversible, so that an equilibrium mixture of reactants and products can be approached from either direction. A catalyst system increases the rate of approach to equilibrium but, of course, cannot affect the equilibrium distribution.

Figure 2.1 shows the formation of only one product. Most catalytic systems produce more than one product. Each additional product requires an additional loop in the mechanism, though generally the loops are joined by reaction pathways or by common catalytic intermediates. As we shall see, some of these systems can give complex networks of reactions.

In addition to substrates and products which may coordinate and dissociate, there are usually other ligands, designated $L$ in Figure 2.1, which also enter or leave the loops but whose role is less apparent. The primary interest of this chapter is in systems where $L$ is a phosphorus donor ligand. The behavior of $L$ will generally depend on both its electronic and steric character. For monodentate ligands, these can be defined in terms of an electron accepting character $\nu\dagger$ (based on the $A_1$ carbonyl stretching frequency $\nu$ of $Ni(CO)_3L$) and a ligand cone angle $\theta$.[4] Understanding how to "fine tune" the ligand effects to make catalyst systems more active, more selective, and longer lasting is a major challenge facing chemists in homogeneous catalysis. The development of a rational approach to this fine tuning depends on another major challenge—determining the mechanisms of homogeneous catalytic systems.

## 1.2. Mechanisms, Rate Laws, and Dominant Species

A mechanism is the "road map" of a chemical reaction. It reflects the sequence of reaction steps and the composition of the intermediates. Mechanisms are postulated based on spectroscopic studies, rate studies, product analyses, and isotopic labeling. Certain types of reactions may be unlikely based on energy considerations (e.g., the Woodward–Hoffman Rules[5] or the 16 and 18 Electron Rule[6]). There is *never* enough information to prove one mechanism to the exclusion of all others, nor is there usually enough data to uniquely determine all the rate constants for the elementary steps of a proposed mechanism. Some possible paths can be shown to be unimportant. In a typical case, a few rate or equilibrium

---

† An alternative and sometimes more convenient electronic parameter $\chi$ is often used, where $\chi = \nu(L) - \nu[P(t\text{-Bu})_3]$ where $\nu[L]$ is the frequency in cm$^{-1}$ of the $A_1$ carbonyl stretching mode of $Ni(CO)_3L$ in $CH_2Cl_2$, from Reference 4. Since $P(t\text{-Bu})_3$ gave the lowest frequency (2056.1 cm$^{-1}$) observed for a phosphorus ligand, $\chi$ gives a convenient scale of positive numbers with values ranging from 0 for $P(t\text{-Bu})_3$ to about 50 cm$^{-1}$ for $PF_3$.

constants can be determined, while others can only be set above or below certain limits.

A mechanism may be represented by a list of chemical reactions (each of which is unimolecular or bimolecular), or more graphically by a diagram, like Figure 2.1, which more clearly shows the flow and cyclic nature of the reactions.

A rate law describes the dependence of the rate of product formation (or substrate disappearance) at steady state on the concentrations of catalyst, cocatalysts, substrates, and ligands. A fact that is not generally appreciated is that the functional form of the rate law often depends on the range of experimental variables investigated, as will be shown subsequently.

Several mechanisms may be (and usually are) consistent with a given rate law, so that other experiments are necessary before we can have any confidence in a detailed mechanism. In particular, parallel spectroscopic experiments on the system—preferably on solutions under actual catalytic conditions—are usually necessary before one interprets the rate law.[7] As a simple example of a stoichiometric reaction, it is found that the rate of oxidative addition of $CH_3I$ to $(C_2H_4)Pt(PPh_3)_2$ is inhibited in an inverse first-order manner by the addition of excess $C_2H_4$ to the system.[8] Without spectroscopic experiments one doesn't know whether the inhibition is the result of association of excess $C_2H_4$ to form an inactive $(C_2H_4)_2PtL_2$ complex, decreasing the concentration of reactive $(C_2H_4)PtL_2$ [reactions (1) to (3)], or whether $C_2H_4$ suppresses formation of a reactive $PtL_2$ complex from an inactive $(C_2H_4)PtL_2$ ((4) and (5)).

$$C_2H_4 + (C_2H_4)PtL_2 \overset{K}{\rightleftharpoons} (C_2H_4)_2PtL_2 \qquad (1)$$

$$RX + (C_2H_4)PtL_2 \overset{k}{\longrightarrow} RPtL_2X(C_2H_4) \qquad (2)$$

$$RPtL_2X(C_2H_4) \overset{\text{fast}}{\longrightarrow} RPtL_2X + C_2H_4 \qquad (3)$$

$$(C_2H_4)PtL_2 \overset{1/K}{\rightleftharpoons} PtL_2 + C_2H_4 \qquad (4)$$

$$RX + PtL_2 \overset{k}{\longrightarrow} RPtL_2X \qquad (5)$$

The rate law in either case is:

$$-d[Pt(0)]/dt = k[Pt(0)][RX]/(1 + K[C_2H_4]) \qquad (6)$$

which reduces to:

$$-d[Pt(0)]/dt = k[Pt(0)][RX]/K[C_2H_4] \qquad (7)$$

provided that $(C_2H_4)_2PtL_2$ is the major species in the first case or $(C_2H_4)PtL_2$ is the major one in the second (i.e., $K[C_2H_4] \gg 1$). Spectrophotomeric measurements and $^1H$ nmr spectra show that $(C_2H_4)_2PtL_2$ does indeed form in solution in the presence of added $C_2H_4$ and that it provides the intermediate for the relatively rapid exchange of free and coordinated ethylene which occurs in solution at room temperature.[9]

In a stoichiometric reaction, the rate law tells the difference between the composition of the major initial species in solution and the composition of the transition state of the rate-determining step (abbreviated CTSRDS). Thus, under conditions where $(C_2H_4)_2PtL_2$ is the major Pt(0) species,

$$\text{CTSRDS} = (C_2H_4)_2PtL_2 - C_2H_4 + RX = (RX)(C_2H_4)PtL_2 \qquad (8)$$

We know nothing, of course, about the structure of the transition state, only its composition. Owing to mass balance, this approach is valid even if rapid equilibria are involved prior to the rate-determinining step; however, there are situations where deceptively simple rate laws can lead to difficulty. These possibilities will be discussed after this approach is illustrated.

In a catalytic system, the rate law distinguishes between the composition of the dominant species present and the CTSRDS†. With 1-butene isomerization with $NiL_4$ [$L = P(OEt)_3$], in the early stages of the reaction and with small concentrations of added acids, the rate law is

$$-d[1\text{-B}]/dt = k[\text{Ni}][H^+][1\text{-B}]/[L] \qquad (9)$$

Since spectroscopic studies show that $NiL_4$ is the major Ni-containing species under these conditions,[10] the CTSRDS is obtained as in Equation 10.

$$\text{CTSRDS} = NiL_4 + H^+ + 1\text{-}B - L = (C_4H_9)NiL_3^+ \qquad (10)$$

This does not, of course, tell us the order in which $H^+$ and 1-B coordinate and $L$ dissociates, or whether the $H^+$ is bound to Ni or carbon in the transition state. Such detailed questions are best answered by examining the rates and equilibria of individual elementary steps in the reaction, by breaking the system up and studying its component parts. Thus, one must address the problem of identifying intermediates on the loop by physical or chemical methods, as well as suggesting reasonable compositions for all the species present.

† If there is more than one active species over the range of variables explored, the rate law will contain more than a simple multiplicative term.

Spectroscopic studies at low temperature can sometimes permit one to trap and characterize reactive intermediates that might be unstable at room temperature; often their structures can be deduced, especially by nmr. In favorable cases even thermally sensitive complexes can be isolated and X-ray crystal structures determined, at low temperatures if necessary. More often, X-ray structures are determined on model systems which are stable enough to be isolated and handled at room temperature. (See Chapter 2.) In this connection, model compounds containing third-row metals are often considerably more stable than their congeners in the first or second rows; however, their catalytic activity may also be substantially lower.

It is sometimes said facetiously that if a species can be observed directly, it probably is not a key intermediate in a catalytic reaction. More indirect methods of inferring the composition and structure of intermediates will be discussed subsequently (Section 1.3); however, it is appropriate to consider some guidelines for postulating "reasonable" intermediates in the catalytic cycle.

One rule of thumb that is often very useful in studying catalytic mechanisms (or stoichiometric reactions, for that matter) is the 16 and 18 Electron Rule.[6] Simply stated, it says that diamagnetic transition-metal complexes undergo reactions, including catalytic ones, which proceed by elementary steps involving only intermediates with 16 or 18 metal valence electrons. Other pathways are usually unfavorable because of the relatively high energy of 14- or 20-electron species. Exceptions are known involving both 14- and odd-electron species; however, 20-electron species are still extremely rare, and their proposal should be regarded with suspicion unless the evidence for them is compelling. It has proven possible to isolate 14-electron complexes with very bulky ligands, such as $Pd[P(t\text{-}Bu)_3]_2$[11] and $RhCl(PCy_3)_2$.[12] Furthermore, species having stable 16-electron configurations, such as $Rh^I$, $Ir^I$, $Pt^{II}$, and $Pd^{II}$, are more likely to show exceptions. There is good kinetic evidence for a $RhClL_2$ intermediate in the reaction of $H_2$ with $RhCl(PPh_3)_3$[13,14] and for a $Bu_2PtPPh_3$ intermediate in the decomposition of $Bu_2Pt(PPh_3)_2$[15]—though electron counting is difficult in cases where triphenyl phosphine could donate electrons from an aromatic ring as it does in $Rh(PPh_3)_3^+$.[16] Another possibility for $RhCl(PPh_3)_2$, first suggested by Osborn[17] and more recently supported by the work of deCroon and co-workers,[18] is that the apparent 14-electron complex is stabilized in benzene solution as a 16-electron $\eta^2$-benzene solvate. There is growing evidence that many organometallic reactions actually proceed through paramagnetic odd-electron species,[19,20] and so fall outside the 16- and 18-Electron Rule. Nevertheless, it still encompasses a great deal of the chemistry of diamagnetic transition-metal complexes and often provides a useful starting point for studying mechanisms.

The approaches most familiar to organometallic chemists for the inference of mechanisms from kinetics studies have been discussed by Pearson.[21-23] It also seems that organometallic chemists interested in determining mechanisms of homogeneous catalytic reactions can profit a great deal from examining the approaches by enzymologists and biochemists. Lineweaver and Burke[24] showed that kinetic data can often be put in convenient linear form for determining rate constants and rate laws by plotting rate$^{-1}$ against $[S]^{-1}$, where [S] is the concentration of substrate or reactant. King and Altman[25] described a simplified procedure for writing rate equations for a cyclic series of reactions—found in both enzyme and organometallic catalytic systems. Cleland has published extensively on enzyme kinetics and the relationships between rate laws and mechanisms.[26-28] Hopefully, organometallic chemists will begin to apply more of these methods. Segel's book[29] is recommended for the interested reader.

Examination of the history of mechanisms inferred from gas phase data emphasizes the need for caution in interpreting kinetics without the use of ancillary confirmation from other methods. For example, the concept of CTSRDS assumes a single well-defined rate-determining step. Since the rate-determining step may vary with changes in concentrations, difficulties may be encountered, particularly when radical chain processes are involved. Thus, one may need to answer the question: "What is the mechanism of the reaction under this set of conditions?;" rather than simply "What is the mechanism of the reaction?" Furthermore, one assumes he has determined the rate law completely and may be bamboozled by deceptively simple rate laws. For example, kinetics studies do not generally allow detection of solvent involvement. Hence, a $k$ in an equation such as (7) may actually be

$$k = k'[\text{solvent}]^n \qquad (11)$$

and the composition of the transition state in the rate-determining step determined from the observed rate expression might either include or be missing several molecules of solvent.

Kinetics studies of gas phase reactions of $H_2$ and $I_2$ dating from the turn of the century provided the classic textbook example of a bimolecular reaction until 1967. That "the reaction between hydrogen and iodine is known to take place at bimolecular collisions involving a single molecule of each kind"[21] was a cornerstone for the study of reaction mechanisms until Sullivan[30] showed that mechanisms involving radical chain processes were consistent with additional data. Thus the concerted reaction mechanism (12):

$$H_2 + I_2 = 2HI \qquad (12)$$

is no longer the cornerstone of the study of mechanisms that it once was. This reaction, however, serves to illustrate the difficulty of identifying the presence of radical chain pathways by conventional kinetics.

One might think that radical reactions would be generally accompanied by fractional order in the rate law, but often surprisingly simple rate laws can be found. The decomposition of ethane (13), for instance, gives a rate law

$$C_2H_6 = C_2H_4 + H_2 \tag{13}$$

that is simply first order in ethane. More detailed studies reviewed by Laidler[31] show that complex radical paths are involved. Many "first-order" gas phase processes, such as the isomerization of *trans*-1,2,-dideutereoethylene, become second-order processes at low pressure.[32] This follows from a rate law in which a term in concentration becomes insignificant at low concentration. Thus, in these cases the CTSRDS approach might be misleading.

Prior to 1979, the Wacker oxidation of ethylene to acetaldehyde was generally suggested to proceed via a *cis* addition of Pd(II) and coordinated hydroxide based upon kinetic evidence.[33] The original rate law is given in equation (14):

$$d[\text{product}]/dt = k[\text{ethylene}][\text{Pd}]/[\text{H}^+][\text{Cl}^-] \tag{14}$$

The $1/[\text{H}^+]$ dependence was originally interpreted in terms of *cis* attack (15). The stereochemical studies of Baeckvall, Akermark, and Ljunggren[33]

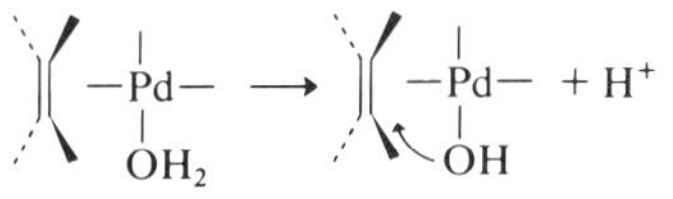

$$\begin{array}{c}\diagup\\ \diagdown\end{array}\Big(\begin{array}{c}|\\ -\text{Pd}-\\ |\\ \text{OH}_2\end{array} \longrightarrow \begin{array}{c}\diagup\\ \diagdown\end{array}\Big(\begin{array}{c}|\\ -\text{Pd}-\\ \diagdown\text{OH}\end{array} + \text{H}^+ \tag{15}$$

showed that *trans* attack of water, (16), on coordinated ethylene was consistent not only with the kinetic data, but also with the stereochemical data.

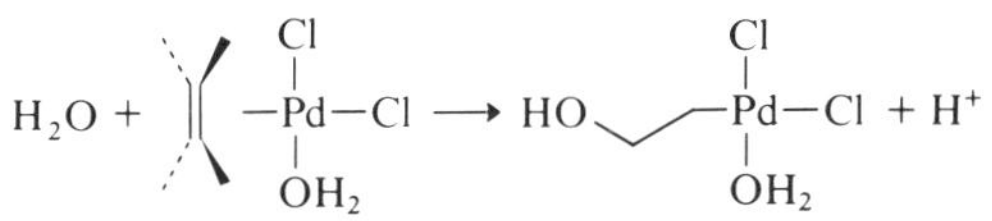

$$\text{H}_2\text{O} + \Big(\begin{array}{c}\text{Cl}\\ |\\ -\text{Pd}-\text{Cl}\\ |\\ \text{OH}_2\end{array} \longrightarrow \text{HO}\diagdown\begin{array}{c}\text{Cl}\\ |\\ -\text{Pd}-\text{Cl}\\ |\\ \text{OH}_2\end{array} + \text{H}^+ \tag{16}$$

Note that this latter mechanism requires nucleophilic attack on a coordinated olefin which is more electrophilic owing to the charge on the complex. Thus, the rate laws derived from mechanisms incorporating (15) or (16) differ in the order of solvent, which illustrates the potential pitfalls and difficulties of dealing with solvent in rate expressions. Thus, as suggested by Atkins,[34] a *proof of mechanism in a chemical reaction is less like a mathematical proof and more like a proof in a court of law.* In general, the greater the number of confirmatory experiments, the better. Nevertheless, as in the courts, the reputation and rhetoric of the lawyer and the persuasiveness of the circumstantial evidence may sway opinions regarding the correctness of a mechanism.

### 1.3. The Use of Isotopic Techniques

Following the caveats above with regard to possible solvent involvement or unusual rate laws, one can make a reasonable determination of the CTSRDS and provide reasonable estimates of intermediates in a catalytic cycle using simple rules, such as the 16- and 18-Electron Rule. Thus, one can create a diagram, such as Figure 2.1, with reasonable species, which is consistent with the observed rate law. Unfortunately, this provides a model, not a proof, of the events involved in the catalysis. The case can be made stronger by testing the mechanism, or individual steps in it, by spectroscopic or other more indirect methods.

#### 1.3.1. Isotopic Labeling

Isotopic labeling can often be very helpful and sometimes indispensable in elucidating the fine details of mechanisms, both from tracing the location of a particular atom or group throughout the course of the series of steps from reactant to product, and by utilizing kinetic isotope effects. As the principles are straightforward and well-known, they will not be detailed here; however, some recent developments will be reviewed.

The lowering of carbonyl stretching frequencies by the introduction of $^{13}CO$ has provided innumerable systems in which the course of carbon monoxide substitution, insertion, and migration could be readily followed by infrared. The advent of Fourier Transform nmr and high field nmr spectrometers has provided straightforward analysis not only of $^{13}C$ labeling, but of $^2H$, and $^{17}O$, as well as direct detection of many metal nuclei.[35-37] The use of $^{17}O$ nmr appears to provide a particularly useful technique for the study of carbonyls in metal systems with $I > \frac{1}{2}$ nuclei, such as cobalt, for which $^{13}C$ resonances are often broad or unobservable.[38]

The large number of catalytic applications of rhodium suggests that use of $^{103}Rh$ nmr spectra should be potentially interesting. Unfortunately, the sensitivity for this nucleus is quite low and relaxation times are quite long, which implies that high concentrations and long collection times would be rquired for the observations. Until recently $^{103}Rh$ chemical shifts were only obtained indirectly through spin-decoupling techniques

(INDOR);[37,39] however, direct detection is practical with modern instrumentation.[40,41] The concentrations and time required for these studies make extensive use of $^{103}$Rh nmr by conventional Fourier Transform methods relatively unattractive for catalytic study applications. The development of a polarization transfer technique (as opposed to an Overhauser enhancement) known as INEPT,[42] however, promises to allow detection under reasonable concentration and time conditions.[43] Perhaps one of the more interesting developments has been the development of sample systems that allow the determination of nmr spectra under high pressure, such as 1000 atm of $CO/H_2$.[44] Even without exceptional equipment, it is practical to collect spectra in heavy-walled tubes ($\sim$10 atm), as in the $^{31}$P nmr studies of rhodium–phosphine hydroformylation systems.[45,46]

GC-mass spectroscopy also provides a straightforward method of analyzing isotopic distributions in product mixtures; however, nmr generally is more useful for evaluating the stereochemistry of isotopic substitution. This ease of assigning position by nmr presumably accounts for the lack of general use of radioactive labels by organometallic chemists; however, this technique is certainly deserving of more attention than it currently receives. On occasion, microwave spectroscopy can provide an extraordinarily unique characterization of diastereomers, particularly of those in which stereoisomers involving chiral centers containing deuterium are involved.[33]

The use of isotopes, as well as the appropriate control experiments required for valid interpretation, is illustrated well by the reductive elimination studies of Norton.[47] Reductive elimination is often a key step in proposed mechanisms and is generally presumed to be intramolecular in nature; however, crossover experiments have shown that intermolecular processes can be involved. Thus, the observation that mixed elimination species and an osmium dimer, $Os_2(CO)_8R_2$, **10**, are found requires an intermolecular reaction (17, 18).

$$Os(CO)_4H_2 + Os(CO)_4D_2 = H_2, HD, D_2 + \mathbf{10} \qquad (17)$$

$$Os(CO)_4H(CH_3) + Os(CO)_4D(CD_3) = CH_4, CH_3D, CD_3H, CD_4 + \mathbf{10} \qquad (18)$$

The absence of $CD_2H_2$ and other control experiments, such as proving that scrambling of products does not occur under the reaction conditions, prove that intermolecular reactions are involved.

### 1.3.2. Spin-State Labeling

When confronted with the problem of labeling a group or an atom, a synthesis of a new compound using isotopically labeled groups often comes to mind. In relatively rapid reactions ($t_{1/2} < 10$ sec), it is often possible to use the effect of spins on the nuclei already present to label a position. This is particularly useful with phosphorus-containing ligands and with metals having nuclei with spin one-half.

In essence, within the magnetic field there are two separate types of phosphorus nuclei, those with spin($+\frac{1}{2}$) and those with spin ($-\frac{1}{2}$). Interchange of P($+\frac{1}{2}$) with P($-\frac{1}{2}$) can often be detected by nmr. This can frequently be used to detect rapid intermolecular interchange of phosphine ligands. Thus, if a group attached to a metal is coupled to a $^{31}$P nucleus in a phosphine, the resonance for that group is split into a doublet. One-half of the doublet arises from the P with spin $+\frac{1}{2}$ and the other from P $-\frac{1}{2}$. Essentially one-half arises from one set of molecules and the other from another set. Consequently, no matter what intramolecular rearrangements may be taking place within the species containing P $+\frac{1}{2}$, that resonance should remain distinct from the other with P $-\frac{1}{2}$. Once P $+\frac{1}{2}$ and P $-\frac{1}{2}$ exchange, the two resonances average, just as the case of a proton exchange between water and an alcohol gives a single resonance. Thus, the loss of coupling to phosphorus indicates intermolecular exchange on the nmr time scale. The rates can be calculated from line–shape analysis[48,49], if an intermediate case is found; however, exchange rates of greater than 1 sec$^{-1}$ are required. This is an important point, nevertheless, because rates of over $\sim 1$ sec$^{-1}$ are required to induce a line–shape change. Thus if the temperature is lowered and the splitting, or a "static" spectrum, is observed, it does not mean that exchange has stopped, merely that it has slowed below a rate of 1 sec$^{-1}$.

This approach can be useful in distinguishing when a ligand, such as ethylene, is exchanging rapidly between complexes, because coupling of the protons of ethylene to both the metal and the phosphine ligands will be averaged and effectively removed.

Intramolecular rearrangements in metal clusters can also be detected using the effects of different spin-state interactions. Platinum clusters are particularly bizarre in that one not only considers the effects of various $^{195}$Pt isotopomers and their relative populations, but also the effects of nonequivalences resulting from having a coupling not only to the $^{195}$Pt that is directly attached, but also to a remote $^{195}$Pt.[50] The static $Pt_3(\mu\text{-}CO)_3(PR_3)_3$ systems are informative in this regard. The isomeric distribution for platinum has significant concentrations of several even mass nuclei with I = 0, but only $^{195}$Pt with a natural abundance of 33.7% has a significant nonzero spin with I = $\frac{1}{2}$. Thus the analysis of spectra of platinum clusters requires consideration of a system with approximately one-third of the nuclei with spin $\frac{1}{2}$ and the remainder with spin zero. The statistical distribution of $^{195}$Pt among the isotopomers **11** to **14** is shown opposite.

The observed spectra are a superposition of the spectra arising from all four isotopomers. Portions of the $^{31}$P spectra can be analyzed for **12** by

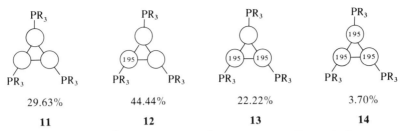

considering **12**-Pt($+\frac{1}{2}$) and **12**-Pt($-\frac{1}{2}$) separately. The relative ease of determining $^{195}$Pt spectra and the potential use of platinum in catalytically active systems suggests that the use of nmr in these systems will continue to increase. The coupling constants of $^{195}$Pt to other metal nuclei appear to be some of the largest known, e.g., $^1J(^{195}$Pt-$^{119}$Sn) in trans-[PtCl(SnCl$_3$)(PEt$_3$)$_2$ is 28,954 Hz,[51] and certain aspects of the spectra are simplified. One should note, however, that with extremely large coupling constants $J/\delta$ may be significantly greater than zero, and splittings arising from couplings between different metals may be large enough relative to the observed frequencies that the spectra may not be first order. This gives the unusual feature that $^{195}$Pt satellites in $^{119}$Sn spectra may not be centered about the zero-spin platinum resonances.

The use of spin-state labeling requires that interchanges of coupled nuclei occur on a time scale rapid compared to the relaxation time of the nuclei. Ordinarily, the rates necessary to average resonances are sufficiently fast, and this is not a problem; however, there are two situations where difficulties may be encountered. The coupling between a proton and another nucleus in two different environments may be of opposite sign; hence, the averaged coupling constant may be zero if appropriate ratios of the two environments are populated. This might occur in an intramolecular exchange which involved a free and bound phosphorus ligand. A situation can also develop where coupling to a nucleus can be lost owing to rapid relaxation of that nucleus. This can occur in systems which contain paramagnetic species.[52] For example, a small fraction of nickel (II) (from a syringe needle, for instance) can relax $^{31}$P nuclei sufficiently and rapidly to effectively decouple them. In this case, the rapid relaxation, when the $^{31}$P coordinates to the nickel, in addition to the rapid exchange between bound and free phosphine, results in minor shifts of the proton resonances but complete decoupling of the phosphorus.

### 1.3.3. Saturation Labeling in nmr

The broadening and eventual coalescence of resonances of nuclei involved in exchange between sites has provided a convenient tool for determining exchange rates[48,49] for relatively rapid reactions. The additional width of the resonances, W, as exchange begins to proceed rapidly enough to be observed as line broadening is related to the rate for leaving that site by a simple relationship (19). In some cases, different mechanisms

$$k = \pi W \qquad (19)$$

can be distinguished by the lifetimes of nuclei in various environments, and since 1966, when line-shape arguments were first used for an organometallic mechanistic study,[53] many insights into mechanisms have been provided by line-shape studies. As these methods are now fairly well known,[48,49] they are mentioned here only to indicate their potential utility in studying exchange of nuclei which proceed so rapidly that isotopic labeling would be impractical. For example, a situation may occur where an isotopic-labeling study of the mechanism of interconversion of two tautomeric species may be made difficult owing to relatively rapid equilibrium, which would scramble the labels before observation was possible. The rates and at least a knowledge of the permutations of the nuclei would be forthcoming from an analysis of line shapes as a function of temperature.

Although line-shape analysis is probably the best approach for the study of rates using nmr, the technique of saturation transfer is more valuable for investigating mechanisms and the nature of interchanges of several nuclei.[49] If the rate of exchange is within 10% of the $T_1$ relaxation rate of a nucleus or faster, i.e., $\sim 10^{-1} < k < 10 \sec^{-1}$, the effect can be observed. In the simple case where there is an exchange between nuclei in an A site and B site, "saturation" of the nuclei in the A site will be transferred to the B site when the exchange rate is faster than the relaxation rate. Thus, irradiating nuclei in the A site not only saturates the A resonance and decreases its intensity to nearly zero but also decreases the intensity of the B resonance. The magnitude of the effect on B is a function of the relative magnitudes of relaxation rate and exchange rate. The exchange between a methyl group and a methylene hydride (see Equation 20) has been demonstrated by this technique.[55]

$$HOs_3(CO)_{10}CH_3 = H_2Os_3(CO)_{10}CH_2 \qquad (20)$$
$$\mathbf{15} \qquad\qquad \mathbf{16}$$

Saturation of the methyl group in **15** leads to an observed decrease in the intensity of *only one* of the hydride signals in **16**. This spin saturation transfer experiment conclusively demonstrates the elimination of a hydride from a methyl to produce a methylene and that only one hydride site is involved to a significant degree. It should be noted that it is impractical to detect transfer to the other hydride site if the rate is less than a few percent of that of the primary process.

More complex transfers of sites have been demonstrated by saturation transfer studies used in the elucidation of sigma-pi rearrangements of allyl groups,[56,57] understanding the nature of which is essential for interpreting asymmetric induction in reactions of allyl metal complexes.

The original saturation transfer methods were developed for field-swept spectrometers.[58] With the advent of fixed-field modes of operation, the use of an external oscillator makes it a reasonably convenient technique. While it was straightforward to observe an effect, the demonstration that the relaxation times were comparable in each site or the measurement of relaxation times in the sites was tedious.[54] With the advent of FT nmr spectrometers, the potential of effectively labeling sites by spin saturation has increased owing to the ease of measuring relaxation times. In FT spectrometers the saturation of a given site is usually accomplished in one of two ways: "continuous" irradiation or decoupling, or gated irradiation. The normal mode used for decoupling in an FT instrument is analogous to using an additional frequency source for the decoupling frequency. (This frequency may actually be turned on and off rapidly to prevent wreaking havoc with the detection system.) Although only low power is necessary to saturate resonances, some decoupling of resonances coupled to the one being irradiated may occur leading to confusion about intensities. In gated irradiation, saturation is provided up to the point where a 90° pulse is given to observe the intensities and then turned off during acquisition of the data. The saturation observed in the transformed spectrum is that appropriate to the time when the pulse was given, as the loss of saturation is a $z$-magnetization property, and the data acquisition measures $x, y$-magnetization produced by the 90° pulse.

Most modern spectrometers provide these options for observing protons, hence readily allowing saturation transfer experiments, but they have only recently been used with $^{13}C$[59,60] on modified spectrometers. The addition of another frequency source, as well as a knowledge of mixers, wiring diagrams, and the confidence of other users of the instrumentation are required for these modifications. This often prevents routine use of the method for a variety of nuclei. This can partially be circumvented by transfer of Overhauser enhancement instead of saturation; fortunately, however, Morris and Freeman[61] developed the DANTE pulse sequence which effectively allows homonuclear irradiation of the nucleus being observed without the need of additional external frequency sources. More complex 2-D methods can also provide similar information when many sites are involved.

### 1.3.4. Kinetic Isotope Effects

The introduction of isotopes, particularly deuterium, for labels may also provide the opportunity for further elucidation of the mechanism via kinetic isotope effects.[62-65] Physical organic chemists, in particular, developed the interpretation of these effects in terms of a mechanism for a large number of systems, and their approach and vocabulary should prove useful when applied to catalysis by organometallics.

The differences in zero-point energy cause the dissociation energy for a C–D bond to be about 2.3 kcal/mol greater than for a C–H bond.[64] Hence, if no new bonds were formed in the activated complex of a hydrogen transfer reaction, the activation energy would be 2.3 kcal/mol less for H than for D. This would translate into an isotope effect of $k_H/k_D$ of nearly fifty at 25°. Many primary isotope effects involving deuterium in organic systems lie in the range of four to eight. Since new bonds are usually being formed as old ones are being broken in the activated complex, one must also consider zero point effects in the activated complex. The simplest approach leads one to suggest that a bigger isotope effect means more bond breaking occurring in the transition state. Furthermore, it also suggests that if one proposes CH-bond breaking in the rate-determining step of a reaction, strong evidence in its favor would be a significant isotope effect on the rate. Unfortunately, this is a naive approach, and the absence of a large isotope effect may not obviate the proposal, particularly in cases where metal hydrides are involved. An alternative approach is to describe the degree of bond breaking or making in the activated complex, as an early transition state, symmetrical TS, or late TS.[63,65]

$$\begin{array}{ccc} \text{C-H}-\text{B} & \text{C}-\text{H}-\text{B} & \text{C}-\text{H-B} \\ \text{early TS} & \text{symmetrical TS} & \text{late TS} \end{array} \quad (21)$$

One also might speak of the transition state being more like reactants or products in attempting to arrive at rationalizations for differences in $E_a$ in comparing compounds. The important feature here is that a symmetrical transition state should give the largest isotope effect, because it presents the largest differences in zero point energy in the TS relative to the ground state.

In organic systems where a transfer of hydrogen is often between atoms for which the atom-hydrogen vibrational frequencies are similar, the concept of a "symmetrical" transition state can be readily visualized. It would appear that the most "symmetrical" TS is often viewed as one in which the H–C and H–B bonds in (21) are stretched to the same degree. More esoteric descriptions involving intersections of hypothetical potential surfaces are also used; nevertheless, definitions of "symmetrical" in cases other than those in which C = B are obscure. An empirical approach has been proposed on the basis that the acidity of the C–H and B–H bonds should reflect the dissociation energy and, hence, a measure of the potential surfaces' C–H and B–H bond breaking.[66] The observation that maximum isotope effects are observed when the B–H and C–H acidities are similar gives the approach credibility. Nevertheless, caution is warranted in offering a "symmetrical" or "unsymmetrical" transition state as a rationalization of isotope effects, if the bonding in C–H and B–H are substantially different.

Since metal hydride stretching frequencies are low compared to C–H frequencies (usually 1500–2100 for $M$–H compared to C–H 2900–3100 $cm^{-1}$), the transfer of hydrogen from metal to a carbon can result in a situation with major differences in zero-point energy in reactant and product. More importantly, the likelihood of an "unsymmetrical" TS is high and the possibility that frequencies are even higher in the activated complex than in the reactant exists. This is illustrated in the reduction of $\alpha$-methylstyrene by $HMn(CO)_5$.[67] Transfer of a hydrogen atom from the metal hydride ($\nu Mn$–H = 1780 $cm^{-1}$) has been suggested as the rate-determining step[62].

$$(C_6H_5)C(CH_3)=CH_2 + HMn(CO)_5 = (C_6H_5)(CH_3)_2C^{\bullet} + (CO)_5Mn^{\bullet} \qquad (22)$$

Although the kinetics in this system are complicated by the possibility of forming $(C_6H_5)(CH_3)_2CMn(CO)_5$ initially or by radical cage recombination and subsequent alpha elimination, the fact remains that at some point a Mn–H bond is broken and an inverse isotope effect of $k_H/k_D = 0.4$ is observed.

Inverse isotope effects have been observed for some time with metal hydride reactions, such as tin hydrides,[68] but their observation in transition metal reductions may have a considerable impact on the interpretation of results of catalytic studies. It would appear that a process such as (22) may be involved in reductions with $Co(CO)_4H$, which also show an inverse isotope effect ($k_H/k_D = 0.58$ for 1,1-diphenylethylene; and 0.43 for 9-methylidenefluorene).[69,70] There is now evidence for radical involvement in the cobalt system.[70] Neither the involvement of radicals, nor an unsymmetrical transition state, however, is required for an inverse isotope effect. For example, the radicals taken to have arisen from H atom addition in (22) might have arisen from addition of $HMn(CO)_5$ to the double bond (23) followed by homolytic cleavage (24).

$$(C_6H_5)C(CH_3)=CH_2 + HMn(CO)_5 = (C_6H_5)(CH_3)_2C-Mn(CO)_5 \qquad (23)$$

$$(C_6H_5)(CH_3)_2C-Mn(CO)_5 = (C_6H_5)(CH_3)_2C^{\bullet} + (CO)_5Mn^{\bullet} \qquad (24)$$

Distinguishing radical species involved in the catalytic loop instead of being involved in an unimportant off-loop reaction is difficult but can be accomplished via CIDNP experiments[67] (see Section 1.5.3.). An inverse isotope effect can also arise from the isotope effect on an equilibrium prior to the rate-determining step.[66] An inverse isotope effect of 0.29 is observed in the reduction of carbonyls with $HFe_2(CO)_8^-$,[72] and it has been attributed to an equilibrium isotope effect on (25), which effectively reduces the concentration of the reactive deuterated alkyl-$Fe_2(CO)_8^-$ in the rate-determining step (26).

$$DFe_2(CO)_8^- + RCH=CHCOR = RCH(D)-CH(CO_2R)[Fe_2(CO)_8]^- \quad (25)$$

$$RCH(D)-CH(CO_2R)[Fe_2(CO)_8]^- + D^+ \rightarrow RCHD-CHD(CO_2R) + Fe_2(CO)_8 \quad (26)$$

Consideration of changes in vibrational frequency suggested a $K_H/K_D$ of approximately 0.5. This requires a rather large equilibrium isotope effect in (25) to account for the observed $k_H/k_D$ of 0.29, as well as any effect from a subsequent reaction such as (26). Nevertheless, the point to be made here is that an equilibrium isotope effect on a prior equilibrium can possibly account for the observation of an inverse kinetic isotope effect.

### 1.3.5. Equilibrium Isotope Effects

Saunders[73] developed a very useful technique for studying equilibrium isotope effects on the proton nmr spectra of carbonium ions, which is known as *isotopic perturbation of degeneracy*. It is effective in problems where distinguishing symmetrical from fluxional unsymmetrical compounds is required. This effect is amplified in $^{13}C$ spectra[74] and has been used to distinguish fluxional $\eta^1$-cyclopentadienyl metal complexes from $\eta^5$-Cp complexes.[75]

The process of $\alpha$-elimination from methyl groups bound to metals[76] may be a key step in catalytic alkane activation, and hydride addition to metal-bound methylene has been suggested as the chain propagation step in the Fischer–Tropsch reaction.[77–79] The relationship between symmetrical, **17**, and unsymmetrical bridging methyl groups, **18**, and methylene hydrides, **19** and **20**, can be addressed by isotopic perturbation of degeneracy studies; and other perturbations of nmr spectra by the introduction of isotopes.

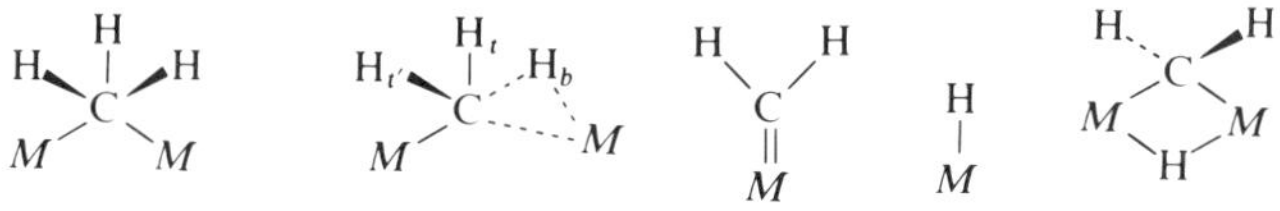

    **17**        **18**        **19**        **20**

The presence of an unsymmetrical bridging methyl as in **18** has been clearly demonstrated in an X-ray structure in which hydrogen atoms were located in $[Fe_2(\mu\text{-}CH_3)(\mu\text{-}CO)(CO)_2(\mu\text{-}dppm)(\eta\text{-}C_5H_5)]PF_6$.[80] The unsymmetrical nature of the C–H–Fe interaction can be implicated by the isotopic perturbation of degeneracy observed in the proton nmr spectrum of a partially deuterated methyl species. This isotopic perturbation method also led to the earlier suggestion of a similar bridging group in $HOs_3(CO)_{10}CH_3$.[55]

The single resonance observed in the proton nmr of the iron complex suggests a low barrier to rotation of the methyl group. Hence, the methyl resonance of $\delta - 2.90$ occurs at a chemical shift which is the weighted average of the two terminal proton environments and the one bridging environment. The averaged shift may be considered as reflecting the percentage of time a nucleus spends in a given environment. That is, it is given by equation (27) where the population of protons is $p$.

$$\delta_{ave} = (p_t\delta_t + p_{t'}\delta_{t'} + p_b\delta_b)/(p_t + p_{t'} + p_b) \tag{27}$$

If the two terminal environments are identical, which they often are not, the equation reduces to (28).

$$\delta_{ave} = (2p_t\delta_t + p_b\delta_b)/(2p_t + p_b) \tag{28}$$

Hence, for a $CH_3$ where $p_t$ and $p_b$ are one, the equation simplifies even further to (29).

$$\delta_{ave} = (2\delta_t + \delta_b)/3 \tag{29}$$

In the mono- and didutereo species, the zero-point energy differences will favor H over a D in the bridging position owing to the lower vibrational frequencies expected for a Fe–H–C bridge relative to a terminal C–H. Therefore, the populations in (27) or (28) will no longer be equal. If $K$ is the equilibrium constant for a single pairwise exchange between a bridging and terminal position (30),

$$K = [18B]/[18T] \tag{30}$$

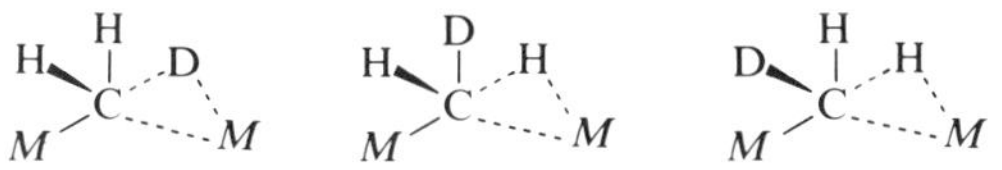

18B      18T      18T'

equations (31) and (32) follow from the approximate equation (28).[55]

$$(CH_2D)_{ave} = (\delta_t + K\delta_t + \delta_b)/(K + 2) \tag{31}$$

$$(CHD_2)_{ave} = (2K\delta_t + \delta_b)/(2K + 1) \tag{32}$$

Thus, according to (31) with $K < 1$ and $\delta_b$ upfield of $\delta_t$, an upfield shift in the proton spectrum of the -CH$_2$D would be expected. The $\delta - 3.47$ shift observed for [Fe$_2$($\mu$-CH$_2$D)($\mu$-CO)(CO)$_2$($\mu$-dppm)($\mu$-C$_5$H$_5$)]PF$_6$ represents an upfield shift of 0.57 ppm. In the deuterium nmr spectrum, the resonance is observed at $\delta - 1.88$, a downfield shift of 1.02 ppm showing approximately twice the effect required by the statistics (Equation (32) with $K^{-1}$ substituted for $K$). Equations (31) and (32) neglect an *intrinsic shift*[73-75] in the proton spectrum arising purely from substitution of $D$ for $H$. This effect is relatively small (~0.01 ppm) in proton spectra but accounts for lack of exact agreement between $^1$H and $^2$H nmr results and may become important in evaluating $K$.

The proton resonance for the CH$_2$D group shows a pronounced temperature dependence. Lowering the temperature 95° changes the shift by 0.25 ppm to $\delta - 3.72$, whereas the CH$_3$ resonance shifts only 0.02 ppm. This temperature dependence is an expected property of the perturbation of an equilibrium by an isotope effect and should always be verified experimentally before an isotopic perturbation of degeneracy rationalization is invoked.

A further indication of a bridging methyl group is provided by the average coupling constants of the protons to $^{13}$C. The $^{13}$C spectrum shows a quartet with $^1J(CH)$ of 114 Hz. This also represents an average of $^1J(CH_b)$, $^1J(CH_t)$, and $^1J(CH_{t'})$ as in equation (27) with $J$'s replacing $\delta$'s. A bridging methylene as in H$_2$Os$_3$(CO)$_{10}$CH$_2$, which contains a type **20** unit,[81-82] has values of $^1J(CH)$ of 140 and 143 Hz.[83] The lower $s$ character in the bonding of methyl would suggest a lower $^1J(CH)$ in a symmetrical methyl bridge, **17**, as in Al$_2$(CH$_3$)$_6$[84] or Y$_2$(CH$_3$)$_2$($\eta^5$-C$_5$H$_5$).[85] The bridging methyl CH would probably have a lower $^1J(CH)$ and the terminal CH a slightly higher $^1J(CH)$ in **18** than observed in **17**. Thus, the averaged $^1J(CH)$ in an unsymmetrically bridging methyl, **18**, should fall in a range between 95 Hz ($J_t = 142$; $J_b = 0$) for the extreme methylene hydride **19** form of a weak $M$-H—C bond and about 125 Hz for the symmetrical form, **17**. Solution of analogs of equations (29), (30), and (31) for the couplings of $^1J(CH)$ observed for 121.1(CH$_3$); 118.9(CH$_2$D) and 116.4(CHD$_2$) for HOs$_3$(CO)$_{10}$CH$_3$[55] yield values of $^1J(CH)_t = 150 \pm 10$ Hz and $^1J(CH)_b = 60 \pm 20$ Hz.[55] The values for the $^1J(CH)$ for the CH$_2$D and CHD$_2$ also show the expected variation with temperature.

Perhaps one might now confidently assume that a moderately high-field three-proton resonance showing an isotopic perturbation of resonance is indicative of an unsymmetrical bridging methyl, **18**. Some caution is advised, however. The presence of a coupling constant, $^1J(CH)$, above about 105 Hz, is a requirement for this situation, as is a variation with temperature of both $^1J(CH)$ and $\delta(CH)$ in the partially deuterated complex. Perhaps the most difficult problem to eliminate is that a fluxional methylene hydride could give rise to many of these observations. *The isotopic perturbation of resonance only indicates the lack of a symmetrical structure, but does not indicate the nature of the asymmetry.* The conversion of methylene hydride to methyl has been observed to occur with a barrier of about 20 kcal/mol.[75] Thus a fluxional structure, such as **19** or **20**, would show all the properties of unsymmetrical bridging methyls with the exception of having a slightly smaller $^1J(CH)$. A rapid equilibrium between a symmetrical methyl and methylene hydride would be virtually impossible to detect by the nmr methods discussed above. For this reason, corroborative evidence for the presence of a single isomer by a technique such as IR should be available.

Without the potential problem of a different structural form being isolated in the solid than is present in solution, X-ray and neutron diffraction studies would be the methods of choice for proving an unsymmetrical bridging methyl. This problem arises in one of the currently claimed cases of unsymmetrical bridging methyls, $HOs_3(CO)_{10}(CH_3)$, where the equilibrium with a methylene hydride results in the crystallization of $H_2Os_3(CO)_{10}(CH_2)$.[74,76] One should note that neutron diffraction experiments also allow a quantitative measurement of the equilibrium isotope effect.[74]

For simplicity, we have neglected the fluxional process in which the stronger bond is interchanged between one metal and the other in this argument. If the chemical shifts of the terminal protons in a static bridging methyl group, **18**, are nonequivalent and there is a reasonably large equilibrium isotope effect, (30), the rotation of the methyl will not make the averaged environments for the terminal protons equivalent. Thus, in both $[Fe_2(\mu\text{-}CH_2D)(\mu\text{-}CO)(CO)_2(\mu\text{-}dppm)(\eta\text{-}C_5H_5)]PF_6$ and $HOs_3(CO)_{10}\text{-}CH_2D$ the two methyl protons should be nonequivalent if only methyl rotation giving equilibria between **18B** = **18T** = **18T'** were involved. The observation of a single signal for these protons implies that the shift difference is either not resolved or that the methyl protons rapidly switch from bridging one metal to the other. This nonequivalence could also be observed in an unsymmetrical methylene hydride system, **19**, if $H$ addition and removal were stereospecific.

The isotopic perturbation method is a powerful one for demonstrating the unsymmetrical nature of equilibrating systems; however, as indicated above, the nature of the dissymmetry requires care in interpretation. In particular, the evaluation of the magnitude of the averaged $^{13}C\text{-}^1H$ coupling and the variation of the shift with temperature are essential. Owing to the possibility of a methylene–hydride equilibrium, it is also prudent to investigate the possible presence of these tautomers by other physical methods.

## 1.4. The Implications of Product Stereochemistry for Mechanisms

One of the main potential advantages of homogeneous catalysts is the ability to control or modify the regioselectivity and stereoselectivity of the catalysed reactions. It follows, therefore, that the stereochemical consequences attending the key steps of a catalytic cycle may be of prime importance in understanding and designing catalyst function. In this section some of the key steps proposed in catalytic cycles will be shown as a vehicle for illustrating some of the stereochemical methods.

The retention, inversion, or loss of stereochemistry at a chiral center is one of the prime objectives of a stereochemical study of a reaction. Thus, displacement reactions, such as a Walden inversion, are classic examples of such studies. Owing to difficulties, such as beta-eliminations and uncertainties with regard to absolute configurations, following the stereochemistry of many key reaction steps was not widely practiced until the seventies. For example, if a M–C(H)(CH$_3$)(C$_6$H$_5$) was stable and did not eliminate styrene forming the metal hydride, the absolute configuration of the metal complex or the stereochemistry might be unknown for the degradations yielding organics of known configuration.

### 1.4.1. NMR Determinations of *Erythro* and *Threo* (CHD)$_2$ Fragments

Snyder[86-88] originated the study of dideutereo-2-phenylethyl groups for determining the stereochemistry of displacement reactions. Whitesides[89] initiated the use of these systems in the study of organometallic systems through examination of the reaction of erythro-3,3-dimethylbutyl-1,2-$d_2$-$p$-bromobenzenesulfonate, **21**, with [(C$_5$H$_5$)Fe(CO)$_2$]$^-$ (33) and the subsequent carbonyl insertion reaction promoted by triphenylphosphine (34).

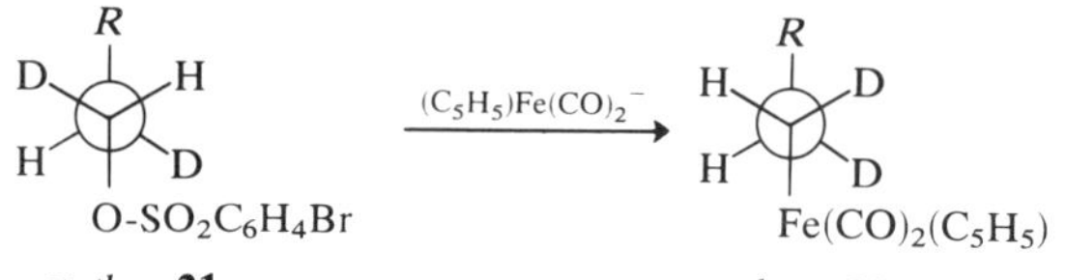

$$\text{erythro-}\mathbf{21} \xrightarrow{(C_5H_5)Fe(CO)_2^-} \text{threo-}\mathbf{22} \quad (33)$$

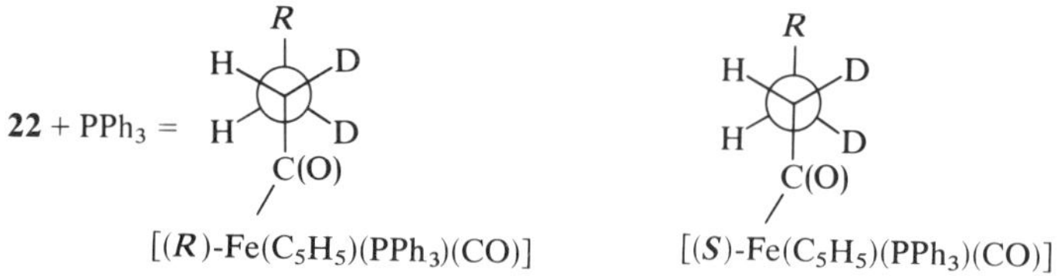

$$\mathbf{22} + PPh_3 = [(R)\text{-Fe}(C_5H_5)(PPh_3)(CO)] \quad [(S)\text{-Fe}(C_5H_5)(PPh_3)(CO)] \quad (34)$$

$$(R, S, R)\text{-}\mathbf{23} \qquad (S, S, R)\text{-}\mathbf{23}$$

Unfortunately, the difficulty of the syntheses and the yields of the *erythro* sulfonates dissuaded many potential users. Although the yields were eventually improved,[90] the synthetic procedures were still lengthy. Alternate routes involving hydroboration[91] and hydroalumination[92] might also be moderately attractive; however, the most straightforward synthesis of relatively pure 21-($R = t$-Bu) appears to be via sequential $Cp_2Zr$ reactions.[93] The most efficient synthesis of *threo*-**22**-($R$ = Ph) is via Rh-catalyzed hydrogenation of commercially available $\beta$-methoxystyrene.[94] Although the brosylate, **21**, contains an effective leaving group for strong nucleophiles, such as $[C_5H_5Fe(CO)_2]^-$ or pyridinebis(dimethylglyoximato)-cobalt(I), the triflate is more generally useful with less nucleophilic organometallic anions.[95]

Fortunately, the stereochemistry of the $(CHD)_2$ moiety can be determined readily by nmr experiments. The *trans* configuration of the protons in the *erythro* isomers implies a larger $^3J(HH)_{vic}$ coupling than in the gauche configuration found in the *threo* isomers. Hence, the coupling in *erythro*-**22**-($R = t$-Bu) is 13.1 Hz, whereas that in *threo*-**22** is 4.5 Hz.[89] Since a chiral iron center is formed in the phosphine addition–carbonyl-insertion reaction (34), the methylene protons are diastereotopic[96] and thus $(R,S,R)$-**23** and $(S,S,R)$-**23** are different, which requires careful interpretation of the observed spectra. Nevertheless, this method has become one of the most straightforward ways of examining changes in chirality in reactions of organometallics.

In equation (34) the ligand-promoted carbonyl-insertion reaction occurs with ~90% retention of configuration. Treatment of **22**-($R = t$-Bu) with bromine in pentane, chloroform, or carbon disulfide gives reaction (35) with >90% inversion of configuration.[90]

$$\text{threo-}\mathbf{22}\text{-}(R = t\text{-Bu}) + Br_2 \longrightarrow \text{erythro-}\mathbf{24} \quad (35)$$

It is tempting to extrapolate these results to all alkyl metal systems, and although carbonyl insertions generally appear to take place with retention in the alkyl group, *mirabile dictu*, halogenations occur via several paths in other organometallics. Thus, bromination of the $R = t$-Bu analog of **22** with $Cp_2Zr$ takes place with retention.[93] Since the mechanism for the iron system appears to occur via oxidation of the complex followed by backside nucleophilic attack of halide ion,[94,97] (36), the highly oxidized Zr(IV) reaction proceeds via a different route owing to the inaccessability of the oxidative route. Halogenations of cis-[*threo*-Ph(CHD)$_2$-Mn(CO)$_4$PEt$_3$] proceed with retentions of from 17% to 77% depending upon the conditions.[98] The more bizarre occurrence, however, is that the bromination

(35) of *threo*-**22**-($R$ = Ph) yields *threo*-**24**.$^{(94,97)}$ Hence, exactly *opposite* conclusions are drawn for the same metal systems with the only difference being $(CHD)_2C(CH_3)_3$ vs. $(CHD)_2C_6H_5$. Baird has used the $(CHD)_2Ph$ functionality extensively for stereochemical studies$^{(94,97-100)}$ and attributes this difference to the potential for formation of the bridged phenonium ion.$^{(94,97)}$ A key indicator of the phenonium ion or radical involvement is the observation that $\alpha$-$^{13}C$-labeled alkyl metal yields an equal mixture of $\alpha$- and $\beta$-labeled product,

$$PhCH_2C^*H_2\text{-}M + I_2 \rightarrow PhCH_2C^*H_2I + PhC^*H_2CH_2I \quad (36)$$

A similar conclusion also resulted from a study of the dideuteriomethylene complex, $(\eta^5\text{-}C_5H_5)Fe(CO)_2\text{-}(CD_2)(CH_2)Ph$.$^{(101)}$ Essentially, the bridged species retains the "cis" orientation of the deuteria in this intermediate, **25**, in equation (37), which accounts for retention in *threo*-**22**.

$$threo\text{-}\mathbf{22}\text{-}(R = Ph) + I_2 \longrightarrow \begin{array}{c} \text{bridged intermediate} \\ \mathbf{25} \end{array} \longrightarrow threo\text{-}\mathbf{24} \quad (37)$$

Iodide attack *trans* to the bridging phenyl at either carbon yields the *threo* isomer, but will yield the ($RS$) or ($SR$) isomer depending on which carbon is attacked. The basic assumption used in the interpretation of the *erythro* and *threo* isomer interconversions is that the configuration of the carbon $\alpha$ to the $R$ group remains constant. Hence, if inversion occurs at both centers, it would not be detected. Thus, an observation interpreted as "retention" with the $Ph(CHD)_2$-system should be viewed with suspicion; nevertheless, racemization or inversion would be interpreted correctly. Studies with optically-active $(CHD)Ph$ corroborate the observed inversion with bromination found with $-(CHD)_2t$-$Bu$;$^{(97)}$ hence, care must be taken with interpretations using the $(CHD)_2Ph$ system, and it would appear that the butyl system provides a more reliable indication of processes at a primary carbon.

Since the stereochemistry of the $(CHD)_2$ group can be determined so conveniently directly from the nmr, it will remain a popular technique, and its utility is not limited to studies of previously prepared $X$-$(CHD)_2R$ compounds. Further evidence for the *trans* attack of water on coordinated olefin in the Wacker reaction (see Equation 16 in Section 1.2) was provided by the isolation of a $\beta$-alkoxyethylpalladium complex from the reaction of a *cis*-dideuterioethylene complex (38) with MeOH.

$$[(C_5H_5)(PPh_3)Pd^+] - \begin{matrix} D & H \\ & \\ D & H \end{matrix} + MeO^- \rightarrow \begin{matrix} & OMe \\ H & D \\ & \\ H & D \\ & Pd(C_5H_5)(PPh_3) \end{matrix} \quad (38)$$

threo-**25**

The vicinal coupling constants for the $(CH_2)_2$ analog of **25** showed vicinal coupling constants of 4.8 and 12.3 Hz. The $^3J(HH)_{vic}$ of 4.8 Hz in the deuterated product proved that it was threo-**25** and the attack of methoxide was trans to the metal.[102]

### 1.4.2. nmr Determinations of Other Diastereomers

The use of threo and erythro alkyl isomers can be extended and in some cases the preparative aspects improved by the use of other groups on the $\alpha$-carbon besides H and D. For example, $(RR, SS)$ and $(RS, SR)$-$C_6H_5(CHF)(CHD)Br$ have been used in studies of oxidative addition.[93] Although a secondary carbon is introduced, $(RS, SR)$-Ph(CHF)-$(CHBrCO_2Et)$ has also been used successfully. Both of these systems, however, have the potential of interpretational problems arising if apparent "retention" is observed. Cis- and trans-substituted cyclohexyl groups provide an opportunity to utilize well-known coupling patterns for the study of stereochemistry.[91,103] Thus, the oxidative addition of cyclohexyl bromides to $[(py)(dmg)_2Co(I)]^-$ proceeds with inversion.[103,37]

$$\text{Br-cyclohexyl-Br} + Co(dmg)_2py \rightarrow \text{Br-cyclohexyl-}Co(dmg)_2py \quad (39)$$

Care is required here, however, as the $^3J(HH)_{vic}$ coupling constants to metal-substituted CH may not be well established; furthermore, C–M bond lengths are sufficiently long that an axial metal substituent is not unexpected.

The inversion or retention of stereochemistry in pseudo tetrahedral complexes of metals, such as $(\eta^5\text{-}C_5H_5)Fe(Me)(CO)(PPh_3)$, **26**, can be informative with regard to the effects of the reaction at the metal center. The absolute configuration at the metal center can be designated by $R$ and $S$ following rules which take polyhapto ligands as pseudo atoms corresponding to the mass (actually the atomic number) of the atoms bound to the metal.[104,105] Hence, $(\eta^5\text{-}C_5H_5)$ would correspond to a mass of 60 or an atomic number of 30. This yields priorities of $(\eta^5\text{-}C_5H_5) > PPh_3 > CO > CH_3$. A $(\eta^5\text{-}1\text{-}CH_3\text{-}3\text{-}C_6H_5\text{-}C_5H_5)Fe$ moiety is also chiral and can be designated by Cahn–Ingold–Prelog nomenclature,[105] as well. Thus, $(RR\text{-}SS)$ and $(RS\text{-}SR)$ diastereomers of $(\eta^5\text{-}1\text{-}CH_3\text{-}3\text{-}C_6H_5\text{-}$

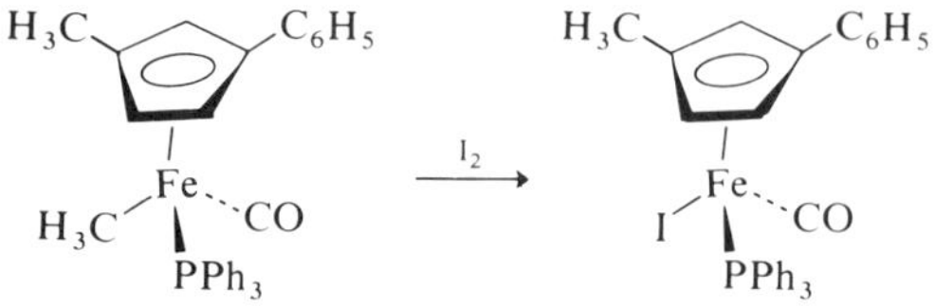

$$\text{(RS)-27} \xrightarrow{I_2} \text{(RR)-28} \quad (40)$$

$C_5H_5)Fe(Me)(CO)(PPh_3)$, **27**, exhibit different nmr properties and can be separated by crystallization.[106] The reaction with iodine, (40), can then be studied by nmr assuming that the chirality of the CpFe group is retained. The nmr results indicate that the reaction proceeds with only partial selectivity. Unfortunately, it is impractical to distinguish an (RS–SR) pair from a (RR–SS) pair by nmr, and X-ray crystallography is required to establish the stereochemistry. Two features are clear, however, without the X-ray results: a) reaction did not occur with pure retention or inversion; and b) partial epimerization of **27** occurred during the reaction. This led to the proposal that initial attack of the iodine resulted in a 5-coordinate pseudo-square pyramidal intermediate, **30**, as shown in Equation 41.

$$(RS)\text{-}27 + I_2 \rightarrow \mathbf{30} \rightarrow (RR)\text{-}28 + CH_3I \quad (41)$$

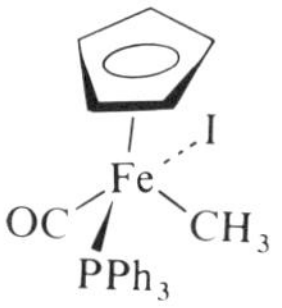

Since these 5-coordinate complexes can readily rearrange,[107] the reverse reaction of **30** → **27** provides a route for the epimerization at the iron center.

### 1.4.3. Enantiomeric Ratio Determination by Polarimetry

The use of an optically active cyclopropyl bromide with $[(py)\text{-}(dmg)_2Co(I)]^-$ gave a complex for which optical rotations could not be measured owing to the high-extinction coefficients.[108] Thus, the possibility of racemization in the formation of the complex could not be investigated. More often, however, one is thwarted by the uncertainty of the relationship between chiroptical properties and absolute configuration. One of the earliest organometallic stereochemical studies[109] showed that decarbonylation of a manganese complex occurred with retention of the "sign" of rotation (42):

$$(+)\text{-}(PhCH_2)(CH)(CH_3)COMn(CO)_5 \rightarrow (+)\text{-}(PhCH_2)(CH)(CH_3)Mn(CO)_5 \quad (42)$$

Unfortunately, this only proved that racemization had not occurred, as there had not been a way to relate sign of rotation to configuration.

In a stereochemical study where one organic product of known absolute configuration is converted into an organic product of known absolute configuration, the interpretation is straightforward. Hence, in the decarbonylation of $(-)$-$(R)$-2-methyl-2-phenylbutanal with $Rh(PPh_3)_3Cl$, (43), the 94% optical purity and relative configuration of the product were readily determined from measurements of optical rotation.[110]

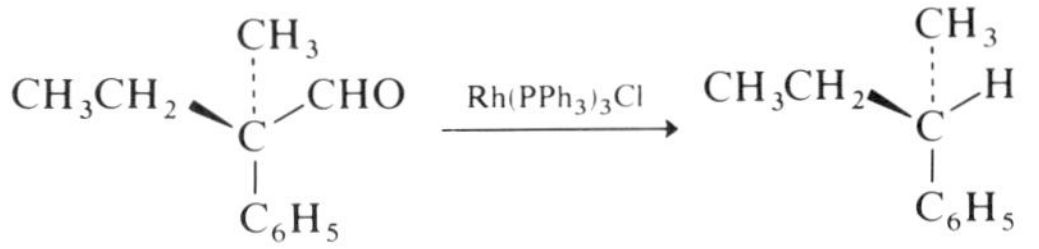

$$CH_3CH_2\underset{C_6H_5}{\overset{CH_3}{\underset{|}{\overset{|}{C}}}}CHO \xrightarrow{Rh(PPh_3)_3Cl} CH_3CH_2\underset{C_6H_5}{\overset{CH_3}{\underset{|}{\overset{|}{C}}}}H \qquad (43)$$

$(-)$-$(R)$-**31** $\qquad\qquad$ $(+)$-$(S)$-**32**

Note that the decarbonylation proceeds with retention, even though the chirality descriptor, $R$, is reversed, as is the sign of the rotation. This reversal of the descriptor is a result of changing relative assignment priorities of the groups attached to the chiral carbon and sometimes leads to confusion in descriptions of stereochemical experiments.

The optical rotations, particularly in the uv, of chiral centers of $R(CHD)R'$ are sufficiently large that it is practical to use polarimetry to measure optical purity.[111] Generally, however, a sequence of steps to produce a known configuration organic product from a known configuration substrate is required, since the chiroptical properties of the metal complexes are not known. For example, the sequence of predominate inversion in oxidative addition and retention in carbonyl insertion has been investigated using $Ph(CHD)Cl$, Equation 44 $(R = D)$.[112]

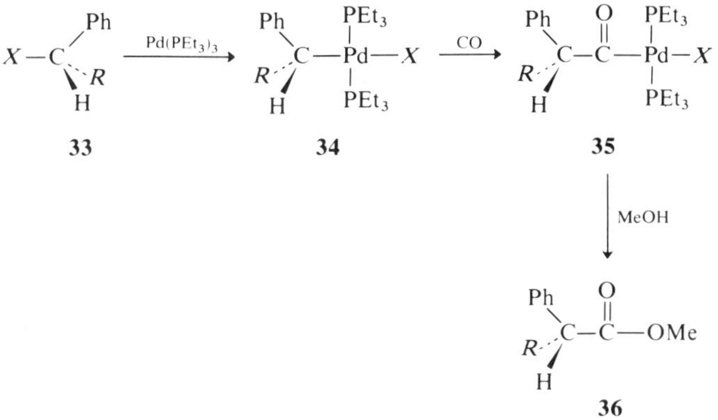

$$X-\underset{H}{\overset{Ph}{\underset{|}{C}}}{\overset{}{\diagdown}}R \xrightarrow{Pd(PEt_3)_3} \underset{H}{\overset{Ph}{R\cdots\overset{|}{C}}}-\underset{PEt_3}{\overset{PEt_3}{\underset{|}{Pd}}}-X \xrightarrow{CO} \underset{H}{\overset{Ph}{R\cdots\overset{|}{C}}}-\overset{O}{\underset{}{\overset{||}{C}}}-\underset{PEt_3}{\overset{PEt_3}{\underset{|}{Pd}}}-X$$

**33** $\qquad\qquad$ **34** $\qquad\qquad$ **35**

$\downarrow$ MeOH $\qquad(44)$

$$\underset{H}{\overset{Ph}{R\cdots\overset{}{C}}}-\overset{O}{\underset{}{\overset{||}{C}}}-OMe$$

**36**

Although the situation is improving, there are still a relatively small number of metal compounds for which the absolute configurations have been firmly established. Although there has been some success with relating the configuration of olefins bound to platinum[113] and allyls bound to palladium[114] to chiroptical properties with quadrant rules, it has proven extremely risky to attempt correlations with chiral metal centers. Absolute configuration studies of $(\eta^5\text{-}C_5H_5)Mo^{(57,115)}$ and $(\eta^5\text{-}C_5H_5)Fe^{(116)}$ compounds have suggested possible correlations, but generally only very minor changes can be tolerated before significant changes in CD or ORD spectra make interpretations unreliable. For example, a change in conformation of an allyl group from $(R)\text{-}[(\eta^5\text{-}C_5H_5)Mo(NO)(CO)\text{endo-(allyl)}]^+$ to $(R)\text{-}[(\eta^5\text{-}C_5H_5)Mo(NO)(CO)\text{exo-(allyl)}]^+$ can reverse the signs of some CD absorptions.[57] Thus, it appears that until a signifiant number of absolute configurations have been determined for metal centers, most examinations of metal-center stereochemical changes in reactions will require X-ray structure determinations. Physical methods providing a measure of diastereomeric purity will be useful, but establishing relative configurations may be difficult.

### 1.4.4. Binding of Chiral Substrates—Chiral Shift Reagents

In attempting to elucidate aspects of reaction mechanisms based on inversion or retention of configuration, the problem of identifying enantiomeric composition or optical purity is encountered. Polarimetric methods are subject to uncertainties of chemical purity, variations with solvent composition, and questionable literature values for the standard rotation of a 100.0% pure enantiomer. One may utilize the approach of making a derivative with an optically-active reagent and detecting the ratio of diastereomers by nmr or gas chromatography.[117] One of the most convenient and reliable techniques is based on the chiral lanthanide shift reagents introduced by Whitesides.[118] Their use has become standard practice for the determination of enantiomeric ratio and enantiomeric excess. When these reagents (usually derived from optically pure camphor) interact with donor atoms in an enantiomeric pair, the diastereomeric interactions yield separate nmr signals for each enantiomer which can then be directly integrated. The use of these reagents in the determination of optical purity has been recently reviewed.[119] The most effective reagents for binding to weak donors often contain fluorinated alkyl groups to increase the coordination efficiency; hence, some of the initial complexes studied were based on trifluoromethylhydroxymethylen-$d$-camphor, such as $Eu(facam)_3$. The europium complex prepared from di[$d$-campholyl]-methanato ligand, $Eu(dcm)_3$, however, appears to generally give the largest diastereomerically induced separations of resonances.[119] These reagents, which are commercially available, provide the most straightforward method of determining the enantiomeric excess in products containing donor groups, such as esters, ketones, and alcohols. Bridging carbonyls in organometallic compounds,[120] as well as cyanides and isocyanates,[121] also bind to lanthanide shift reagents and consequently should allow optical purity determinations via this method.

The "enantiomer shift differences," $\Delta\Delta\delta$, are predominately induced by the differing equilibrium constants between the enantiomers and the paramagnetic shift reagent (45), (46).

$$(R)\text{-substrate} + \text{Eu(dcm)}_3 \xrightleftharpoons{K_R} [(R)\text{-substrate}]\text{-}[\text{Eu(dcm)}]_3 \quad (45)$$

$$(S)\text{-substrate} + \text{Eu(dcm)}_3 \xrightleftharpoons{K_S} [(S)\text{-substrate}]\text{-}[\text{Eu(dcm)}]_3 \quad (46)$$

Since europium-substrate complex itself would have very large shifts and the observed shifts are averages of bound and free substrate, the greatest shifts would usually be observed for the substrate with the largest $K$. *When considering catalytic cycles involving chiral intermediates, it may be necessary to consider differences arising from diastereomeric interactions*, such as equations 45 and 46. While the chiral discrimination in the binding is the major source of the shift, the intrinsic shifts in the diastereomeric complexes are not equal. Thus, it is possible to even use the reagents for optical purity determinations in cases where $K_R$ and $K_S$ are effectively equal, as with Ph(CHD)OH.

### 1.4.5. Binding of Prochiral Substrates—Asymmetric Induction

In the studies of asymmetric hydrogenation mechanisms using rhodium complexes of chiral *bis*-phosphine complexes (see Chapter 4), there are parallels to the binding of substrates by the europium complexes containing chiral-diketone ligands. There are some important features in the process which selects the binding of substrates and their impact on mechanistic paths that need be addressed. Bosnich and his co-workers,[122-123] have considered the design of chiral *bis*-phosphines and have presented evidence that the major source of the discriminatory interaction was the chiral array of phenyl groups. On this basis, as well as on the basis of studies to identify important intermediates, one would presumably be well on the way to rationally designing a catalyst to produce high-optical yields. The importance of allowing the substrate to bind with the proper stereochemistry is reminiscent of the importance attached to "lock-and-key" models of enzymes. Thus, on observing an intermediate in a catalytic cycle, one might proceed to design a system for which the substrate of the appropriate chirality would be bound. The advantages and potential pitfalls of this approach are well-illustrated by studies of asymmetric hydrogenation.

The rhodium-catalyzed homogeneous hydrogenation of olefins using $(PPh)_3RhCl$ has been studied extensively (see Section 2.2), and similar mechanistic pathways might be anticipated in the asymmetric hydrogenation reactions. It should be noted that the most effective asymmetric hydrogenation catalysts[123-125] are based on the cationic bisphosphine hydrogenation catalysts developed by Schrock and Osborn.[126-127] Furthermore, the chelating phosphine, chiraphos, gives some of the highest optical yields. The chiral process has been discussed by Bosnich.[122]

Aspects of mechanism relating to the order of attack of olefin and hydrogen have received considerable attention in consideration of the reduction with $(PPh_3)_3RhCl$ and cationic triphenylphosphine analogs.[128-130]

$$(PPh_3)_2Rh^+ + H_2 \rightarrow (PPh_3)_2RhH_2^+ \xrightarrow{olefin} (PPh_3)_2RhH_2(olefin)^+ \quad (47)$$

$$(PPh_3)_2Rh^+ + olefin \rightarrow (PPh_3)_2Rh(olefin)^+ \xrightarrow{H_2} (PPh_3)_2RhH_2(olefin)^+ \quad (48)$$

The cycle is completed, and the product is obtained by hydride addition from a cis-hydride followed by reductive elimination. The "hydride" route (47) is now generally accepted.[131] The evidence for the $(ROH)_2(PPh_3)_2RhH_2^+$ cation as an intermediate was obtained by direct nmr observation of this cis-dihydride[126-127] after hydrogenation of the precursor in alcohol. The "unsaturate" route (48), however, was implicated with chelating phosphines by the observation that only a $[(diphos)Rh(ROH)_2]^+$ species is observed upon hydrogenation of [(norbornadiene)Rh(diphos)]$^+$.[132] Thus, the change of two $PPh_3$ to a diphos ligand completely changes the principal pathway. This illustrates the care that must be exercised in extending the results from one system to another.

The addition of prochiral substrates, such as esters of $\alpha$-acetamidocinnamic acids, to a solution of the solvate of a chiral diphosphine rhodium complex allowed the observation of at least one of the two diastereomeric complexes, Equation 49.[133-138]

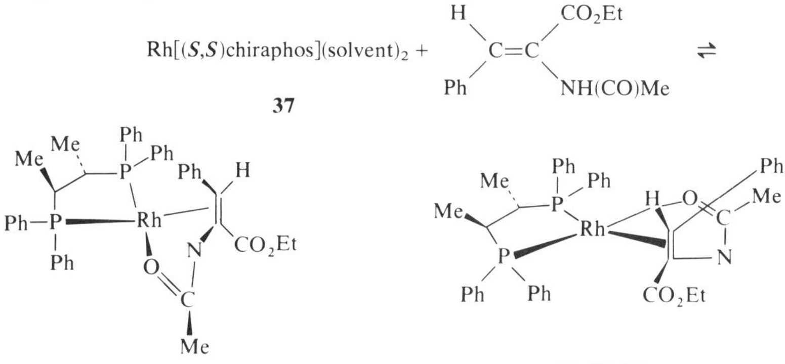

$$Rh[(S,S)chiraphos](solvent)_2 + \underset{Ph}{\overset{H}{\diagdown}}C=C\underset{NH(CO)Me}{\overset{CO_2Et}{\diagup}} \rightleftharpoons$$

**37**

(49)

(S,S)-38        (R,R)-38

Presuming that the minor stereoisomer in solution is not the least soluble,[138] the crystal structure[136] indicates that the major isomer in solution is $(S, S)$-**38**, where $(S, S)$ refers to the configuration at the 1- and 2-carbons of the bound olefin. The rate-determining step is the addition of hydrogen to give **39**, and the equilibrium shown in equation (49) occurs rapidly.[138] After hydrogen addition, *cis*-hydride transfer from the metal to the olefin gives a metal alkyl, **40**, followed by reductive elimination yielding the product, Equations 50–51.

$$37 \underset{K_{SS}}{\overset{\text{olefin}}{\rightleftharpoons}} (S, S)\text{-}38 \xrightarrow{H_2} (S, S)\text{-}39 \rightarrow (R)\text{-}40 \rightarrow PhCH_2C\overset{NH(CO)Me}{\underset{COOEt}{\rule{0pt}{1em}\smash{-}H}} \qquad (50)$$

$$(S)\text{-}41$$

$$37 \underset{K_{RR}}{\overset{\text{olefin}}{\rightleftharpoons}} (R, R)\text{-}38 \xrightarrow{H_2} (R, R)\text{-}39 \rightarrow (S)\text{-}40 \rightarrow PhCH_2C\overset{COOEt}{\underset{NH(CO)Me}{\rule{0pt}{1em}\smash{-}H}} \qquad (51)$$

$$(R)\text{-}41$$

The predominant product (>95% ee) of this reaction, however, is the ester of N-acetyl-$(R)$-phenylalanine, $(R)$-**41**. This requires that the Rh(chiraphos) (olefin) *intermediate*, $(R, R)$-**38**, *which is in lowest concentration, yields most of the product*. This illustrates again that the dominant species observed is not necessarily the most reactive, nor is it necessarily a loop species in the most important catalytic cycle.

The important consideration, therefore, is the relative rate of addition of $H_2$ to the Rh (chiraphos) (olefin) intermediates, $(S, S)$-**38** and $(R, R)$-**38**.[138] Hence, as discussed in more detail by Bosnich,[122] if all prior equilibria ($K_{RR}$ and $K_{SS}$) are fast, the *only* relative energies of any consequence are those of the transition states between **38** and **39**. Under these circumstances the only diastereomeric discriminatory of importance are those in these transition states, and *not* those of the prior equilibria. Thus, in general, when attempting to design catalysts with greater enantioselectivity or stereoselectivity, it is essential that relative transition-state energies be considered, either instead of, or in addition to, those of ground states.

### 1.5. Distinguishing Radical Pathways

Although one might hope that the key steps in catalytic systems, such as oxidative addition or reductive elimination, would proceed via a universally valid mechanism, it is clear that many of these steps may occur by several paths. Several paths may even occur within the same system and minor perturbations may alter the primary route. The success of the 16- and 18-Electron Rule in rationalizing many aspects of organometallic chemistry and catalytic systems has tended to lead to nearly exclusive consideration of reaction mechanisms involving only even numbers of electrons. There are often enough alternate mechanistic paths conceivable, using the 16- and 18-Electron Rule, that the inclusion of additional odd-electron paths produces an unwieldy number of candidates for the mechanism. It is now clear, however, that one-electron processes are increasingly being encountered and cannot be neglected. By the mid-seventies, Kochi[139] had investigated radical participation in nontransition metal systems extensively, but the studies of oxidative addition mechanisms by Osborn[140-141] and Lappert[142-143] pointed out the importance of radical paths in organometallic systems which had been presumed to occur by conventional two-electron $S_N2$ paths.

Although one might expect that an oxidative addition is the most likely candidate for a radical path, even the displacement of carbonyl by phosphine may proceed by a radical path. For example, $HRe(CO)_5$ does not readily undergo phosphine substitution under scrupulously clean conditions, but does react rapidly under conditions in which the $Re(CO)_5$ radical is formed.[144]

Stable organometallic radicals are rare, owing to their proclivity for dimerization, and most that are known owe their stability to steric effects which prevent dimerization. Thus, $V(CO)_6$ and derivatives of $(\eta^3$-allyl)-$Fe(CO)_3$[145] have been known for nearly twenty years. Sublimation of some organometallic dimers with relatively weak bonds onto a liquid nitrogen-cooled probe has provided a method of preparing radicals, such as $Co(CO)_4$[146] and $(\eta^5-C_5H_5)Cr(CO)_3$[147] for ESR study. The major path for carbonyl substitution in some metal–metal bonded dimers is believed to involve breaking of the dimer bond, substitution on the radical, and reformation of the dimer.[148]

Relatively stable organometallic radicals, such as $[(\eta^5-C_5H_5)Fe-(diphos)CH_3]^+$ and $[Mo(CO)_2(diphos)_2]^+$, can be prepared by chemical oxidation,[149,150] and the paramagnetic hydride $FeH(diphos)_2$ is available through reduction of the $[FeHCl(diphos)_2]^+$. The largest number, however, are available through electrochemical techniques.[151] Electron rich systems are oxidized relatively easily and $E_{1/2}$ measurements provide a quantitative measure of the facility of oxidation.

The importance of radical pathways can usually be correlated with the ease of electron transfer.

Generally, ESR spectra of organometallic radicals are unexceptional and are similar to those of organic radicals with $g \sim 2$. For example, trans-$[Mo(CO)_2(dmpe)_2]^{\bullet}$ has a $g = 2.053$ and a coupling of $a_P = 25$ G is resolved.[152] A value of $g = 2.085$ and poorly resolved fine structures are found for $HFe(diphos)_2^{\bullet}$.[153] The relative sharpness of the resonances is important, particularly with iron compounds, because inadvertent oxidation could give rise to a low spin $d^5$ complex. Since these $d^5$ complexes have one unpaired electron and a $g \sim 2$, they might be confused with an organometallic radical. These species, however, often have rapid relaxation times at room temperature so that very broad resonances are observed.[153] Variations in line width with temperature are observed for $[Fe_2(CO)_8]^{\overline{\cdot}}$ with $g = 2.0385$ and a width $\sim 3$ G at $-80°$ and $\sim 20$ G at $25°$.[154,155] The spectrum of $[Fe_3(CO)_{12}]^{\overline{\cdot}}$, $g = 2.0016$, is sharp at $-80°C$ and clearly shows $\sim 3$ G couplings to equivalent $^{13}C$ nuclei.

### 1.5.1. Direct Observation of Radicals

The possibility of a radical path in a reaction is often suggested by an unexpected racemization or an unusual reactivity order (see Section 1.5.4). In an effort to prove that the mechanism involves radical participation, it might appear that direct observation of the radicals by ESR would be the method of choice. Generally, however, one wishes to demonstrate not only the presence of radicals, but that the major path of the reaction proceeds via a radical route. The extreme sensitivity of ESR may make detection of radicals relatively straightforward, but making quantitative estimates to prove that a radical is a loop species is another matter. Nevertheless, direct ESR detection of an intermediate is a quite useful approach.

Treatment of cis-$[Mo(CO)_2(dmpe)_2]$ with $Ph_3CCl$ in the cavity of an ESR spectrometer yielded a spectrum of trans-$[Mo(CO)_2(dmpe)_2]^{\bullet}$, which was identified by its characteristic values of $g$ and $a_P$.[152] Metal-centered radicals have also been observed in oxidative additions to $[Mo(N_2)_2(diphos)_2]^{(156)}$ and $(\eta^5-C_5H_5)Fe(diphos)MgBr$.[157] Organic radical fragments have been detected in the reaction of isopropyl iodide with $[(\eta^5-C_5H_5)Fe(CO)_2]^-$ in THF.[158] In this case the characteristic septet of doublets of the $(CH_3)_2HC^{\bullet}$ radical was observed in a flow cell in which the reactants were mixed at room temperature.

The failure to observe radicals by ESR does not, however, allow them to be discounted from the mechanism. The flow experiments with $i$-PrI and $[(\eta^5-C_5H_5)Fe(CO)_2]^-$ would suggest that $[(\eta^5-C_5H_5)Fe(CO)_2]^{\bullet}$ should have been detected. Presumably the very short lifetime of the radical precludes its observation. In some situations, one might have an organometallic anion acting solely as a one-electron donor and then dimerizing rapidly enough that it is not observed.[159] Unfortunately, owing to the difficulties in quantifying the experiments, the observation of ESR signals does not prove that radicals are involved on the main reaction path.

### 1.5.2. Radical Spin Traps

An alternative to the direct observation of the rather elusive radicals themselves is to provide a substrate with which they will react and produce a radical that is stable indefinitely. This spin-trapping procedure has been used primarily for organic radicals,[160-161] but has been extended to metal-based organometallics by Lappert.[142] The technique usually involves a nitroso compound, which reacts with the transient radical to form a nitroxide (see Equation 52). The resultant nitroxide can then be characterized by

$$R\text{-}N=O + R' \rightarrow \underset{R\phantom{xx}R'}{\overset{\overset{\displaystyle O}{|}}{\dot{N}}} \tag{52}$$

the spin splittings of nuclei two or three bonds removed from the nitroxide nitrogen. The most commonly used trap is 2-methyl-2-nitrosopropane.[162] The spectra are simplified with this trap as the $t$-butyl protons in the resulting trapped radicals do not couple to the electron. Furthermore, it exists in solution primarily as the monomer, whereas many nitroso compounds are unreactive dimers.[163] The major drawback is that it tends to decompose by loss of NO forming $(t\text{-Bu})_2\dot{N}O$ at a rate sufficient to interfere in the spectra of the trapped radical. To overcome this problem, other traps such as nitrosodurene were developed. They often exist in solution predominantly as the dimer, but sufficient monomer is usually present for most applications[164]; complications may result, however, if the equilibrium is not established rapidly.

It is important to establish by appropriate control experiments that the trapped radicals were produced exclusively by the reaction under consideration. Lappert[143] has discussed these precautions, such as avoiding the effects of exposure to light and making certain that the reagents themselves do not react with the spin trap. The importance of these controls is demonstrated by the study of the oxidative addition of tosyl halides to $Pt^0$ species, where it was found that tosyl iodide + $R$NO gave a clean ESR signal for $[(MeC_6H_4SO_2)(R)\dot{N}O]$.[143]

### 1.5.3. CIDNP

*C*hemically *i*nduced *d*ynamic *n*uclear *p*olarization (CIDNP) arises in some radical pair processes and is characterized by intensity inversion and enhancement of some of the resonances in the nmr spectra of radical combination products during a reaction.[165] This provides definitive evidence for the involvement of radicals, but as with ESR, it is difficult to quantitatively estimate the fraction of the total reaction proceeding via the radical path. The conditions under which a CIDNP effect will be observed are rather limited, and it has only been observed in a few cases with organometallics.[166] In particular, situations in which a metal-based radical can be implicated are extremely rare.[67,167] Nevertheless, with quantitative measurements of enhancement factors, convincing evidence can be provided for a reaction proceeding predominantly by a radical path.

Two cases of particular relevance are the CIDNP effects observed in the oxidative addition of isopropyl iodide to $Pt(PEt_3)_3$[167] and the reduction of $\alpha$-methylstyrene with $HMn(CO)_5$:[67]

$$(CH_3)_2CHI + Pt(PEt_3)_3 \rightarrow \text{propane} + \text{propene} + 2,3\text{-dimethylbutane}$$
$$+ PtHI(PEt_3)_2 + PtI_2(PEt_3)_2 \qquad (53)$$

$$PhC(CH_3)=CH_2 + 2HMn(CO)_5 \rightarrow PhCH(CH_3)_2 + Mn_2(CO)_{10} \qquad (54)$$

The $i$-PrI reaction produces the expected products of an isopropyl radical reaction, and CIDNP enhancements and multiplet effects were observed in the propene resonances and isopropyl iodide resonances. In the $HMn(CO)_5$ reduction, enhancements were observed in the Mn–H resonance and in the methyl resonance of the product.

As reaction (54) was studied quantitatively and is simpler, its results will be used to illustrate the method. CIDNP effects can be interpreted in terms of a radical pair mechanism. At some stage a geminate radical pair, **42**, is formed and the competition among electronic relaxation times, the

$$PhC(Me)=CH_2 + HMn(CO)_5 \rightleftarrows \overline{Ph(Me)_2CC^\bullet, \,^\bullet Mn(CO)_5} \qquad (55)$$
$$\mathbf{42}$$
$$\mathbf{42} \xrightarrow{\text{cage escape}} Ph(Me)_2C^\bullet + \,^\bullet Mn(CO)_5 \qquad (56)$$

back reaction to starting materials, and the escape of the radicals from the solvent cage around the pair produces the unusual polarizations of the protons observed in the reactants and products. Measurement of the relaxation times and the rates of decay of the intensities of the CIDNP-enhanced resonances provides a measure of the absolute intensity enhancements. The magnitude of the enhancements, ~300, supports the view that most of the product is being formed by the radical path. The calculation of the CIDNP effect involves the dephasing rate of the radical pair, which is related to the relative $g$ values of the radicals. Thus, one can obtain some estimate of the value of $g$ of one of the radicals, if the other is known. This provides an additional tool for estimating the involvement of a metal-based radical.

### 1.5.4. Indirect Methods

Although loss of stereochemistry in a reaction might provide an indication that radical reactions are occurring, retention or inversion does not necessarily preclude radical reactions. There are numerous examples in the organic literature[168-170] where stereoselectivity is observed in radical reactions. For example, homolytic decomposition necessarily generates two radicals in close proximity. These fragments may diffuse away or react within the solvent cage. If reactions take place before the radicals diffuse away from each other, high stereoselectivity can be observed. The lifetime of a radical pair is estimated to be $\sim 10^{-11}$ s, and after escape from the cage racemization usually occurs.

Inference of a radical path can also be made on the basis of structure and reactivity arguments. For example, two-electron $S_N 2$ processes generally show a maximum rate in halide displacements from alkyls of Me > Et > $i$-Pr > $t$-Bu. The reverse reactivity order is generally characteristic of radicals. $R$I > $R$Br > $R$Cl reactivity is also expected for radical reactions. Initiation of polymerization of styrene or acrylonitrile is also an indicator of the presence of radicals.

In addition to a radical pair process, a radical chain process may be involved in some oxidative additions.[140,171]

$$Pt^0 + R^{\bullet} \rightarrow Pt\text{-}R \qquad (57)$$

$$Pt\text{-}R + RX \rightarrow X\text{-}Pt\text{-}R + R^{\bullet} \qquad (58)$$

In such cases, a chain initiation step is required; hence, if the reaction can be initiated with light, AIBN, or benzoyl peroxide, a radical chain process is indicated. Since the chain involves the continual presence of a radical after initiation, inhibition of the reaction by a radical trap, such as galvanoxyl or duroquinone, also provides evidence for a radical chain mechanism. Because some radical traps react with organometallics, however, lack of inhibition may be the result of deactivation of the trap.

An excellent measure of radical involvement for certain types of studies relies upon the rearrangement rates of organic radicals that might be formed under the reaction conditions.[172] Thus, if a free 5-hexenyl radical is formed in the oxidative addition of 5-hexenyl bromide, the product will contain a cyclopentylmethyl group. The rearrangement of the radical occurs at a rate of $10^5$ s$^{-1}$, so that if the radical has a lifetime of more than about $10^{-4}$ s, it will rearrange. This technique for the demonstration of radical paths has been used in a number of cases.[154,155] A similar approach using cyclopropylcarbinyl halides[156,173] ($k_{\text{rearrangement}} \sim 10^8$ s$^{-1}$) suggests that approximately 30% of the reaction follows a radical path with iodide.

$$\triangleright\!\!\!\diagdown\!\!X \xrightarrow{(\eta^5\text{-}C_5H_5)Fe(CO)_2^-} \triangleright\!\!\!\diagdown\!\!Fe(CO)_2(\eta^5\text{-}C_5H_5) \quad 70\%$$

$$+ \qquad (59)$$

$$\diagdown\!\!\diagup\!\!\diagdown\!\!Fe(CO)_2(\eta^5\text{-}C_5H_5) \quad 30\%$$

## 2. APPLICATIONS ILLUSTRATING THE METHODS

### 2.1. Butene Isomerization by $NiL_4$ and $H_2SO_4$

This system[10] nicely illustrates many of the principles outlined in the first section. Most of the work was done using $Ni[P(OEt)_3]_4$ or $Ni[P(OMe)_3]_4$ as catalysts in methanol solvent. In the presence of a strong acid cocatalyst, such as $H_2SO_4$ or HCl (but not without cocatalyst), butenes can be isomerized within minutes at room temperature to an equilibrium mixture of the linear isomers. This is probably the most active olefin isomerization catalyst system known.

Using 1-butene (the least stable isomer, abbreviated 1-B) as the starting olefin, the rate law (at the beginning of the reaction) is given by Equation 60,

$$-d[1\text{-B}]/dt = kK[Ni]_0[H^+][1\text{-B}]/(1 + K[H^+])[L] \qquad (60)$$

where $[Ni]_0$ is the initial concentration of $NiL_4$, and $k = 0.7 \text{ s}^{-1}$ and $K = 50 \text{ M}^{-1}$ † at 25° for $L = P(OEt)_3$.

A plot of $k_{obs}^{-1}$ against $[H^+]^{-1}$ will give a straight line since

$$k_{obs}^{-1} = [L](1 + 1/K[H^+])/k[Ni]_0 \qquad (61)$$

The intercept on the ordinate as $[H^+]^{-1} \to 0$ ($[H^+] \to \infty$) gives the maximum rate attainable for fixed $[L]$ and $[Ni]_0$—when all the nickel is converted to $HNiL_4^+$.

Spectroscopic studies (vis/uv and $^1H$ and $^{31}P$ nmr) of solutions under catalytic olefin isomerization conditions show $NiL_4$ and $HNiL_4^+$, with more of the latter as the acid concentration is increased. Gradually the catalyst dies as these species are converted to Ni(II), in a reaction which is inverse first order in $[L]$. In the absence of olefin, $H_2$ is evolved, while in the presence of butene substrate, butane is formed. The system gives about 300 catalytic cycles at 25° and about 3000 at 0°, indicating that catalyst degradation has a higher activation energy than olefin isomerization.

In an experiment at room temperature using $D_2SO_4$ in $CH_3OD$, 39% of the original 1-butene was isomerized in 15 s. There was some $d_1$-1-butene (0.5%), but more than 99% of the 2-butene products were undeuterated. The ratio of isomerization to deuteration is about 170. In spite of this, we are confident that the reaction involves olefin insertion into a nickel hydride to form a nickel alkyl intermediate, followed by de-insertion of

---

† We have chosen to use $M$ to represent $\text{mol l}^{-1}$ and mM to represent $\text{mmol l}^{-1}$ in hopes that this usage will be more familiar to readers.

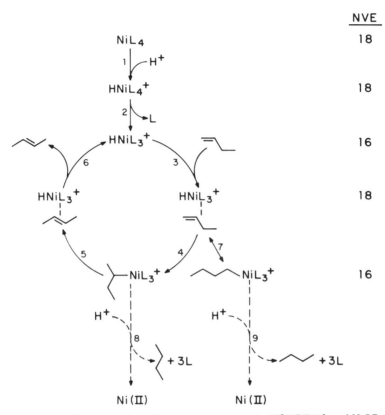

Figure 2.2. The mechanism of butene isomerization by Ni[P(OEt)$_3$]$_4$ and H$_2$SO$_4$.

isomerized olefin. Additional supporting evidence is provided by the quantitative analysis by microwave spectroscopy of various $d_1$-propylene isomers formed on exposure of $d_0$-propylene to solutions of Ni[P(OEt)$_3$]$_4$ and DCl in CH$_3$OD.[174] The analysis indicates that insertion into the NiD bond occurs with relative rates of 4:1 to form DCH$_2$CH(CH$_3$)Ni and CH$_3$CHDCH$_2$Ni intermediates. The presence of deuterium on C$_2$ in the product propylene cannot be accounted for by a $\pi$-allyl hydride mechanism.

On long exposure of 1-butene to a similar solution, one obtains a thermodynamic mixture of the isomers (about 6% 1-butene, 25% *cis*-2-butene, and 69% *trans*-2-butene) along with statistical scrambling of the deuterium originally present in solution with all of the hydrogens of the C$_4$H$_8$.[10]

The mechanism shown in Figure 2.2 satisfactorily accounts for the observations and is consistent with the 16- and 18-Electron Rule.[6] The number of metal valence electrons (NVE) is written to the right of each species.

Ignoring the back reaction of product 2-B (letting $k_{-6} \to 0$) and applying the method of Cleland,[27] we obtain the rate law given in Equation 60, where $K = K_1$ and

$$k = k_3 k_4 k_5 k_6 / \{K_2 k_4 k_5 k_6 + k_{-3}[k_5 k_6 + k_{-4}(k_6 + k_{-5})]\}, \tag{62}$$

where the rate constants are defined in Figure 2.2. Equation 62, clearly shows the impossibility of determining each of the individual rate constants from the overall rate law. We also cannot tell which of steps 3, 4, 5, or 6 is rate determining. In principle, a choice could be made from the deuterium isotope effect. The more rapid isomerization in $CH_3OD$ suggests that olefin insertion is rate determining (breaking an Ni–H bond and forming a stronger C–H bond), but the situation is complicated by the more rapid degradation of $L$ which also occurs in $CH_3OD$.

This is a system where none of the loop species is present in sufficient concentration to be spectroscopically detectable. No alkyl is observed even when $C_2H_4$ is added to the $HNiL_4^+$ solutions. With butadiene, however, the additional stability provided by coordination of the double bond makes the trihapto ($\pi$-crotyl)$NiL_3^+$ stable enough to be isolated.[175]

Spectroscopic and kinetic studies give considerable insight into the sequence and rates of the individual steps of the isomerization mechanism. $Ni[P(OEt)_3]_4$ itself does not dissociate to the $NiL_3$ complex to a detectable extent, even at 70°. Using the very sensitive vis/uv method, $K_d$ is estimated to be less than $10^{-10} M^{-1}$ at 70°,[176] and must be considerably less than this at 25°. The $^{31}P$ NMR singlet for $Ni[P(OEt)_3]_4$ indicates that the complex has the expected tetrahedral structure. Adding $P(OEt)_3$ gives a separate sharp signal, showing that ligand exchange—in the absence of added acid—is slow on the NMR time scale. Though there appear to be no X-ray structures of $NiL_4$ or $NiL_3$ complexes, $Pt(PF_3)_4$[177] and $Pt(PPh_3)_3$[178] have been done and shown to be tetrahedral and trigonal, respectively.

The half-life for dissociation of the first ligand from $Ni[P(OEt)_3]_4$, measured by capturing the $NiL_3$ intermediate by isonitrile[179] or CO,[180] is several hours at 25°—far too slow to allow ligand dissociation to be the first step. Studies in the reaction of $Ni[P(OEt)_3]_4$ with olefins[180] show that these reactions proceed at the same slow rates, and that highly activated olefins (such as maleic anhydride) must be used before any olefin complexes are detectable. The reaction of $H^+$ with $NiL_4$, on the other hand, is so fast that stop-flow techniques must be used. For reaction (63) at 25° using $HClO_4$, $k_1 = 1550\ M^{-1} s^{-1}$, $k_{-1} = 45\ s^{-1}$.[181] The calculated $K_1$ of $35\ M^{-1}$ is in satisfactory agreement with the value of $48 \pm 14\ M^{-1}$, determined

$$H^+ + NiL_4 \underset{k_{-1}}{\overset{k_1}{\rightleftharpoons}} HNiL_4^+ \tag{63}$$

with room temperature equilibrium measurements using $H_2SO_4$, and the $K$ of 50 $M^{-1}$ in the rate equation 60 for butene isomerization.

$^1H$ and $^{31}P$ NMR spectra of $HNiL_4^+$ at all accessible temperatures show that all of the P nuclei are equivalent on the NMR time scale, in spite of the fact that $HNiL_4^+$ probably has a trigonal bipyramidal structure in its lowest energy form. This rapid exchange of axial and equatorial ligands, quite common for 5-coordinate complexes[182] must be part of the reason for the catalytic activity of nickel, since it readily provides a mutually *cis* orientation of the hydride and olefin in the $HNiL_3(Ol)^+$ complexes suggested by the $[L]^{-1}$ dependence of the rate law.

While hydrido-olefin complexes are not observed in this system, a related platinum complex *trans*-$HPt(PEt_3)_2(C_2H_4)^+$ is quite stable.[183] In this case, at least part of the stability is attributable to the mutually *trans* orientation of H and $C_2H_4$; a *cis* orientation appears to be required for insertion.

$HNiL_3^+$ is not detectable in solutions of $HNiL_4^+$, and an upper limit of $4 \times 10^{-5} M$ can be set on $K_2$.[181]

$$HNiL_4^+ \underset{k_{-2}}{\overset{k_2}{\rightleftharpoons}} HNiL_3^+ + L \tag{64}$$

The value of $k_2$ can be determined to be $0.015\ s^{-1}$ at 25° by trapping the $HNiL_3^+$ with butadiene to form $\pi$-crotyl complexes.[175] Thus, the lower limit on $k_{-2}$ is about 400 $M^{-1} s^{-1}$. Comparison of $k_2$ with $k_1$ shows that protonating the $NiL_4$ accelerates $L$ dissociation by a factor of more than $10^2$.

Figure 2.2 is oversimplified in that it does not show the formation of *cis*-2-butene, the reaction of $H^+$ with $HNiL_3^+$ to liberate $H_2$, Ni(II), and $L$ (the dominant path for nickel oxidation in the absence of substrate), nor the acid catalyzed decomposition of $P(OEt)_3$ to $HPO(OEt)_2$[181] which also occurs.

While the kinetics of olefin isomerization have only been examined with $L = P(OEt)_3$ and $P(OMe)_3$, the equilibrium constants for protonation of $NiL_4$ complexes in Table 1 indicate that $K$ decreases rapidly as $\nu$ increases, suggesting that $NiL_4$ complexes with strongly electron withdrawing ligands are likely to be poor catalysts. $Ni[Ph_2PCH_2CH_2PPh_2]_2$ is also expected to be a poor catalyst because of the difficulty in obtaining a coordination site for olefin binding. It is interesting in this respect that $HNi[Ph_2PCH_2CH_2PPh_2]_2^+$ reacts only very slowly with butadiene ($t_{1/2}$ = 8 h at 50° *vs.* less than 0.1 h for $HNi[P(OEt)_3]_4^+$ at 25°). $HCo[P(OEt)_3]_4$ does not react with butadiene at all.[184]

The $HNiL_4^+$ complexes are effective catalysts for isomerization of a variety of olefins. 1,4-pentadiene is isomerized to the 1,3-isomer, which then reacts rapidly and irreversibly with the nickel hydride to form 1,3-

Table 1. Equilibrium Constants for Protonation of $NiL_4$ Complexes

| Ligand | $K(M^{-1})^a$ | $\nu(cm^{-1})^b$ |
|---|---|---|
| $Ph_2PCH_2CH_2PPh_2$ | 410.0 | 2066.7 |
| $PPh(OEt)_2$ | 107.0 | 2074.2 |
| $P(OEt)_3$ | 33.0 | 2076.3 |
| $P(OMe)_3$ | 35.0 | 2079.5 |
| $P(OCH_2CH_2Cl)_3$ | 1.2 | 2084.0 |
| $P(OCH_2CCl_3)_3$ | <0.1 | 2091.7 |

$^a$ Measured in most cases at 0° in methanol using $H_2SO_4$, from C. A. Tolman, *Inorg. Chem.* **11**, 3128 (1972).
$^b$ Carbonyl stretching frequencies used to characterize the electronic character of the ligands. See reference 4.

dimethyl-$\pi$-allyl nickel complexes.[112] Functionally substituted olefins, such as methylpenteneoates, can be readily isomerized at 25° to an equilibrium mixture of the isomers.[180]

## 2.2. Olefin Hydrogenation with Wilkinson's Catalyst

### 2.2.1. Background and Rate Laws

This fascinating catalyst system, which has been studied in considerable detail,[185] is based on $RhCl(PPh_3)_3$ and was apparently discovered independently in Wilkinson's laboratory by J. A. Osborn[17] and at ICI by R. S. Coffey;[186] nevertheless, Wilkinson's name has stuck. Olefins can be smoothly hydrogenated in minutes at room temperature with millimolar Rh, and an $H_2$ pressure of 1 atm. Hydrogenation rates increase with both olefin concentration and $H_2$ pressure as shown by plots of reciprocal rates of cyclohexene hydrogenation against 1/(cyclohexene) and $1/P_{H_2}$ in Figures 2.3A and 2.3B. The dependence of rate on catalyst concentration is complex, and in the absence of added $L$ is approximately half-order in [Rh].[12,187] The rate becomes first-order in [Rh] and nearly inverse first-order in [L] at moderate (~0.01 M) concentrations of added $PPh_3$. Several rate laws advanced by various workers have been summarized by Halpern[188] (Table 2), who reports that his Equation 65 (for cyclohexene) is consistent with rates over wide ranges of concentrations, though he has not published the experimental measurements of catalytic hydrogenations under steady-state conditions.

$$-d[S]/dt = [Rh]_{tot}\{a + b[L]/[H_2] + c[L]/[S]\}^{-1} \qquad (65)$$

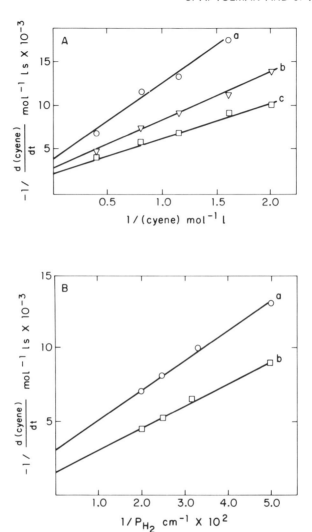

Figure 2.3. Reciprocal plots of the rate of hydrogenation of cyclohexene (abbreviated cyene) by Wilkinson's catalyst in benzene at 25°: A, plotted against reciprocal olefin concentration at 0.625(a), 1.25(b) and 1.875(c) × $10^{-3}$ $M$ RhCl(PPh$_3$)$_3$; B, plotted against reciprocal hydrogen pressure at 2.5(a) and 1.25(b) × $10^{-3}$ $M$ RhCl(PPh$_3$)$_3$, from Reference 17.

Spectroscopic studies under catalytic olefin hydrogenation conditions without added $L$ show the presence of RhCl$L_3$ (**43**), H$_2$RhCl$L_3$ (**44**), (RhCl$L_2$)$_2$ (**45**), and H$_2$(RhCl$L_2$)$_2$ (**46**). (In the case of ethylene (C$_2$H$_4$)RhCl$L_2$ (**47**) is also observed.)[14] The amount of the hydrides increases with H$_2$ pressure, while the dimeric species are increased by

Table 2. Rate Laws[a] for RhCl(PPh$_3$)$_3$ Catalyzed Olefin Hydrogenation

| No. | Rate law | Reference |
|---|---|---|
| 1 | Rate = $[Rh]_{tot}\{a/[H_2][S] + b/[H_2] + c/[S]\}^{-1}$ | 17 |
| 2 | $[Rh]_{tot}\{(a + b[L])/[H_2][S] + c/[H_2] + d/[S]\}^{-1}$ | 193 |
| 3 | $[Rh]_{tot}\{a + (b + c[L])/[H_2][S] + (d + e[L])/[H_2] + (f + g[L])/[S]\}^{-1}$ | 187 |
| 4 | $[Rh]_{tot}\{a + b[L]/[H_2] + c[L]/[S]\}^{-1}$ [b] | 188 |
| 5 | $[Rh]_{tot}\{(a[S] + b[L])[L]/(d[S] + e[L])[S] + (e[S] + f[L])/[H_2]\}^{-1}$ [c] | |

[a] Taken from reference 118. Here [L] and [S] are the concentrations of PPh$_3$ and substrate olefin.
[b] For cyclohexene.
[c] For styrene, based on unpublished results of J. Halpern and T. Okamoto.

temperature and suppressed by added $L$. Structures **43** to **47** are established by $^{31}$P, $^1$H, and $^{13}$C NMR spectra of the $L$ = PPh$_3$ and P($p$-tolyl)$_3$ complexes in solution[14] and by the X-ray crystal structures of RhCl(PPh$_3$)$_3$ and (C$_2$F$_4$)RhCl(PPh$_3$)$_2$ in the solid state.[189] RhCl(PPh$_3$)$_3$ can crystallize in two crystalline modifications with different orientations of the crowded phenyl rings—a red $\alpha$-form with a tetrahedral distortion in which the r.m.s. deviation of the coordinated atoms from the mean coordination plane is 0.43 Å, and an orange $\beta$-form with an r.m.s. deviation of 0.28 Å.[190] A

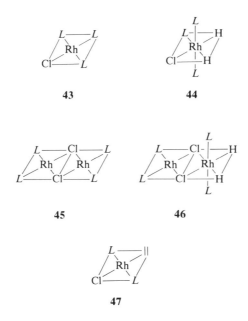

related, less crowded complex with a chelating tridentate ligand, RhCl[Ph$_2$P(CH$_2$)$_3$PPh(CH$_2$)$_3$PPh$_2$], shows an r.m.s. deviation of only 0.08 Å, and Rh–P bond lengths which are systematically shorter by 0.02 Å than those in RhCl(PPh$_3$)$_3$.[191,192]

The dimer itself is quite a good hydrogenation catalyst, though it is not very soluble in the case of $L$ = PPh$_3$, and is subject to poisoning by small amounts of adventitious O$_2$. Addition of small amounts of free $L$ have an accelerating affect on the hydrogenation rate, while larger amounts inhibit with a $[L]^{-1}$ concentration dependence as shown in Figure 2.4 for $L$ = P($p$-tolyl)$_3$.

Isotopic labeling studies using D$_2$ and 1,4-dimethylcyclohexene show that although the $d_2$-cyclohexane is the major product (about 90%) significant amounts of $d_0$, $d_1$, $d_3$, and $d_4$ products can also be detected, indicating stepwise addition and reversible formation of (alkyl)RhH intermediates,[193] which are also indicated by the occurrence of olefin double-bond migration under hydrogenation conditions.[194] Kinetic isotope effects ($k_H/k_D$) in cyclohexene hydrogenation ranging from 0.90 to 1.17 have been reported, but Siegel and Ohrt[195] have pointed out that the value observed depends on the particular conditions of the experiment.

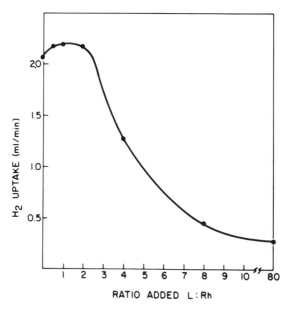

Figure 2.4. Rate of hydrogen uptake at ambient temperature and 1 atm by 100 ml 1 $M$ cyclohexene in toluene containing $1.3 \times 10^{-3} M$ Rh added as [RhCl[P($p$-tolyl)$_3$]$_2$]$_2$ and various amounts of added P($p$-tolyl)$_3$, from Reference 14.

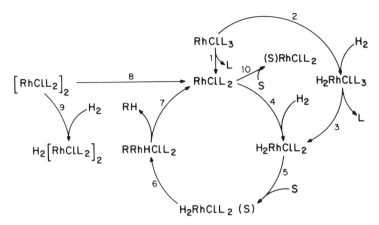

Figure 2.5. The mechanism of cyclohexene (S) hydrogenation by Wilkinson's catalyst. $L$ is $PPh_3$ or $P(p\text{-tolyl})_3$.

### 2.2.2. The Mechanism

Considerable insight into the mechanism shown in Figure 2.5 has been achieved by studying individual steps in the system; rate and equilibrium constants measured so far are given in Table 3. Halpern and Wong[13] used stop-flow techniques to measure the rapid reaction of $H_2$ with $RhCl(PPh_3)_3$ to form $H_2RhClL_3$—a reaction which is inhibited by $L$ but becomes independent of $[H_2]$ at low $[L]$, when $k_1$ becomes rate determining. The value of $k_1 = 0.71\ s^{-1}$ implies a half-life of $L$ dissociation from $RhClL_3$ of 1 s. $RhClL_2$ is never present in significant quantities, but reacts so rapidly with $H_2$ that below ligand concentrations of about $0.15\ M$ it provides the major pathway for $H_2$ absorption; at higher $[L]$ the direct reaction of $H_2$ with $RhClL_3$ dominates[14] and $k_2$ can be determined. While $RhCl(PPh_3)_2$ cannot be observed spectroscopically, the 14-electron complex $RhCl(PCy_3)_2$ with bulkier ligands has been isolated.[12]

Recently deCroon and co-workers[18] explored the rate of cyclohexene hydrogenation by $RhCl(PPh_3)_3$ in benzene over a wide range of substrate concentrations (0.2 to 8.2 $M$) and found that the rate increases with increasing substrate concentration up to about 3 $M$ then decreases by a factor of 2 on increasing the cyclohexane concentration to 8.2 $M$! Their explanation is that coordination of the benzene is important in stabilizing $RhCl(PPh_3)_2$ as $RhCl(PPh_3)_2(\eta^2\text{-benzene})$. Very high substrate concentrations decrease the concentration of available benzene. Thus $RhClL_2$ in Figure 2.5 should perhaps be written as $RhClL_2B$. Note that the kinetic measurements of Halpern and Wong were made in benzene, and that they

Table 3. Values of Individual Rate and Equilibrium Constants Determined[a] in Wilkinson's Hydrogenation System

| Reaction | $k_i$ | $k_{-i}$ | $K_i$ | Combinations | Reference |
|---|---|---|---|---|---|
| 1 | $0.71 \text{ s}^{-1}$ | | $<10^{-5} M$ | | 197 |
| 2 | $4.8 M^{-1}\text{s}^{-1}$ | | $9 \times 10^3 M^{-1}$ [b] | | 13, 14 |
|   | $(12.5 M^{-1}\text{s}^{-1})^b$ | | $(40 \times 10^3 M^{-1})^b$ | | |
| 3 | $500 \text{ s}^{-1}$ | | $\ll 1 M^c$ | | 14 |
| 4 | $>7 \times 10^4 M^{-1}\text{s}^{-1}$ | | | $k_4/k_{-1} = 0.9$ | 13 |
| 5 | | | | $K_5 K_3 = 3.4 \times 10^{-4}$ [d] | 197 |
| 6 | $0.20 \text{ s}^{-1}$ [e] | | | $k_{-6}/k_7\ 0.1^f$ | 197 |
| 7 | | $0^g$ | | | |
| 8 | | | $<6 \times 10^{-6} M$ | $K_1^2 K_8 = 3.3 \times 10^{-4} M^h$ | 14 |
|   | | | | $(2.5 \times 10^{-4} M)$ | |
| 9 | $5.4 M^{-1}\text{s}^{-1}$ | | | | 14 |
|   | $(12.5 M^{-1}\text{s}^{-1})^b$ | | $(5.5 \times 10^3 M^{-1})^b$ | | |
| 10 | | | | $K_1 K_{10} \ll 1;\ 0.4$ for $C_2 H_4$ | 14 |
|   | | | | $(\ll 1)$ | |
|   | | | | $(1.7)$ for $C_2 H_4$ | |

[a] At 25°, with L = PPh$_3$ (or P(p-tolyl)$_3$ in parentheses) in benzene, with S = cyclohexene, unless listed otherwise. Reaction numbers refer to Figure 2.5, with forward reactions in the directions indicated by arrows.
[b] Taken from reference 14 but converted units assuming an H$_2$ solubility in benzene or toluene at 25° of $2.0 \times 10^{-3} M/\text{atm}$.
[c] Rapid and reversible by nmr. See Figure 2.6.
[d] Both steps 5 and 3 were assumed to be rapid and reversible.
[e] This could be the slow step in the inner loop, in accord with the deuterium isotope effect.
[f] We estimate this value in order to account for the non-$d_2$ products when D$_2$ is used.
[g] Step 7 is irreversible.
[h] $2\text{RhCl}L_3 \xrightleftharpoons{K_1^2/K_8} [\text{RhCl}L_2]_2 + 2L$.

would have no way to exclude this possibility, based on their data. One way to check deCroon's hypothesis would be to use mesitylene or durene as the solvent, rather than benzene. The methyl groups should block $\eta^2$ coordination, severely retarding the rate of H$_2$ uptake.

The rapid and reversible nature of step 3 in Figure 2.5 is shown by $^{31}$P NMR studies. Figure 2.6 shows spectra before and after addition of H$_2$ to RhCl(PPh$_3$)$_3$. At the high [Rh] used, the dimer and free PPh$_3$ were not observed. The absence of P–P coupling and Rh–P coupling for the upfield P nucleus in the spectrum (B) at 30° after adding H$_2$ shows that the unique phosphine (the one *trans* to H in structure **44**) undergoes rapid

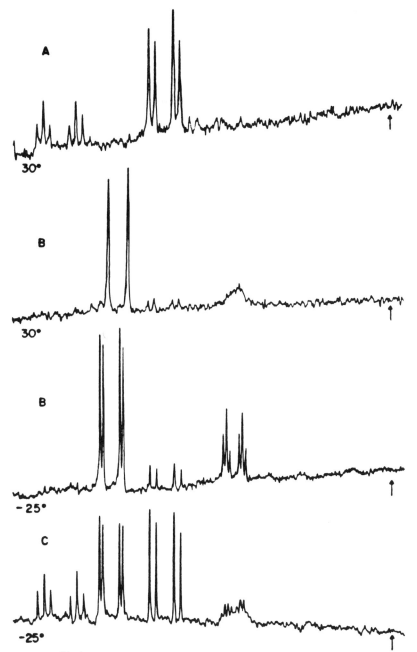

Figure 2.6. $^{31}P\{^1H\}$ nmr spectra of 0.14 $M$ RhCl(PPh$_3$)$_3$ in CH$_2$Cl$_2$ at the temperatures shown: A, before adding hydrogen; B, after adding hydrogen; and C, after sweeping nitrogen through the solution, the arrow indicates the position of free PPh$_3$. From Reference 196.

intermolecular exchange at this temperature.$^{(196)}$ A value for $k_3$ at 25° of $500 \, M^{-1} \, s^{-1}$ can be estimated by line-shape analysis. A kinetic study of the reaction of cyclohexene with preformed $H_2RhCl(PPh_3)_3$ serves to support steps 3, 5, and 6.$^{(197)}$ The value of $k_6$ (and therefore also of $K_5$ in Reference 13) is questionable, however, since it depends on an extrapolation of a plot of $k_{obs}^{-1}$ against $[L]/[S]$ at low concentrations of added $L$ (0.75 mM) where about 25% of the Rh must have been present as $H_2(RhClL_2)_2$, which itself reacts rapidly with cyclohexene.$^{(14)}$

In a system with enough added $L$ present to keep dimeric species insignificant (above about $0.01 \, M$ $[L]$ dimers are less than 2% of the Rh at 25°), a weakly binding olefin like cyclohexene, and 1 atm $H_2$, and assuming that steps 1, 3, and 5 in the mechanism in Figure 2.5 are rapid and reversible relative to the other steps, and that step 7 is irreversible, one can use the treatment of Cleland$^{(27)}$ to derive the rate law in Equation 66.

$$k^{-1} = [Rh]_{tot}/Rate = [L]/(k_2[L] + K_1k_4)[H_2] + [L]/k_6'K_3K_5[S] \qquad (66)$$

where $k_6' = k_6k_7/(k_7 + k_{-6})$.

The two terms arise from the two principal species present under catalytic cyclohexene hydrogenation conditions—$RhClL_3$ and $H_2RhClL_3$, respectively. An additional constant term, like that shown in Rate Law No. 4 in Table 2.2, is not necessary unless a significant amount of another intermediate $[H_2RhClL_2(S)]$ is also present in the system under conditions where the rate law applies.†

With cyclohexene or 1-hexene and no dimeric species, the relative amounts of $RhClL_3$ and $H_2RhClL_3$ depend on the rate at which $RhClL_3$ is converted to $H_2RhClL_3$ by $H_2$ compared to the rate at which $H_2RhClL_3$ can be converted to $RhClL_3$ by substrate olefin, indicated schematically in Figure 2.7. When the two concentrations are equal, the two reactions contribute equally to the overall resistance of the system. The $1/k'$ in Figure 2.7 can be thought of as the system resistance (reciprocal of the overall conductivity) and the two terms as the resistances of the two reactions in series.

The complexity of the Wilkinson hydrogenation system emphasizes the need to specify the ranges of variables over which a rate law applies, and it illustrates the power of combining spectroscopic studies with kinetic studies of the individual steps.

The rate law of Equation 66 becomes inadequate at low $[L]$ concentrations, when $[RhClL_2]_2$ and $H_2[RhClL_2]_2$ become significant, or in the presence of ethylene, when $(C_2H_4)RhClL_2$ can be a major species. Each

---

† Equation (66) reduces to Rate Law 4 in Table 2 if $a = 0$ and $[L]$ is small enough so that most of the reaction goes through Step 4 in Figure 2.5 rather than through Step 2.

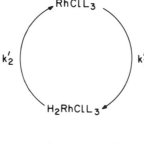

Figure 2.7. A simplified kinetic mechanism for Wilkinson's catalyst with added $L$. Here $k'_1$ and $k'_2$ are functions of the true rate constants $k_i$, [S], [$H_2$], and [$L$]. (See Reference 27).

additional significant species will add a term to the rate law written in reciprocal form.

The accelerating effect of adding small amounts of $L$ to the dimer (Figure 2.4) suggests a step such as reaction (67) in a more complete mechanism, which might also include an additional catalytic cycle based

$$H_2[RhClL_2]_2 + L \rightarrow H_2RhClL_2 + RhClL_3 \qquad (67)$$

on dimer.[14] The question of whether an olefin route exists with ethylene, with a path involving direct reaction of $H_2$ with $(C_2H_4)RhClL_2$, remains to be resolved.

### 2.2.3. Extensions to Related Systems

Changing the phosphorus ligand has a marked effect on hydrogenation rate, as seen in Table 4. Both electronic and steric effects are evident. Putting an electron-donating methyl group in the *para* position of $PPh_3$ increases the rate of 1-hexene hydrogenation by 2.2. Independent studies of the rate of reaction of $H_2$ with $RhCl[P(pC_6H_4Me)_3]_3$ show that the rate is 2.4 times as fast as with the $PPh_3$ analog.[14] Thus, better donors increase $k'_1$ in Figure 2.7. While $PEtPh_2$ is a better donor than $PPh_3$, the overall hydrogenation rate decreases by more than a factor of 2. In this case $k'_1$ probably increases while $k'_2$ substantially decreases, owing no doubt to the decreased equilibrium constant $K_3$ for $PEtPh_2$ dissociation from $H_2RhClL_3$, which is probably the major species in solution under those conditions. A smaller value of $K_3$ is attributable to the reduced steric crowding in the $H_2RhClL_3$ complex relative to the case where $L = PPh_3$. In the case of $P(oC_6H_4Me)_3$ the very slow overall reaction may be the result not of a

Table 4. Phosphorus Ligand Dependence of Relative Rates of Hydrogenation of 1-Hexene by $RhClL_3$[a] Complexes

| Ligand | Rate (ml/min)[b] |
|---|---|
| $P(pC_6H_4OMe)_3$ | 99.5 |
| $P(pC_6H_4Me)_3$ | 85.3 |
| $PPh_3$ | 38.9 |
| $PEtPh_2$ | 17.5 |
| $P(pC_6H_4Cl)_3$ | 1.58 |
| $P(oC_6H_4Me)_3$ | 0.11 |
| $P(2,4,6-C_6H_2Me_3)$ | 0.09 |
| $P(OPh)_3$ | 0.02 |

[a] Catalyst prepared *in situ* by the addition of $L$ to [RhCl(cyclooctene)]$_2$.
[b] 1.25 mM Rh, 0.6 $M$ 1-hexene in benzene, and 50 cm Hg $H_2$ pressure at 25°, from C. O'Connor and G. Wilkinson, *Tetrahedron Lett.* **18**, 1375 (1969).

small $K_3$, but of a very small $K_5$—the equilibrium constant for olefin complexation to $H_2RhClL_2$. Indeed $RhCl(PCy_3)_2$ reacts readily with $H_2$ to form $H_2RhCl(PCy_3)_2$; however, as a hydrogenation catalyst for cyclohexene, it has only 1/40 the activity of a $RhCl(PPh_3)_3$ under similar conditions.[198] $RhCl[Ph_2P(CH_2)_3PPh(CH_2)_3PPh_2]$ should be electronically very similar to $RhCl(PEtPh_2)_3$. It forms a very stable dihydride at room temperature, but does not act as a hydrogenation catalyst.[199] The reason given by Dubois and Meek[200] is that the coordinatively saturated dihydride, with structure **48** analogous to **44**, is unable to dissociate the P *trans* to H without breaking two chelate chains. The most active catalyst of the Wilkinson type appears to be $RhCl\{PhP[N(CH_2)_5]\}_3$. The turnover rate of $n$-alkenes at 30° is 1.7 $s^{-1}$ compared to 0.2 $s^{-1}$ for $RhCl(PPh_3)_3$.[201]

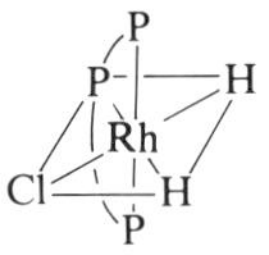

**48**

Rates of olefin hydrogenation also depend on olefin structures, as seen in Table 5. Steric inhibition is indicated by the much slower rate of

*Table 5. Olefin Structure Dependence of Hydrogen Rates Using RhCl(PPh$_3$)$_3$*

| Olefin | $k \times 10^2\ (M^{-1}\ s^{-1})^a$ |
|---|---|
| Styrene | 93.0 |
| Cyclopentene | 34.3 |
| Cyclohexene | 31.6 |
| 1-Hexene | 29.1 |
| cis-2-Pentene | 23.2 |
| 4-Methyl-cis-2-pentene | 9.9 |
| 4-Methyl-trans-2-pentene | 1.8 |
| 1-Methylcyclohexene | 0.6 |

$^a$ 1.25 in $M$ Rh in benzene at 25°, from F. H. Jardine, J. A. Osborn, and G. Wilkinson, *J. Chem. Soc. (A)*, 1574 (1967).

1-methylcyclohexene compared to the similar rates of cyclohexene and 1-hexene. Styrene hydrogenates considerably more rapidly than either, but Halpern has found a different rate law in this case and suggests a pathway involving a H$_2$RhCl$L$(S)$_2$ intermediate.[188]

Hydrogenation rates also depend on the halogen in RhXL$_3$ complexes, with rates increasing in the series $X = $ Cl $<$ Br $<$ I.[17] Both steric and electronic factors are probably involved, but not enough is currently known about the bromide and iodide systems to say more.

A theoretical study has been carried out on the olefin insertion reaction (68) by Dedieu.[202]

$$H_2RhCl(PH_3)_2(C_2H_4) \rightarrow C_2H_5RhHCl(PH_3)_2 \qquad (68)$$

It is found that the reaction is best described as an olefin insertion, rather than a hydrogen migration, and that it is promoted by good $\pi$-donor ligands in the coordination sphere. Note that from the rate law (66) one cannot say whether the slowest step in the inner loop of Figure 2.5 is due to olefin insertion $k_6$ or reductive elimination $k_7$.

The Co and Ir analogues of Wilkinson's catalyst are completely inactive, but for quite different reasons. The Co complex does not react with H$_2$,[203] while the Ir analog reacts irreversibly to give H$_2$IrCl(PPh$_3$)$_3$. This 18-electron dihydride, however, is stable to ligand dissociation and thus does not provide a site for olefin coordination.[204] Other iridium complexes can, however, be extremely active.

[Ir(COD)(PMePh$_2$)$_2$]PF$_6$ in CH$_2$Cl$_2$ at 0° will give 3800 cycles/hr of cyclohexene hydrogenation compared to only 70 for RhCl(PPh$_3$)$_3$ (in benzene/ethanol).[205] It has been possible to identify the [H$_2$Ir(COD)(PMePh$_2$)$_2$]PF$_6$ intermediate by $^1$H NMR at −80°.[206]

### 2.3. The Nickel-Catalyzed Cyclooligomerization of Butadiene

#### 2.3.1. Background

The nickel-catalyzed cyclooligomerization of butadiene (BD) is one of the most thoroughly studied homogeneous catalytic reactions.[207,208] Extensive $^{13}$C NMR studies of intermediates and model compounds have recently been reported.[209] It provides one of the best examples of the ability of phosphorus ligands to control both rates and product distributions in homogeneous catalysis and shows just how complex catalytic systems can become.

In the absence of added phosphorus ligands, the principal product (80–90%) is *trans, trans, trans*-1,5,9-cyclododecatriene (*ttt*-CDT **49**), with smaller amounts of *trans, trans, cis*-1,5,9-cyclododecatriene (*ttc*-CDT **50**), *trans, cis, cis*-1,5,9-cyclododecatriene (*tcc*-CDT **51**), 1,5-cyclooctadiene (COD **52**), 4-vinylcyclohexene (VCH **53**), and *cis*-1,2-divinylcyclobutane (DVCB **54**). With the addition of phosphorus ligands, especially bulky phosphites, COD can become 96% of the product, and the rate of butadiene consumption can be increased by a factor of 10 (at 80°).[211]

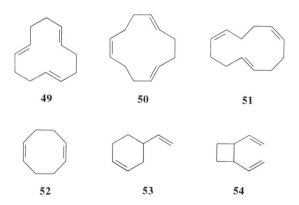

Ethylene can be readily cooligomerized with BD to produce *cis, trans*-1,5-cyclododecadiene (**55**) in good yield,[212] but propylene reacts only poorly in this way and 2-butene not at all. The strained olefin norbornene, however, gives **56** in high yield,[213] and dimethyl acetylene can be used to prepare **57**.[214]

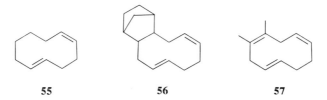

**55**   **56**   **57**

A variety of nickel catalysts may be used, including Ni(II) with reducing agents, $(\pi\text{-}C_3H_5)_2Ni$, $Ni(COD)_2$, $Ni(CH_2=CHCN)_2$, $Ni(CO)_4$, and even atomic Ni (by metal atom evaporation).† Phosphorus ligands may be added separately or coordinated in an added nickel complex such as $(CDT)NiL$, $(COD)NiL_2$, or $NiL_4$.

### 2.3.2. Phosphorus Ligand Effects and Intermediates Involved

Rates and product distributions depend on $L$ : Ni ratio, as well as ligand type, temperature, and BD conversion. Table 6 shows the effect of varying the ligand : Ni ratio for $PPh_3$. The rate of CDT formation (sum of all isomers **49–51**) is reduced from 25 cycles/hr‡ with no added $PPh_3$, to 4 at 1:1, and to less than 0.4 with $PPh_3$ : Ni of 2:1 or greater. The maximum rate of $C_8$ products (COD + VCH) is realized at a 3:1 ratio (not shown) but decreases from 105 cycles/hr§ to 45 as the $L$ : Ni ratio is increased to 8:1.

Table 6. Dependence of Product Distribution[a] on $L$ : Ni Ratio for $PPh_3$ at 80°

|  | 0:1 | 1:1 | 2:1 | 4:1 | 8:1 |
|---|---|---|---|---|---|
| % CDT | 87.2 | 6.0 | 0.5 | 0.6 | — |
| % COD | 8.2 | 64.0 | 62.0 | 56.0 | 5.0 |
| % VCH |  | 27.0 | 36.0 | 41.0 | 48.0 |
| % > $C_{12}$ | 3.6 | 2.8 | 1.9 | 1.5 | 1.6 |
| g BD/g Ni hr | 75.0 | 180.0 | 185.0 | 165.0 | 80.0 |
| Mol BD/mol Ni hr | 85.0 | 200.0 | 205.0 | 185.0 | 90.0 |

[a] Data at high conversion, from Reference 211 except for 0:1 data from reference 210.

† See references cited on p. 136 of Reference 207. $Ni(CH_2=CHCN)_2$ might appear to be a 14-electron complex, but its insolubility in noncoordinating solvents suggests a polymeric structure with nitrile bridges. Attempts (with L. J. Guggenberger) to determine a single crystal X-ray structure of the solid failed because of disorder problems.
‡ The turnover rate for CDT is one-third the moles of BD consumed per mol of Ni per hour, times the fraction of BD going to CDT, or $1/3(85 \times 0.872)$.
§ The turnover rate for cyclodimers is one-half the moles of BD consumed per mol of Ni per hour, times the fraction of BD going to cyclodimers, or $1/2(865 \times 0.991)$.

Table 7. Dependence of Product Distribution[a] on L:Ni Ratio for $P(OoC_6H_4Ph)_3$ at 80°

|  | 0:1 | 0.5:1 | 1:1 | 2:1 | 3:1 |
|---|---|---|---|---|---|
| % CDT | 87.2 | 0.4 | 0.2 | — | — |
| % COD | 8.2 | 96.0 | 96.0 | 96.0 | 96.0 |
| % VCH |  | 3.1 | 3.1 | 3.5 | 3.7 |
| % > $C_{12}$ | 36.0 |  | 0.2 | 0.3 | 0.3 |
| gBD/gNi hr | 75.0 | 380.0 | 780.0 | 550.0 | 230.0 |
| Mol BD/mol Ni hr | 85.0 | 420.0 | 865.0 | 610.0 | 255.0 |

[a] Data at high conversions, from same sources as Table 6.

The ratio COD/VCH decreases from 2.4 to 1.0 as the $PPh_3$:Ni ratio is increased from 1:1 to 8:1—but actually increases again to nearly 2.0 at 100:1.[215]

Table 7 shows the effect of varying the $L$:Ni ratio for $P(OoC_6H_4Ph)_3$. In this table, the selectivity to COD is very high at all ratios of 1:1 or greater, but the high rate of COD formation (415 cycles/hr) decreases as more $L$ is added above 1:1. With the phosphite, the $L$ is so strongly bonded to Ni that the trimerization reaction is effectively cut off. The at first puzzling result that CDT is only a minor product when the $L$:Ni ratio is 0.5:1 (when only half the nickel can be tied up by $L$) is explained by the fact that the other half is tied up as $Ni(COD)_2$; at the temperature of the experiment and in the presence of excess COD, the BD cannot effectively compete for coordination. Note that the BD consumption rate in Table 7 in this case is only half the rate at a 1:1 ratio. COD is a more severe inhibitor of cyclotrimerization than is CDT, because $Ni(COD)_2$ is considerably more stable than $Ni(CDT)$.[216]

Table 8. Dependence of Product Distribution[a] on Conversion, with 1:1 $P(OoC_6H_4Ph)_3$:Ni at 20°[b]

| % Conversion: | 4 | 8 | 55 | 85 | 95 | 100 |
|---|---|---|---|---|---|---|
| % COD | 55.0 | 56.0 | 59.0 | 60.0 | 84.0 | 98.0 |
| % VCH | 8.7 | 5.3 | 1.9 | 1.8 | 2.2 | 2.0 |
| % DVCB | 36.0 | 38.0 | 39.0 | 38.0 | 14.0 | 0.0 |

[a] In 50% benzene, 50% butadiene, from Reference 211. Less than 1% of CDT was observed.
[b] The rate was about 7 g BD/g Ni (4 mol dimer/mol Ni hr). Less than 1% was >$C_8$.

Divinylcyclobutane (DVCB) is also formed in these systems as a kinetically controlled product (as much as 40% of the cyclodimers) but is isomerized to the more thermodynamically stable COD and VCH. The isomerization rate is enhanced by higher temperatures and inhibited by BD, so that isomerization becomes significant under low-temperature catalytic conditions only at high conversions, as seen in Table 8. (The results in Tables 6 and 7 were under high conversion conditions, where DVCB was no longer present.)

The *cis* orientation of the vinyl groups in DVCB is consistent with its formation by coupling of carbons $C_3$ and $C_6$ in an intermediate of structure **58a** (or its *anti, anti*-isomer, **58b**, below).

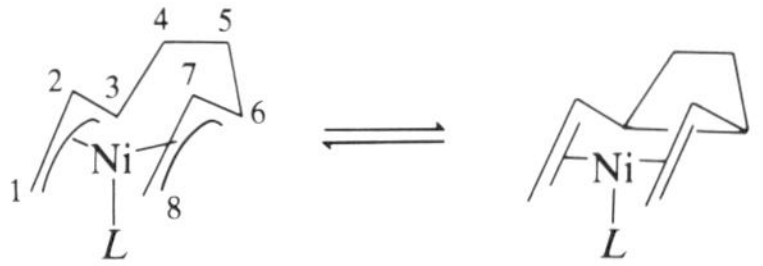

$$\text{58a} \rightleftharpoons \text{59} \tag{69}$$

This *cis* (head-to-head) orientation of $\pi$-allyl groups has been established by X-ray crystallography for $(\pi\text{-}C_3H_5)_2\text{NiPMe}_3$, which retains its *bis*-$\pi$-allyl structure in solution, as shown by NMR and Raman spectra.[217] Reversal of *rxn* (69) by oxidative addition of Ni(0) in **59** can regenerate the *bis*-$\pi$-allyl. The inhibition of DVCB decomposition by excess BD can be attributed to a competition of BD and DVCB for coordination to the available Ni(0). Coupling of BD in *rxn* (70) provides a pathway to **58**, and, by further coupling of $C_1$ and $C_8$, to COD.

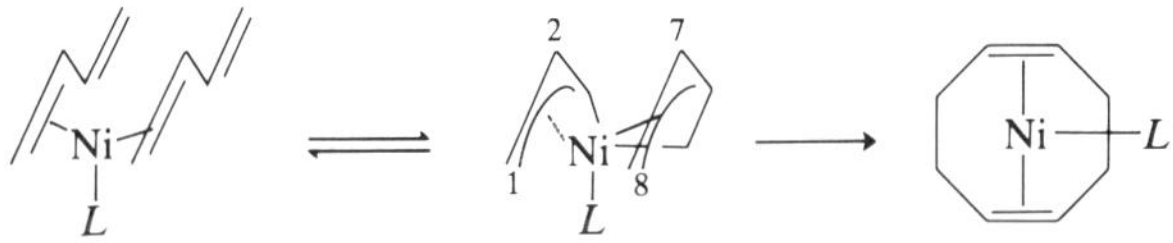

$$\rightleftharpoons \text{58b} \longrightarrow \tag{70}$$

Since the $C_1$-$C_8$ coupling in **58** fixes the stereochemistry of the double bonds, the fact that they are both *cis* in COD suggests that the $CH_2$ substituents on the $\pi$-allyl groups, as in **58b**, are *anti* with respect to the hydrogens on $C_2$ and $C_7$ at the moment of coupling. NMR studies[217] on solutions of $(\eta^3\text{-}1\text{MeC}_3H_4)_2\text{NiPMe}_3$, however, show that the only isomer observed in solution has both methyl groups *syn*.

With sufficient steric crowding (and $L$ = phosphine), both $\eta^1$ and $\eta^3$-allyls can be observed in solution, for example with **60** and **61**.[217]

Coupling of $C_1$ and $C_8$ in **61** provides a likely path to VCH, the major $C_8$ product with this ligand. Structure **62**, the analog of **61** produced by isoprene cyclodimerization, has been established by X-ray diffraction.[219]

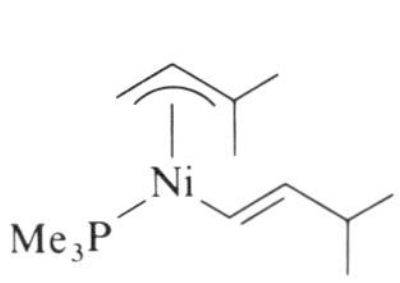

**60** 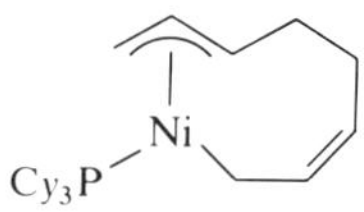   **61** 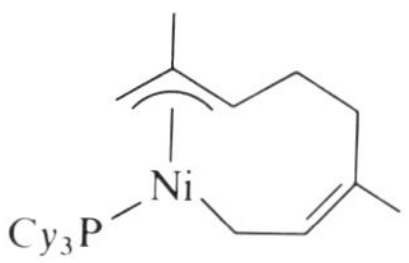   **62**

In the absence of added phosphorus ligand, BD can take the place of $L$ in intermediate **58**. In that case, however, in addition to coupling $C_1$ and $C_8$ (or $C_3$ and $C_8$), there is the possibility of coupling $C_1$ with a terminal carbon of the coordinated butadiene. This leads to a new *bis*-$\pi$-allyl complex **63**, an intermediate in the cyclotrimerization of BD which can be isolated and is stable at low temperature in the absence of free BD or $L$.[210] Recent $^{13}$C [$^1$H] and $^1$H NMR studies[219] show that **63** exists as two isomers in solution, both with coordinated *trans*-olefinic double bonds. The major one is assigned structure **64**, where the double bond is parallel to the planes

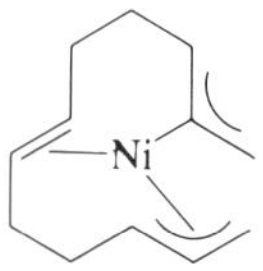

**63**    **64**

defined by the $\pi$-allyl groups, while the minor one is obtained by rotating the double bond 90° about the Ni-olefin bond. A structure analogous to **64** has been found by X-ray diffraction for the $RuCl_2$ analog **65**.[220]

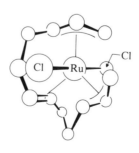

**65**

In the absence of a 2-electron donor, such as the double bond in **64** or the $L$ in **58**, *bis*-$\pi$-allyl nickel complexes are found with the $\pi$-allyl groups in nearly parallel planes and *trans* to each other (tail-to-tail), as shown by the X-ray structures of $(\eta^3\text{-2MeC}_3\text{H}_4)_2\text{Ni}$ (**66**)[221] and $(\eta^3\text{-cyclooctatrienyl})_2\text{Ni}$.[219] In toluene-$d_8$ solution $(\eta^3\text{-C}_3\text{H}_6)_2\text{Ni}$ and $(\eta^3$-

**66**

$2\text{MeC}_3\text{H}_4)_2\text{Ni}$ are found by $^{13}\text{C}$ NMR to exist as a mixture of *cis* and *trans* isomers—30% of one and 70% of the other.[219]

With addition of a 2-electron donor to **63** at room temperature, coupling of the terminal carbons occurs to give CDT. With one $\text{PE}t_3$ per nickel, $(\text{CDT})\text{NiPE}t_3$ can be isolated.[210] Excess phosphine displaces CDT from Ni. The X-ray structure of (*ttt*-CDT)Ni shows the expected ruffled symmetrical 3-coordinate structure.[222]

The presence of a *trans* double bond in trimerization intermediate **63** explains why the observed cyclotrimers **49–51** all contain at least one *trans* double bond. The two *cis* double bonds in **51** or the *cis* and second *trans* bonds in **50** must form during the final $C_{12}$ ring closure—implying that the precursors to **51** and **50** have $\pi$-allyl groups whose substituents (the chain) on $C_3$ and $C_{10}$ are *anti, anti* or *anti, syn*, respectively. (The possibility that BD couples with the terminal carbon of an $\eta^1$-allyl containing a *cis* double bond (as **61**) is regarded as less likely.) The greater thermodynamic stability of *syn*-$\pi$-allyls accounts for the predominance of the *ttt*-CDT among the trimeric products.

Cyclooligomerization reactions of $d_0$- and $d_6$-BD mixtures have been carried out and found to give $d_0$, $d_6$, $d_{12}$, or $d_{18}$ CDTs (**49–51**) and $d_0$, $d_6$, and $d_{12}$ cyclodimers (**52–54**) in nearly statistical ratios—the small deviations from statistic being attributable to secondary isotope effects.[223] This work confirms the idea that the mechanism involves C–C-bond formation without breaking or making C–H bonds.

### 2.3.3. Ligand-Concentration Control Maps and Mechanisms

Schenkluhn and co-workers[208,215,224] have made extensive studies of cyclooligomer concentrations as a function of $L:\text{Ni}$ ratios. By plotting percentages of various products against $\log(L/\text{Ni})$, they get "titration curves"[224] or "ligand-concentration control maps"[215] like that shown in

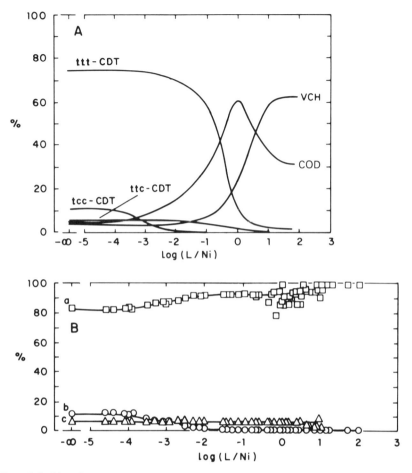

Figure 2.8. Ligand control charts or titration curves for butadiene oligomerization by nickel with $L = PPh_3$: A, product distribution among cyclotrimers and cyclodimers; B, distribution among the cyclotrimers. From Reference 215.

Figure 2.8. it is fascinating to see with either pyridine[224] or $PPh_3$[215] that increasing $L$:Ni above about $10^{-4}$ starts shutting off the formation of *tcc*-CDT and that above $10^{-2}$ ($L$ at 1% of the Ni concentration) this trimer is essentially gone. Increasing $L$:Ni over this range increases the *ttt*-CDT but has virtually no effect on the percentage of the *ttc*-isomer! The explanation of the disappearance of the *tcc*-isomer is that the added ligand decreases the steady-state concentration of an *anti, anti-bis-π*-allyl $C_{12}$ intermediate to a more thermodynamically stable *syn, syn-bis-π*-allyl by accelerating the rate of isomerization. Kinetic preference for an *anti-π*-allyl complex, which subsequently isomerizes to a more stable *syn* isomer, is also seen in the reaction of BD with $HNiL_4^+$ [$L = P(OEt)_3$]. The equilibrium constant Equation 71 is about 20 at 25°.[175]

$$\underset{NiL_3^+}{\text{(anti)}} \rightleftharpoons \underset{NiL_3^+}{\text{(syn)}} \qquad (71)$$

The virtual elimination of CDT between $\log(L/Ni) = -2$ and 0 indicates that the major catalytic trimerization intermediate (**63**) changes to one (**58**) which contains one $L$ per Ni. The fact that COD reaches a maximum near $\log(L/Ni) = 0$ and then declines at higher $L/Ni$ ratios in favor of VCH shows that VCH is formed by an intermediate (**67**) which has two $L$s.

**67**

The switch from COD to VCH can be understood in terms of the 16- and 18-Electron Rule. Coupling in an intermediate like **58** (to COD or DVCB) is allowed, because it converts an 18-electron complex into a 16. A 16-electron intermediate like **61** would drop to 14 electrons if coupling occurred in the $\eta^1$, $\eta^3$ form. On the other hand **67** is an 18-electron complex and should couple much more rapidly, giving VCH. This probably accounts for why **62** can be isolated whereas a complex like **67** has not been reported—though we have identified the analogous ($syn$-$\eta^3$-1MeC$_3$H$_4$)NiL$_2$CN intermediate in the hydrocyanation of BD.[225] At very high concentrations of some phosphorus ligands an 18-electron $\eta^1$, $\eta^1$ intermediate **68** can become important, as we shall see, providing an additional route to COD.

$$L_3Ni\text{—}\bigcirc$$

**68**

The thermodynamic stability of the 18-electron complex **64** (in the absence of added donors) may be the result of the unfavorable *trans* orientation across the nickel of the terminal carbon atoms. Addition of another ligand probably displaces the *trans* double bond and allows the complex to rearrange to the more favorable **69**, which then leads to *ttt*-CDT.

**69**

With a catalytic system as complex as this, no one has attempted to write a rate law. Using the 16- and 18-Electron Rule one can, however, write a plausible mechanism for cyclotrimerization, shown in Figure 2.9. Here S is used to represent substrate BD, bound as a simple olefin ($\eta^2$) when attached to Ni. A *bis*-$\pi$-allyl intermediate with an 8-carbon chain (as in **58**) is written as $\eta^6$-$C_8$ and the one with a twelve carbon chain (as in **69**) as $\eta^6$-$C_{12}$. No attempt is made to show the detailed stereochemistries of the intermediates or the paths leading to isomeric cyclotrimers.

The catalytic cycle in Figure 2.9 takes up three molecules of substrate BD and produces one CDT. Step 1, in which the first two BDs are coupled, is a type of oxidative addition, since the oxidation state, the number of coordinated atoms, and the number of valence electrons all increase by 2.[6] Coupling of the third BD in step 2, a type of olefin insertion reaction, fixes the stereochemistry of the first formed double bond in the $C_{12}$ chain [the $\eta^2$ in Ni($\eta^3$, $\eta^2$, $\eta^1$-$C_{12}$)] *trans*. Step 3 gives the 18-electron complex

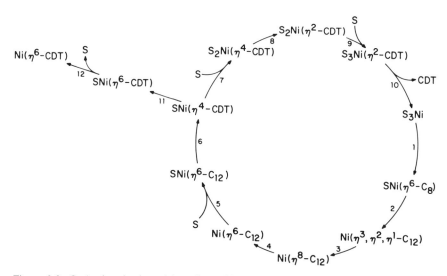

Figure 2.9. Cyclotrimerization of butadiene (S) to 1, 5, 9-cyclododecatriene (CDT). The multiplicity of Ni-carbon bonds is shown by the superscripts on $\eta$. $C_8$ and $C_{12}$ show the lengths of carbon chains.

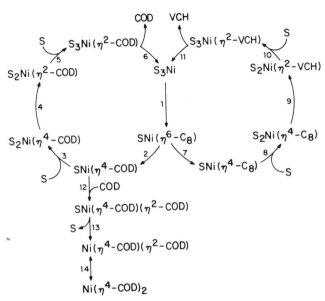

Figure 2.10. Cyclodimerization of butadiene (S) to cyclooctadiene (COD) and vinylcyclohexene (VCH) in the absence of added phosphorus ligand $L$.

**63.** Note that this is one of the few examples in this chapter where a catalytic loop species has been sufficiently stable to isolate. In step 4, the double bond dissociates to provide a coordination position for S in step 5. Step 6 is the final irreversible ring closure. Steps 7–10 then accomplish the stepwise replacement of CDT by S to regenerate $S_3Ni$. Reversible steps 11 and 12 give the isoluble Ni(CDT) complex. It is clear that Ni(CDT), $Ni(\eta^8\text{-}C_{12})$, or $(\eta^3\text{-}C_3H_5)_2Ni$ [an analog of $Ni(\eta^6\text{-}C_{12})$] can be used as catalysts for the cyclotrimerization.

In addition to adding a third BD to the chain, $SNi(\eta^6\text{-}C_8)$ can undergo $C_1$–$C_8$ ring closure to give COD, or go to a 16-electron $\eta^1$, $\eta^3$ complex, written $SNi(\eta^4\text{-}C_8)$ in Figure 2.10, where the left-hand loop produces COD and the right, VCH. Steps 12–14 show how $Ni(COD)_2$ can be used as a catalyst, and how the concentration of loop species can be decreased at high COD concentrations (See Table 7 and the associated discussion.)

In the presence of added phosphorus ligand, loops exactly like those in Figure 2.8 are possible, except that one S in each loop species is replaced by $L$ (Figure 2.11). In addition, there is a path (steps 12–17), where ring closure to VCH occurs in a complex with two $L$s (step 13). This additional path can be expected to contribute at higher $L$ concentrations. At very high $[L]$ yet another path (steps 18–22 in Figure 2.12) could

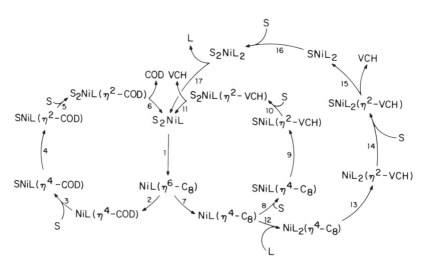

Figure 2.11. Cyclodimerization of butadiene in the presence of added $L$.

become important, in which COD forms (step 20) by reductive elimination in a dialkyl intermediate like **68**.

The effect of these multiple cycles on product distribution is seen in the distribution of dimers, plotted against $\log(L/\text{Ni})$ for $L = \text{PPh}_3$ and $\text{PPh(OPh)}_2$ in Figure 2.13. At very low ligand concentrations, the dimer distribution (about 48% COD and 42% VCH) is determined primarily by the ligand-free cycles in Figure 2.10. At $\log(L/\text{Ni})$ between $-2$ and $0$, the Ni$L$ cycles (Steps 1–11 in Figure 2.11) dominate. Even with only 1% of

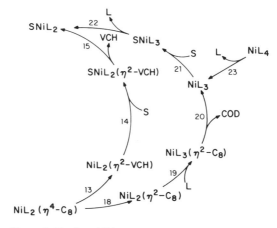

Figure 2.12. An additional pathway to COD at high $[L]$.

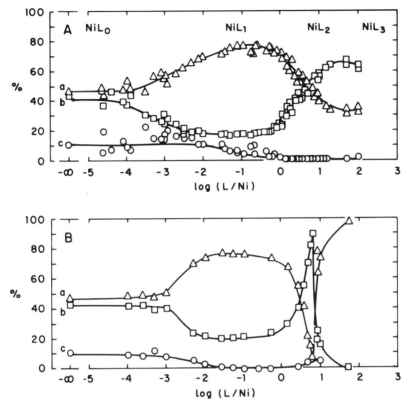

Figure 2.13. Product distribution among the dimers for: A, $L$ = PPh; and B, $L$ = PPh(OPh)$_2$: COD($a$), VCH($b$), and linear octatrienes($c$). The Ni$L_n$ show the regions where different nickel species dominate the dimer distribution. From References 208 and 215.

the Ni in the Ni$L$ cycles, most of the COD and VCH are produced here, both because of the more rapid C$_8$ ring closure induced by $L$ compared to S and because of a reduced inhibiting effect of COD in the Ni$L$ system.

The crossovers in Figure 2.13 at about log $(L/\text{Ni})$ = 0.5 indicated that at higher $L$ : Ni ratios more VCH is forming by step 13 in Figure 2.11 than by step 9. The second crossover in Figure 2.13B with $L$ = PPh(OPh)$_2$ at log $(L/\text{Ni})$ = 0.9 indicates that at higher $L$ : Ni ratios more reaction is going through step 20 in Figure 2.12 than through step 13—once again making COD the major cyclodimer. That a second crossover does not occur with PPh$_3$ but does with PPh(OPh)$_2$ is probably a consequence of the large cone angle of PPh$_3$ ($\theta$ = 145°), which precludes a favorable equilibrium to form **68**, which contains three $L$s.

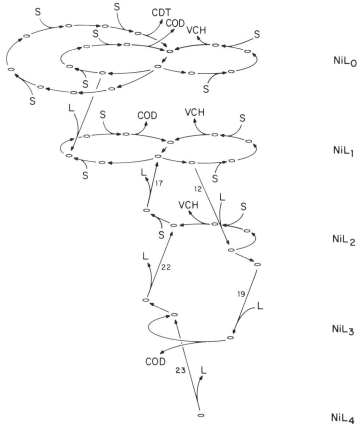

Figure 2.14. A complete mechanism for butadiene cyclooligomerization at all $L:\text{Ni}$ ratios.

The various cycles shown in Figures 2.9–2.12 can be combined to show the entire BD cyclooligomerization system in a three-dimensional representation in Figure 2.14. The reactions are shown in different planes ($\text{Ni}L_0$, $\text{Ni}L_1$, ... $\text{Ni}L_4$), according to the number of coordinated phosphorus ligands on each complex in that plane. Species in the $\text{Ni}L_1$ and $\text{Ni}L_2$ planes are the same as in the $\text{Ni}L_0$, except for replacing one or two $S$s by $L$s. Some of the reactions connecting planes are shown. Others, indicated in Equation (72), where Ol represents an olefinic double bond, no doubt occur also. Adding $L$ shifts the steady-state distribution of

$$(\text{Ol})_4\text{Ni} \underset{+\text{Ol}}{\overset{-\text{Ol}}{\rightleftarrows}} (\text{Ol})_3\text{Ni} \underset{-L}{\overset{+L}{\rightleftarrows}} (\text{Ol})_3\text{Ni}L \underset{+\text{Ol}}{\overset{-\text{Ol}}{\rightleftarrows}} (\text{Ol})_2\text{Ni}L$$

$$\underset{-L}{\overset{+L}{\rightleftarrows}} (\text{Ol})\text{Ni}L_2 \underset{-L}{\overset{+L}{\rightleftarrows}} (\text{Ol})\text{Ni}L_3 \underset{+\text{Ol}}{\overset{-\text{Ol}}{\rightleftarrows}} \text{Ni}L_3 \underset{-L}{\overset{+L}{\rightleftarrows}} \text{Ni}L_4 \qquad (72)$$

Table 9. Examples[a] of $(Ol)_xNiL_y$ Species

| NVE | Compd | Example | Reference |
|---|---|---|---|
| 18 | $(Ol)_4Ni$ | $Ni(COD)_2$ | 216 |
| 16 | $(Ol)_3Ni$ | $(C_2H_4)_3Ni$ | b |
| 18 | $(Ol)_3NiL$ | $(CDT)NiPEt_3$ | 216 |
| 16 | $(Ol)_2NiL$ | $(CH_2=CHCN)_2NiPPh_3$ | c |
| 18 | $(Ol)_2NiL_2$ | $(1\text{-Hexene})_2Ni[P(Oo\text{-tolyl})_3]_2$ | 233 |
| 16 | $(Ol)NiL_2$ | $(C_2H_4)Ni[P(Oo\text{-tolyl})_3]_2$ | d |
| 18 | $(Ol)NiL_3$ | $(C_2F)_4Ni(Ph_2PCH_2CH_2)_3CMe$ | e |
| 16 | $NiL_3$ | $Ni[P(Oo\text{-tolyl})_3]_3$ | f |
| 18 | $NiL_4$ | $Ni[PMe_3]_4$ | g |

[a] For further examples and for $^{13}C$ NMR data on several of these, see reference 209, Table IV, p. 270.
[b] K. Fischer, K. Jonas, and G. Wilke, *Angew. Chem. Int. Ed. Engl.* **12**, 565 (1973).
[c] G. N. Schrauzer, *J. Amer. Chem. Soc.* **81**, 5310 (1959).
[d] L. J. Guggenberger, *Inorg. Chem.* **12**, 499 (1973).
[e] J. Browning and B. R. Penfold, *J. Chem. Soc., Chem. Commun.*, 198 (1973).
[f] L. W. Gosser and C. A. Tolman, *Inorg. Chem.* **9**, 2350 (1970).
[g] C. A. Tolman, *J. Amer. Chem. Soc.* **92**, 2956 (1970).

nickel species to lower and lower planes and eventually cuts off the BD reactions.

Examples are known of every type of species shown in Equation 72, beginning with $Ni(COD)_2$ in Table 9 as an example of $(Ol)_4Ni$. All known $(Ol)_xNiL_y$ complexes have 16- or 18-valence electrons.

Calorimetric studies[226] of the reaction of $Ni(COD)_2$ with phosphorus ligands show why the coupling reactions are cut off, particularly for small $L$. With excess $P(OMe)_3$ ($\theta = 107°$) COD is quantitatively displaced to give $NiL_4$, in a reaction that is exothermic by 51 kcal/mol. The average Ni–P bond strength in $Ni[P(OMe)_3]_4$ (37.5 kcal) is stronger than the average Ni–Ol bond in $Ni(COD)_2$ (24.6) by 13 kcal/mol. With the bulkier ligand $P(Oo\text{-}C_6H_4Me)_3$ ($\theta = 141°$), the final product is $(COD)NiL_2$ ($\Delta H = -20$ kcal/mol), while with $P(Oo\text{-}C_6H_4\text{-}t\text{-Bu})_3$ ($\theta = 175°$), no reaction with $Ni(COD)_2$ is observed.

### 2.3.4. Steric and Electronic Control Surfaces

Both steric and electronic properties of the ligands are very important in determining product distributions, as seen in the dimer distributions in Figure 2.13. Figure 2.15 shows the percentage of CDT at $1:1$ $L:Ni$ in a large series of experiments,[215] where the response surface is plotted as a

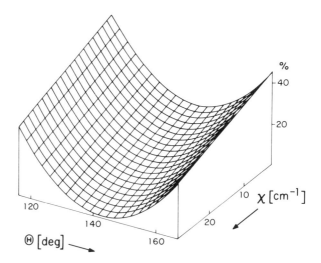

Figure 2.15. The response surface showing the effects of varying ligand steric and electronic character on the percentage of cyclotrimers in butadiene cyclooligomers for $L/Ni = 2$. The lowest point on the surface represents the maximum cyclodimer formation. From Reference 215.

function of $\theta$ and $\chi$. The minimum CDT (max COD) is found at large $\chi$ (Electronegative $L$ favors reductive elimination and $C_8$ ring closure) and moderate $\theta$. With very large ligands much of the nickel may remain in the $NiL_0$ plane, while with very small ones "disproportionation" to $NiL_0$ and $NiL_2$ planes may occur. By analyzing the surface, Schenkluhn and co-workers[215] have concluded that the control of oligomer distribution in Figure 2.15 is 75% steric and 25% electronic.

The first application of multiple regression analysis to define a response surface as a function of $\theta$ and $\chi$ was reported by Schenkluhn and co-workers using data from a calorimetric study of the reaction of 22 phosphorus ligands with $[(1,3-Me_2-\eta^3-C_3H_4)NiMe]_2$ at $0°$. In that case, the heat evolved in forming ($\pi$-allyl)NiMe$L$ complexes was found to be 40% steric and 60% electronic, the greatest exotherms being given by small electronegative ligands.[227]†

### 2.3.5. Further Extensions

The cyclotrimerization of BD seems to be peculiar to nickel. Thus, while $bis$-$\pi$-allyl nickel is an excellent catalyst for producing CDT, the Pd analog gives only linear $n$-dodecatetraenes and higher oligomers; $bis$-$\pi$-allyl

† References 228–230 are recommended for those interested in reading further.

Pt shows no reaction with BD up to the temperature of its decomposition with precipitation of Pt metal.[231] The absence of catalytic activity with Pt does not appear to be the result of very different Pt–Ol bond strengths compared to Ni. Calorimetric studies of the reaction of excess $P(OMe)_3$ with $Pt(COD)_2$ gave an exotherm of 51 kcal/mole[232]—the same value found using $Ni(COD)_2$.[226] The formation of linear oligomers does, of course, require transfer of a hydride at one point in the mechanism. N-octatrienes form in the nickel system (Figure 2.13B), but they are minor by-products.

Additional ring compounds can be prepared using substituted dienes, or by cooligomerizing olefins or acetylenes with dienes. One intriguing result involves the stereochemistry of the DVCB derivative (**70**) prepared from *cis*-piperylene and a nickel catalyst containing $P(Oo\text{-}C_6H_4Ph)_3$. With *trans*-piperylene only **71** and **72** were isolated.† The regioselectivity—in particular the absence of head-to-head coupling—suggests that prior to coupling at least one of the olefins is coordinated to Ni by its least-substituted

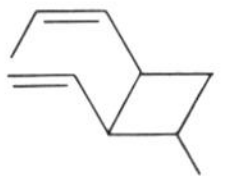

**70** 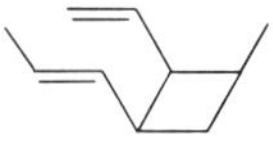  **71**

double bond. This is in accord with measurements of equilibrium constants for reactions of olefins with $Ni[P(Oo\text{-}tolyl)_3]_3$, where substitution of a methyl group on the terminal carbon of a 1-alkene decreases $K$ in Equation 73

$$Ol + NiL_3 \xrightleftharpoons{K} (Ol)NiL_2 + L \qquad (73)$$

at 25° by a factor of 200.[233] The stereoselectivity shown in products **70–72** is not in accord with a simple 2 + 2 cycloaddition reaction, but is consistent with application of the Woodward–Hoffmann rules to a stepwise process in which one C–C bond forms at a time.[218,234] Formation of DVCB and its methyl-substituted derivatives from dienes is reversible at ambient temperature. With *trans*-piperylene, for example, the $K$ for Equation 74 is $2.3 \times 10^{-2} M^{-1}$ at 30°, giving a maximum conversion to **72** of only 20% in neat diene:[234]

$$2 \, trans\text{-piperylene} \xrightleftharpoons{K} \qquad (74)$$

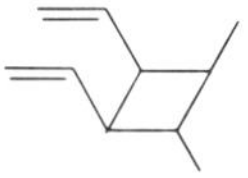

**72**

Cooligomerization of propylene with butadiene gives two derivatives of *cis, trans*-cyclodecatriene: 76% **73** and 24% **74**.

† See p. 150 in reference 207 and references cited therein.

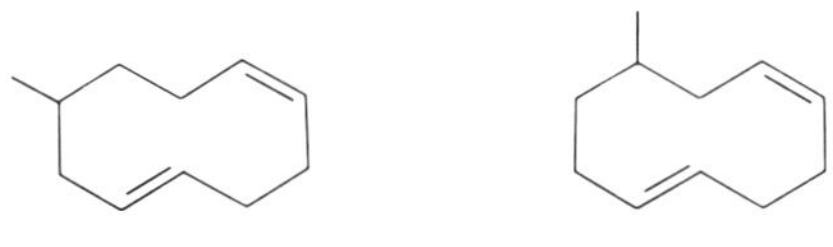

**73**    **74**

The relative amounts can be explained in terms of the interaction between the LUMO of propylene and the HOMO on the $\pi$-allyl group to which it couples to give intermediates like **75** and **76**.[235]

**75**    **76**

While propylene gives primarily **73**, styrene gives exclusively the phenyl-substituted derivative corresponding to **74**. It is noteworthy in this connection that hydrocyanation of propylene (using a $Ni[P(Ootolyl)_3]_3$ catalyst) gives primarily n-butyronitrile, while styrene gives predominantly the branched product 2-phenyl-propionitrile.[225,236]

Notice that **75** and **76** contain a *trans*-double bond and are strictly analogous to the $Ni(\eta^3, \eta^2, \eta^1-C_{12})$ intermediate for cyclotrimerization shown in Figure 2.9. The lack of dependence of the ratio **73**:**74** on the type and concentration of added ligand indicates that the coupling occurs in an intermediate $(S)Ni(\eta^6-C_8)$ in which propylene takes the place of BD as S.

An intermediate analogous to **75** (or **76**) has been isolated from the reaction of $(CDT)NiPPh_3$ with butadiene and the diethylester of acetylenedicarboxylic acid. Structure **77** has been established by assignment of the $^1H$ NMR spectrum. On treatment with CO at $-20°$, **78** is produced along with $Ni(CO)_4$.[237] Note that the *syn*-$\pi$-allyl in **77** must isomerize to *anti* in order to give the unsubstituted *cis* double bond observed in the product.

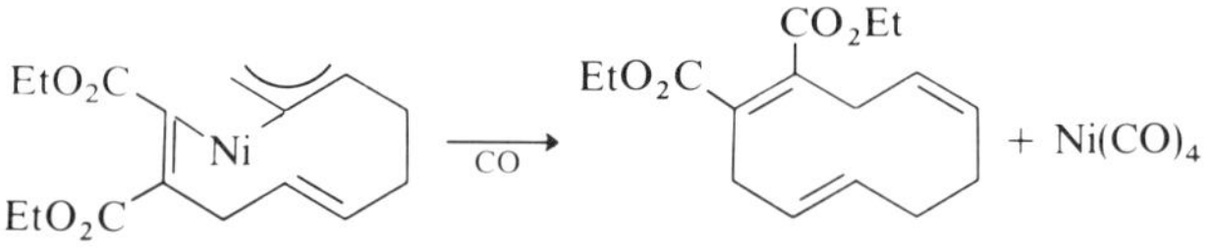

$$\text{77} \xrightarrow{CO} \text{78} + Ni(CO)_4 \qquad (75)$$

As the reader may imagine, the co-oligomerization of various dienes, olefins, and acetylenes can be elaborated to prepare a vast number of ring compounds, some including heteroatoms. References 208 and 234 are recommended to the interested reader.

## 2.4. Rhodium-Catalyzed Olefin Hydroformylation

### 2.4.1. Background and Rate Law

Hydroformylation, shown in Equation 76 for a terminal olefin, involves the addition of hydrogen and carbon monoxide to produce

$$RCH=CH_2 + H_2 + CO \rightarrow RCH_2CH_2CHO \tag{76}$$

an aldehyde and is exothermic by about 28 $k$cal/mol.[239a] The formal addition of a hydrogen atom and a formyl group across the double bond provides the origin of the name "hydroformylation." It has also been called the "Oxo" synthesis. Since its discovery using cobalt catalysts by Rolen[238] in 1938, the reaction has been very thoroughly studied with both Co and Rh catalysts. A review by Cornils in Falbe's recent book[239] contains nearly 2000 references, and has an extensive discussion of industrial processes and applications. An excellent recent review has also been published by Pruett.[240] Hydroformylation is now one of the largest-scale processes based on homogeneous catalysis by transition metal complexes.[1]

Industrially, the rhodium-catalyzed hydroformylation is normally operated at about 100°, at pressures up to 50 atm; and in the presence of a large excess of added phosphorus ligand, it can be carried out in molten $PPh_3$. Under these conditions, a terminal olefin can be converted in over 90% yield to linear aldehyde. By-products include branched aldehydes as well as small amounts of alkanes and isomerized olefins. Advantages over the more conventional cobalt catalysts include lower temperatures and pressures, higher ratios of linear to branched products, and less hydrogenation of aldehyde products to alcohols.[239b]

As in other homogeneous catalyst systems, many different catalysts can be used. These include Rh metal on carbon,[241] $Rh_4(CO)_{12}$,[242] $Rh_2O_3$,[242] $RhCl_3 \cdot 3H_2O$,[243] $(acac)Rh(CO)_2$,[244] $RhCl(CO)(PPh_3)_2$,[243] and $HRh(CO)(PPh_3)_3$.[243] In cases where one starts with a higher oxidation state of Rh, reduction by $H_2$ and CO takes place under reaction conditions. An induction period may be observed. In the case of $RhCl(CO)(PPh_3)_2$, it can be eliminated by adding a base to the system which removes HCl in reactions like (77).†

$$RhCl(CO)(PPh_3)_2 + NEt_3 + H_2 + CO \rightarrow HRh(CO)_2(PPh_3)_2 + HNEt_3^+Cl^- \tag{77}$$

With $HRh(CO)(PPh_3)_3$ or $(acac)Rh(CO)_2$, hydroformylation can be carried out directly at ambient temperature and pressure with no induction

---
† Evans and coworkers[243] write this reaction without CO to produce $HRh(CO)(PPh_3)_2$; however, this species has never been directly observed.

period.[243,244] Spectroscopic studies show that $(acac)Rh(CO)_2$ is rapidly converted to $HRh(CO)(PPh_3)_3$ in the presence of $PPh_3$, $H_2$, and $CO$.[245] An empirical rate law has been written in the form,

$$d[\text{aldehyde}]/dt = k[S]^x[Rh]^y[H_2]/[CO] \qquad (78)$$

where $x$ can be less than 1.0 and depends on the substrate olefin and $y$ can be less than 1.0.[239c] While CO inhibits the reaction at high pressures (>40 atm), it can accelerate the reaction at low pressures,[242] and, of course, is one of the reactants necessary for hydroformylation—illustrating the limited range of applicability of Equation 78. A similar rate expression for cobalt has been proposed for high pressures (100 atm), where the rate is linear in [Co], [S], and $[CO]^{-1}$.[242]

### 2.4.2. Spectroscopic Studies

High-temperature and pressure IR studies with the original phosphine-free cobalt system under conditions for catalytic hydroformylation of cyclohexene (150°, 250 atm) normally show $Co_2(CO)_8$ and $HCo(CO)_4$ as the only spectroscopically detectable species, with the hydride as the major species. With 1-octene or ethylene, the hydride is not observed and the only detectable species are $Co_2(CO)_8$ and $RCOCo(CO)_4$.[246,247] With excess $PBu_3$ (190°, 80 atm), the dimer species $Co_2(CO)_7L$ and $Co_2(CO)_6L_2$ and hydrides $HCo(CO)_3L$ and $HCo(CO)_2L_2$ are observed, and there is no evidence for the alkyl or acyl complexes, $RCo(CO)_3L$ or $RCOCo(CO)_3L$, which can be prepared by other routes.[248]

In recent work with ethylene and a phosphine-free rhodium system using a high-pressure IR cell, King and co-workers have assigned bands at $2115\,m$, $2037\,s$, and $2019\,s\,\text{cm}^{-1}$ to $C_2H_5Rh(CO)_4$[249] and bands at $2075\,w$ and $2026\,s\,\text{cm}^{-1}$ (in $n$-tetradecane) to $HRh(CO)_3(C_2H_4)$.[250] While these assignments must be regarded as tentative in the absence of confirming NMR data, they are very interesting. Many of the experiments were done at 35° and less than 100 atm; the new species are quite unstable and decompose on reducing the pressure to 1 atm or removing the solvent. The studies show that equilibria (79) and (80) lie to the right under the conditions

$$2HRh(CO)_4 \rightleftharpoons H_2 + Rh_2(CO)_8 \qquad (79)$$

$$2Rh_2(CO)_8 \rightleftharpoons Rh_4(CO)_{12} + 4CO \qquad (80)$$

used, in the absence of ethylene, but that the clusters are effectively broken down to monomeric species by the addition of ethylene. $HRh(CO)_4$ has only recently been unambiguously characterized, using FTIR.[251] Unfortunately, there appear to be no spectroscopic studies involving hydroformylation of other olefins,† or using phosphorus ligands with rhodium under the catalytic conditions of high pressures, where the rate laws have been determined. Attempts were made by Wilkinson and co-workers[252] to determine steady-state concentrations of rhodium species during hydroformylation at near ambient conditions, using an IR cell and a pump to circulate solution from the reaction vessel, but the equipment was not good enough to get useful results. Extensive spectroscopic studies employing both IR and $^1$H nmr at ambient conditions using $HRh(CO)(PPh_3)_3$ have, however, been made by Wilkinson and co-workers[243,252-254] and provide considerable insight into individual steps of the mechanism. $^{31}$P and $^{13}$C nmr studies have been reported by Kastrup and co-workers[244,245,45] at various temperatures with solutions of $HRh(CO)(PPh_3)_3$ containing added $PPh_3$, $H_2$, CO, olefins, and chelating diphosphines.

The $^{31}$P spectrum of a solution containing $HRh(CO)(PPh_3)_3$ and added $PPh_3$ in a 1:6 ratio shows separate slightly broadened resonances of complex and free ligand at 25° in a 1:2 intensity ratio, the former appearing as a doublet due to Rh–P coupling. On heating, the resonances broaden and at 60° the Rh–P coupling is lost. At 105° a single broad resonance is seen about 2/3 of the way between the original resonances of $HRh(CO)(PPh_3)_3$ and $PPh_3$—consistent with rapid exchange and a very small $K_d$ for Equation 81 at this temperature.

$$HRh(CO)(PPh_3)_3 \underset{}{\overset{K_d}{\rightleftharpoons}} HRh(CO)(PPh_3)_2 + PPh_3 \qquad (81)$$

The fact that the shape of the $HRh(CO)(PPh_3)_3$ doublet is independent of the concentration of added $PPh_3$ shows that the ligand-exchange process is indeed dissociative. Fitting of the line shapes gives $k_d = 100\,s^{-1}$ at 25° and $E_d = 20 \pm 1\,kcal/mol$.[45]

The $^1$H nmr spectrum of $HRh(CO)(PPh_3)_3$ shows a single broad line,‡ which sharpens to the expected quartet of doublets on cooling to $-30°$. The trigonal bipyramidal structure indicated in solution by nmr is also found in the solid state by X-ray.[255] In solution the IR shows only two bands around $2000\,cm^{-1}$ independent of temperature from $-45$ to $+30°$: a stronger one at $2000\,cm^{-1}$ assigned to $\nu$Rh–H and a weaker one at 1920

---

† King and Tanaka[251a] have identified a new species $RRh(CO)_4$ on treatment of $\eta^3$-$C_3H_5Rh(CO)_2$ with CO at higher pressure; $R$ is either $CH_2{=}CHCH_2-$ or $CH_2{=}CHCH_2CO-$.

‡ The observation of only a single broad line without resolvable H–P coupling in the $^1$H spectrum, at the same temperature at which the $^{31}$P spectrum shows P–Rh coupling, is a consequence of the fact that $J(H-P) \ll J(P-Rh)$.

assigned to $\nu_{CO}$;† the higher frequency band is not seen in the RhD analog.$^{(245)}$ The simple spectrum is consistent with no measurable dissociation of the complex by Equation 81. Evans and co-workers$^{(321)}$ proposed extensive phosphine dissociation to $HRh(CO)(PPh_3)_2$ and even to $HRh(CO)PPh_3$ (a 14-electron complex) based on low-molecular weights;‡ however, measuring reliable molecular weights is notoriously difficult, especially with air-sensitive compounds, and extensive dissociation is not supported by more recent spectroscopic studies.$^{(244,245,45)}$

On bubbling CO through a solution of $HRh(CO)L_3$ ($L$ = $PPh_3$) equilibrium (82) is rapidly established, the yellow solution becoming pale yellow.

$$HRh(CO)L_3 + CO \rightleftharpoons HRh(CO)_2L_2 + L \qquad (82)$$

With further bubbling for about 30 minutes, the color changes to yellow orange via *rxn* (83), the CO serving to sweep out $H_2$.

$$2\,HRh(CO)_2L_2 \rightleftharpoons H_2 + [Rh(CO)_2L_2]_2 \qquad (83)$$

Reintroducing $H_2$ reforms the hydride. Under 1 atm $H_2$ and CO in a 1:1 ratio, the equilibrium is heavily on the left-hand side. This can be compared with reaction (79) in the absence of phosphine where reduced electron density on the metal favors $H_2$ loss. $HRh(CO)_2(PPh_3)_2$ has not been isolated, since it is stable only in an atmosphere of $H_2$ and CO. The Ir analog is, however, well-known, and its X-ray structure has been done.$^{(258)}$

On bubbling $N_2$ or oxygen through solutions of the yellow dimer, CO is lost, and red compounds of composition $Rh(CO)(PPh_3)_2$(solvent) can be isolated. These are presumed to be dimeric, showing bridging carbonyl bands in the IR, but in the absence of other evidence could contain $Rh_4$ or higher clusters. The red compounds absorb $H_2$ in the presence of added $PPh_3$, and $HRh(CO)(PPh_3)_3$ is re-formed.$^{(254)}$

Treatment of $HRh(CO)L_3$ with ordinary olefins (even ethylene at 40 atm) does not give detectable concentrations of alkyls (or acyls). $C_2F_4$, however, gives $HCF_2CF_2Rh(CO)L_2$, which is stable enough to be isolated as a crystalline solid; with $H_2$ and CO (or CO alone) $HCF_2CF_2Rh(CO)_2L_2$ and $HCF_2CF_2Rh(CO)_3L$ are formed, but no acyl complexes.$^{(252)}$ Allene and butadiene give stable ($\pi$-allyl)$Rh(CO)L_2$ complexes, which do not react with $H_2$ or CO at 25° and 1 atm.$^{(259)}$ In the absence of dienes, normal

---

† The strange relative intensities are due to kinetic coupling of the two stretching modes, possible when the H and CO are mutually *trans*. The coupling also causes '$\nu_{CO}$' to increase from 1920 to 1960 cm$^{-1(254)}$ on replacing H by D.$^{(256)}$

‡ By believing the MW data, the authors were forced to assume that the various $HRh(CO)$-$(PPh_3)_n$ complexes ($n$ = 1 to 3) have accidentally similar IR spectra. We have found differences of about 10 cm$^{-1}$ in $\nu_{CN}$ in $HNiL_nCN$ complexes, with higher frequencies in $n$ = 2 than in $n$ = 3.$^{(257)}$

olefins are isomerized. In one elegant experiment,[243] $DRh(CO)L_3$ in benzene was treated with 1-pentene. The high field resonance of $HRh(CO)L_3$ grew in with a 20 s half-life at 25°, which increased to 45 min when $PPh_3$ (added $L:Rh = 10:1$) was added. Isomerization of 1-pentene to 2-pentenes also occurred in these solutions, but was much slower than exchange (in the absence of added ligand $t_{1/2} > 1$ hr), and was still slower with ligand added (13% isomerization in 1 day). These experiments are consistent with $L$ dissociation from the 18-electron $HRh(CO)L_3$ as the first step preceding olefin coordination in both exchange and isomerization reactions, which proceed through low concentrations of alkyl rhodium intermediates. The fact that exchange is about 200 times faster than isomerization indicates that the linear alkyl $RRh(CO)L_2$ in Figure 2.16 forms about 200 times as often (by reversible steps 2 and 6) as loop cycling which passes through the branched alkyl $R'Rh(CO)L_2$ formed in step 3. S and S' represent 1-alkene and 2-alkene (*cis* and *trans*), respectively. Note the similarity of Figures 2.16 and 2.2. In the rhodium system, the activity is gradually lost as an orange compound proposed to be $[Rh(CO)(PPh_3)_2]_2$ forms, even in the absence of added olefin.

On treatment of $HRh(CO)(PPh_3)_3$ with $H_2$ (even at 30 atm), one does not observe a reaction.[260] Though apparently no $H_3RhL_3$ complexes are

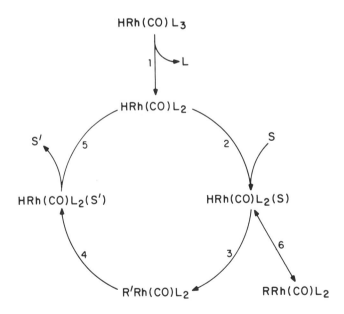

Figure 2.16. A mechanism for isotope exchange (beginning with $DRh(CO)L_3$) and isomerization of terminal olefin (S) to internal olefin (S').

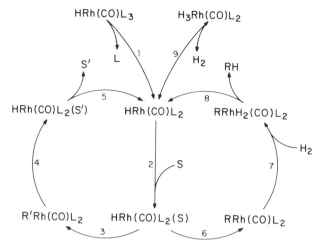

Figure 2.17. A mechanism which includes terminal olefin (S) isomerization and hydrogenation, and conversion of $HRh(CO)L_3$ to $DRh(CO)L_3$ (using $D_2$). $R$ and $R'$ are linear and branched alkyls.

known,†[262] several of the type $H_3IrL'L_2$ are, including $H_3Ir(CO)(PEt_2Ph)_2$.[263] With $D_2$, however, $DRh(CO)(PPh_3)_3$ forms rapidly from $HRh(CO)(PPh_3)_3$, with a half-life of about 2 minutes[254]—presumably via a $HRhD_2(CO)L_2$ intermediate. Analogous $RRhH_2(CO)L_2$ complexes are believed to be intermediates in the hydrogenation of 1-alkenes catalyzed by $HRh(CO)L_3$.[260]

Figure 2.17 shows a mechanism which includes paths for isomerization of 1-alkene (via steps 2–5), hydrogenation (2 and 6–8), and deuterium exchange (9), using $HRh(CO)L_3$. Of the species shown, only $HRh(CO)L_3$ is sufficiently stable to be used as the catalyst. (The dimeric complexes described earlier could also be used but are not shown in the interest of simplicity.) Detailed studies by Yagupsky and Wilkinson[264] show that isomerization and hydrogenation of 1-pentene occur at about the same rate (3 turnovers/min at 27° and 1 atm $H_2$ pressure). Both reactions are suppressed by adding $L$,‡ which will decrease the concentrations of species in both loops. The dependence of rate on catalyst concentration in the absence of added $L$ is approximately half-order,§ consistent with a larger fraction of Rh in the loops (via equilibrium 1 in Figure 2.17) as

---

† $H_3RhL_2$ complexes have been prepared with the very bulky electron donating ligands $PCy_3$, $PPh(tBu)_2$, and $P(tBu)_3$.[261]
‡ Though the experiment has not yet been reported, we can anticipate that adding $L$ will also suppress formation of $DRh(CO)L_3$ for $HRh(CO)L_3$ and $D_2$.
§ See Figure 2 in Reference 264.

the concentration is decreased. In the presence of sufficient added $L$, the rate should become first-order in [Rh], since the ratio of [HRh(CO)$L_2$]/[HRh(CO)$L_3$] will be fixed by [$L$] in Equation 6, independent of [Rh].

Figure 2.17 shows oxidative addition of $H_2$ only to the linear alkyl complex RRh(CO)$L_2$. Hydrogenation via R'Rh(CO)$L_2$ could also occur, but must be very minor because of the high selectivity for hydrogenation of terminal alkenes. The selectivity could be due to an unfavorable oxidative addition of $H_2$ to R'Rh(CO)$L_2$ because of steric crowding in the branched alkyl complex,† but a more important factor may be a very small steady-state ratio of R'Rh:RRh due to the lower stability of the more crowded complex.

With added CO, but not in its absence, ethylene reacts rapidly with HRh(CO)$L_3$ solutions to give the acyl complex EtCORh(CO)$_2L_2$.[252] With 1 atm of $H_2$ and CO (in a 2:1 ratio) propionaldehyde forms with a half-life of 5 minutes. Using a 1:2 mixture of $H_2$ and CO instead of 2:1 increases the half-life to an hour, suggesting that CO dissociation from the 18-electron acyl complex is necessary before oxidative addition of $H_2$. Isomerization and hydrogenation are much slower under hydroformylation conditions (with $H_2$ and CO) than under $H_2$ alone. These reactions are typically only 1–2% of the hydroformylation rate. The reason must be high rates of capture of 16-electron alkyl rhodium complexes by CO compared to rates of $H_2$ capture or $\beta$-hydride abstraction to form isomerized hydrido-olefin complexes.

### 2.4.3. Catalytic Hydroformylation with HRh(CO)(PPh$_3$)$_3$

With olefins, CO, and $H_2$ catalytic hydroformylation takes place even at 25° and subatmospheric pressure. Rates and product distributions depend on substrate type, [S], [$H_2$], [CO], ligand type, [$L$], [Rh], and temperature. Rates with selected olefins are given in Table 10. Note that 2-pentenes react about 25 times slower than 1-pentene, and that 2-methyl-1-pentene (a hindered terminal olefin) is slower still. Cyclooctene is much faster than cyclohexene, presumably because of ring strain effects on olefin coordination.‡ Butadiene reacts rapidly with the catalyst to form an inert ($\pi$-crotyl)Rh(CO)$L_2$ complex and no gas uptake occurs at 25°. 1,5-Hexadiene can be successfully hydroformylated, because the hydroformylation rate (to primarily linear dialdehyde) is fast compared to the rate of isomerization

---

† Decreasing rate and equilibrium constants for oxidative addition of $H_2$ and HCl to IrCl(CO)$L_2$ complexes with increased steric crowding by $L$ have been reported by Vaska and coworkers.[265]

‡ The equilibrium constant for olefin coordination to Ni(0) displacing P(O$o$tolyl)$_3$ is larger for cyclooctene than for cyclohexene by a factor of 180.[233]

Table 10. Hydroformylation Rates[a] of Various Olefins Using HRh(CO)(PPh$_3$)$_3$ as Catalyst at 25°

| Substrate | Gas uptake (ml/min) | Turnover rate[b] (cycles/hr) |
|---|---|---|
| Allyl alcohol | 7.05 | 4.5 |
| Styrene | 4.32 | 2.8 |
| 1,5-Hexadiene | 4.26 | 2.7 |
| 1-Pentene | 3.74 | 2.4 |
| Allyl cyanide | 3.27 | 2.4 |
| Ethylene | 4.55 | 2.4[c] |
| 1-Hexene | 3.52 | 2.3 |
| 1-Dodecene | 3.18 | 2.0 |
| Cyclooctene | 0.26 | 0.17 |
| 2-Pentene(cis & trans) | 0.15 | 0.10 |
| 2-Methyl-1-pentene | 0.06 | 0.04 |
| Cyclohexene | <0.05 | <0.03 |
| 1,3-Butadiene | 0 | 0[d] |
| C$_2$F$_4$ | 0 | 0[e] |

[a] With 2.5 mM HRh(CO)(PPh$_3$)$_3$, 50 cm Hg gas pressure of H$_2$ and CO (1:1), and usually 1 $M$ substrate in 50 ml benzene. With ethylene 1:1:1 mixtures of C$_2$H$_4$, H$_2$ and CO were used at 60 cm total pressure. From reference 253, except as noted otherwise.
[b] Turnover rate = gas uptake (ml/min) × 50 cm Hg × 0.5 × 60 min/hr·[24.5 ml atm/m mol × 760 cm Hg/atm × 2.5 m mol/1 × 0.050 l] = gas uptake × 0.64 except for C$_2$H$_4$.
[c] For ethylene, turnover rate = gas uptake (ml/min) × 60 cm Hg × 0.33 × 60 min/hr·[24.5 ml atm/m mol × 760 cm Hg/atm × 2.5 m mol/1 × 0.050 l] = gas uptake × 0.52.
[d] ($\eta^3$-1MeC$_3$H$_4$)Rh(CO)PPh$_3$)$_2$ forms, but no hydroformylation occurs at 25°.
[e] HCF$_2$CF$_2$Rh(CO)(PPh$_3$)$_2$ forms on reaction of C$_2$F$_4$ with the HRh(CO)$L_3$ complex. Under H$_2$ and CO no hydroformylation occurs, but HCF$_2$CF$_2$Rh(CO)$_2L_2$ and HCF$_2$CF$_2$Rh(CO)$_3L$ can be observed.

of the double bonds into conjugation. The fact that the turnover rate for ethylene is not significantly greater than for many of the other olefins is no doubt because the ethylene concentration in solution in these experiments was only that in equilibrium with 20 cm Hg C$_2$H$_4$ gas, while the concentration of liquid olefins was 1 $M$. Styrene is unusual in giving both a high-reaction rate and a predominance of branched aldehyde product. (Normal/branched ratios observed are in the range of 0.1 to 0.5.)[253] The reader may recall (Section 2.3.5) that unusual regioselectivities for C–C coupling at the carbon adjacent to the phenyl ring have been observed in both the cooligomerization of styrene and butadiene (to give solely structure

**74**, with Ph in place of Me) and in styrene hydrocyanation (to give primarily branched 2-phenyl-propionitrile). We attribute this effect to an enhanced steady-state ratio **80**:**79** in these systems relative to **82**:**81**.

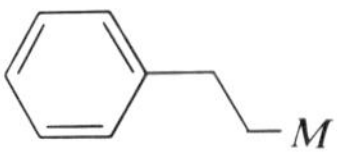

|  79  |  80  |  81  |  82  |

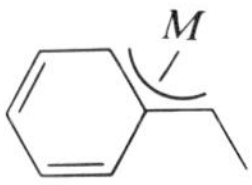

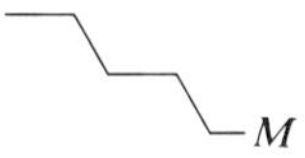

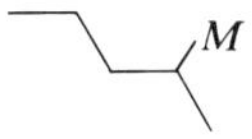

The higher ratio is achieved by $\eta^2$ electron donation from the benzene ring to the otherwise 16-electron alkyl complex. The resulting $\eta^2$ bonding is, however, not nearly as complete as in a normal $\pi$-allyl, because of the benzene resonance stabilization effect.

The high rate of hydroformylation of allyl alcohol in Table 10 could also involve some stabilization of intermediate alkyl complexes by $\eta^2$ electron donation from the oxygen.

The effects of various reaction variables on rates and product distributions using $HRh(CO)(PPh_3)_3$ and 1-hexene near ambient temperature and pressure have been reported by Wilkinson and co-workers,[253] and using 1-butene at higher temperatures (90–160°) and pressures (25–50 atm) by Kastrup and co-workers;[244] results of both groups are given in Table 11. The four products observed are linear aldehyde ($R$ CHO), branched aldehyde ($R'$CHO), alkane ($R$ H), and cis and trans-2-alkene (S'). The designations

Table 11. Effects of Reaction Variables on Rates and Product Distributions in Hydroformylations Using $HRh(CO)PPh_3)_3{}^a$

| Variables | Rate | RCHO/R'CHO | (RH + S')/Σ Products | RH/S' |
|---|---|---|---|---|
| [H$_2$] | + | + | + | + |
| [CO] | −(+) | −$^b$ | − − | − |
| [S] | + | 0 | $^c$ | $^c$ |
| T | ++ | ±$^d$(−) | + | 0(+) |
| [Rh] | + | +(0) | $^c$ | $^c$ |
| [L] | − | + | − − (0) | 0 |

$^a$ Using 1-hexene in benzene,[253] or 1-butene in 2-ethylhexylacetate[244] (in parentheses if different). For the meanings of the ++, +, 0, −, and − − entries, see the text. Abbreviations: [H$_2$] and [CO], concentrations (or pressures) of H$_2$ and CO; [S], concentration of olefin substrate; $T$, temperature; [Rh], initial catalyst concentration; [L] concentration of ligand PPh$_3$; RCHO/R'CHO, ratio of normal to branched aldehyde; RH, alkane; S', cis and trans-2-alkene.
$^b$ In this absence of added CO no appreciable hydroformylation occurs, but isomerization and hydrogenation can be fast.
$^c$ Not reported.
$^d$ The RCHO/R'CHO ratio can increase or decrease with temperature.[253]

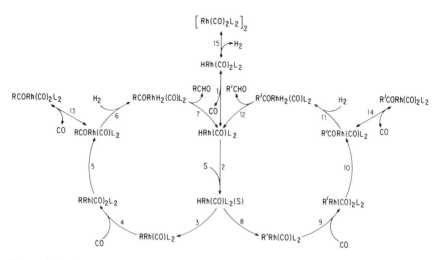

Figure 2.18. A mechanism for hydroformylation of a 1-alkene, involving only $RhL_2$ species. $R$CHO and $R'$CHO are linear and branched aldehyde products.

++ to −− express both the direction and magnitude of the effects, with 0 meaning no significant effect observed. The differences in the two studies emphasize the dependence of some of the effects on the region explored. For example, increasing the [Rh] does not affect product linearity[254] with a large excess of added $L$, but does increase linearity without added $L$.[244]

Figure 2.18 shows a mechanism for the formation of the aldehyde products which is consistent with the spectroscopic experiments described earlier and with the 16- and 18-Electron Rule. Steps are numbered starting from $HRh(Co)_2L_2$ in step 1, both because this is the principal species in solution under 1 atm $H_2$ and CO (1:1) before addition of olefin, and for the sake of simplicity, one could equally well begin with the dimer because of the reversibility of step 15. Other dimers, tetramers, and $HRh(CO)_nL_{4-n}$ complexes may be present under some conditions. The rapid formation of $R\text{CORh}(CO)_2L_2$ by steps 1–5 and 13 (or of $R'\text{CORh}(CO)_2L_2$ by similar steps) observed when 1-hexene is added to a $HRh(CO)_2L_2$ solution suggests that these steps are fast, and that steps 6 and 11 are rate determining.† This accounts for the positive effect of $[H_2]$ on rate in Table 11. The inhibiting effect of [CO] (at low $P_{CO}$) is attributable to the large equilibrium constants $K_{13}$ and $K_{14}$ (and also to the small values of $K_1$), which cause a decrease in a concentration of loop species as [CO] is increased. Though CO is taken up in steps 4 and 9, these reactions are so fact that they do

---

† The alternate possibility, that 7 and 12 are rate determining but that equilibrium constants $K_6$ and $K_{11}$ are small, cannot be excluded.

not affect the overall rate. Very rapid capture of (alkyl)Rh(CO)$L_2$ by CO is consistent with the severe inhibiting effect that [CO] has on the rates of hydrogenation and isomerization, which pass through the same intermediates (Figure 2.17). Competition of $H_2$ and CO for these 16 electron alkyls explains why hydrogenation can become faster than hydroformylation at high $H_2$:CO ratios.[254]

Isomerization probably occurs primarily via $R'$Rh(CO)$L_2$ prior to capture by CO in step 9. After $R'$CORh(CO)$_2L_2$ forms by further steps 10 and 14, getting back to $R'$Rh(CO)$L_2$ by the reverse of steps 14, 10, and 9 seems unlikely, especially under CO and $H_2$ pressure. This conclusion is supported by isotopic-labeling studies by Pino.[266] If steps 9 and 4 are essentially irreversible, then the $R$CHO/$R'$CHO ratio will primarily depend on the steady-state concentration ratio of $R$Rh(CO)$L_2$ to $R'$Rh(CO)$L_2$. If $k_4$ and $k_9$ are similar†, then $k_3$, $k_8$, and the relative stabilities ($K_3/K_8$) of $R$Rh(CO)$L_2$ and $R'$Rh(CO)$L_2$ will determine the aldehyde product distribution. Steric crowding in the branched alkyl complex is no doubt a major factor in destabilizing it compared to the linear alkyl as early recognized by Evans, Osborn, and Wilkinson;[243] the increased crowding of phosphines relative to carbonyls accounts for the increased yields of desired linear products in both the Rh and Co hydroformylation systems. Electronic factors no doubt also play a role, and in fact most dominate the product distribution with styrene, where $R'$Rh(CO)$L_2$ capture by CO is favored over $R$Rh(CO)$L_2$. The question is, to what extent is the ratio [$R$Rh(CO)$L_2$]/[$R'$Rh(CO)$L_2$] kinetically or thermodynamically controlled under steady-state conditions?

Treatment of HRh(CO)$L_3$ with styrene and CO (without $H_2$) gave a mixture of acyl complexes $R$CORh(CO)$_2L_2$ (>70%) and $R'$CORh(CO)$_2L_2$ (<30%), which could be distinguished by their $^1$H nmr spectra.[254] Bleeding $H_2$ into the nmr tube gave a mixture of aldehydes in a similar ratio. The $R$CHO/$R'$CHO ratio of >2 in the nmr experiment was much larger than observed under any *catalytic* hydroformylation conditions, where the maximum value with styrene was about 0.6. This indicates that under the conditions of stepwise addition in the nmr experiments, where the alkyls and acyls presumably had time to equilibrate, the $R$Rh/$R'$Rh and $R$CORh/$R'$CORh ratios were higher than at steady state. Thus with styrene $K_3 > K_8$ but $k_8 > k_3$.

Increasing 1-hexene concentration has no effect on the product ratio (Table 11), as expected from Figures 2.18 and 2.17, since all four products depend on the same step 2. The overall rate does, however, depend on

---

† The rate constants for steps 4 and 9 are probably very similar. Reduced steric crowding should slightly favor $k_4$ over $k_9$, while a slightly higher electron density on the metal in $R'$Rh(CO)$L_2$ should favor $k_9$. The same can be said for $k_6$ and $k_{11}$.

[S]. Lineweaver–Burke plots[24] of rate$^{-1}$ against [S]$^{-1}$ give straight lines with positive intercepts on the $y$-axis, indicating that at very high olefin concentrations the rate becomes independent of [S], and the composition at the transition state of the rate-determining step (CTSRDS, Section 1.1) contains the olefin (as acyl complexes). At low [S] the rate becomes proportional to [S], and the CTSRDS does not contain olefin, suggesting that the RDS changes under these conditions to steps 1 or 2.

Not surprisingly, the total rate is strongly temperature dependent, the rate of 1-hexene consumption increasing by about a factor of 10 on increasing the temperature from 15° to 30°.† This corresponds to an activation energy of about 25 kcal/mol. Studies by the Exxon group[244] gave a similar activation energy for 1-butene hydroformylation near 100°, but the very small effect of temperature between 145° and 160° corresponds to an apparent activation energy of less than 5 $k$ cal/mole, indicating a different rate-determining step.‡ The fraction of $(RH + S')/\Sigma$ products increases with temperature, indicating a higher activation energy for hydrogenation and isomerization than for hydroformylation, but this undesirable effect can be largely overcome by increasing [$L$]. Good results can be achieved by operating near 100° in molten $PPh_3$.[254] The temperature effect on $R$CHO/$R'$CHO was positive with 1-hexene under catalytic hydroformylation conditions, but negative under stepwise reaction conditions.

Figure 2.18 is drawn with two phosphorus ligands in each one of the complexes, in what we shall call the Rh$L_2$ plane, by analogy with the Ni$L_2$ plane of Section 2.3.3. Similar figures could be drawn for Rh$L_0$, Rh$L_2$, and Rh$L_3$ planes, which differ from Figure 2.18 only in the number of $L$ (and CO) ligands in each complex, keeping the total electron count the same for complexes similarly located in the different planes. Figure 2.19 shows a three-dimensional view of the four planes stacked one above the other. There are no catalytic cycles in the Rh$L_4$ plane, since it contains only 18-electron species: $R$CORh$L_4$, $R$Rh$L_4$, HRh$L_4$, $R'$Rh$L_4$, and $R'$CORh$L_4$ (from left to right). The figure is oversimplified in not showing dimer or cluster species, or the paths for olefin isomerization and hydrogenation indicated in Figure 2.17. It does, however, serve to explain many of the remaining effects in Table 11, as well as the effects of changing phosphorus ligands.

Increasing the concentration of HRh(CO)(PPh$_3$)$_3$ (in the absence of excess added $L$) increases the hydroformylation rate, but in a nonlinear way, indicating that the turnover rate is greater at higher dilution; the $R$CHO/$R'$CHO ratio is, however, lower in dilute solutions. This is readily

---
† See Figure 6 in reference 253.
‡ Great care must be exercised with rapid reactions involving gases to be sure that the reaction rate is not mass transfer limited. A simple test for chemical rate control is to show that the rate does not increase with increasing stirring speed.

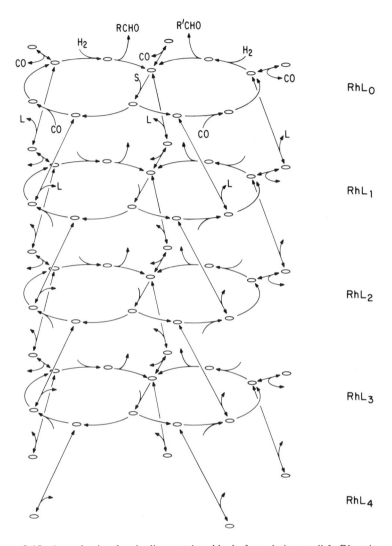

Figure 2.19. A mechanism for rhodium catalyzed hydroformylation at all $L:Rh$ ratios.

explained, since higher dilution will tend to favor species with fewer $L$s as CO and $L$ compete for coordination, i.e., reaction will tend to shift from the $RhL_2$ to the $RhL_1$ and $RhL_0$ planes at high dilution. The turnover rates are higher in the higher planes, but the product distribution is worse because of diminished steric crowding as $L$ is replaced by CO. In the $RhL_0$ plane—as in a phosphine-free system using $Rh_2(CO)_8$ or $Rh_4(CO)_{12}$ as catalysts— the product ratio $RCHO/R'CHO$ is typically less than one. (See Table 12 below.)

Table 12. Effects of Olefin Structure on Hydroformylations[a] Using AcacRh(CO)$_2$ in 1 M PPh$_3$ at 145°

| Olefin | [Rh] (mM) | $\kappa/[\text{Rh}]$[b] (min$^{-1}$ M$^{-1}$) | Percent linear aldehyde |
|---|---|---|---|
| CH$_3$CH=CH$_2$ | 1.0 | 91.0 | 88.1 |
| CH$_3$CH$_2$CH=CH$_2$ | 0.25 | 200.0 | 91.8 |
| CH$_3$(CH$_2$)$_2$CH=CH$_2$ | 0.25 | 260.0 | 91.4 |
| CH$_3$(CH$_2$)$_3$CH=CH$_2$ | 0.25 | 416.0 | 91.8 |
| CH$_3$CHCH$_2$CH=CH$_2$<br>    &#124;<br>    CH$_3$ | 0.6 | 240.0 | 90.5 |
| CH$_3$CH$_2$CHCH=CH$_2$<br>    &#124;<br>    CH$_3$ | 0.5 | 284.0 | 95.5 |
| CH$_3$CH$_2$C=CH$_2$<br>    &#124;<br>    CH$_3$ | 5.0 | 6.0 | 100.0 |
| CH$_3$CH=CHCH$_3$ | 5.0 | 5.6 | 41.9 |

[a] Twenty percent by weight olefin in 2-ethylhexylacetate under 5:1 H$_2$:CO at 25 atm. From Reference 244.
[b] Observed first-order rate constant for CO (not olefin) consumption divided by the rhodium concentration.

Adding $L$ has effects which are the opposite of diluting catalyst; shifting more of the reaction to lower planes reduces the turnover rate, but reduces hydrogenation and isomerization rates even more, while increasing the $R$CHO/$R'$CHO ratio. The negative effect of [CO] on product isomer ratio can be understood in terms of competition between CO and $L$ for coordination.

It is unlikely, for ligands as large as PPh$_3$, that much of the rhodium will be found under hydroformylation conditions in the Rh$L_3$ or Rh$L_4$ planes of Figure 2.19. Compounds in these planes, such as HRh(CO)- (PPh$_3$)$_3$ and HRh(PPh$_3$)$_4$, can, however, be used as catalysts. In the presence of CO, even in a large excess of PPh$_3$, HRh(PPh$_3$)$_4$ no doubt rapidly loses PPh$_3$ in favor of CO. The strain energy in this highly crowded molecule is indicated by the severe distortion of the three equatorial phosphines of the trigonal bipyramid away from the axial phosphine to give a nearly tetrahedral arrangement of P atoms in the X-ray crystal structure.[255]

The importance of substrate steric effects on hydroformylation rate and product linearity is shown by the results in Table 12. There is essentially no difference in product linearity for the normal alkenes longer than propylene, or for 4-methyl-1-pentene; the linearity with propylene is a little lower due to the slightly reduced crowding in its R'Rh intermediate. (The lower rate with propylene is no doubt due to the reduced solubility of this olefin under the high-temperature conditions of the experiment.) With 3-methyl-1-pentene the additional crowding significantly increases the product linearity but does not affect the overall rate. The more crowded 2-methly-1-butene reacts very slowly but gives explusively linear aldehyde. 2-Butene reacts very slowly and gives a poor product distribution because some of the $R'$Rh intermediate is trapped by CO; some goes on to isomerize to 1-butene (detected at the end of the run)[244] and give $n$-aldehyde.

### 2.4.4. Ligand Steric and Electronic Effects

With phosphorus ligands smaller than PPh$_3$ the steric strain will be less, as will the tendency to be replaced by CO. Studies of the competition between CO and phosphorus ligands for coordination to Ni(O), starting with Ni(CO)$_4$, showed that the degree of substitution of CO by $L$ decreases linearly with increasing cone angle $\theta$.[4] PPh$_3$ ($\theta = 145°$) was unable to displace more than two COs. Smaller ligands like P(OMe)$_3$ ($\theta = 107°$) were able to displace all four. On this basis we can anticipate that the distribution of Rh among the RhL$_n$ planes will depend on $\theta$, and that systems containing a very small ligand like P(OMe)$_3$, especially if it is present in excess, should show little catalytic activity.

Important studies of rhodium-catalyzed hydroformylation with ligands other than PPh$_3$ were carried out by Pruett and Smith.[241] Table 13 shows some of their results with varying concentrations of P(OPh)$_3$, using rhodium on carbon as the catalyst charged. The metal dissolved under the reaction conditions.† The increase in percentage of linear aldehyde with added $[L]$

Table 13. Dependence of Product Distribution on $L$:Rh Ratio for P(OPh)$_3$ in 1-Octene Hydroformylation[a]

| Wt. P(OPh)$_3$ (g) | $L$:Rh[b] | Percent linear aldehyde |
|---|---|---|
| 0 | 0.0 | 31 |
| 5 | 2.2 | 74 |
| 15 | 6.6 | 86 |
| 30 | 13.3 | 87 |
| 60 | 26.6 | 89 |

[a] 112 g of 1-octene and 15 g of 5% Rh/C in 200 ml toluene at 90° with 80-100 psig of 1:2 H$_2$:CO. Data from Reference 241.
[b] Mole ratio, assuming that all the Rh metal dissolved.

† When $L$ was PPh$_3$, the same rates and products were observed starting with HRh(CO)(PPh$_3$)$_3$.

Table 14. Dependence of Product Distribution on Pressure in 1-Octene Hydroformylation$^a$

| Pressure (psig)$^b$ | Percent linear aldehyde |
|---|---|
| 80–100 | 86 |
| 280–300 | 80 |
| 560–600 | 74 |
| 2500 | 69 |

$^a$ 112 g of 1-octene, 15 g of 5% Rh/C, and 15 g P(OPh)$_3$ in 200 ml toluene at 90°. $L$:Rh = 6.6:1.0. Data from Reference 241.
$^b$ Total pressure of H$_2$ and CO in a 1:1 ratio.

is striking, but this trend can be partially reversed by increasing the total pressure, as shown in Table 14. The authors recognized that the effect was due to a competition of $L$ and CO for coordination, and wrote Equation 84:†

$$\text{HRh(CO)}_4 \underset{\text{CO}}{\overset{L}{\rightleftarrows}} \text{HRh(CO)}_3 L \underset{\text{CO}}{\overset{L}{\rightleftarrows}} \text{HRh(CO)}_2 L_2 \underset{\text{CO}}{\overset{L}{\rightleftarrows}} \text{HRh(CO)} L_3 \underset{\text{CO}}{\overset{L}{\rightleftarrows}} \text{HRh} L_4$$

(84)

theorizing that each species results in an individual reaction rate and a characteristic product distribution. The species in Equation 84 are represented in Figure 2.19 by the small circles which are furthest away from the viewer in each plane. Here we see that each plane has an individual reaction rate and characteristic product distribution, with the ratio of $R$CHO/$R'$CHO increasing—due to more steric crowding—as the reaction shifts to successively lower planes.

Table 15 shows the effects of changing the steric and electronic character of $L$ on the product distribution when other variables are kept constant. For the $p$-substituted phenyl phosphites, which are sterically similar, the percentage of linear aldehyde is greatest for $p$-Cl. Without spectroscopic data on these systems under hydroformylation conditions, we cannot know whether the electronic effects are due to changing the distribution among the planes of Figure 2.19, or to changing the relative rates of the two loops within a plane, though the latter seems more likely. Though the authors do not state that the reactions were all carried to approximately the same

---

† Pruett and Smith[241] obtained IR evidence for the middle three of the five species in Equation 84 by using $L$ = PPh$_3$ under simulated hydroformylation conditions.

Table 15. Dependence of Product Distribution on Phosphorus Ligand in Hydroformylation of 1-Octene[a]

| Ligand | Percent linear aldehyde | Reaction time (min) | $\theta$ (deg)[b] | $\chi$ (cm$^{-1}$)[b] |
|---|---|---|---|---|
| P(O$p$C$_6$H$_4$Cl)$_3$ | 93 | 55 | 128 | 33.2 |
| P(OPh)$_3$ | 86 | 50 | 128 | 29.2 |
| P(O$p$C$_6$H$_4$Ph)$_3$ | 85 | 70 | 128 | 28.9[c] |
| P(O$p$C$_6$H$_4$OMe)$_3$ | 83 | 270 | 128 | 28.0 |
| PPh$_3$ | 82 | 35 | 145 | 12.8 |
| P(OBu)$_3$ | 81[e] | 60 | 107 | 20.2[d] |
| P(OC$_6$H$_4$Me)$_3$ | 78 | 52 | 141 | 28.0 |
| PBu$_3$ | 71 | 225 | 132 | 4.2 |
| P(O$o$C$_6$H$_4$Ph)$_3$ | 52 | 95 | 132 | 28.9 |
| P(O-2, 6-Me$_2$C$_6$H$_3$)$_3$ | 47 | 80 | 190 | 27.1 |

[a] 112 g 1-octene, 10 g 5% Rh/C, and 0.05 mol $L$ ($L$:Rh = 10:1) with 80-100 psig of 1:1 H$_2$:CO at 90° (unless noted otherwise). From Reference 241.
[b] Taken from ref. 239b unless noted otherwise. $\chi = \nu - 2056.1$ cm$^{-1}$.
[c] The value for P(O$o$C$_6$H$_4$Ph)$_3$.
[d] The value for P(OEt)$_3$.
[e] At 110°.

conversion, we assume that they were and that the reaction times reported give a measure of the relative reaction rates. The best electron donor of the $p$-substituted phenyl phosphites gives the lowest rate (perhaps due to reduced CO dissociation constants from 18-electron acyls). PBu$_3$, the best donor in Table 15, is also very slow.

Electronically similar $o$-substituted phenyl phosphites show a decreasing percentage of linear aldehyde as the cone angle is increased; this is no doubt due to a shifting of the Rh to higher and higher planes. Though the rate of formation of $R$CHO relative to $R'$CHO within a plane probably increases with $\theta$,† due to increased crowding, the shifting of Rh to higher planes (through competition of CO and $L$) dominates, and the net result is less linear product.

---

† This hypothesis can be tested by carrying out experiments with $o$-substituted phenyl phosphites, adjusting the $L$:Rh ratio to maintain the Rh distributions the same among the Rh$L_n$ planes (as determined spectroscopically) as $\theta$ is varied.

The reaction times in Table 15 suggest that hydroformylation of 1-octene is relatively fast using $PPh_3$. Ogata and co-workers[267] report that hydroformylation of styrene (at 50°) is 2.5 times as fast using $HRh(DBP-Ph)_4$ as the catalyst as with $HRh(PPh_3)_4$. DBP-Ph is 5-phenyl-5-H-dibenzophosphole (**83**).

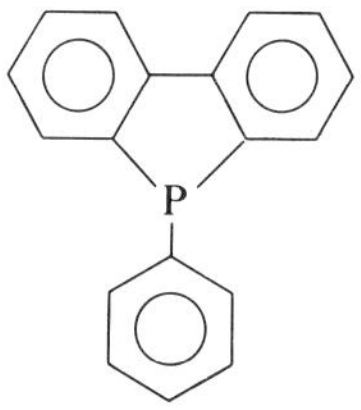

**83**

The percentage of linear aldehyde was not reported for these experiments, but in others at 140° the percentage increased from 23% to 26% on going from $PPh_3$ to DBP–Ph. We cannot say to what extent the differences in behavior are due to changing steric size and to what extent to changing electron donor character. BDP–Ph has a smaller $\theta$ but a larger $\chi$ than $PPh_3$. Replacing $PPh_3$ by $Ph_2P(CH_2)_2PPh_2$ or $Ph_2P(CH_2)_3PPh_2$ decreases both rates and percentages of linear aldehyde in the hydroformylation of 1-butene.[45]

The mechanisms shown in Figures 2.18 and 2.19 are consistent with the 16- and 18-Electron Rule and are able to account qualitatively for the behavior described in Table 11. They correspond to the "dissociative mechanism" of hydroformylation of Evans and co-workers.[243] These authors also proposed an "associative mechanism" in which the first step in the reaction of 1-alkene with $HRh(CO)_2L_2$ was coordination to form a 20 electron $HRh(CO)_2L_2(S)$ complex. They were led to this because of the different selectivities using $HRh(CO)L_3$ ($L = PPh_3$) for 1-alkene $vs$ 2-alkene in the hydrogenation and hydroformylation reactions, where the rate ratios were about 200:1 and about 25:1, respectively. They wrote, "The objection can be raised that for a $d^8$ species, the effective atomic number rule is thus exceeded, but we do not regard this as serious for a short-lived species."[243] The associative pathway has been widely quoted, including in hydroformylations using cobalt.† We regard the associative pathway as both unlikely and unnecessary. The necessary experiments to show this would involve studying the kinetics of reaction of 1-alkene and 2-alkene with $HRh(CO)_2L_2$ to form acyls as a function of CO pressure, varying $[L]$ to keep $HRh(CO)_2L_2$ as the major initial Rh species in solution.

† See, for example, Figures 1.6. and 1.7. in Reference 239.

Table 16. Comparison of Cobalt and Rhodium in Position to which Formyl Group is Bound[a]

| Catalyst | $P_{CO}$ (atm) | Percent distribution of formyl placement | | | | |
|---|---|---|---|---|---|---|
| | | $CD_3-CH_2-CH_2-CH=CH_2$ | | | | |
| $Co_2(CO)_8$ | 100 | 9 | 3 | 3 | 12 | 72 |
| $Rh_4(CO)_{12}$ | 100 | 0 | 4 | 6 | 35 | 54 |
| | 4-5 | 8 | 19 | 12 | 25 | 36 |
| | | $CD_3-CH_2-CH=CH-CH_3$ | | | | |
| $Co_2(CO)_8$ | 400 | 29 | 19 | 12 | 20 | 39 |
| | 100 | 25 | 7 | 7 | 14 | 47 |
| $Rh_4(CO)_{12}$ | 100 | 1 | 16 | 28 | 48 | 7 |
| | 4-5 | 11 | 24 | 14 | 26 | 25 |

[a] At 100° with 80–100 atm $H_2$, from Reference 270.

The 16- and 18-Electron Rule predicts a rate law given by Equation 85

$$-d[HRh(CO)_2L_2]/dt = k[HRh(CO)_2L_2][S]/[CO] \qquad (85)$$

while a combination of dissociative and associative paths would give (86).

$$-d[HRh(CO)_2L_2]/dt = k[HRh(CO)_2L_2][S](a/[CO] + b) \qquad (86)$$

### 2.4.5. Comparison of Rhodium with Cobalt and Iridium Catalysts

Cobalt and iridium complexes can also function as hydroformylation catalysts, but of course the cobalt catalysts are much better known.† The reduced activity of Co and Ir compared to Rh has been attributed to the greater stability of their 18-electron complexes to ligand dissociation.[5] The Co and Ir systems do provide some interesting comparisons with Rh,[269] both in the behavior of the catalytic systems and in the identification of catalytic intermediates.

Table 16 shows some interesting work by Pino and co-workers[270] comparing the position of formyl placement by Co and Rh in the hydroformylation of isotopically-labeled $d_3$-pentenes. The greater percentage of linear aldehyde products with phosphine-free Co than with Rh has been attributed to greater steric crowding in Co complexes as a consequence of its smaller atomic radius.[266] In addition to the tendency for terminal

† A review of $HCo[CO]_4$ chemistry has recently been published by Onchin.[268]

formylation with Co, there is an increased tendency to isomerization, seen in the relatively large amounts of formylation at the labeled end of the molecule in reactions at 100 atm CO. Even when the starting olefin is 2-pentene, the major product is linear aldehyde. With Rh there is a much greater tendency to add the formyl to the carbons which were originally on the double bond. Increased CO pressure, with either metal, has the effect of increasing the extent of formylation of the original double-bond carbons. This is clearly a consequence of trapping by CO of the alkyl complexes first formed by metal hydride addition across the double bond, before $\beta$-hydride abstraction can occur to move the double bond down the chain.

High-temperature and pressure IR spectroscopic studies by Penninger and co-workers[271] have accurately defined equilibrium constants for Equation 87:

$$H_2 + Co_2(CO)_8 \rightleftharpoons 2HCo(CO)_4 \quad (87)$$

At 100°, they report $K = 0.20$. For the forward reaction, the pseudo first-order rate constant $\kappa = 8.8 \times 10^{-3}$ min$^{-1}$† at $P_{H_2} = 25$ atm (and $P_{CO} = 25$ atm). $\Delta H = 6.6$ $k$cal/mole and $E_A = 11.3$ $k$cal/mol. The fact that both the equilibrium constant and forward rate constant increase with increasing temperature explains why high temperatures are used with cobalt to achieve high rates. Under these conditions, however, higher CO pressures are also required to stabilize the carbonyls against decomposition to metallic Co.

Ungvary and Marko[272] have more recently reported that the kinetics of (87) are complex, and that the back reaction, while second-order in [HCo(CO)$_4$] as expected, is inhibited by CO and accelerated by Co$_2$(CO)$_8$, with a concentration dependence of [Co$_2$(CO)$_8$]$^{0.5}$. They propose a mechanism involving ·Co(CO)$_4$ radicals.

Penninger and co-workers[273] studied the behavior of a system of Co$_2$(CO)$_8$ and HCo(CO)$_4$ at equilibrium at 125° and 100 atm of H$_2$ and CO (1:1) to which excess 1-pentene was injected. They observed a rapid decrease ($t_{1/2} < 2$ min) in the concentration of HCo(CO)$_4$ from about 3.8 mM to a new steady-state concentration of about 1.3 mM, with formation of a mixture of (acyl)Co(CO)$_4$ complexes,‡ (about 1.3 mM) and an *increase* in Co$_2$(CO)$_8$ from about 0.25 to 1.5 mM. They explained the fact that [Co$_2$(CO)$_8$] was higher at steady state than at equilibrium in terms of Equation 88. Hydrogenolysis of acyls was said to be by reaction with

---

† The authors[271] report both forward and reverse rate constants in units of min$^{-1}$. This is okay for a pseudo first-order rate constant for the forward reaction (if [H$_2$] is essentially constant), but cannot be correct for the reverse reaction.

‡ Characterized by a new band at 2010 cm$^{-1}$.

HCo(CO)$_4$, rather than with H$_2$, implying that Co$_2$(CO)$_8$ is a loop species.

$$RCOCo(CO)_{3,4} + HCo(CO)_4 \rightarrow RCHO + Co_2(CO)_{7,8} \qquad (88)$$

The extent to which hydrogenolysis takes place by reaction with HCo(CO)$_4$ rather than with H$_2$, as in the conventional Heck and Breslow mechanism[274] and under what conditions, and whether it plays a role at all in rhodium hydroformylations, remains to be established by kinetic and spectroscopic experiments. The situation with cobalt is also complicated by the occurrence under some conditions of odd-electron pathways. Ungvary and co-workers[275,276] have shown that reactions of HCo(CO)$_4$ with olefins can be catalyzed by Co$_2$(CO)$_8$, implying a significant role for ˙Co(CO)$_4$.

While iridium is less active than rhodium as a catalyst, it does provide insight into the mechanism, because several intermediates are stable enough to be isolated or observed when the rhodium analogs cannot. Examples include HIr(CO)$_2$(PPh$_3$)$_2$, whose X-ray structure[258] was mentioned earlier, and the EtIr(CO)$_2L_2$ (alkyl), EtCOIr(CO)$L_2$ (16-electron acyl), and EtCOIr(CO)$_3L$ (tricarbonyl acyl) with $L$ = PPh$_3$ reported by Yagupsky and co-workers.[252] Formation of propionaldehyde from EtCOIr(CO)$_2L_2$ or EtCOIr(CO)$_3L$ with H$_2$ is inhibited by CO pressure, indicating that the oxidative additions proceed via the 16-electron acyls as shown for Rh in Figure 2.18. While the acyldihydrides have not been seen, HCl and EtCOIr(CO)$L_2$ give the EtCOIrHCl(CO)$L_2$ analog, whose formation is also inhibited by CO.

While HRh(CO)$L_3$ does not react with C$_2$H$_4$ under pressure, HIr(CO)$L_3$ reacts under 10 atm C$_2$H$_4$ at 35° in 15 minutes to give about 15% conversion to a new species with $^1$H nmr signals at $\delta$ 1.26 (triplet) and 1.78 (broad) assigned to CH$_3$ and CH$_2$ protons in EtIr(CO)$L_2$. While HRh(CO)$_2L_2$ (formed from HRh(CO)$L_3$ under CO) reacts rapidly and quantitatively with C$_2$H$_4$ to give EtCORh(CO)$_2L_2$, HIr(CO)$_2L_2$ undergoes a complex series of reactions over 1–2 h at 35°, giving resonances assigned to EtIr(CO)$L_2$, EtCOIr(CO)$_2L_2$, EtCOIr(CO)$_2$(C$_2$H$_4$)$L$, and EtCOIr(CO)$L_2$.[252]

By successive treatment of HIr(CO)$_3L'$ [$L'$ = P($i$Pr)$_3$] with C$_2$H$_4$, CO, and H$_2$ in a high pressure IR cell, Whyman[277] has been able to identify C$_2$H$_5$Ir(CO)$_3L'$, C$_2$H$_5$COIr(CO)$_3L'$, and C$_2$H$_5$COH + HIr(CO)$_3L'$, and follow the stepwise conversion of one to the next—each reaction requiring about half an hour at 50°. He thus has direct spectroscopic evidence for many of the steps shown in Figure 2.18, except in the Ir$L'_1$ plane. The absence of Ir$L'_2$ species is no doubt due to the large size of P($i$Pr)$_3$ ($\theta$ = 160°)[4]

The greatly reduced activity of Ir compared to Rh can be seen in the following experiments. Using 1-pentene and HIrCO(PPh$_3$)$_3$ in a 540:1

ratio in benzene under 100 atm $H_2$ and CO (1:1), less than 10% conversion to aldehyde occurred in 18 h at 70° (<3 cycles/h). Under similar conditions at 70° even $RhCl(CO)(PPh_3)_2$ gave complete conversion.[243] $HRh(CO)(PPh_3)_3$ is faster still and gives nearly 3 cycles/h at 25° (Table 10).

## ACKNOWLEDGMENTS

We are indebted to Drs. H. Schenkluhn, P. Heimbach, J. Halpern, R. Kastrup, and R. B. King for sending information prior to its publication, to Ms. Ruth McFarlane for technical information assistance, and to Ms. Carol Farber and Ms. Sue Koblitz for painstaking typing of the manuscript, and to the Journals Department of the American Chemical Society for permission to reproduce some of the figures.

## REFERENCES

1. G. W. Parshall, *Homogeneous Catalysis, The Applications and Chemistry of Catalysis by Soluble Transition Metal Complexes* (Wiley-Interscience of John Wiley and Sons, New York, 1980).
2. J. P. Collman and L. S. Hegedus, *Principles and Applications of Organotransition Metal Chemistry* (University Science Books, Mill Valley, California, 1980) a) p. 316.
3. C. Masters, Homogeneous Transition-Metal Catalysis—A Gentle Art (Chapman and Hall, London, 1981).
4. C. A. Tolman, *Chem. Rev.* **77**, 319 (1977).
5. R. B. Woodward and R. Hoffman, *Angew. Chem. Internat. Ed. Engl.* **8**, 781 (1969).
6. C. A. Tolman, *Chem. Soc. Revs.* **1**, 337 (1972).
7. C. A. Tolman and J. P. Jesson, *Science* **181**, 501 (1973).
8. J. P. Birk, J. Halpern, and A. L. Pickard, *J. Amer. Chem. Soc.* **90**, 4491 (1968).
9. C. A. Tolman, W. C. Seidel, and D. H. Gerlach, *J. Amer. Chem. Soc.* **94**, 2669 (1972).
10. C. A. Tolman, *J. Amer. Chem. Soc.* **94**, 2994 (1972).
11. S. Otsuka, T. Yoshida, M. Matsumoto, and K. Nakatsu, *J. Amer. Chem. Soc.* **98**, 5850 (1976).
12. W. L. M. Van Gaal and F. L. A. Van Den Bekerom, *J. Organomet. Chem.* **134**, 237 (1977).
13. J. Halpern and C. S. Wong, *J. Chem. Soc. Chem. Commun.* 629, (1973).
14. C. A. Tolman, P. Z. Meakin, D. L. Lindner, and J. P. Jesson, *J. Amer. Chem. Soc.* **96**, 2762 (1974).
15. G. M. Whitesides, J. F. Gaasch, and E. R. Stedronski, *J. Amer. Chem. Soc.* **94**, 5258 (1972).
16. Y. W. Yared, S. L. Miles, R. Bau, and C. A. Reed, *J. Amer. Chem. Soc.* **99**, 7076 (1977).
17. J. A. Osborn, F. H. Jardine, J. F. Young, and G. Wilkinson, *J. Chem. Soc.* **A**, 1711 (1966).
18. M. H. J. M. de Croon, P. F. M. T. van Nisselrooij, H. J. A. M. Kuipers, and J. W. E. Coenen, *J. Mol. Catalysis* **4**, 325 (1978).
19. T. L. Brown, *Anal. N. Y. Acad. Sci.*, edited by D. W. Slocum and O. R. Hughes, **333**, 80 (1980).

20. J. Halpern, *Fundamental Research in Homogeneous Catalysis*, edited by M. Tsutsui (Plenum, New York, 1978), Vol. 3, pp. 25-40.
21. A. A. Frost and R. G. Pearson, *Kinetics and Mechanism*, 2nd ed. (John Wiley and Sons, New York, 1961) a) p. 2.
22. J. W. Moore and R. G. Pearson, *Kinetics and Mechanism*, 3rd ed. (John Wiley and Sons, New York, 1981).
23. F. Basolo and R. G. Pearson, *Mechanisms of Inorganic Reactions*, 2nd ed. (John Wiley and Sons, New York, 1967).
24. H. Lineweaver and D. Burke, *J. Amer. Chem. Soc.* **56**, 658 (1934).
25. E. L. King and C. Altman, *J. Phys. Chem.* **60**, 1375 (1956).
26. W. W. Cleland, *Biochem. Biophys. Acta.* **67**, 104 (1963).
27. ———, *Biochem.* **14**, 3220 (1975).
28. D. F. Cook and W. W. Cleland, *Biochem.* **20**, 1790, 1797, 1805 (1981).
29. I. H. Segel, *Enzyme Kinetics* (John Wiley and Sons, New York, 1975).
30. J. H. Sullivan, *J. Chem. Phys.* **46**, 73 (1967).
31. K. J. Laidler, *Chemical Kinetics* (McGraw-Hill Book Co., New York, 1965), pp. 400-406.
32. M. J. Pilling, *Reaction Kinetics* (Clarendon Press, Oxford, 1975).
33. J. E. Baeckvall, B. Akermark, and S. O. Ljunggren, *J. Am. Chem. Soc.* **101**, 2411 (1979).
34. P. W. Atkins, *Physical Chemistry*, 2nd ed. (W. H. Freeman & Co., San Francisco, 1982).
35. R. K. Harris and B. E. Mann (Eds), *NMR and the Periodic Table* (Academic Press, London, 1978).
36. P. S. Pregosin and R. W. Kunz, *NMR Basic Principles and Progress* (Springer-Verlag, Heidelberg, 1979) Vol. 16.
37. R. G. Kidd, *Annu. Rep. NMR Spectrosc.* **10A**, 1 (1980).
38. S. Aime and D. Osella, L. Milone, G. E. Hawkes, and E. W. Randall, *J. Am. Chem. Soc.* **103**, 5920 (1981).
39. C. Brown, B. T. Heaton, L. Longhetti, W. T. Povey, and D. O. Smith, *J. Organometal. Chem.* **192**, 93 (1980).
40. K. D. Brueninger, A. Schwenk, B. E. Mann, *J. Magn. Reson.* **41**, 354 (1980).
41. O. A. Gansow, D. S. Gill, F. J. Bemis, J. R. Hutchinson, J. L. Vidal, R. C. Shoening, *J. Am. Chem. Soc.* **102**, 2449 (1980).
42. G. A. Morris and R. Freeman, *J. Am. Chem. Soc.* **101**, 760 (1979).
43. C. Brevard, G. C. van Stein, and G. van Koten, *J. Am. Chem. Soc.* **103**, 6746 (1981).
44. B. T. Heaton, J. Jonas, T. Eguchi, and G. Hoffman, *J. Chem. Soc., Chem. Commun.*, 313 (1981).
45. R. V. Kastrup, J. S. Merola, and A. A. Oswald, *Advances in Chemistry* **196**, 78 (1981), edited by E. C. Alyea and D. W. Meek, American Chemical Society.
46. A. A. Oswald, R. V. Kastrup, J. S. Merola, E. J. Mozeleski, and J. C. Reisch. *ACS Symp. Ser.* **171**, 503 (1981).
47. J. Norton, *Accts. Chem. Res.* **12**, 139 (1977).
48. L. M. Jackman and F. A. Cotton, eds., *Dynamic Nuclear Magnetic Resonance Spectroscopy* (Academic Press, New York, 1975).
49. J. W. Faller, *Adv. Organometal. Chem.* **16**, 211 (1977).
50. A. Moor, P. S. Pregosin, and L. M. Venanzi, *Inorg. Chim. Acta* **48**, 153 (1981).
51. K-H. A. O. Starzewski and P. S. Pregosin, *Angew. Chem., Int. Ed. Engl.* **19**, 316 (1980).
52. J. W. Faller and G. N. LaMar, *Tetrahedron Lett.* **16**, 1381 (1973).
53. M. A. Bennett, F. A. Cotton, A. Davison, J. W. Faller, S. J. Lippard, and S. M. Morehouse, *J. Am. Chem. Soc.* **88**, 4371 (1966).
54. J. W. Faller, *Determination of Organic Structures by Physical Methods*, edited by F. C. Nachod and J. I. Zuckerman (Academic Press, New York, 1973), Vol. 5, p. 75.
55. R. Bruce Calvert and J. R. Shapley, *J. Am. Chem. Soc.* **100**, 7726 (1978).

56. J. W. Faller, M. E. Thomsen, and M. J. Mattina, *J. Am. Chem. Soc.* **93,** 2642 (1971).
57. J. W. Faller, Y. Shvo, K. Chao, and H. H. Murray, *J. Organometal. Chem.* **226,** 251 (1982).
58. R. A. Hoffman and S. Forsen, *Progr., Nucl., Magn., Reson., Spectrosc.* **1,** 15 (1966).
59. F. W. Dahlquist, K. J. Longmuir, and R. B. DuVernet, *J. Magn. Resonance* **17,** 406 (1975).
60. B. E. Mann, *Prog. NMR. Spectros.* **11,** 95 (1977).
61. G. A. Morris and R. Freeman, *J. Magn. Resonance* **29,** 433 (1978).
62. L. Melander, *Isotope Effects on Reaction Rates* (Ronald Press, New York, 1960).
63. L. Melander and W. H. Saunders, Jr., *Reaction Rates of Isotopic Molecules* (John Wiley and Sons, New York, 1980).
64. K. B. Wiberg, *Physical Organic Chemistry* (John Wiley and Sons, New York, 1964), pp. 351–363.
65. F. H. Westheimer, *Chem. Rev.* **61,** 265 (1961).
66. R. P. Bell and D. M. Goodall, *Proc. Roy. Soc., Ser. A.* **294,** 273 (1966).
67. R. L. Sweany and J. Halpern, *J. Am. Chem. Soc.* **99,** 8335 (1977).
68. A. J. Leusink, H. A. Budding, and W. Drenth, *J. Organometal. Chem.* **9,** 295 (1967).
69. T. E. Nalesnik and M. Orchin, *J. Organometal. Chem.* **199,** 265 (1980).
70. ———, *Organometallics* **1,** 223 (1982).
71. J. P. Collman, R. G. Finke, P. L. Matlock, R. Wahren, R. G. Komoto, and J. I. Brauman, *J. Am. Chem. Soc.* **100,** 1119 (1978).
72. R. P. Bell, *Chem. Rev.* **3,** 513 (1974).
73. M. Saunders, M. H. Jaffe, and P. Vogel, *J. Am. Chem. Soc.* **93,** 2558 (1971).
74. M. Saunders, L. Telkowski, and M. R. Kates, *J. Am. Chem. Soc.* **99,** 8070 (1979).
75. J. W. Faller, H. H. Murray, and M. Saunders, *J. Am. Chem. Soc.* **102,** 2306 (1980).
76. N. J. Cooper and M. L. H. Green, *J. Chem. Soc., Chem. Commun.,* 761 (1974).
77. R. C. Brady and R. Pettit, *J. Am. Chem. Soc.* **103,** 1287 (1981).
78. J. G. Ekerdt and A. T. Bell, *J. Catal.* **62,** 19 (1980).
79. L. E. McCandlish, *J. Catal.*, to be published.
80. G. M. Dawkins, M. Green, A. Guy Orpen, and F. G. A. Stone, *J. Chem. Soc., Chem. Commun.* 41 (1982).
81. R. B. Calvert, J. R. Shapley, A. J. Shultz, Jack M. Williams, S. L. Sluib, and G. D. Stucky, *J. Am. Chem. Soc.* **100,** 6240 (1978).
82. A. J. Schultz, J. M. Williams, R. B. Calvert, J. R. Shapley, and G. D. Stucky, *Inorg. Chem.* **18,** 319 (1979).
83. R. B. Clavert and J. R. Shapley, *J. Am. Chem. Soc.* **99,** 5226 (1977).
84. J. C. Huffman and W. E. Streib, *J. Chem. Soc., Chem. Commun.,* 911 (1971).
85. J. Holton, M. F. Lappert, D. G. H. Ballard, R. Pearce, J. L. Atwood, and W. E. Hunter, *J. Chem. Soc., Dalton Trans.,* 54 (1979).
86. R. J. Jablonski and E. I. Snyder, *Tetrahedron Lett.* 1103, (1968).
87. R. G. Weiss and E. I. Snyder, *J. Chem. Soc., Chem. Commun.,* 1358 (1968).
88. R. J. Jablonski and E. I. Snyder, *J. Am. Chem. Soc.* **91,** 4445 (1969).
89. G. M. Whitesides and D. J. Boschetto, *J. Am. Chem. Soc.* **91,** 4313 (1969).
90. P. L. Bock, D. J. Boschetto, J. R. Rasmussen, J. P. Demers, and G. M. Whitesides, *J. Am. Chem. Soc.* **96,** 2814 (1974).
91. J. A. Labinger and J. A. Osborn, *Inorg. Chem.* **19,** 3230 (1980).
92. G. Zweifel and C. C. Whitney, *J. Am. Chem. Soc.* **89,** 2753 (1967).
93. J. A. Labinger, D. W. Hart, W. E. Seibert, and J. Schwartz, *J. Am. Chem. Soc.* **97,** 3851 (1975).
94. D. A. Slack and M. C. Baird, *J. Am. Chem. Soc.* **98,** 5539 (1976).
95. P. L. Bock and G. M. Whitesides, *J. Am. Chem. Soc.* **98,** 5539 (1976).
96. J. W. Faller and A. S. Anderson, *J. Am. Chem. Soc.* **91,** 1550 (1969).

97. W. N. Rogers, J. A. Page, and M. C. Baird, *Inorg. Chem.* **20,** 3521 (1981).
98. D. Dong, B. K. Hunter, and M. C. Baird, *J. Chem. Soc., Chem. Commun.*, 11 (1978).
99. D. Dong, D. A. Slack, and M. C. Baird, *J. Organometal. Chem.* **153,** 219 (1978).
100. ———, *Inorg. Chem.* **18,** 188 (1979).
101. T. C. Flood and F. J. DiSanti, *J. Chem. Soc.* 18 (1975).
102. T. Majima and H. Kurosawa, *J. Chem. Soc., Chem. Commun.*, 610 (1977).
103. F. R. Jensen, V. Madan, and D. H. Buchanan, *J. Am. Chem. Soc.* **92,** 1414 (1970).
104. K. Stanley and M. C. Baird, *J. Am. Chem. Soc.* **97,** 6598 (1975).
105. R. S. Cahn, C. Ingold, and V. Prelog., *Angew. Chem. Int. Ed. Eng.* **5,** 385 (1966).
106. T. G. Attig, R. G. Teller, S-M. Wu, R. Bau, and A. Wojcicki, *J. Am. Chem. Soc.* **101,** 619 (1979).
107. J. W. Faller and A. S. Anderson, *J. Am. Chem. Soc.* **92,** 5852 (1970).
108. C. Dodd and M. D. Johnson, *J. Chem. Soc.* **D,** 571 (1971).
109. F. Calderazzo and K. Noack, *Coord. Chem. Rev.* **1,** 118 (1966).
110. H. M. Walborski and L. E. Allen, *J. Am. Chem. Soc.* **93,** 5465 (1971).
111. L. Verbit, *Progr. Phys. Org. Chem.* **7,** 51 (1970).
112. J. K. Stille and S. Y. Lau, *Accts. Chem. Res.* **10,** 434 (1977).
113. A. I. Scott and A. D. Wrixon, *Tetrahedron* **27,** 2339 (1971).
114. J. W. Faller and M. T. Tully, *J. Am. Chem. Soc.* **94,** 2676 (1972).
115. H. Brunner, *Adv. Organometal. Chem.* **18,** 151 (1980).
116. C. K. Chou, D. L. Miles, R. Bau, and T. C. Flood, *J. Am. Chem. Soc.* **100,** 7271 (1978).
117. J. A. Dale, D. L. Dull, and H. S. Mosher, *J. Org. Chem.* **34,** 2543 (1969).
118. M. D. McCreary, D. W. Lewis, D. L. Wernick, and G. M. Whitesides, *J. Am. Chem. Soc.* **96,** 1038 (1974).
119. G. R. Sullivan, *Topics in Stereochemistry*; edited by E. L. Eliel and N. L. Allinger (Wiley-Interscience of John Wiley and Sons, New York, 1978), p. 288.
120. T. J. Marks, J. S. Kristoff, A. Alich, and D. F. Shriver, *J. Organometal. Chem.* **33,** C35 (1971).
121. J. W. Faller and B. V. Johnson, *J. Organometal. Chem.* **96,** 99 (1975).
122. P. A. MacNeil, and N. K. Roberts, and B. Bosnich, *J. Am. Chem. Soc.* **103,** 2273 (1981).
123. M. D. Fryzuk and B. Bosnich, *J. Am. Chem. Soc.* **99,** 6262 (1977).
124. J. D. Morrison, W. F. Masler, M. K. Neuberg, *Adv. Catal.* **25,** 81 (1976).
125. D. S. Valentine, Jr and J. W. Scott, *Synthesis*, 329 (1978).
126. R. R. Schrock and J. A. Osborn, *J. Am. Chem. Soc.* **98,** 2134 (1976).
127. ———, *J. Am. Chem. Soc.* **98,** 4450 (1976).
128. J. Halpern, T. Okamoto, and A. Zakhariev, *J. Mol. Catal.* **2,** 65 (1977).
129. C. Rousseau, M. Evard, and F. Petit, *J. Mol. Catal.* **3,** 309 (1978).
130. M. H. J. M. DeCroon, P. F. M. T. van Nisselrooij, H. J. A. M. Kuipers, and J. W. E. Coener, *J. Mol. Catal.* **4,** 325 (1978).
131. F. A. Cotton and G. Wilkinson, *Advanced Inorganic Chemistry* (John Wiley and Sons, New York, 1980), a) p. 1266.
132. J. Halpern, D. P. Riley, A. S. C. Chan, and J. J. Pluth, *J. Am. Chem. Soc.* **99,** 8055 (1977).
133. J. M. Brown and D. Parker, *J. Chem. Soc., Chem. Commun.* 342 (1980).
134. J. M. Brown and P. A. Chaloner, *J. Chem. Soc., Chem. Commun.*, 344 (1980).
135. ———, *J. Am. Chem. Soc.* **102,** 3040 (1980).
136. A. S. C. Chan, J. J. Pluth, and J. Halpern, *Inorg. Chim. Acta* **37,** L477 (1979).
137. A. S. C. Chan, and J. Halpern, *J. Am. Chem. Soc.* **102,** 838 (1980).
138. A. S. C. Chan, J. J. Pluth, and J. Halpern, *J. Am. Chem. Soc.* **102,** 5952 (1980).
139. J. K. Kochi, *Accts. Chem. Res.* **7,** 351 (1974).
140. J. A. Osborn, *Organotransition-Metal Chemistry*, edited by Y. Ishii and M. Tsutsui (Plenum Press, New York, 1975), p. 65.
141. J. A. Labinger, J. A. Osborn, and N. J. Coville, *Inorg. Chem.* **19,** 3236 (1980).

142. M. F. Lappert and P. W. Lednor, *Adv. Organometal. Chem.* **14,** 345 (1976).
143. T. L. Hall, M. F. Lappert, and P. W. Lednor, *J. Chem. Soc., Dalton,* 1448 (1980).
144. B. H. Byers and T. L. Brown, *J. Am. Chem. Soc.* **97,** 947 (1975).
145. H. D. Murdock and E. A. C. Lucken, *Helv. Chim. Acta* **47,** 1517 (1964).
146. H. J. Keller and H. Wawersik, *Z. Naturforsch.* **20b,** 938 (1965).
147. H. J. Keller, *Z. Naturforsch.* **23b,** 133 (1968).
148. J. P. Fawcett and A. Poe, *J. Chem. Soc., Dalton Trans.,* 2039 (1976).
149. P. M. Treichel, K. P. Wagner, and H. J. Much, *J. Organometal. Chem.* **86,** C13 (1975).
150. A. M. Bond, R. Colton, and J. J. Jackowski, *Inorg. Chem.* **14,** 2526 (1975).
151. R. E. Dessy and L. A. Bares, *Accts. Chem. Res.* **5,** 415 (1972).
152. J. A. Connor and P. I. Riley, *J. Chem. Soc., Chem. Commun.,* 634 (1976).
153. M. Gargano, P. Giannocaro, M. Rossi, G. Vassapollo, and A. Sacco, *J. Chem. Soc., Dalton,* 9 (1975).
154. P. J. Krusic, J. San Filippo, Jr., B. Hutcins, R. L. Hance, and L. M. Daniels, *J. Am. Chem. Soc.* **103,** 2129 (1981).
155. P. J. Krusic, *J. Am. Chem. Soc.* **103,** 2131 (1981).
156. J. Chatt, R. A. Head, and G. J. Leigh, *J. Chem. Soc., Dalton Trans.,* 1638 (1978).
157. H. Felkin and B. Meunier, *Nouv. J. Chimie* **1,** 231 (1977).
158. P. J. Krusic, P. J. Fagan, and J. San Filippo, *J. Am. Chem. Soc.* **99,** 250 (1977).
159. P. J. Krusic, U. Klabunde, C. P. Casey, and T. F. Block, *J. Am. Chem. Soc.* **98,** 2015 (1976).
160. C. Lagercrantz, *J. Phys. Chem.* **75,** 3466 (1971).
161. E. G. Janzen, *Accts. Chem. Res.* **4,** 31 (1971).
162. S. Terabe and R. Konaka, *J. Chem. Soc., Perkin* **II,** 2136 (1972).
163. B. G. Gowerlock and J. Trotman, *J. Chem. Soc.,* 4190 (1955).
164. S. Terabe, K. Kuruma, and R. Konaka, *J. Chem. Soc., Perkin* **II,** 1252 (1973).
165. A. R. Lepley and G. L. Closs, *Chemically Induced Magnetic Polarization* (John Wiley and Sons, New York, 1973).
166. H. R. Ward, *Free Radicals,* edited by J. K. Kochi (John Wiley and Sons, New York, 1973), Vol. 1, Chapt. 8.
167. A. V. Kramer and J. A. Osborn, *J. Am. Chem. Soc.* **96,** 7832 (1974).
168. F. D. Green, M. A. Berwick, and J. C. Stowell, *J. Am. Chem. Soc.* **92,** 867 (1970).
169. D. J. Carlsson and K. U. Ingold, *J. Am. Chem. Soc.* **90,** 7047 (1968).
170. P. D. Bartlett and J. M. McBride, *Pure Appl. Chem.* **15,** 89 (1967).
171. J. A. Labinger, J. A. Osborn, and N. J. Coville, *Inorg. Chem.* **19,** 3236 (1980).
172. F. R. Jensen, L. H. Gale, and J. E. Rodgers, *J. Am. Chem. Soc.* **90,** 5793 (1968).
173. J. S. Filippo, Jr., J. Silberman, and P. J. Fagan, *J. Am. Chem. Soc.* **100,** 4834 (1978).
174. C. A. Tolman and L. H. Scharpen, *J. Chem. Soc. Dalton Trans.,* 584 (1973).
175. C. A. Tolman, *J. Amer. Chem. Soc.* **92,** 6777 (1970).
176. C. A. Tolman, W. C. Seidel, and L. W. Gosser, *J. Amer. Chem. Soc.* **96,** 53 (1974).
177. J. C. Marriott, J. A. Salthouse, M. J. Ware, and J. M. Freeman, *J. Chem. Soc., Chem. Commun.,* 575 (1970).
178. V. Albano, P. L. Bellon, and V. Scatterin, *J. Chem. Soc., Chem. Commun.,* 507 (1966).
179. M. Meier, F. Basolo, and R. G. Pearson, *Inorg. Chem.* **8,** 795 (1969).
180. C. A. Tolman, unpublished results.
181. ———, *J. Am. Chem. Soc.* **92,** 4217 (1970).
182. P. Meakin and J. P. Jesson, *J. Amer. Chem. Soc.* **96,** 5751 (1974) and references therein.
183. A. J. Deeming, B. F. G. Johnson, and J. Lewis, *J. Chem. Soc.* (D) 1848 (1973).
184. C. A. Tolman, *J. Am. Chem. Soc.* **92,** 6785 (1970).
185. B. R. James, *Homogeneous Hydrogenation* (Wiley-Interscience of John Wiley and Sons, New York, 1973).

186. R. S. Coffey, British Patent No. 1 121 642 (18 February 1965).
187. S. Siegel and D. Ohrt, *Inorg. Nucl. Chem. Lett.* **8,** 15 (1972).
188. J. Halpern, *Trans. Am. Cryst. Assoc.* **14,** 59 (1978).
189. P. B. Hitchcock, M. McPartlin, and R. Mason, *J. Chem. Soc., Chem. Commun.*, 1367 (1969).
190. M. J. Bennett, P. B. Donaldson, P. B. Hitchcock, and R. Mason, *Inorg. Chim. Acta* **12,** L9 (1975).
191. T. E. Nappier, D. W. Meek, R. M. Kirchner, and J. A. Ibers, *J. Am. Chem. Soc.* **95,** 4197 (1973).
192. R. Mason and D. W. Meek, *Angew. Chem. Int. Ed. Engl.* **17,** 183 (1978).
193. A. S. Hussey and Y. Takeuchi, *J. Amer. Chem. Soc.* **91,** 672 (1969).
194. G. C. Bond and R. H. Hillyard, *Discuss. Farad Soc.* **46,** 20 (1968).
195. S. Siegel and D. W. Ohrt, *J. Chem. Soc. Chem. Commun.*, 1529 (1971).
196. P. Meakin, J. P. Jesson, and C. A. Tolman, *J. Am. Chem. Soc.* **94,** 3240 (1972).
197. J. Halpern, T. Okamoto, and A. Zakhariev, *J. Mol. Cat.* **2,** 65 (1976).
198. H. L. M. Van Gaal, F. G. Moers, and J. J. Steggerda, *J. Organomet. Chem.* **65,** C43 (1974).
199. T. E. Nappier, Jr., Ph.D. Dissertation, The Ohio State University, 1972.
200. D. L. Dubois and D. W. Meek, *Inorg. Chim. Acta* **19,** L29 (1976).
201. R. Ugo, *Chem. Ind.* (Milan) **58,** 631 (1976), quoting results in L. Sajus, *Rev. Inst. Franc. Petrole* **24,** 1477 (1969).
202. A. Dedieu, *Inorg. Chem.* **20,** 2803 (1981).
203. N. Aresta, M. Rossi, and A. Sacco, *Inorg. Chim. Acta* **3,** 227 (1969).
204. M. A. Bennett and D. A. Milner, *J. Am. Chem. Soc.* **91,** 6983 (1969).
205. R. H. Crabtree, *Accounts Chem. Res.* **12,** 331 (1979).
206. R. H. Crabtree, H. Felkin, and G. E. Morris, *J. Chem. Soc., Chem. Commun.*, 716 (1976).
207. P. W. Jolly and G. Wilke, *The Organic Chemistry of Nickel* (Academic Press, London, 1975) Vol. II, Chapt. III.
208. P. Heimbach and H. Schenkluhn, *Topics in Current Chemistry* (Springer-Verlag, Berlin Heidelberg, 1980), Vol. 92, p. 46.
209. P. W. Jolly and R. Mynott, *Adv. Organomet. Chem.* **19,** 257 (1981).
210. B. Bogdanovic, P. Heimbach, M. Kroner, G. Wilke, E. G. Hoffmann, and J. Brandt, *Liebigs Ann. Chem.* **727,** 143 (1969).
211. W. Brenner, P. Heimbach, H. Hey, E. W. Muller, and G. Wilke, *Liebigs Ann. Chem.* **727,** 161 (1969).
212. P. Heimbach and G. Wilke, *Liebigs Ann. Chem.* **727,** 183 (1969).
213. P. Heimbach, R. V. Meyer, and G. Wilke, *Liebigs Ann. Chem.* 743 (1975).
214. W. Brunner, P. Heimbach, and G. Wilke, *Liebigs Ann. Chem.* **727,** 194 (1969).
215. P. Brille, P. Heimbach, J. Kluth, and H. Schenkluhn, *Angew. Chem. Int. Ed. Engl.* **18,** 400 (1979). a) H. Schenkluhn and P. Heimbach, lecture notes used in Essen, 1978.
216. B. Bogdanovic, M. Kroner, and G. Wilke, *Liebigs Ann. Chem.* **699,** 1 (1966).
217. B. Henc, P. W. Jolly, R. Salz, S. Stobbe, G. Wilke, R. Benn, R. Mynott, K. Seevogel, R. Goddard, and C. Kruger, *J. Organomet. Chem.* **191,** 449 (1980).
218. B. Barnett, B. Bussemeier, P. Heimbach, P. W. Jolly, C. Kruger, I. Tkatchenko, and G. Wilke, *Tetrahedron Lett.* **15,** 1457 (1972).
219. B. Henc, P. W. Jolly, R. Salz, G. Wilke, R. Benn, E. G. Hoffmann, R. Mynott, G. Schroth, K. Seevogel, J. C. Sekutowski, and C. Kruger, *J. Organomet. Chem.* **191,** 425 (1980).
220. J. E. Lydon, J. K. Nicholson, B. L. Shaw, and M. R. Truter, *Proc. Chem. Soc.* (London), 421 (1964).
221. R. Uttech and H. H. Dietrich, *Z. Kristalloger* **122,** 60 (1965).

222. H. Dietrich and H. Schmidt, *Naturwissenchaften* **52**, 301 (1965).
223. G. Schomburg, D. Henneberg, P. Heimbach, E. Janssen, H. Lehmkuhl, and G. Wilke, *Liebigs Ann. Chem.*, 1667 (1975).
224. F. Brille, J. Kluth, and H. Schenkluhn, *J. Mol. Catalysis* **5**, 27 (1979).
225. C. A. Tolman *et al.*, *Organometallics*, accepted for publication.
226. C. A. Tolman, D. W. Reutter, and W. C. Seidel, *J. Organomet. Chem.*, C30 (1976).
227. H. Schenkluhn, W. Scheidt, B. Weimann, and M. Zahres, *Angew. Chem. Int. Ed. Engl.* **18**, 401 (1979).
228. R. Berger, H. Schenkluhn, and B. Weimann, *Trans. Met. Chem.* **6**, 272 (1981).
229. H. Schenkluhn, R. Burger, B. Pinttel, and M. Zahres, *Trans. Met. Chem.* **6**, 277 (1981).
230. H̄. Schenkluhn, H. Bandmann, R. Berger, and E. Hubinger, *Trans. Met. Chem.* **6**, 287 (1981).
231. G. Wilke, B. Bogdanovic, P. Hardt, P. Heimbach, W. Kleim, M. Bruner, W. Oberkirch, K. Tamaka, E. Steinrucke, D. Walter, and H. Zimmerman, *Angew Chem. Int. Ed. Engl.* **5**, 151 (1966).
232. A. Butler and C. A. Tolman, unpublished results.
233. C. A. Tolman, *J. Am. Chem. Soc.* **96**, 2780 (1974).
234. P. Heimbach, *Angew. Chem. Int. Ed. Engl.* **12**, 975 (1973).
235. P. Heimbach, A. Roloff, and H. Schenkluhn, *Angew. Chem. Int. Ed. Engl.* **16**, 252 (1977).
236. C. M. King, W. C. Seidel, and C. A. Tolman, United States Patent No. 3 925 445 (1975).
237. B. Bussemeier, P. W. Jolly, and G. Wilke, *J. Am. Chem. Soc.* **96**, 4726 (1974).
238. O. Rolen, German Patent No. 849 548 to Ruhrchemie A. G. (1938).
239. B. Cornils *New Syntheses with Carbon Monoxide* edited by J. Falbe, (Springer-Verlag, New York, 1980), pp. 1–225. a) p. 156. b) p. 174. c) p. 17.
240. R. L. Pruett, *Adv. Organomet. Chem.* **17**, 1 (1979).
241. R. L. Pruett and J. A. Smith, *J. Org. Chem.* **34**, 327 (1969).
242. P. Pino, *J. Organomet. Chem.* **200**, 223 (1980) and references therein.
243. D. Evans, J. A. Osborn, and G. Wilkinson, *J. Chem. Soc.* (**A**), 3133 (1968).
244. A. A. Oswald, D. E. Hendriksen, R. V. Kastrup, J. S. Merola, and J. C. Reisch, presented at the Lubrizol Award Sympoium of the 1982 Spring ACS meeting, Las Vegas, Nevada.
245. A. A. Oswald, J. S. Merola, E. J. Mozeleski, R. V. Kastrup, and J. C. Reisch, *Adv. Chem. Series* **104**, 503 (1981).
246. R. Whyman, *J. Organomet. Chem.* **66**, C23 (1974).
247. R. Whyman, *J. Organomet. Chem.* **81**, 97 (1974).
248. F. Piacenti, M. Bianchi, and E. Benedetti, *Chim. Ind. Milan* **49**, 245 (1969).
249. R. B. King, A. D. King, Jr., M. Z. Iqbal, *J. Am. Chem. Soc.* **101**, 4893 (1979).
250. R. B. King, A. D. King, Jr., M. Z. Iqbal, and K. Tanaka, *Ann. N. Y. Acad. Sci.* **333**, 74 (1980).
251. J. L. Vidal and W. E. Walker, *Inorg. Chem.* **20**, 249 (1981).
251a. R. B. King and K. Tanaka, *J. Indian Chem. Soc.*, in press.
252. G. Yagupsky, C. K. Brown, and G. Wilkinson, *J. Chem. Soc.* (**A**) 1392 (1970).
253. C. K. Brown and G. Wilkinson, *J. Chem. Soc.* (**A**), 2753 (1970).
254. D. Evans, G. Yagupsky, and G. Wilkinson, *J. Chem. Soc.* (**A**), 2660 (1968).
255. S. J. LaPlaca and J. A. Ibers, *Acta Cryst.* **18**, 511 (1965).
256. L. Vaska, *J. Amer. Chem. Soc.* **88**, 4100 (1966).
257. J. D. Druliner, A. D. English, J. P. Jesson, P. Meakin, and C. A. Tolman, *J. Am. Chem. Soc.* **98**, 2656 (1976).
258. M. Ciechanowicz, A. C. Skapski, and P. G. H. Troughton, *Acta Cryst.* **B32**, 1676 (1976).
259. C. K. Brown, W. Mowat, G. Yagupsky, and G. Wilkinson, *J. Chem. Soc.* (**A**), 850 (1971).
260. C. O'Connor and G. Wilkinson, *J. Chem. Soc.* (**A**), 2665 (1968).
261. T. Yoshida, T. Okano, D. L. Thorn, T. H. Tulip, S. Otsuka, and J. A. Ibers, *J. Organomet. Chem.* **181**, 183 (1979).

262. J. P. Jesson *Transition Metal Hydrides*, edited by E. L. Meutterties (Marcel Dekker, New York, 1972), p. 129.
263. B. E. Mann, C. Masters, and B. L. Shaw, *J. Chem. Soc., Chem. Commun.*, 846 (1970).
264. M. Yagupsky and G. Wilkinson, *J. Chem. Soc.* **(A)**, 941 (1970).
265. R. Brady, W. H. deComp, B. R. Flynn, M. L. Schneider, J. D. Scott, L. Vaska, and M. F. Werneke, *Inorg. Chem.* **14**, 2669 (1975).
266. P. Pino, *J. Organomet. Chem.* **200**, 223 (1980).
267. T. Hayashi, M. Tanaka, and J. Ogata, *J. Mol. Catal.* **6**, (1979).
268. M. Orchin, *Accounts Chem. Res.* **14**, 209 (1981).
269. F. E. Paulik, *Cat. Rev.* **6**, 49 (1972).
270. D. A. von Bezard, G. Consiglio, F. Morandini, and P. Pino, *J. Mol. Cat.* **7**, 431 (1980).
271. N. H. Alemdaroglu, J. M. L. Penninger, and E. Oltay, *Monats. Chem.* **107**, 1043 (1976).
272. F. Ungvary and L. Marko, *J. Organomet. Chem.* **193**, 303 (1980).
273. N. H. Alemdaroglu, J. M. L. Penninger, and E. Oltay, *Monats. Chem.* **107**, 1153 (1976).
274. R. F. Heck and D. S. Breslow, *J. Am. Chem. Soc.* **83**, 4023 (1961).
275. J. Csizmadia, F. Ungvary, and L. Marko, *Trans. Met. Chem.* **1**, 170 (1976).
276. F. Ungvary, *Acta Chim. Hung*, **111**, 117 (1982).
277. R. Whyman, *J. Organomet. Chem.* **94**, 303 (1975).

# 3

# Structurally Characterized Transition-Metal Phosphine Complexes of Relevance to Catalytic Reactions

*Nancy L. Jones and James A. Ibers*

## 1. INTRODUCTION

A catalytic reaction that occurs under homogeneous conditions can be probed by spectroscopic and kinetic methods, but hardly by crystallographic techniques. Why then include in a book concerned with homogeneous catalysis a chapter on the structures of isolable transition-metal phosphine complexes? There are several answers. One is that such structures provide details, albeit indirectly, on the substrate-metal interaction that occurs in such a reaction. How this interaction is affected by the nature of the metal and its ligands can be probed macroscopically by kinetic measurements; it can also be probed at the molecular level in a qualitative way by spectroscopic methods. But if one seeks metrical details about the interaction, then one must turn to diffraction studies on stable, isolable analogues of the original system. Of course, one must employ spectroscopic methods to establish that the analogue is indeed a good model for the temporal species in the catalytic reaction. A second answer is that such structural information is an aid to the formulation of reaction schemes, some of which are based

---

*Nancy L. Jones and James A. Ibers* • Department of Chemistry, Northwestern University, Evanston, Illinois 60201.

on minimal data. In such schemes there frequently are proposed intermediates of new or unusual structures. Known structures may serve as a basis for such speculation. Sometimes, of course, a scheme may precede the isolation and characterization of a new structural type. In such instances the ensuing characterization of a useful analogue supports the credibility (or perhaps minimizes the incredibility) of the scheme. As an example, transition-metal allyl hydride complexes have been proposed as intermediates in a number of catalytic reactions, including olefin isomerization,[1] olefin metathesis,[2] and rearrangements of strained rings.[3] But it was not until 1978 that a stable transition-metal allyl hydride was synthesized and characterized.[4] This isolation and characterization lends credibility to the various reaction schemes.

In this chapter we tabulate structurally characterized transition-metal phosphine complexes that are analogues of some of the intermediates suggested in various catalytic schemes. While a tabulation of metrical details may prove useful for a given class of compounds, e.g., metallacycles, the range of metal phosphine complexes and the types of reactions in which they enter are so vast that tabulations of metrical information appear to serve little purpose in general. Rather, we have chosen to tabulate compounds by their type and, except for metallacycles, to provide minimal metrical details. We believe that these tabulations are reasonably complete. They are derived from computer searches of the Cambridge Crystallographic Data File (January 1981 version) along with manual searches of the more recent literature up to January 1982. The literature is indeed vast, there being approximately 2500 known structures of transition-metal phosphine complexes. Nevertheless, it will be apparent from the tabulations that there is a critical lack of concerted, systematic, high-quality studies on metal phosphine complexes of relevance to catalytic reactions. Perhaps this chapter will encourage such studies.

## 2. ALKENE HYDROGENATION

The most active metal phosphine complexes for the selective catalytic hydrogenation of alkenes and alkynes in the presence of other functional groups are $RhCl(PPh_3)_3$,[5] $RuCl_2(PPh_3)_3$,[6] and $RhH(CO)(PPh_3)_3$.[7] The structures of these three complexes are known.[8-10] Scheme 1[5,6] displays typical reaction pathways proposed for these catalytic reductions. Although the specific details of the reactions are not known in all cases, and are particularly sketchy for the Ru system, there is general agreement that the key intermediates are $Tr(P)(H)(olefin)$ and $Tr(P)(H)(alkyl)$, where $P$ = phosphine and $Tr$ = transition-metal (Sc-Zn, Y-Cd, Hf-Hg, La-Lu, Ac- ).

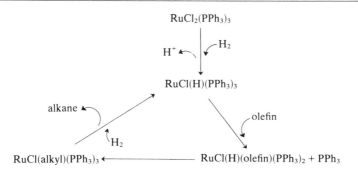

Scheme 1. Alkene Hydrogenation

It is presumed that formal hydride migration to the $\pi$-bound olefin to afford the $\sigma$-bound alkyl is facilitated if hydride and olefin ligands are initially cis to one another.

What can we learn from the structural literature on isolable compounds related to these intermediates? It turns out that very little can be learned.

Table 1. $Tr(P)(H)(olefin)$ and $Tr(P)(H)(\pi\text{-}allyl)$[a]

| Compound | Geom. | $d^n$ | CN | Description | Reference |
|---|---|---|---|---|---|
| 1. $[MoH(H_2C{=}CH_2)_2(Ph_2PCH{=}CHPPh_2)_2][CF_3CO_2]$ | pent bp | $d^4$ | 7 | H not located one olefin in eq plane, one ⊥ to plane 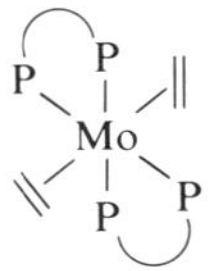 | 11 |
| 2. $Ir(H)[(CN)_2C{=}C(CN)_2](CO)(PPh_3)_2$ | tbp | $d^8$ | 5 | H not located, but cis to olefin; olefin lies in eq plane 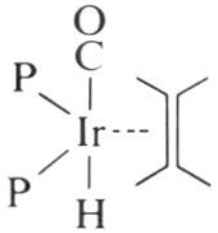 | 12 |
| 3. $Ir(H)(CH_2{=}CH{-}CH{=}CH_2)(P(i\text{-}Pr)_3)_2$ | dist tbp | $d^8$ | 5 | H located, Ir-H 1.8 Å, $\eta^4$-butadiene | 13 |
| 4. $[Ru(H)(CH_2{=}CH{-}CH{=}CH_2)(PMe_2Ph)_3][PF_6]$ | oct | $d^6$ | 6 | H not located, P's facial, $\eta^4$-butadiene | 14 |
| 5. $[Pt(H)(CH_2{=}C{=}C(CH_3)_2)(PCy_3)_2][PF_6]$ | sq pl | $d^8$ | 4 | H not located, but trans to allene | 15 |

6. IrCl(H)($\eta^3$-H$_2$C⋯CH⋯CHPh)(PPh$_3$)$_2$    oct    $d^6$    6    H located, Ir-H 1.5 Å    4

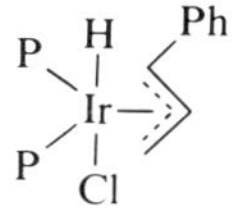

[Structure diagram showing Ir center with H, Cl, two P ligands, and η³-CH$_2$CHCHPh]

---

[a] In this and ensuing tables the following conventions and abbreviations are used:
$d^n$ represents the $d$-electronic configuration of the metal.
CN is the coordination number of the metal. In defining CN a $\pi$-bound olefin or $\pi$-allyl is counted as a monodentate ligand and a cyclopentadienyl as a tridentate ligand.

| | | | |
|---|---|---|---|
| pent | pentagonal | eq | equatorial |
| bp | bipyramid | fac | facial |
| tbp | trigonal bipyramid | dist | distorted |
| oct | octahedral | sq pl | square planar |
| rect | rectangular | sq py | square pyramid |

Ph = C$_6$H$_5$
i-Pr = CH(CH$_3$)$_2$
Cy = cyclohexyl
Me = CH$_3$
mnt = S(CN)C=C(CN)S
P$_3$N = N(CH$_2$CH$_2$PPh$_2$)$_3$
n-Bu = CH$_2$CH$_2$CH$_2$CH$_3$
Et = C$_2$H$_5$
p-tol = p-C$_6$H$_4$CH$_3$
dmpe = (CH$_3$)$_2$PCH$_2$CH$_2$P(CH$_3$)$_2$
PPN$^+$ = [Ph$_3$P=N=PPh$_3$]$^+$

There appear to be no known structures of $Tr(P)(H)(alkyl)$ complexes. Table 1 tabulates $Tr(P)(H)(olefin)$ complexes. Even for these the tabulation is very small and the results are unsatisfactory. There are only two complexes involving a mono-olefin, **1** and **2**. In neither of these structures was the position of the hydride ligand located. But on the basis of the stereochemistry it is clear that H is cis to olefin in **2**. In fact, compound **2** represents the best (and only) model for the generally proposed intermediate. That it contains an activated olefin, tetracyanoethylene, and Ir rather than Rh may be the reason for its stability.

Also included in Table 1 are known structures that involve diolefins. Complexes **3** and **4** both have coordinated butadiene. Compound **4** is rather poorly defined: the butadiene ligand appears to be $\eta^4$ coordinated; the hydride ligand was not located. Another diolefin, 1,1-dimethylallene, is coordinated to a $PtHP_2$ center in compound **5**. The H ligand, though not located, is presumed to be trans to the allene, which is coordinated to the Pt center through the C(2)–C(3) bond.

Several transition-metal phosphine complexes, including $RuCl_2(PPh_3)_3$, are effective for the selective reduction of dienes to mono-olefins.[16] It is generally believed that this selectivity arises from the exceptional stability of an intermediate $\pi$-allyl complex. The key intermediate in this process may be a $\pi$-allyl hydride complex. Complex **6**, as we indicated in the Introduction, is the only known example of such a complex. The hydride ligand is cis to the allyl group.

## 3. HYDROFORMYLATION

Hydroformylation, the addition of H and CHO to an olefinic double bond, is the oldest and largest industrial process that involves homogeneous transition-metal catalysts. The original catalyst was derived from $Co_2(CO)_8$.[17,18] Later, phosphine modification of the system was introduced to afford better product selectivity and lower operating pressures.[19] Still later, related Rh systems were introduced.[20]

The considerable mechanistic work on these systems leads, for example, to Scheme 2 for $RhH(CO)(PPh_3)_2$.[20a] With minor modifications a similar scheme applies to the $Co_2(CO)_8$ and phosphine-modified Co systems.[17–19] Thus the important intermediates are: (i) $M(H)(CO)L_2$; (ii) $M(H)(CO)(olefin)L_2$; (iii) $M(CO)(R)L_2$; (iv) $M(CO)(COR)L_2$; and (v) $MH_2(COR)L_3$, where $M$ = Co or Rh and $L$ = CO or phosphine. We will now point out the paucity of structural analogues for these various species.

Table 2 summarizes the known structures of $Tr(P)(H)(CO)$ complexes, among which could occur (i) $M(H)(CO)L_2$. In fact, the presumed

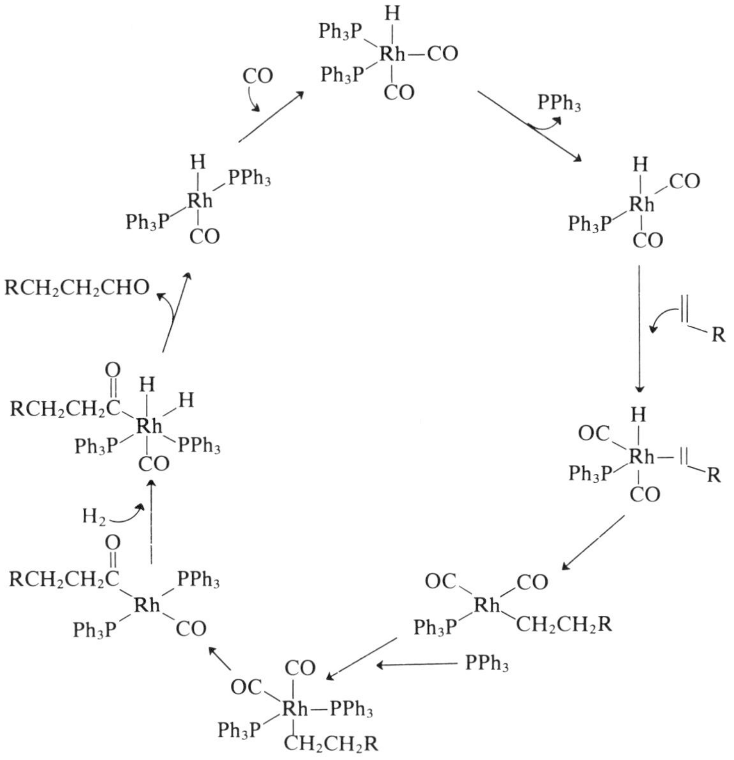

Scheme 2. Hydroformylation

precursors to $M(H)(CO)L_2$ do occur as the $d^8$ 5-coordinate species $M(H)(CO)L_3$, $M$ = Co (**7**),[21] Rh (**8**),[10] and Ir (**9**),[22] $L$ = CO or $PR_3$. In each of these structures the position of the hydride ligand was determined. As may be seen from **7** and **9**, two different trigonal bipyramidal arrangements about the metal center occur, depending upon $L$. The sensitivity of trigonal bipyramidal coordination to ligand type is well-known;[35] it may be that the sensitivity of the geometries of some of the proposed intermediates (Scheme 2) to phosphine substitution is an important aspect of such substitution. Presumably loss of the fifth ligand (either CO or P) in these $M(H)(CO)L_3$ systems results in all instances in the desired $d^8$

Table 2. Tr(P)(H)(CO)

| | Compound | Geom. | $d^n$ | CN | Description | | Reference |
|---|---|---|---|---|---|---|---|
| 7. | Co(H)(CO)(PPh$_3$)$_3$ | tbp | $d^8$ | 5 | P's equatorial, Co-H 1.4 Å 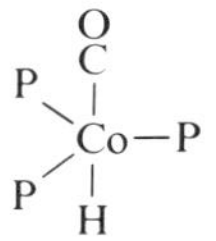 | $\begin{array}{c}\text{O}\\\text{P}-\overset{\|\|}{\underset{\|}{\text{C}}}-\text{Co}-\text{P}\\\text{P}\quad\text{H}\end{array}$ | 21 |
| 8. | Rh(H)(CO)(PPh$_3$)$_3$ | tbp | $d^8$ | 5 | P's equatorial, Rh-H 1.6 Å | | 10 |
| 9. | Ir(H)(CO)$_2$(PPh$_3$)$_2$ | tbp | $d^8$ | 5 | H, P axial, Ir-H 1.6 Å 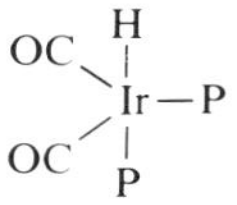 | $\begin{array}{c}\text{OC}\diagdown\overset{\text{H}}{\underset{\|}{\text{Ir}}}-\text{P}\\\text{OC}\diagup\quad\text{P}\end{array}$ | 22 |
| 10. | Os(H)(N$_2$C$_6$H$_5$)(CO)(PPh$_3$)$_3$ | dist tbp | $d^8$ | 5 | Os-H ≈ 1.2 Å, P's trans | | 23 |
| 11. | Ir(H)(Br)(C$_6$H$_5$)(CO)(PEt$_3$)$_2$ | oct | $d^6$ | 6 | P's trans, H trans to Br | | 24 |
| 12. | Ir(H)(Si(CH$_3$)$_2$OSi(CH$_3$)$_2$)(CO)(PPh$_3$)$_2$ | oct | $d^6$ | 6 | H not located, P's cis, eq, H trans to CO | | 25 |
| 13. | Os(H)(Br)(CO)(PPh$_3$)$_3$ | oct | $d^6$ | 6 | P's meridional, H not located, Br trans to CO | | 26 |

| | | | | | |
|---|---|---|---|---|---|
| 14. | $\overline{Os(H)[C(S)SCN(Me)(p\text{-}tol)]}$ $-(CO)(PPh_3)_2$ | oct | $d^6$ | 6 | P's trans, H assumed trans to CS | 27 |
| 15. | $Os(H)(CS_2CH_3)(CO)_2(PPh_3)_2$ | oct | $d^6$ | 6 | Os-H 1.6 Å, P's trans, H, CO trans | 28 |
| 16. | $Os(H)(Cl)(CO)(SO_2)(PCy_3)_2$ | oct | $d^6$ | 6 | H trans to $SO_2$, CO trans to Cl | 29 |
| 17. | $\overline{Ru(H)[N(p\text{-}tol)NN(p\text{-}tol)]}$ $-(CO)(PPh_3)_2$ | dist oct | $d^6$ | 6 | P's axial, Ru-H (disordered), 1.8 Å | 30 |
| 18. | $\overline{Ru(H)[N(p\text{-}tol)CHN(p\text{-}tol)]}$ $-(CO)(PPh_3)_2$ | dist oct | $d^6$ | 6 | P's axial, H, CO cis | 31 |
| 19. | $Mn(H)(CO)_3(PMePh_2)_2$ | dist oct | $d^6$ | 6 | P's trans, Mn-H, 1.5 Å | 32 |
| 20. | $[Ir(H)(CO)(P(OMe)_3)]_2$ $-(\mu\text{-}SMe)_2$ | rect py | $d^7$ | 5 | P axial, Ir-H, 1.7 Å (assumed) 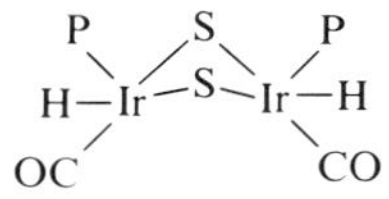 | 33 |
| | | | | | $\begin{array}{c} P\quad\quad S\quad\quad P \\ H-Ir-S-Ir-H \\ OC\quad\quad\quad CO \end{array}$ | |
| 21. | $Ta(H)(CO)_2(dmpe)_2$ | dist capped oct | $d^4$ | 7 | H capping | 34 |

Table 3. Tr(P)(CO)(olefin)

| | Compound | Geom. | $d^n$ | CN | Description | Reference |
|---|---|---|---|---|---|---|
| 22. | $Ir(Br)[(CN)_2C=C(CN)_2](CO)(PPh_3)_2$ | tbp | $d^8$ | 5 | Olefin in eq plane with P's | 36 |
| 23. | $Ir[(CN)_2C=C(CN)_2]$ $-(N=C=C(CN)CH(CN)_2](CO)(PPh_3)_2$ | tbp | $d^8$ | 5 | Olefin in eq plane, CO, N apical | 37 |
| 24. | $\overline{Co[\eta^2-CH_2=CHCH_2CF(CF_3)CF(CF_3)]}$ $-(CO)_2(P(OMe)_3)$ | tbp | $d^8$ | 5 | Olefin lies in eq plane 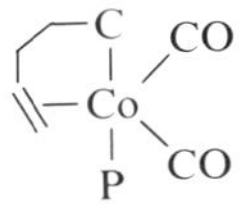 | 38 |
| 25. | $\overline{Fe[PF_2N(Me)C(POF_2)=C(Me)CH=CHPh]}$ $-(CO)[PF_2N(Me)(PF_2)]$ | 3-legged piano stool | $d^8$ | 5 | 4C chain nearly planar 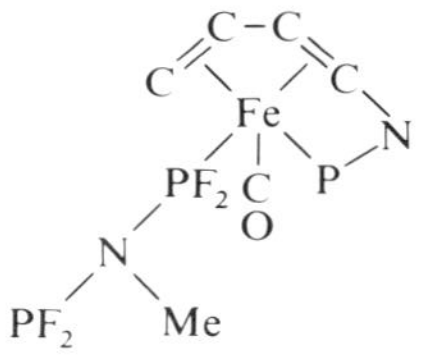 | 39 |
| 26. | $Fe(\eta^4-PhCH=CHCH=O)(CO)_2(PPh_3)$ | sq py | $d^8$ | 5 | P apical, 2 CO's and $\eta^4$-diene in basal plane | 40 |

| | | | | |
|---|---|---|---|---|
| 27. Fe(CO)$_2$(PPh$_3$)($\mu$-CH$_2$=C=C=CH$_2$)—Fe(CO)$_3$ | dist oct 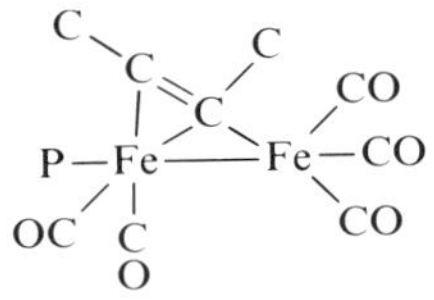 | $d^6$ | 6 | 2 CO, P facial, P trans to Fe–Fe $\pi$-allyl bonding to each Fe 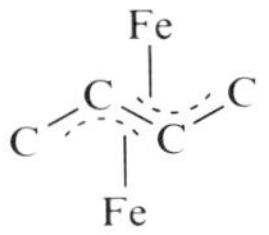 | 41 |
| 28. Ru(Cl)$_2$(H$_2$C=CH$_2$)(CO)(PMe$_2$Ph)$_2$ | oct | $d^6$ | 6 | Cl trans to olefin, trans P's, olefin parallel to P's | 42 |
| 29. Mo($\eta^2$-CH$_3$CH=CHC$_6$H$_4$PPh$_2$)(CO)$_4$ | oct | $d^6$ | 6 | 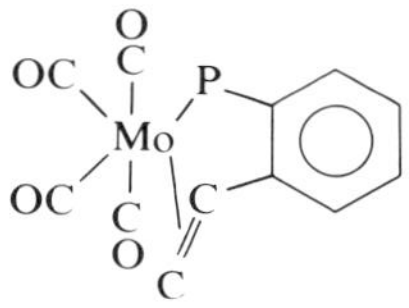 | 43 |
| 30. [Mo(CO)$_2$(P(n-Bu)$_3$)$_2$]$_2$—($\mu$-CH$_2$=CHCN)$_2$ | oct | $d^6$ | 6 | Olefin $\perp$ Mo(CO)$_2$ plane 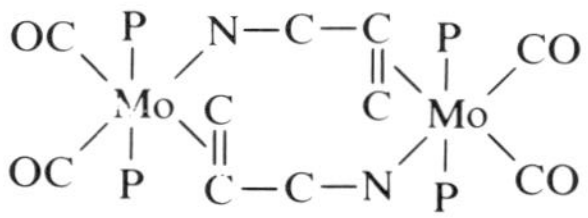 | 44 |

4-coordinate $M(H)(CO)L_2$ (i) species. The other $Tr(P)(H)(CO)$ complexes of Table 2 have no direct relevance to hydroformylation.

The general structural type $Tr(P)(H)(CO)(\text{olefin})$ encompasses intermediate (ii), $M(H)(CO)(\text{olefin})L_2$. There is only one such example, namely $Ir(H)[(CN)_2C=C(CN)_2](CO)(PPh_3)_2$, **2**, a structure known since 1969! This complex has the H and olefin ligands cis, as is proposed in Scheme 2.

Not only is the structural class $Tr(P)(H)(CO)(\text{olefin})$ extremely limited (one example!), but the less restricted class $Tr(P)(CO)(\text{olefin})$ is also very limited, with only nine examples. For the sake of completeness we tabulate this class in Table 3. Among these only **22**, a $d^8$ 5-coordinate species that contains Br rather than H, is even vaguely related to the proposed intermediates in hydroformylation.

There are no known structural analogues of (iii), $M(CO)(R)L_2$, $M =$ Co, Rh. The structural class $Tr(P)(CO)(R)$, Table 4, is very limited. The three known examples are all $d^6$ 6-coordinate complexes with R = Me.

Intermediate (iv), $M(CO)(COR)L_2$ is represented by Table 5, which tabulates complexes of the type $Tr(P)(CO)(COR)$. But as can be seen, none of the known structures involves Co, Rh, or Ir. Even if we turn to a less restricted class, namely $Tr(P)(COR)$ compounds (Table 6), we find a suprisingly small tabulation in view of the interest in transition-metal acyl complexes. Although no $d^8$ 4-coordinate complexes of the Co triad are found, several $d^6$ complexes of the triad occur, compound **44** being perhaps the closest (but distant) analogue of intermediate (iv).

There are no known structures of the type $Tr(P)(H)(COR)$ and hence no analogues of intermediate (v), $MH_2(COR)L_2$. However, a number of $d^6$ 6-coordinate complexes of the Co triad are found in the various tables. These complexes are generally unremarkable; the propensity of the Co triad to form such complexes is well known.

## 4. OLEFIN OLIGOMERIZATION

Although olefin oligomerization, especially polymerization, is an exceedingly high-volume industrial process, most of the commercial catalysts do not contain phosphines. However, one of the most active catalysts is obtained by treating $NiX_2L_2$ ($X$ = anionic ligand, $L$ = phosphine) with an aluminum alkyl.[64,65] Such systems are believed to involve $Ni(H)(Y)L$, where $Y$ is a complex Lewis-acid anion of the type $XAlR_xX_{3-x}$. A possible reaction scheme is shown in Scheme 3.[66] A key intermediate is of the type $Tr(P)(R)(\text{olefin})$ where olefin and R are cis to one another. Table 7 tabulates the known structures of this type. The relevance to the Ni-mediated catalysis is slight. The three compounds do

Table 4. Tr(P)(CO)(R)

| | Compound | Geom. | $d^n$ | CN | Description | Reference |
|---|---|---|---|---|---|---|
| 31. | [Rh($C_5H_5$)($CH_3$)(CO)[(S) – (P($C_6H_5$)$_2$ –NHCH($CH_3$)($C_6H_5$)]][$BF_4$] | oct | $d^6$ | 6 | $CH_3$, CO, P fac | 45 |
| 32. | [Rh($\mu$-X)(X)($CH_3$)(CO)(PMe$_2$Ph)]$_2$, X = Cl, Br | oct | $d^6$ | 6 | 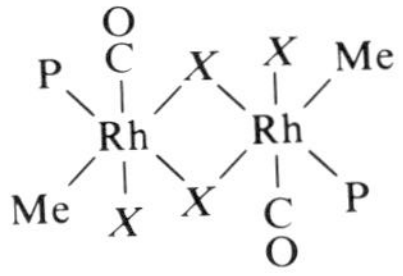 (dimer structure) | 46 |
| 33. | Mn($CH_3$)(CO)$_4$(PPh$_3$) | oct | $d^6$ | 6 | P, $CH_3$ cis | 47 |

Table 5. Tr(P)(CO)(COR)

| | Compound | Geom. | $d^n$ | CN | Description | Reference |
|---|---|---|---|---|---|---|
| 34. | [PPN][Fe(COCH$_3$)(CO)$_2$-($\mu$-PPh$_2$)$_2$Fe(CO)$_3$] | sq py | $d^6$ | 6 | 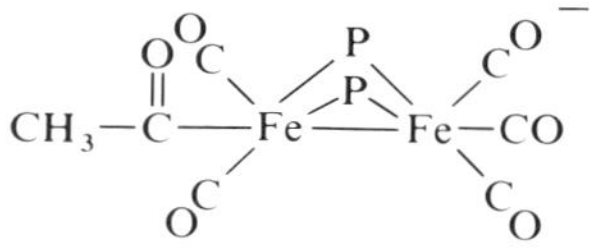 (structure diagram) | 48 |
| 35. | [Na][Fe(COCH$_3$)(CO)$_2$($\mu$-PPh$_2$)$_2$Fe(CO)$_3$] | sq py | $d^6$ | 6 | | 48 |
| 36. | Fe($\eta^5$-1-Me-3-Ph-C$_5$H$_3$)(COCH$_3$)(CO)-(PPh$_3$) | oct | $d^6$ | 6 | CO, P, acyl fac | 49 |
| 37. | Fe[C(OCH$_3$)CH(CO$_2$CH$_3$)CH$_2$CO](CO)$_3$(PPh$_3$) | oct | $d^6$ | 6 | Cyclopentenone, acyl, CO in ring plane, 3 CO's fac 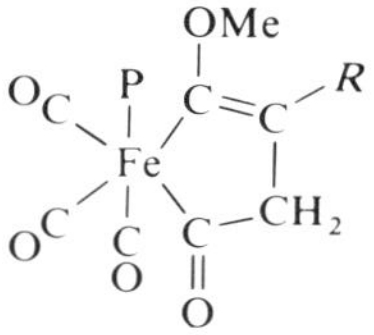 | 50 |
| 38. | Ru(I)($\eta^2$-COCH$_3$)(CO)(PPh$_3$)$_2$ | sq py | $d^6$ | 5 | P's trans, $\eta^2$-acyl apical | 51 |
| 39. | [Mo($\mu$-Cl)($\eta^2$-OC(CH$_2$SiMe$_3$))-(CO)$_2$(PMe$_3$)]$_2$ | oct | $d^4$ | 6 | $\eta^2$-acyl CO, P trans to $\eta^2$-acyl, 2 CO's, 2 $\mu$-Cl's in eq plane | 52 |
| 40. | Mo(C$_5$H$_5$)(COCH$_3$)(CO)$_2$(PPh$_3$) | 4-legged piano stool | $d^4$ | 7 | P trans to acetyl, C$_5$H$_5$ is stool seat | 53 |
| 41. | V($\eta^2$-OC(C$_3$H$_2$Ph$_3$))(CO)$_3$-(Ph$_2$PCH$_2$CH$_2$AsPh$_2$) | oct | $d^3$ | 6 | $\eta^2$-acyl CO, 3 CO's facial | 54 |

## Table 6. Tr(P)(COR)

| | Compound | Geom. | $d^n$ | CN | Description | Reference |
|---|---|---|---|---|---|---|
| 42. | Rh(I)(COCH$_3$)(PMe$_2$Ph)($\mu$-SCH$_3$)$_2$Rh(CO)—(PMe$_2$Ph) | sq py | $d^6$ | 5 | Acyl apical, sq py Rh joined to sq pl Rh through S $\cdots$ S edge | 55 |
| 43. | Rh(COCH$_2$CH$_2$CH$_3$)(mnt)(PEt$_3$)$_2$ | sq py | $d^6$ | 5 | Acyl apical | 56 |
| 44. | [Rh(Cl)(COCH$_3$)(PMe$_2$Ph)$_3$][PF$_6$] | sq py | $d^6$ | 5 | Acyl apical | 57 |
| 45. | [AsPh$_4$][Rh(I)(mnt)(COCH$_2$CH$_3$)(PPh$_3$)] | sq py | $d^6$ | 5 | Acyl apical | 58 |
| 46. | [[Rh(COCH$_3$)(PMe$_2$Ph)$_2$]$_2$($\mu$-Cl)$_3$][PF$_6$] | dist oct | $d^6$ | 6 | 3 Cl's fac | 59 |
| 47. | Rh(C$_5$H$_5$)(I)(COCH$_3$)—[(S)—(P(C$_6$H$_5$)$_2$NHCH(Me)(C$_6$H$_5$)] | oct | $d^6$ | 6 | C$_5$H$_5$ occupies fac sites | 45 |
| 48. | Ir(C$_5$H$_5$)[$\mu$-C(C$_6$H$_5$)O][$\mu$-C(CH$_3$)O]—[$\mu$-P(C$_6$H$_5$)$_2$]Mn(CO)$_3$ | Mn oct<br>Ir oct | $d^4$<br>$d^6$ | 6<br>5 | 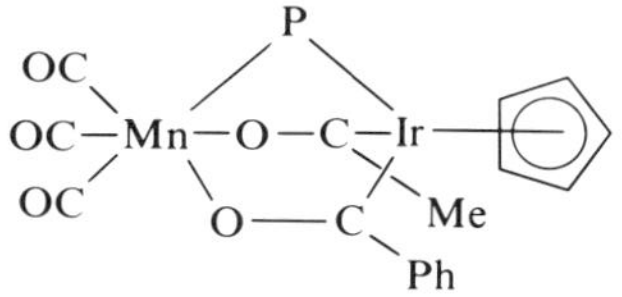 | 60 |
| 49. | Ni(Cl)(COCH$_3$)(PMe$_3$)$_2$ | sq pl | $d^8$ | 4 | P's trans | 61 |
| 50. | Pd(Cl)(COCH$_2$CH$_2$CH$_3$)(PPh$_3$)$_2$ | sq pl | $d^8$ | 4 | P's trans | 62 |
| 51. | Pt(Cl)(COCH$_2$CH$_2$CH$_3$)(PPh$_3$)$_2$ | sq pl | $d^8$ | 4 | P's trans | 62 |
| 52. | [Ni(COCH$_3$)(p$_3$N)][BPh$_4$] | sq pl | $d^8$ | 5 | P's eq, N trans to acyl 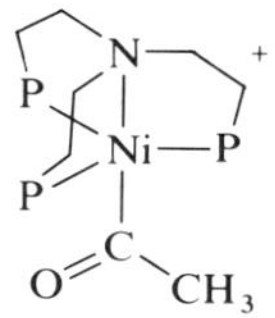 | 63 |

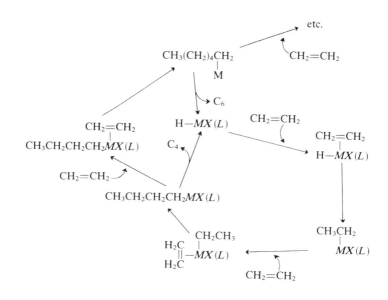

Scheme 3. Oligomerization of Olefins

not contain Ni. However, each has the olefin and R ligands cis to one another.

## 5. METALLACYCLES

Metallacycles have been suggested as intermediates in many transition-metal catalyzed reactions of olefins, acetylenes, and cyclopropanes. Metallacyclobutane complexes are invoked in olefin and cyclopropane isomerization schemes[3] (Scheme 4[4]), as well as in olefin metathesis schemes.[70,71] Metallacyclopentane, -pentene, and -pentadiene complexes can all be invoked in olefin and acetylene dimerization and polymerization[72–78] (Scheme 5[77]). Many of these involve early transition metals and do not include phosphine ligands.

Several platinacyclobutane complexes have been isolated from the reaction of platinum-phosphine complexes with cyclopropanes. Those structurally characterized are listed in Table 8. No other structures of metallacyclobutane-phosphine complexes have been reported, but an

## Table 7. Tr(P)(R)(olefin)

| | Compound | Geom. | $d^n$ | CN | Description | Reference |
|---|---|---|---|---|---|---|
| 53. | Co(C$_6$H$_5$)(H$_2$C=CH$_2$)(PMe$_3$)$_3$ | tbp | $d^8$ | 5 | Olefin and C$_6$H$_5$ cis and coplanar, 2 P's axial | 67 |
| 54. | $\overline{\text{Ir}[\text{P}(\text{CH}(\text{CH}_3)\text{CH}_2)(i\text{-Pr})_2]}$(C$_2$H$_4$)$_2$ (P(i-Pr)$_3$)$_2$ | tbp | $d^8$ | 5 | Olefin and CH$_2$ cis, olefins and metallated P in eq plane | 68 |
| 55. | cis-$\overline{\text{Pt}(\text{CH}_3)_2[\text{P}(o\text{-C}_6\text{H}_4\text{CH}=\text{CH}_2)(\text{Ph})_2]}$ | sq pl | $d^8$ | 4 | Olefin and CH$_3$ cis, olefin ⊥ to sq pl Pt | 69 |

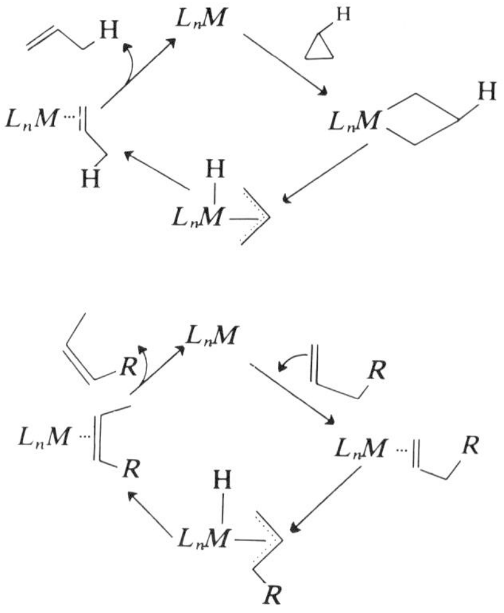

**Scheme 4.** Olefin and Cyclopropane Isomerization

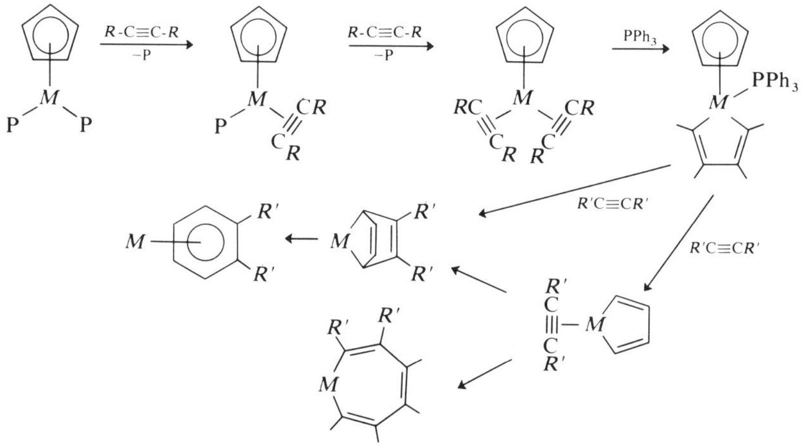

**Scheme 5.** Acetylene Trimerization

iridiacyclobutane arsine complex, $\overline{\text{Ir(H)}(\text{CH}_2\text{C}(\text{CH}_3)_2\text{CH}_2)}(\text{AsMe}_3)_3$, is known.[94] As may be seen in Scheme 4, migration of the $\beta$-hydrogen atom from the ring to the metal is believed to occur upon isomerization of the metallacycle to the corresponding transition-metal allyl hydride complex. It may be that transfer of the $\beta$-hydride atom from ring to metal is facilitated as the pucker angle increases.[73,74] For the platinacyclobutanes this angle ranges from 22.4° to 49.7° and generally increases as the substitution on the ring increases. Of course, in these solid state structures the pucker angle may be merely a manifestation of packing forces. In solution the barrier to ring inversion is small[71] and greater puckering may be possible.

The formation of a metallacyclopentane from two moles of ethylene has been observed with nickel phosphine complexes.[72] Metallacyclopentanes are often included in mechanisms for the dimerization of olefins. Table 8 contains four Pt complexes, as well as one each of Co, Ni, and Ir.

Metallacyclopentene and -pentadiene phosphine complexes that have been structurally characterized are also included in Table 8. These may be relevant to the dimerization of two acetylenes or acetylenes with olefins[75,76] and to the cyclotrimerization of acetylene with an olefin.[78]

## 6. SUMMARY

This chapter has presented a tabulation of transition-metal phosphine complexes bearing some relation to proposed intermediates in a number of catalytic reactions. It is an interesting commentary on structural inorganic chemistry that, despite the determination of approximately 2500 crystal structures of such complexes, there are so few structures of direct relevance to such processes. For example, there is but one direct structural analogue (compound 2) for any of the intermediates proposed in the hydroformylation reaction, the most important and oldest industrial process based on a homogeneous transition-metal phosphine complex. Similarly, the structure of only one transition-metal allyl hydride is known,[4] despite the fact that allyl hydrides are frequently proposed intermediates. This paucity of structural information does not necessarily result from the inherent reactivity of such intermediates—for example the synthesis of suitable analogues for the hydroformylation reaction would appear to be straightforward. There appear to be several potential routes to complexes of the type $M\text{H}_2(\text{COR})L_3$, and one would expect that complexes of the type $M(\text{H})(\text{CO})L_2$ would be stable, especially for $L$, a bulky phosphine.

There are several examples of metallacycle structures of direct importance to catalytic processes, but such structures only serve to emphasize another problem, that of the relation of structure to reactivity.

Table 8. Tr(P)(metallacycle)

| | Compound | Geom. | $d^n$ | CN | M-C(1)[a] | M-C(3) | C(1)-C(2) | C(2)-C(3) | C(3)-C(4) | Description[b] | Reference[c] |
|---|---|---|---|---|---|---|---|---|---|---|---|
| A. Metallacyclobutanes $\overline{M\text{-}C(1)\text{-}C(2)\text{-}C(3)}$ | | | | | | | | | | | |
| 56. | $\overline{Pt[C(CN)_2CH_2C(CN)_2]}(PPh_3)_2$ | sq pl | $d^8$ | 4 | 2.137(6)[d] | 2.139(6) | 1.545(9) | 1.584(9) | | 24.4° | 79 |
| 57. | $\overline{Pt[CH(CO_2CH_3)C(O)CH(CO_2CH_3)]}$ —$(PPh_3)_2$ | sq pl | $d^8$ | 4 | 2.149(6) | 2.128(6) | 1.496 | 1.456 | | 49.7° | 80 |
| 58. | $\overline{Pt[C(CN)(CO_2CH_2CH_3)CH(C_6H_5)C(CN)_2]}$ —$(PPh_3)_2$ | sq pl | $d^8$ | 4 | 2.158(14) | 2.200(14) | 1.556(19) | 1.509(14) | | 29.7° | 81 |
| 59. | $\overline{Pt[C(CN)_2CH(C_6H_5)C(CN)_2]}(PPh_3)_2$ | sq pl | $d^8$ | 4 | 2.137(6) | 2.159(6) | 1.557(9) | 1.548(10) | | 28.6° | 81 |
| 60. | $\overline{Pt[CH_2C(CH_3)_2CH_2]}(PEt_3)_2$ | sq pl | $d^8$ | 4 | 2.086(6) | 2.080(6) | 1.535(9) | 1.536(9) | | 22.4° | 82* |
| B. Metallacyclopentanes $\overline{M\text{-}C(1)\text{-}C(2)\text{-}C(3)\text{-}C(4)}$ | | | | | | | | | | | |
| 61. | $\overline{Pt(I)_2(CH_2CH_2CH_2CH_2)}(PMe_2Ph)_2$ | dist oct | $d^6$ | 6 | 2.15(1) | 2.15(1) | 1.54(2) | 1.49(4) | 1.54(2) | | 83 |
| 62. | $\overline{Pt[CH(COCH_3)CH_2CH(COCH_3)CH_2]}$ —$(PPh_3)_2$ | sq pl | $d^8$ | 4 | 2.139(7) | 2.094(9) | 1.540(11) | 1.579(23) | 1.430(20) | | 84 |
| 63. | $\overline{Pt[CH(CH=CH_2)CH_2CH_2CH(CH=CH_2)]}$ —$(PMe_3)_2$ | sq pl | $d^8$ | 4 | 2.142(8) | 2.131(5) | 1.52(1) | 1.52(1) | 1.54(1) | | 85* |
| 64. | $\overline{Pt(CH_2CH_2CH_2CH_2)}(PPh_3)_2$ | sq pl | $d^8$ | 4 | 2.12(2) | 2.05(2) | 1.57(3) | 1.45(3) | 1.48(3) | | 86 |
| 65. | $\overline{Ni[CH_2C(CH_2)C(CH_2)CH_2]}$ —$[(C_6H_{11})_2P(CH_2)_2P(C_6H_{11})_2]$ | sq pl | $d^8$ | 4 | 1.975[d] | 1.965 | 1.47 | 1.47 | 1.48 | Ring planar | 87 |
| 66. | $\overline{Co(C_5H_5)(CH_2CH_2CH_2CH_2)}(PPh_3)$ | dist oct | $d^6$ | 6 | 2.025(6) | 2.024(7) | 1.509[d] | 1.466 | 1.561 | | 88 |
| 67. | $\overline{Ir[C_5(CH_3)_5](CH_2CH_2CH_2CH_2)}$ $(PPh_3)$ | dist oct | $d^6$ | 6 | 2.10(4) | 2.14(5) | 1.30[d] | 1.45 | 1.67 | | 88 |

## STRUCTURALLY CHARACTERIZED COMPLEXES

### C. Metallacyclobutene ($\overline{M\text{-}C(1)=C(2)\text{-}C(3)}$)

| | | | | | | | | | | | |
|---|---|---|---|---|---|---|---|---|---|---|---|
| 68. | Pt[C(C$_6$H$_5$)=C(C$_6$H$_5$)$\overline{C(O)}$](PPh$_3$)$_2$ | sq pl | | 4 | 2.09(4) | 2.08(6) | 1.31(8) | 1.45(7) | | | 89 |

### D. Metallacyclopent-2-enes $\overline{M\text{-}C(1)=C(2)\text{-}C(3)\text{-}C(4)}$

| | | | | | | | | | | | |
|---|---|---|---|---|---|---|---|---|---|---|---|
| 69. | $\overline{\text{Co(C}_5\text{H}_5)[\text{C(C}_6\text{H}_5)=\text{C(CO}_2\text{CH}_3)\text{CH}}$ $-(\text{CO}_2\text{CH}_3)\overline{\text{CH(CO}_2\text{CH}_3)}](\text{PPh}_3)$ | dist oct | $d^6$ | 6 | 1.947(11) | 2.079(12) | 1.375(17) | 1.527(13) | 1.513(15) | Co 0.583 Å out of CoC$_3$ plane | 90 |
| 70. | $\overline{\text{Fe}[\text{C(OCH}_3)=\text{C(CO}_2\text{CH}_3)\text{CH}_2\overline{\text{CO}}](\text{CO})_3}$ $-(\text{PPhMe}_2)$ | oct | $d^6$ | 6 | 2.001(4) | 2.005(4) | 1.349(5) | 1.499(5) | 1.509(5) | | 50 |

### E. Metallacyclopent-3-ene $\overline{M\text{-}C(1)\text{-}C(2)=C(3)\text{-}C(4)}$

| | | | | | | | | | | | |
|---|---|---|---|---|---|---|---|---|---|---|---|
| 71. | $\overline{\text{RhCl}[\text{C(O)C(Cl)=C(Cl)}\overline{\text{C}}(\text{O})]}$(H$_2$O) $-(\text{PMe}_2\text{Ph})_2$ | oct | $d^6$ | 6 | 1.970(6) | 1.970(6) | 1.50(1) | 1.32(1) | 1.52(1) | P's trans | 91 |

### F. Metallacyclopentadienes $\overline{M\text{-}C(1)=C(2)\text{-}C(3)=C(4)}$

| | | | | | | | | | | | |
|---|---|---|---|---|---|---|---|---|---|---|---|
| 72. | $\overline{\text{Co(C}_5\text{H}_5)[\text{C(C}_6\text{F}_5)=\text{C(C}_6\text{F}_5)\text{C(C}_6\text{F}_5)=\overline{\text{C}}}$ $-(\text{C}_6\text{F}_5)](\text{PPh}_3)$ | dist oct | $d^6$ | 6 | 1.995(11) | 1.993(11) | 1.326(15) | 1.467(16) | 1.335(15) | Co 0.203 Å out of C$_4$ plane | 92 |
| 73. | $\overline{\text{Rh}C_5H_5)[C(C_6F_5)=C(C_6F_5)C(C_6F_5)=\overline{\text{C}}}$ $-(\text{C}_6\text{F}_5)](\text{PPh}_3)$ | dist oct | $d^6$ | 6 | 2.060(12) | 2.067(11) | | 1.457(16) | 1.354(15) | Rh 0.237 Å out of C$_4$ plane | 93 |

[a] Distances in Å.
[b] The pucker angle given for metallacyclobutanes is the dihedral angle between C(1)-C(2)-C(3) and C(1)-$M$-C(3).
[c] An asterisk on the reference indicates a low-temperature structure determination (−70°C or below).
[d] Estimated standard deviation, if available, is in parentheses.

From the structures of the platinacyclobutanes and the one iridiacyclobutane it is not at all evident why the synthesis of the Pt complex, **60**, requires forcing conditions,[82] whereas the Ir analogue is formed easily.[94] Moreover, even among the platinacyclobutanes the relation, if any, of solid-state structure, especially the pucker angle, to reactivity is unclear. Of course, structure-reactivity relationships are at the very foundation of our potential understanding of such catalytic processes. At least in the area of transition-metal phosphine complexes it seems clear that such an understanding will not emerge without considerably increased effort in the isolation and structural characterization of carefully tailored, model complexes.

## ACKNOWLEDGMENTS

We thank Ms. Jan Goranson for skillful typing of the manuscript, Ms. Jean Wisner for thorough proofreading, and Dr. Paul Swepston for assistance with the structural search. JAI thanks the Sherman Fairchild Distinguished Scholars Program of the California Institute of Technology under whose auspices a portion of this review was written.

## REFERENCES

1. For reviews of transition metal-catalyzed olefin isomerizations see (a) C. A. Tolman, *Transition Metal Hydrides*, edited by E. L. Muetterties (Marcel Dekker, New York, 1971) pp. 271–312; (b) A. J. Hubert and H. Reimlinger, *Synthesis* **1**, 405–430 (1970); (c) N. R. Davies, *Rev. Pure Appl. Chem.* **17**, 83–93 (1967); (d) M. Orchin, *Adv. Catal.* **16**, 1–47 (1966).
2. M. Ephritkhine, M. L. H. Green, and R. E. MacKenzie, *J. Chem. Soc., Chem. Comm.* **1976**, 619–621.
3. For a review, see K. C. Bishop III, *Chem. Rev.* **76**, 461–486 (1976).
4. T. H. Tulip and J. A. Ibers, *J. Am. Chem. Soc.* **101**, 4201–4211 (1979).
5. J. A. Osborn, F. H. Jardine, J. F. Young, and G. Wilkinson, *J. Chem. Soc.* **A 1966**, 1711–1732.
6. P. S. Hallman, B. R. McGarvey, and G. Wilkinson, *J. Chem. Soc.* **A 1968**, 3143–3150.
7. C. O'Connor and G. Wilkinson, *J. Chem. Soc.* **A 1968**, 2665–2671.
8. P. B. Hitchcock, M. McPartlin, and R. Mason, *J. Chem. Soc., Chem. Comm.* **1969**, 1367–1368; M. J. Bennett and P. B. Donaldson, *Inorg. Chem.* **16**, 655–660 (1977).
9. S. J. La Placa and J. A. Ibers, *Inorg. Chem.* **4**, 778–783 (1965).
10. ———, *Acta Crystallogr.* **18**, 511–519 (1965).
11. J. W. Byrne, J. R. M. Kress, J. A. Osborn, L. Ricard, and R. E. Weiss, *J. Chem. Soc., Chem. Comm.* **1977**, 662–663.
12. K. W. Muir and J. A. Ibers, *J. Organomet. Chem.* **18**, 175–187 (1969).

13. G. Del Piero, G. Perego, and M. Cesari, *Gazz. Chim. Ital.* **105,** 529–537 (1975).
14. T. V. Ashworth, E. Singleton, M. Laing, and L. Pope, *J. Chem. Soc., Dalton Trans.* **1978,** 1032–1036.
15. H. C. Clark, M. J. Dymarski, and N. C. Payne, *J. Organomet. Chem.* **165,** 117–128 (1979).
16. (a) J. Tsuji and H. Suzuki, *Chem. Lett.* **1977,** 1083–1084; (b) A. Andreetta, F. Conti, and G. F. Ferrari, *Aspects on Homogeneous Catal.* **1,** 203–267 (1970).
17. R. F. Heck and D. S. Breslow, *J. Am. Chem. Soc.* **83,** 4023–4027 (1961).
18. N. H. Alemdaroglu, J. L. M. Penninger, and E. Oltay, *Monatshefte für Chemie* **107,** 1153–1165 (1976).
19. M. van Boven, N. H. Alemdaroglu, J. L. M. Penninger, *Ind. Eng. Chem., Prod. Res. Dev.* **14,** 259–264 (1975); L. H. Slaugh and R. D. Mullineaux, *J. Organomet. Chem.* **13,** 469–477 (1968); E. R. Tucci, *Ind. Eng. Chem., Prod. Res. Dev.* **7,** 32–38 (1968).
20. (a) C. K. Brown and G. Wilkinson, *J. Chem. Soc.* **A 1970,** 2753–2764; (b) R. L. Pruett and J. A. Smith, *J. Org. Chem.* **34,** 327–330 (1969).
21. J. M. Whitfield, S. F. Watkins, G. B. Tupper, and W. H. Baddley, *J. Chem. Soc., Dalton Trans.* **1977,** 407–413.
22. M. Ciechanowicz, A. C. Skapski, and P. G. H. Troughton, *Acta Crystallogr., Sect.* **B 32,** 1673–1680 (1976).
23. M. Cowie, B. L. Haymore, and J. A. Ibers, *Inorg. Chem.* **14,** 2617–2623 (1975).
24. U. Behrens and L. Dahlenburg, *J. Organomet. Chem.* **116,** 103–111 (1976).
25. M. D. Curtis, J. Greene, and W. M. Butler, *J. Organomet. Chem.* **164,** 371–380 (1979).
26. P. L. Orioli and L. Vaska, *Proc. Chem. Soc., London* **1962,** 333.
27. G. R. Clark, T. J. Collins, D. Hall, S. M. James, and W. R. Roper, *J. Organomet. Chem.* **141,** C5–C9 (1977).
28. J. M. Waters and J. A. Ibers, *Inorg. Chem.* **16,** 3273–3277 (1977).
29. R. R. Ryan and G. J. Kubas, *Inorg. Chem.* **17,** 637–641 (1978).
30. L. D. Brown and J. A. Ibers, *Inorg. Chem.* **15,** 2788–2793 (1976).
31. L. D. Brown, S. D. Robinson, A. Sahajpal, and J. A. Ibers, *Inorg. Chem.* **16,** 2728–2735 (1977).
32. M. Laing, E. Singleton, and G. Kruger, *J. Organomet. Chem.* **54,** C30–C32 (1973).
33. J. J. Bonnet, A. Thorez, A. Maisonnat, J. Galy, and R. Poilblanc, *J. Am. Chem. Soc.* **101,** 5940–5949 (1979).
34. P. Meakin, L. J. Guggenberger, F. N. Tebbe, and J. P. Jesson, *Inorg. Chem.* **13,** 1025–1032 (1974).
35. B. A. Frenz and J. A. Ibers, "M.T.P. International Review of Science, Physical Chemistry," Series One, **11,** 33–72 (1972).
36. J. A. McGinnety and J. A. Ibers, *J. Chem. Soc., Chem. Comm.* **1968,** 235–237.
37. J. S. Ricci and J. A. Ibers, *J. Am. Chem. Soc.* **93,** 2391–2397 (1971).
38. M. Bottrill, R. Goddard, M. Green, and P. Woodward, *J. Chem. Soc., Dalton Trans.* **1979,** 1671–1678.
39. M. G. Newton, R. B. King, M. Chang, and J. Gimeno, *J. Am. Chem. Soc.* **101,** 2627–2631 (1979).
40. M. Sacerdoti, V. Bertolasi, and G. Gilli, *Acta Crystallogr. Sect.* **B 36,** 1061–1065 (1980).
41. J. N. Gerlach, R. M. Wing, and P. C. Ellgen, *Inorg. Chem.* **15,** 2959–2964 (1976).
42. L. D. Brown, C. F. J. Barnard, J. A. Daniels, R. J. Mawby, and J. A. Ibers, *Inorg. Chem.* **17,** 2932–2935 (1978).
43. H. Luth, M. R. Truter, and A. Robson, *J. Chem. Soc.* **A 1969,** 28–41.
44. F. Hohmann, H. T. Dieck, C. Krüger, and Y.-H. Tsay, *J. Organomet. Chem.* **171,** 353–364 (1979).
45. S. Quinn, A. Shaver, and V. W. Day, *J. Am. Chem. Soc.* **104,** 1096–1099 (1982).
46. M. J. Doyle, A. Mayanza, J.-J. Bonnet, P. Kalck, and R. Poilblanc, *J. Organomet. Chem.* **146,** 293–310 (1978).

47. A. Mawby and G. E. Pringle, *J. Inorg. Nucl. Chem.* **34,** 877–883 (1972).
48. R. E. Ginsberg, J. M. Berg, R. K. Rothrock, J. P. Collman, K. O. Hodgson, and L. F. Dahl, *J. Am. Chem. Soc.* **101,** 7218–7231 (1979).
49. T. G. Attig, R. G. Teller, S.-M. Wu, R. Bau, and A. Wojcicki, *J. Am. Chem. Soc.* **101,** 619–628 (1979).
50. T.-A. Mitsudo, T. Sasaki, Y. Watanabe, Y. Takegami, K. Nakatsu, K. Kinoshita, and Y. Miyagawa, *J. Chem. Soc., Chem. Comm.* **1979,** 579–580.
51. W. R. Roper, G. E. Taylor, J. M. Waters, and L. J. Wright, *J. Organomet. Chem.* **182,** C46–C48 (1979).
52. E. C. Guzman, G. Wilkinson, J. L. Atwood, R. D. Rogers, W. E. Hunter, and M. J. Zaworotko, *J. Chem. Soc., Chem. Comm.* **1978,** 465–466.
53. M. R. Churchill and J. P. Fennessey, *Inorg. Chem.* **7,** 953–959 (1968).
54. U. Franke and E. Weiss, *J. Organomet. Chem.* **165,** 329–340 (1979).
55. A. Mayanza, J.-J. Bonnet, J. Galy, P. Kalck, and R. Poilblanc, *J. Chem. Res.* **146,** 2101–2133 (1980).
56. C.-H. Cheng, D. E. Hendriksen, and R. Eisenberg, *J. Organomet. Chem.* **142,** C65–C68 (1977).
57. M. A. Bennett, J. C. Jeffery, and G. B. Robertson, *Inorg. Chem.* **20,** 323–330 (1981).
58. C.-H. Cheng, B. D. Spivack, and R. Eisenberg, *J. Am. Chem. Soc.* **99,** 3003–3011 (1977).
59. M. A. Bennett, J. C. Jeffery, and G. B. Robertson, *Inorg. Chem.* **20,** 330–335 (1981).
60. J. R. Blickensderfer, C. B. Knobler, and H. D. Kaesz, *J. Am. Chem. Soc.* **97,** 2686–2691 (1975).
61. G. Huttner, O. Orama, and V. Bejenke, *Chem. Ber.* **109,** 2533–2536 (1976).
62. R. Bardi, A. M. Piazzesi, G. Cavinato, P. Cavoli, and L. Toniolo, *J. Organomet. Chem.* **224,** 407–420 (1982).
63. P. Stoppioni, P. Dapporto, and L. Sacconi, *Inorg. Chem.* **17,** 718–725 (1978).
64. G. LeFebvre and Y. Chauvin, *Aspects of Homogeneous Catal.* **1,** 107–201 (1970).
65. R. G. Miller, P. A. Pinke, R. D. Stauffer, H. J. Golden, and D. J. Baker, *J. Am. Chem. Soc.* **96,** 4211–4220 (1974); P. A. Pinke and R. G. Miller, *J. Am. Chem. Soc.* **96,** 4221–4229 (1974); P. A. Pinke, R. D. Stauffer, and R. G. Miller, *J. Am. Chem. Soc.* **96,** 4229–4234 (1974); H. J. Golden, D. J. Baker, and R. G. Miller, *J. Am. Chem. Soc.* **96,** 4235–4243 (1974).
66. B. Bogdanović, B. Henc, H.-G. Karmann, H.-G. Nüssel, D. Walter, and G. Wilke, *Ind. Eng. Chem.* **62** (12), 34–44 (1970).
67. H.-F. Klein, R. Hammer, J. Gross, and U. Schubert, *Angew. Chem. Int. Ed. Eng.* **19,** 809–810 (1980).
68. G. Perego, G. Del Piero, M. Cesari, M. G. Clerici, and E. Perrotti, *J. Organomet. Chem.* **54,** C51–C52 (1973).
69. M. A. Bennett, H.-K. Chee, J. C. Jeffery, and G. B. Robertson, *Inorg. Chem.* **18,** 1071–1076 (1979).
70. J.-L. Herisson and Y. Chauvin, *Makromol.* Chem. **141,** 161–176 (1970).
71. R. J. Puddephatt, *Coord. Chem. Rev.* **33,** 149–194 (1980).
72. R. H. Grubbs, A. Miyashita, M. Liu, and P. Burk, *J. Am. Chem. Soc.* **100,** 2418–2425 (1978); R. H. Grubbs and A. Miyashita, *J. Am. Chem. Soc.* **100,** 1300–1302 (1978); R. H. Grubbs and A. Miyashita, *J. Am. Chem. Soc.* **100,** 7416–7418 (1978).
73. J. X. McDermott, J. F. White, and G. M. Whitesides, *J. Am. Chem. Soc.* **98,** 6521–6528 (1976).
74. J. X. McDermott, M. E. Wilson, and G. M. Whitesides, *J. Am. Chem. Soc.* **98,** 6529–6536 (1976).
75. J. P. Collman, J. W. Kang, W. F. Little, and M. F. Sullivan, *Inorg. Chem.* **7,** 1298–1303 (1968).

76. A. J. Chalk, *J. Am. Chem. Soc.* **94**, 5928–5929 (1972).
77. D. R. McAlister, J. E. Bercaw, and R. G. Bergman, *J. Am. Chem. Soc.* **99**, 1666–1668 (1977).
78. L. D. Brown, K. Itoh, H. Suzuki, K. Hirai, and J. A. Ibers, *J. Am. Chem. Soc.* **100**, 8232–8238 (1978).
79. D. J. Yarrow, J. A. Ibers, M. Lenarda, and M. Graziani, *J. Organomet. Chem.* **70**, 133–145 (1974).
80. D. A. Clarke, R. D. W. Kemmitt, M. A. Mazid, M. D. Schilling, and D. R. Russell, *J. Chem. Soc., Chem. Comm.* **1978**, 744–745.
81. J. Rajaram and J. A. Ibers, *J. Am. Chem. Soc.* **100**, 829–838 (1978).
82. J. A. Ibers, R. DiCosimo, and G. M. Whitesides, *Organometallics* **1**, 13–20 (1982).
83. A. K. Cheetham, R. J. Puddephatt, A. Zalkin, D. H. Templeton, and L. K. Templeton, *Inorg. Chem.* **15**, 2997–2999 (1976).
84. M. Green, J. A. K. Howard, P. Mitrprachachon, M. Pfeffer, J. L. Spencer, F. G. A. Stone, and P. Woodward, *J. Chem. Soc., Dalton Trans.* **1979**, 306–314.
85. G. K. Barker, M. Green, J. A. K. Howard, J. L. Spencer, and F. G. A. Stone, *J. Chem. Soc., Dalton Trans.* **1978**, 1839–1847.
86. C. G. Biefeld, H. A. Eick, and R. H. Grubbs, *Inorg. Chem.* **12**, 2166–2170 (1973).
87. P. W. Jolly, C. Krüger, R. Salz, and J. C. Sekutowski, *J. Organomet. Chem.* **165**, C39–C42 (1979).
88. P. Diversi, G. Ingrosso, A. Lucherini, W. Porzio, and M. Zocchi, *J. Chem. Soc., Chem. Comm.* **1977**, 811–812.
89. W. Wong, S. J. Singer, W. D. Pitts, S. F. Watkins, and W. H. Baddley, *J. Chem. Soc., Chem. Comm.* **1972**, 672–673.
90. Y. Wakatsuki, K. Aoki, and H. Yamazaki, *J. Am. Chem. Soc.* **101**, 1123–1130 (1979).
91. P. D. Frisch and G. P. Khare, *J. Am. Chem. Soc.* **100**, 8267–8269 (1978).
92. R. G. Gastinger, M. D. Rausch, D. A. Sullivan, and G. J. Palenik, *J. Am. Chem. Soc.* **98**, 719–723 (1976).
93. ———, *J. Organomet. Chem.* **117**, 355–364 (1976).
94. T. H. Tulip and D. L. Thorn, *J. Am. Chem. Soc.* **103**, 2448–2450 (1981) and private communication.

# 4

# Asymmetric Hydrogenation Reactions Using Chiral Diphosphine Complexes of Rhodium

*John M. Brown and Penny A. Chaloner*

## 1. INTRODUCTION

The archetypal reaction in asymmetric homogeneous hydrogenation is reduction of a z-acyldehydroamino acid to the corresponding amino acid derivative, which represents one of the most efficient routes to introduction of a chiral center, with optical yields often in excess of 98%. Soon after Wilkinson's discovery that *tris*(triphenylphosphine)rhodium chloride was a homogeneous catalyst for alkene hydrogenation,[1] several authors realized the potential for asymmetric induction. Early attempts using simple resolved chiral phosphines were not impressive,[2] and the first major development was due to Kagan and Dang.[3] They demonstrated that the use of a chiral chelating biphosphine, (**1**) (see Figure 4.1), bearing the now-famous acronym of DIOP, gave good optical yields, particularly in the reduction of dehydroamino acid derivatives. The most startling aspect of

---

*Dr. John M. Brown* • Dyson Perrins Laboratory, South Parks Road, Oxford OX1 3QY, England.
*Dr. Penny A. Chaloner* • School of Molecular Sciences, University of Sussex, Falmer, Brighton, U.K.

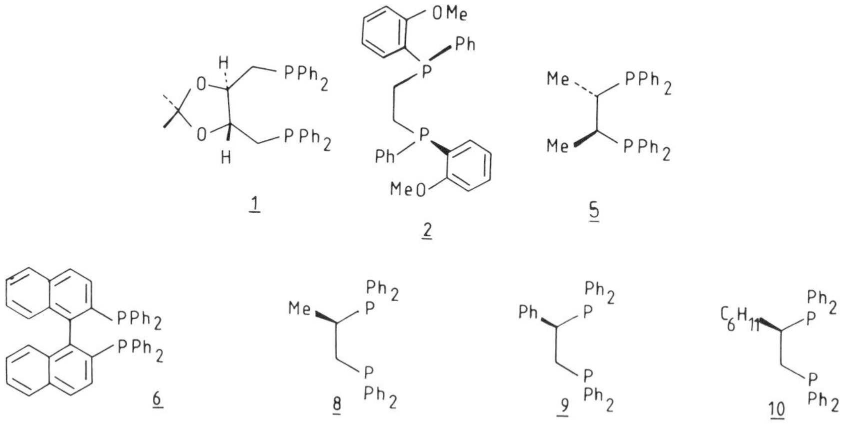

Figure 4.1. Some of the commoner biphosphines which have been successfully employed in asymmetric catalysis.

this result is that the chiral center in the catalyst which controls asymmetric induction is three bonds removed from the metal and spatially remote from it. This leads to an immediate suspicion that the chelate ring is in some way controlling the orientation of the P-Ph rings, which in turn determines the stereochemical outcome of the rate-determining stage in catalysis. A second development, one of considerable commercial significance, was due to W. S. Knowles and co-workers at Monsanto.[4] Their chelating biphosphine, DIPAMP, (**2**), whose synthesis was inspired by Mislow's pioneering work on phosphine oxide resolution,[5] effected the reduction of (**3**) to (**4**)

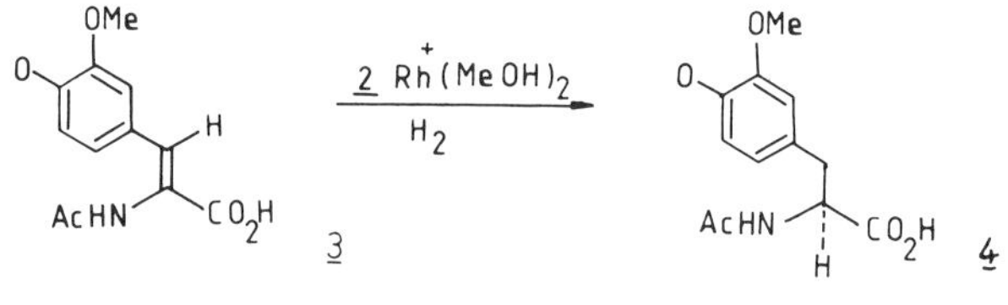

in 95% optical yield and thus afforded a viable catalytic synthesis of L-DOPA. Many catalysts of comparable efficiency have been developed subsequently, all based on the principle of a rigid, cationic rhodium cis-chelated biphosphine complex. Complexes of CHIRAPHOS (**5**)[6] and BINAP (**6**)[7] are among the most successful, but none surpass Knowles's contribution (Figure 4.1).

The simplicity and utility of asymmetric hydrogenation has inevitably attracted a number of reviews[8-14] in which different aspects are emphasized. The article by Markó and Bakos is particularly useful, as it is intended to be a comprehensive source of all applications of asymmetric hydrogenation up to the end of 1978. Because of this background, the present chapter is not intended to be comprehensive, but will draw as far as possible on developments within the last two or three years which have not yet appeared in other reviews. Our intention is to survey the field, present its successes and limitations, and show how mechanistic studies have helped explain the origins of stereoselection.

## 2. THE VARIATION OF REACTION CONDITIONS

The success of asymmetric hydrogenation as a route to amino acid derivatives does not find more general applicability. To appreciate why this is so, an analysis of the various factors involved is required. The limitations will only be overcome when a full appreciation of mechanism and of the source of stereoselectivity has been assimilated.

### 2.1. Changing the Catalyst

Almost without exception[15] monophosphines are ineffective in asymmetric hydrogenation, and flexible chelate biphosphines[16] give rise to poor optical yields. The Monsanto phosphine, DIPAMP, is an apparent exception, but the lack of rigidity in the chelate backbone may be countermanded by a preferred orientation of the *ortho*-methoxy groups during key stages of the catalytic cycle. There are many variables within a rigid *cis*-chelate, and these give rise to subtle changes in characteristics.

The simplest case is that of a 5-ring chelate based on 1,2-*bis*(diphenylphosphino)ethane. Inorganic stereochemists have long appreciated the chiral conformations adopted by 2,3-diaminobutane derivatives, but it was not until 1977 that Fryzuk and Bosnich applied this knowledge to phosphine chemistry. The synthesis of CHIRAPHOS and its cationic rhodium complex (**7**) was most fruitful, and optical yields of nearly

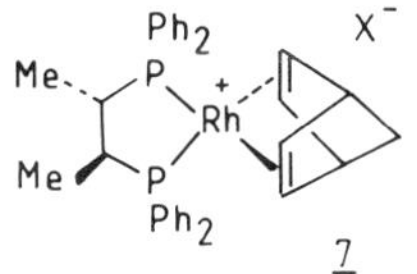

100% were obtained in the reduction of z-acylaminocinnamic acids and esters. In X-ray analysis[17] of complex (**7**), it was shown that the methyl-groups occupy pseudoequatorial sites in a nonplanar chelate that maintains $C_2$ symmetry. Once this had been recognized, an obvious experiment was to prepare the lower homologue PROPHOS, (**8**),[17] and to determine whether a single methyl group was capable of maintaining conformational rigidity.

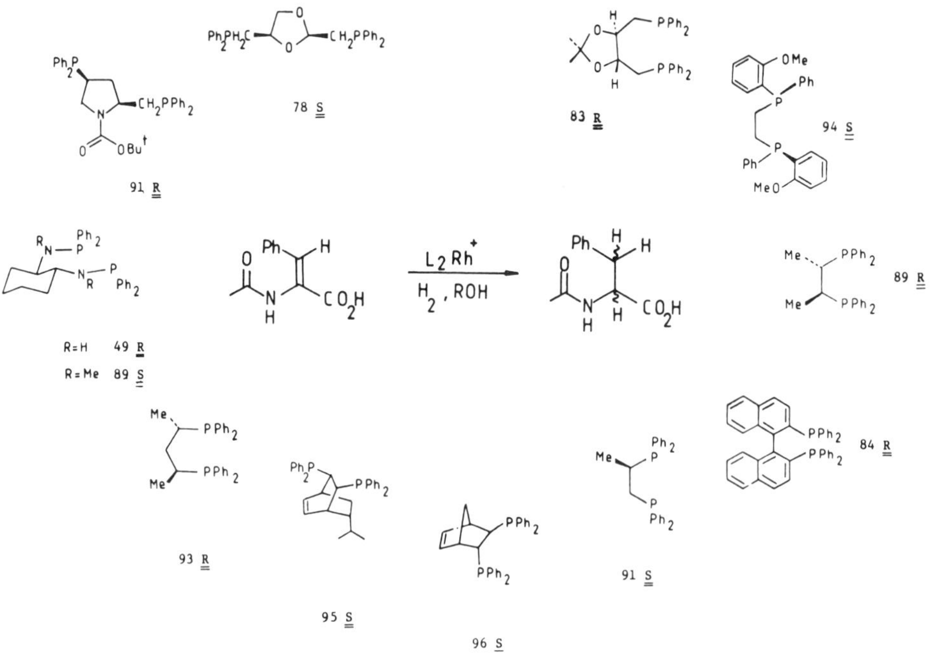

Figure 4.2. Typical optical yields in asymmetric hydrogenation of z-dehydroamino acids by typical catalysis.

Optical yields were only slightly lower (Figure 4.2), and comparable results have subsequently been obtained using the ligands PHEPHOS, (**9**),[18,19] and CYPHOS, (**10**).[20] Chiral 1,2-bis(diphenylphosphino)-ethane units may be derived by resolution procedures as in the case of NORPHOS, (**11**),[21,22]

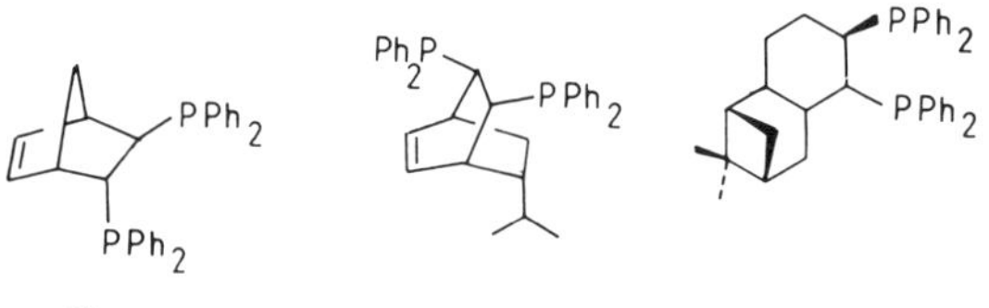

where Brunner and co-workers exploited the selective crystallization of one diastereomer of the complex dibenzoyl tartrate with the *bis*-phosphine oxide of (**11**). The synthesis of PHELLANPHOS, (**12**), involved reaction of *trans*-1,2-*bis*(diphenylphosphino)ethylene disulfide with (−)-phellandrene,[23] whereas NOPAPHOS, (**13**), was prepared by the same Diels Alder reaction with (+)-nopadiene.

The chair-form of a 6-ring biphosphine chelate has an achiral local orientation of $PPh_2$-groups. This makes it predictable that phosphine (**14**) should be a poor ligand in asymmetric hydrogenation,[24,25] but the efficiency of (**15**) is more surprising at first sight. Bosnich and co-workers attribute this to the existence of a preferred $C_2$ twist conformation in complexes of biphosphine (**15**), which is thus dubbed SKEWPHOS. They support their conclusions by showing that square-planar rhodium complexes of (**15**) are five times more CD-active than comparable complexes of (**14**). The catalyst apparently sits on a conformational knife-edge, and z-dehydroisoleucine derivative (**16**) is out of line in giving a poor optical yield on hydrogenation.

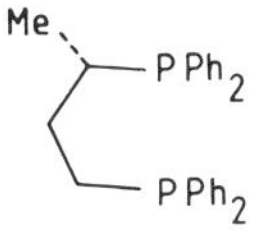

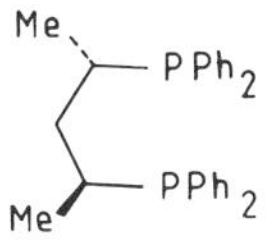

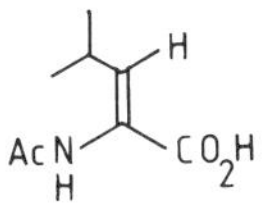

Seven-ring chelates were the subject of much early effort following the initial successes of Kagan and co-workers. It should suffice to comment that the nonrigid chelate, (**17**), does not effect asymmetric hydrogenation, that a range of *meta*- and *para*-substituted arylphosphine analogues of DIOP and a range of stereochemically equivalent bisphosphinobutanes, such as (**18**), have comparable efficiency to the parent,[26] but that the *ortho*-methoxyphenyl analogue, (**19**), gives lowered reactivity and *reversed* chirality in hydrogenation of z-dehydroamino acids.[27]

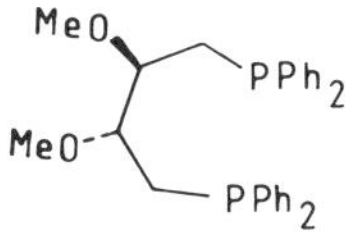

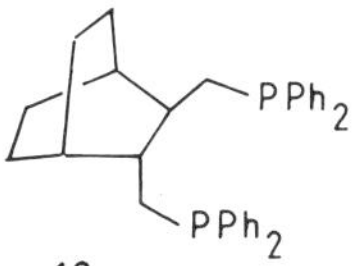

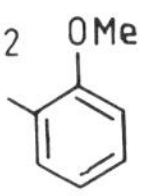

A number of crystal structures of DIOP complexes have been described[28,29,30] (*vide infra*), from which it is apparent that the geometry may vary from an ideal $C_2$-twist-chair to a twist-boat in accord with conformational variations within *bis*(DIOP)platinum.[31] Kagan[32] has carried out a detailed conformational analysis of the ligand in its chelate complexes, Figure 4.3 representing a simplified version.

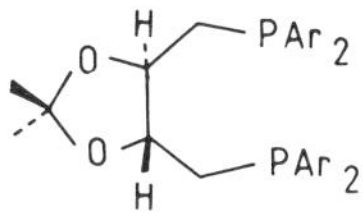

The basic framework of DIOP, which is a $C_2$ chiral 1,4-*bis*(diphenylphosphino)butane derivative, is subject to wide variation. The

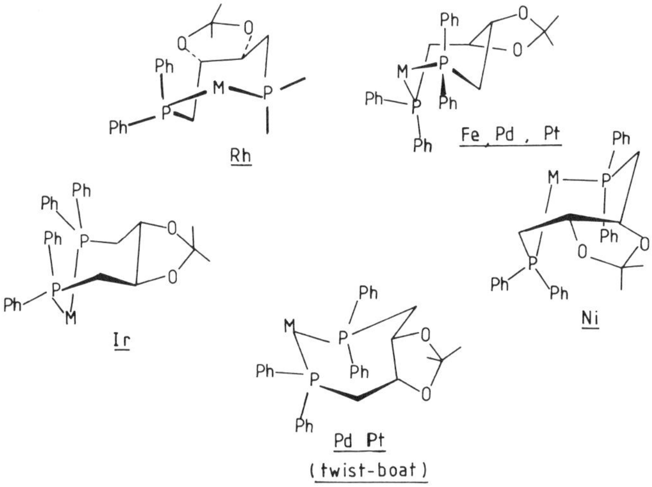

Figure 4.3. Conformational analysis of DIOP complexes.

heteroatom analogues, (**20**) and (**21**), are examples of a range studied by Italian,[33,34,35] Japanese,[36,37] and East German[36] workers. Chiral diaminobutanes are readily available and potentially modified with wide variation by N-substitution. While the rhodium complex of (**20**) gives high optical yields in reduction of z-dehydrophenylalanines, the simple change by N-methylation to (**21**) causes reduction of reaction rate and inversion in the configuration of product. This provides an interesting clue towards the origin of stereoselectivity.

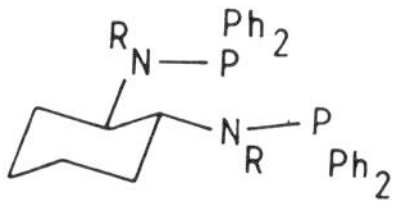

20  R=H
21  R=Me

Two further variations on the theme of 7-ring chelates deserve comment. The ingenious synthesis of a series of chiral biphosphines based on 4-hydroxyproline provided first Achiwa[39] and then Ojima and co-workers[40] with an efficient and reactive catalyst. The N-*t*- butyloxycarbonyl derivative BPPM, (**22**), has been most widely used. If the disadvantage of 7-ring chelates is their flexibility and the availability of alternative conformations of comparable energy, then BINAP complexes deserve attention. The axial chirality in (**6**) ensures that the chelate is rigid, and it does indeed provide very high optical yields in dehydroamino acid reduction. The simplicity of its structure makes it an almost ideal model for mechanistic studies.

Larger-ring chelate biphosphines are rare and may suffer from a number of disadvantages in catalysis. The tendency to *trans*-chelation means that DIOXOP (**23**) only functions as an effective catalyst under certain conditions[41] where the reaction sequence may involve (**24**) as well as (**25**). In the simple 8-ring rhodium chelate derived from 1,5-*bis*(diphenylphosphino)pentane, ready C–H insertion occurs under hydrogenation conditions, so that complex (**26**) is observed. The *cis*-chelate biphosphine unit is apparently essential to successful asymmetric hydrogenation, and purely *trans*-chelating chiral biphosphines, such as (**27**), show poor selectivity.[42]

## 2.2. Changing the Substrate

The previous section summarizes how various biphosphines catalyze asymmetric hydrogenation as their cationic rhodium complexes, illustrated by reduction of z-acetamido- or benzamidocinnamic acids. Generally speaking, 5-ring chelate complexes offer predictable results, and a range of z-dehydroamino acid derivatives give optical yields in excess of 85% with catalysts based on (**2**), (**5**), or (**8**)–(**13**). On lesser evidence, the same may be true of rigid chelates based on (**6**). With 7-ring chelates, the reaction

Table 1. Asymmetric Hydrogenation of $\underset{H}{\overset{R_1}{\diagdown}}\!\!=\!\!\underset{COOR_2}{\overset{NHCOR_2}{\diagup}}$ by R,R-DIOP Rhodium Complexes[a]

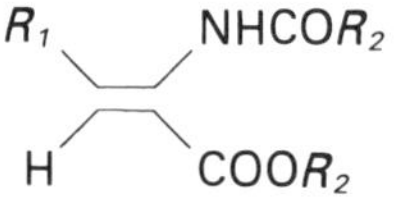

| $R_1$ | $R_2$ | $R_3$ | Enantiomer excess | Configuration | Reference |
|---|---|---|---|---|---|
| H | Me | H | 73 | R | 16 |
| H | Me | Me | 60 | R | 27 |
| iPr | Me | H | 74 | R | 43 |
| iPr | Me | Me | 21 | R | 43 |
| Ph | Me | H | 82 | R | 16 |
| Ph | Me | Me | 69 | R | 43 |
| Ph | Ph | H | 64 | R | 16 |
| Ph | Ph | Me | 55 | R | 16 |
| Ph | iPr | H | 57 | R | 43 |
| Ph | iPr | Me | 15 | R | 43 |
| Ph | tBu | H | 52 | R | 43 |
| Ph | tBu | Me | 0 | — | 43 |
| Ph | Me | Et | 72 | R | 43 |
| Ph | Me | iPr | 76 | R | 43 |
| Ph | Me | tBu | 77 | R | 43 |

[a] The catalyst used was formed *in situ* by reaction of R, R-DIOP with (Rh(cyclooctene)$_2$Cl)$_2$ in benzene/ethanol.

is very sensitive to the precise pattern of substitution in the dehydroamino acid derivative, as the results in Table 1 indicate. Glaser and co-workers[43] have made a systematic study of variations in the ester and amide moiety and find that the latter posseses a much more powerful stereodirecting effect, explicable if there are conformational variations in the biphosphine chelate induced by alteration of the substrate.

Changing the stereochemistry of the dehydroamino acid has a profound effect. While z-α- benzamidocinnamic acid usually gives optical yields comparable to α- acetamidoacrylic acid, the e-isomer gives greatly lowered optical yields, and isomerization[44,45] may compete with hydrogenation. Not so for DIPAMP in benzene solution[45] (Table 2), nor for BINAP, both of which give high optical yields in the reduction of E-α- benzamidocinnamic acid. While z-acids and esters frequently reduce with equal facility, there is no case of an E-acylaminocinnamic acid ester giving effective asymmetric

Table 2. Enantiomer Excesses in Hydrogenation of α-Benzamidocinnamic Acids and Methyl Esters

| Phosphine | Solvent | z-Acid | z-Ester | e-Acid | e-Ester |
|---|---|---|---|---|---|
| R,R-DIPAMP | EtOH | 94 s | | 30 s | |
| | $C_6H_6$ | 16 s | | 80 s | |
| R,R-DIOP | 3:1 EtOH:$C_6H_6$$^a$ | 70 R | 37.5 R | 25 R | 5 R |
| | $C_6H_6$$^b$ | | | 62 R | |
| s-BINAP | EtOH | 96 R | 93 R | 87 s$^c$ | |

$^a$ Neutral RhCl (DIOP) used as catalyst.
$^b$ Cationic Rh (norbornadiene) DIOP used as catalyst.
$^c$ Reaction carried out in THF.

reduction. A recent and very extensive paper$^{(45a)}$ has shown that both E- and z-acylamino-3-alkylacrylates are hydrogenated with high stereospecificity to s-amino acid derivatives with RR DISAMP-derived catalysts.

The fundamental and apparently indispensable structural unit in asymmetric hydrogenation is the enamide group (*vide infra* for structural studies), and this led several authors to examine related reactants. Kagan showed that simple enamides such as (**28**) were reduced in the presence of DIOP complexes in up to 90% optical yield.$^{(46)}$ Reduction of itaconic acid and its esters has been extensively studied and two groups of workers$^{(47,48)}$ report high optical yields. DIPAMP complexes are effective catalysts for the reduction of a range of itaconate derivatives (**29**),$^{(47)}$ the β-ester being the least successful, while the proline-derived complexes$^{(48)}$ (such as BPPM) only work well in the presence of triethylamine, a point to which we shall return later. Enol esters possess the same relative disposition of carbonyl and olefinic groups and so would be expected to be reduced with high stereoselectivity. The Monsanto group$^{(49)}$ showed that α-aryl vinyl acetates, (**30**), are reduced with moderate optical efficiency by DIPAMP

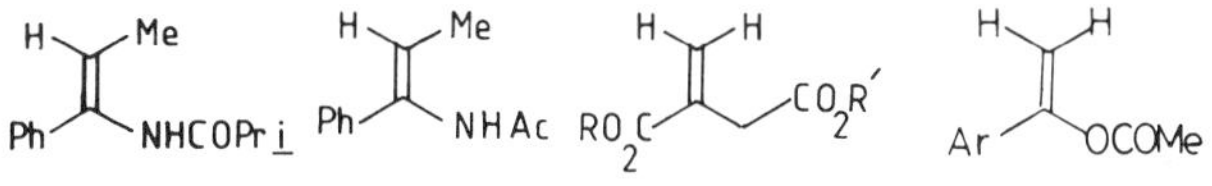

28a     28b     29 R,R′=H,Me     30

complexes, the enantiomer excess increasing with the electron withdrawing power of the aryl residue. The example provided by Bosnich and Fryzuk demonstrates a simple synthesis of (chiral-methyl) lactic acid (Figure 4.4).$^{(17)}$

A final example of the chelating substrate principle is given by reduction of enol phosphates, an interesting feature being the high (and not

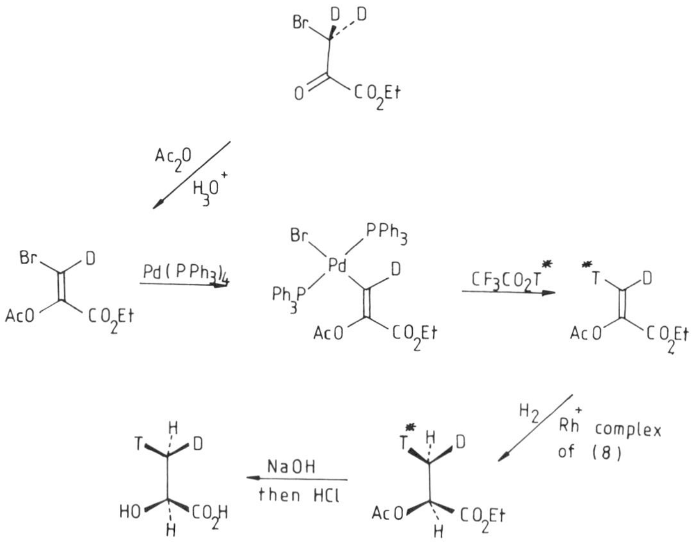

Figure 4.4. The synthetic route to $CH_3$-chiral lactic acid.

readily predicted) specificity afforded by phosphine (**31**) in reduction of (**32**),[50] where DIOP complexes are both chemically and optically less efficient.

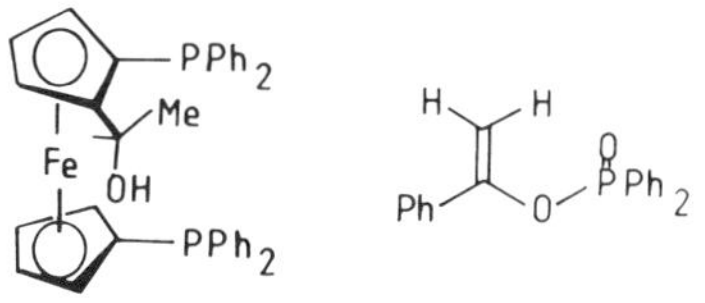

Enamide hydrogenation has found application in dipeptide synthesis, since dehydrodipeptides are quite readily available.[51] Three groups of workers[52-55] have examined the reduction of dehydrophenylalanyldipeptides, with an interesting divergence of results. The asymmetric center in dehydrodipeptide (**33**) (see Figure 4.5) has very little effect on the configuration of the new asymmetric center when the catalyst is derived from DIPAMP,[52] DIOP,[54] or the prolinebiphosphine (**34**),[52] and the major diastereomer formed is that predicted by analogy with simple enamides. Thus high yields of either RS- or SS-dipeptides may be obtained at will. With DIOXOP-derived catalysts, the existing chiral center plays an important part,[55] and high optical yields are only obtained when the chiral amino acid residue in the starting material has the s-configuration (Figure 4.5).

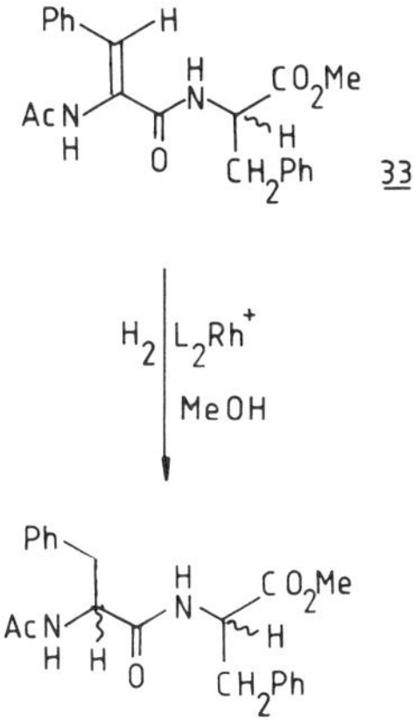

# ASYMMETRIC HYDROGENATION REACTIONS

|  | R → RS + RR |  | S → SS + SR |  | L₂ |
|---|---|---|---|---|---|
|  | 90 | 10 | 95 | 5 | DIOP **1** |
|  |  |  | 95 | 5 | DIPAMP **2** |
|  | 27 | 73 ◆ | 93 | 7 ◆ | DIOXOP **23** |
|  | 0.4 | 99.6 ✱ | 1 | 99 | 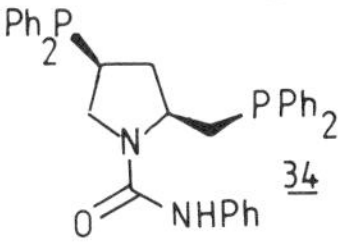 **34** |
|  | 38 | 62 | 62 | 38 | Ph$_2$P(CH$_2$)$_4$PPh$_2$ |

✱ Ligand 22
◆ CO$_2$H (NEt$_3$)

Figure 4.5. Diastereomer yields in the reduction of dehydroamino acid derivatives.

Prochiral $\alpha\beta$-unsaturated carboxylic acids present an anomaly, since some of the best examples of their asymmetric hydrogenation require monophosphine-derived catalysts. An early example using Morrison's neomenthyldiphenylphosphine, (**35**),[56] is shown and in a wide-ranging study[57] respectable optical yields were obtained. In reduction of (**36**) and

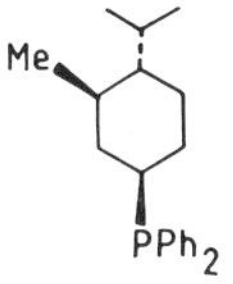

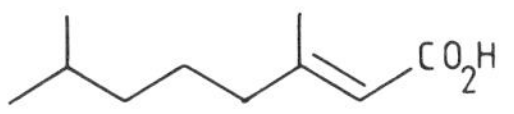

other carboxylic acids, triethylamine seems to be necessary, so that the true substrate is likely to be the $\alpha\beta$-unsaturated carboxylate, which binds more effectively[58] to the rhodium center.

The reader is referred to Reference 12 for a more exhaustive range of substrates, but one class is worth mentioning since it has received little attention. If the chiral center is in the substrate rather than the catalyst, then internal asymmetric induction may occur during hydrogenation. A simple example is in the reduction of dehydrovaline[59] by HT catalyzed by *tris*(triphenylphosphine)rhodium chloride, where the tritium is introduced with ⩾90% stereoselectivity giving (chiral-methyl)valine. Allylic alcohols possessing an asymmetric center (e.g., **37** in Figure 4.6) are reduced with high stereoselectivity by a cationic rhodium complex derived from

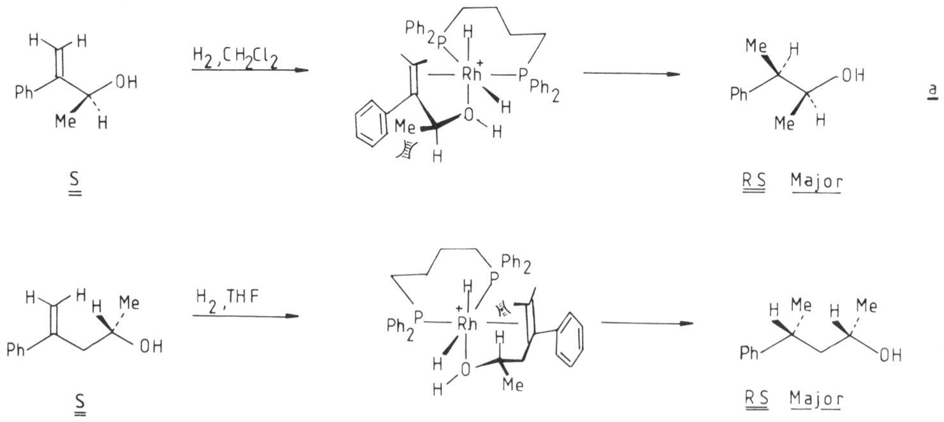

Figure 4.6. Internal asymmetric induction in homogeneous hydrogenation of allylic and homoallylic alcohols.

1,4-bis(diphenylphosphino)butane.[60] The course of reduction is inverted for the homolog (**38**) (Figure 4.6), and stereoselectivity appears to be controlled by nonbonded interactions within a chelated substrate.

### 2.3. Changing the Solvent

Most reductions of z-dehydroamino acids are carried out using cationic biophosphine rhodium diene complexes in polar media, the most popular being methanol. The course of reduction is not much altered by changing the solvent to other polar coordinating media, such as higher alcohols or tetrahydrofuran, although E-dehydroamino acids seem to give much better results in aprotic solvents, particularly benzene.

Some early results by Kagan[46] suggest that the nature of the complex involved in reduction of simple enamides may change with solvent polarity. Substrate (**39**) gives s-product on reduction by "DIOP rhodium chloride"

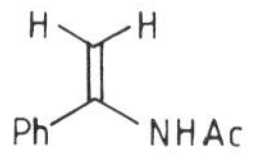

in ethanol, but increasing amounts of benzene in the medium lead to an increased amount of R-enantiomer, which eventually predominates. This implies that the true catalyst is cationic in ethanol, but hydrogenation by a neutral complex (which exhibits opposite selectivity) becomes increasingly important as the medium polarity is reduced.

## 2.4. Changing the Pressure

Most asymmetric hydrogenation is carried out at ambient temperature and pressure, but several workers have explored the relationship between optical yield and overpressure of hydrogen. The most thorough studies are due to Ojima[61] and to Sinou,[62] representative data being presented in Table 3. The general trend is to lower optical yields at higher pressures, but the sensitivity is greatly dependent on both catalyst and substrate. It should be possible in principle to derive mechanistic information from these observations; either the increased hydrogen pressure leads to a change in rate-determining stage, or, alternatively, there is a second mechanism which comes into play at high hydrogen pressures, possibly involving prior coordination of hydrogen to the metal, so that the order of addition (and hence the stereoselectivity) is reversed. Despite one authoritative statement in the literature,[63] there seems no basis for preferring one alternative over the other on present evidence.

## 2.5. Summary of the Features of Effective Catalysts

So many examples of the asymmetric hydrogenation of dehydroamino acid derivatives are known which have sufficiently similar features that a common mechanism is probable. It seems that the rhodium complex must be cationic, preferably in a weakly coordinating solvent, and carry a single chelating biphosphine occupying adjacent coordination sites throughout the catalytic cycle. Only rhodium complexes are genuinely successful, although ruthenium DIOP complexes[64] show some promise in the asymmetric hydrogenation of $\alpha\beta$-unsaturated carboxylic acids. *cis*-Chelate iridium complexes would be likely to be deactivated by $\mu$-hydride formation in the course of catalysis.[65]

Enamides are successful substrates on account of their closely defined coordination geometry, since both the olefin and amide carbonyl group are available to bind at *cis*-related sites. This must be necessary during the rate-determining stage, for species lacking the amide group (or a closely related functionality similarly sited) react much more slowly and with lower stereoselectivity.

## 3. STUDIES OF REACTION MECHANISM

It is probably true to say that more is known about the mechanism of asymmetric hydrogenation than any other process in homogeneous catalysis. This is partly because the complexity of the reaction and its

Table 3. Variation in Optical Yield in Reduction with Hydrogen Pressure

| Substrate | Catalyst | Et$_3$N/Rh | Solvent | pH$_2$ | | | | |
|---|---|---|---|---|---|---|---|---|
| | | | | 1 | 5 | 20 | 50 | 100 |
| z-α-Benzamido-cinnamic acid | (BPPM)Rh$^\oplus$cod | — | EtOH | 83.8 R | — | 21.2 R | 4.7 S | 8.4 S |
| Cinnamic acid | (BPPM) Rh$^\oplus$cod | 2 | EtOH | 93.3 R | — | 78.7 R | 66.2 R | 64.2 R |
| α-Acetamido acrylic acid | (BPPM) Rh$^\oplus$cod | — | MeOH | 95.2 R | | 21.8 S | | |
| α-Acetamido acrylic acid | (BPPM) Rh$^\oplus$cod | 2 | MeOH | 98.5 R | | 7.3 S | | |
| Itaconic acid | (BPPM) Rh$^\oplus$cod | — | MeOH | 91.3 S | | 85.3 S | | |
| Itaconic acid | (BPPM) Rh cod Cl | — | MeOH/C$_6$H$_6$ | 92.3 S | | 79.1 S | | |
| Itaconic acid | (BPPM) Rh cod Cl | 2 | MeOH/C$_6$H$_6$ | 94.5 S | | 91.3 S | | |
| z-α-Acetamido-cinnamic acid | DIOXOP Rh$^\oplus$cod | — | EtOH | 13.0 S | 10.0 R | 32.0 R | 30.0 R | |
| z-α-Acetamido-cinnamic acid | DIOXOP Rh$^\oplus$cod | 3 | EtOH | 84.0 S | 69.0 S | 43.0 S | 35.0 S | |
| Methyl z-α-acetamidocinnamate | DIOXOP Rh$^\oplus$cod | — | EtOH | 1.0 S | 10.0 R | 23.0 R | 33.0 R | |
| Methyl z-α-acetamidocinnamate | DIOXOP Rh$^\oplus$cod | 3 | EtOH | 6.0 S | 10.0 S | 6.0 S | 5.0 S | |

stereochemical component provide many factors that can be systematically varied and partly because several intermediates on the reaction pathway may be isolated or defined in solution, and the stereochemical course of each step thus established.

## 3.1. Reaction Rates

There are only two controlled kinetic studies of asymmetric hydrogenation, one of which[66] was carried out using neutral DIOP complexes before the significance of ionization in polar solvents was fully appreciated, and, hence, the pathway studied is not necessarily the most efficient one. In the second, Halpern and Chan[67,68] demonstrate that the kinetic form of the hydrogenation of methyl z-$\alpha$-acetamidocinnamate catalyzed by the 1,2-bis(diphenylphosphino)-ethane rhodium cation in methanol is:

$$\frac{dP}{dt} = k_2[H_2][Rh_{tot}]$$

with $k_2 = 10^2$ mol L$^{-1}$ s$^{-1}$ at 298 K for a *formal* hydrogen concentration of M/22.4 at STP. The kinetic isotope effect is 1.22.[69] This corresponds to a turnover number of about 0.07 M s$^{-1}$, which is 60 times faster than the corresponding CHIRAPHOS complex. Most of the catalyst under these conditions is bound to the substrate with K = $5 \times 10^3$. These experiments are not informative of the rate-determining stage, but it has recently been shown that this particular reaction occurs irreversibly, since *ortho*- and *para*-hydrogen are not equilibrated in the course of hydrogenation by a *para*-enriched sample.[70] In contrast, reaction in the presence of Wilkinson's catalyst leads to equilibration of nuclear spin-states as reaction proceeds.

A more general qualitative survey[71] of reactivity shows that 5-ring chelate complexes are much less reactive than 6- or 7-ring chelates where turnovers in excess of 0.3 M s$^{-1}$ may be observed. The nature of the substrate is also important, and in DIPAMP complex catalyzed hydrogenations almost any structural change in the substrate leads to a depression of reactivity. This includes N-methylation (which also profoundly affects the stereoselectivity) conversion of the carboxylic acid moiety into a nitrile and changing from z- to E-stereochemistry about the double bond. The latter further indicates the stringent structural requirements of asymmetric catalysis.

## 3.2. X-Ray Structural Analysis of Catalysts and Reaction Intermediates

A fortunate feature of asymmetric catalysis is that chiral biphosphine rhodium diene complexes, especially with norbornadiene, crystallize well

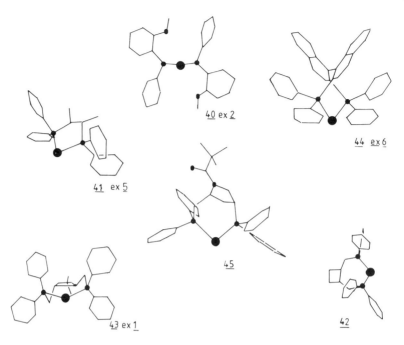

Figure 4.7. Approximate crystal structures of chelating biphosphine rhodium complexes (other ligands omitted for clarity).

and are amenable to X-ray study. The structures will be discussed in detail elsewhere in the book (see Chapter 2), but it is useful to analyze some general features here. Pertinent structures are shown in Figure 4.7; the DIPAMP complex, (**40**), appears in the original work of Knowles and co-workers while the CHIRAPHOS complex structure, (**41**), was published shortly after the description of its use in catalysis.[72] The 1,2-*bis*(diphenylphosphinomethyl) cyclobutane structure (**42**) is due to Townsend[73] and the DIOP structure (**43**) again due to Knowles and co-workers.[74] The BINAP complex (**44**) with its highly constrained geometry was solved by Ito and co-workers.[75]

All of these structures with $C_2$ symmetry have a special feature, first recognized by Knowles. If the P-phenyl groups are viewed from the remote side of the coordination plane, then the axial pair appear edge-on, and the equatorial pair are approximately face-on. Viewed from above, a pair of the *ortho*-hydrogens belonging to opposed axial phenyl groups is in proximity to the metal, within 3 Å. Whether this plays any role in stabilizing the structure is unclear but the P-phenyl orientation is sufficiently well defined to permit prediction of the sense of asymmetric hydrogenation. If the relationship is drawn in (**40**), then the reaction will give rise to an s-amino

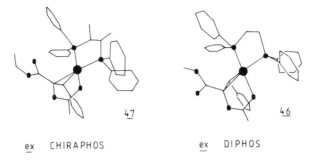

Figure 4.8. Crystal structures of dehydroamino acid complexes.

acid derivative. All catalysts which are structurally defined conform to this rule.

A further important development is due to Halpern and co-workers.[63,67] They managed to crystallize two enamide complexes, one derived from 1,2-*bis*(diphenylphosphino)ethane and the other from CHIRAPHOS. They are almost isostructural with regard to the binding of the dehydroamino acid ester, as indicated in Figure 4.8. Complex (**46**) shows the olefin $\pi$-bonded and rotated somewhat from orthogonality with respect to the coordination plane. The Rh-C distance to the heteroatom-substituted carbon is less than that to the benzylidene carbon, and the N-acetyl group is bonded to rhodium through a $\sigma$-interaction with the lone pair of the carbonyl group. The ligand backbone is skewed so that it is chiral. Interestingly, the CHIRAPHOS complex, (**47**), has the opposite configuration to that of the amino acid ester produced on its hydrogenation by *cis*-transfer of $H_2$ from rhodium. The solid-state structure has been shown to persist in solution by analysis of its CD spectrum.[76]

## 3.3. Structure and Dynamics in Solution

### 3.3.1. Solvates

In a classic paper[77] Tolman and co-workers at Dupont demonstrated that the dihydride (**48**) was observable by $^1H$ and $^{31}P$ nmr when *tris*(triphenyl-phosphine) rhodium chloride was sealed in solution under hydrogen. The stereochemistry of this intermediate was thus established, and at that time it was suggested that stereospecific PPh$_3$ loss preceded alkene coordination in hydrogenation. Later work[78] suggests that at high concentrations of alkene and low concentrations of PPh$_3$, (**48**) may be bypassed in the true catalytic cycle. Thus, early attempts to explain the mechanism and stereoselectivity of asymmetric hydrogenation were influenced by these results and modeled on them, notwithstanding the fact that H$_2$Rh(PPh$_3$)$_2$Cl has two *trans*-related PPh$_3$ ligands and efficient chiral catalysts possess a *cis*-chelated biphosphine. Observations by Halpern[79] and Baird[80] and co-workers clarified the issue. It was shown,[79] for example, that complex (**49**) reacted with three moles of hydrogen in methanol solution giving (**50**), whereas complex (**51**) reacted with two moles of hydrogen in methanol solution giving (**52**), which had no apparent affinity for hydrogen. Further work has established that the division between monophosphines (dihydride formation) and chelating biphosphines (solvate formation) is not quite so clear-cut.[81] Direct evidence for the binding of methanol to rhodium in (**52**) is lacking, but the structure is consistent with spectral properties, and $^{31}$P nmr indicates that the phosphorus nuclei are *trans* to an electronegative ligand. In the corresponding DIPAMP solvate, (**53**), EXAFS measurements[82] indicate that oxygen ligands are *trans* to phosphorus in the inner coordination sphere. Attempts to crystallize complex (**52**) led to the formation of an arene-bridged dimer, characterized as (**54**) by X-ray structure determination.[79] Dimers of a different structure may be formed on hydrogenation of the PHEPHOS complex, (**55**), since it gives a monomeric and two dimeric solvates (four in the case of racemic phosphine) on hydrogenation in methanol.[81]

Related solvate complexes have been formed from many other asymmetric chelate biphosphine complexes and with other coordinating solvents.

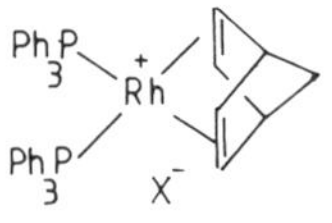

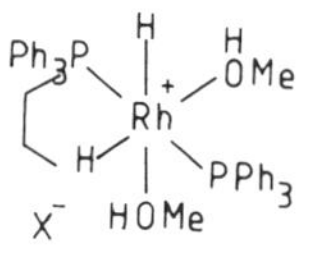

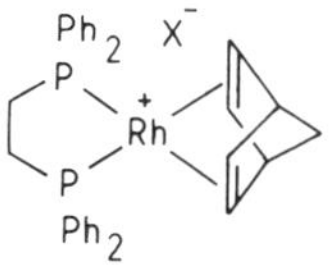

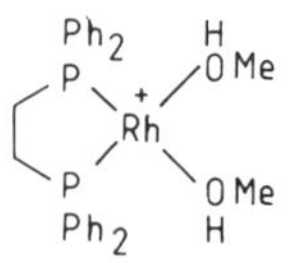

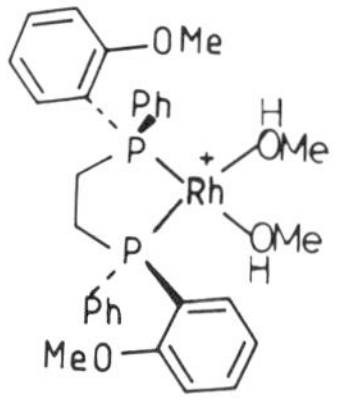

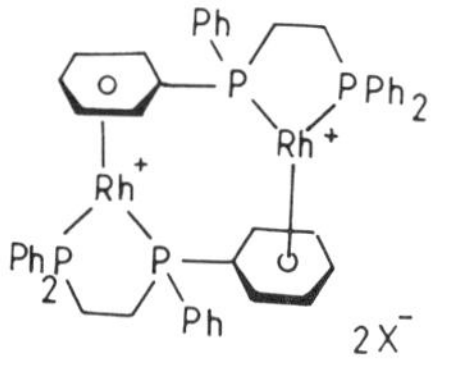

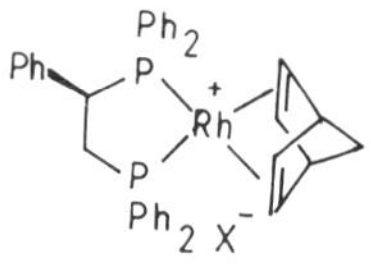

## ASYMMETRIC HYDROGENATION REACTIONS

The apparent formation of arene complexes on hydrogenation in benzene (e.g., in Reference 61) seems to be an anomaly, since their solution nmr spectra are very different from authentic rhodium biphosphine tetraphenylborates.[81]

### 3.2.2. Substrate Complexes

Under an inert atmosphere, solvate complexes of the type described above react with dehydroamino acid derivatives to produce new species, the solution structure of which is comparable to (**46**) or (**47**). Binding of the olefin and amide groups was established by $^{13}C/^{31}P$ nmr employing specifically labeled dehydroamino acids, which additionally demonstrated that the carboxylate group is not involved in coordination in most cases. A range of enamide complexes derived from rhodium *bis*(diphenylphosphino)ethane all exhibit very similar $^{31}P$ nmr spectra which are sharp at room temperature.[83] With larger ring chelates, the corresponding complexes give $^{31}P$ spectra which may be broadened by dynamic processes at room temperature and require cooling before the characteristic sharp 8-line spectrum is observed. The nature of this dynamic process is revealed by saturation transfer experiments employing the DANTE pulse sequence.[84] This demonstrates that an *inter*molecular exchange is involved (Figure 4.9),

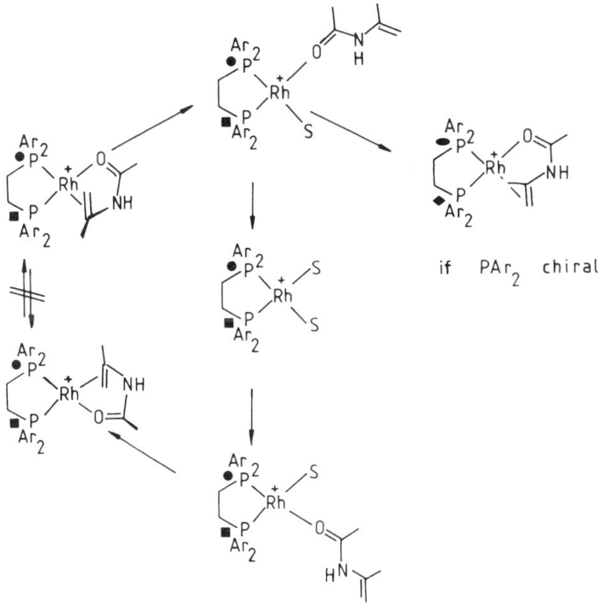

Figure 4.9. The mechanism of ligand exchange in enamide complexes.

which is fast in relation to the time-scale of catalysis but strongly affected by the nature of the amide-group, with the pivalamide exchanging an order of magnitude more slowly.

Of a number of enamide complexes studied by the present authors, those derived from DIPAMP proved to be the most rewarding.[81,83,85] If the solvate, (**53**), is reacted with z-α-benzamidocinnamic acid or its methyl ester, then *two* enamide complexes of similar structure are formed in a ratio of *ca* 10:1 at room temperature. Since $^{13}$C nmr of labeled enamides shows that these have very similar solution structures, they are presumed to be the two diastereomeric complexes (**56**) and (**57**), although at this stage the experiment gives no information on their configuration. The DANTE experiment[86] provides the intriguing result that their *intra*molecular exchange (by coordination/decoordination of the olefinic bond) is considerably faster than dissociative exchange and again rapid on the time-scale of catalysis.

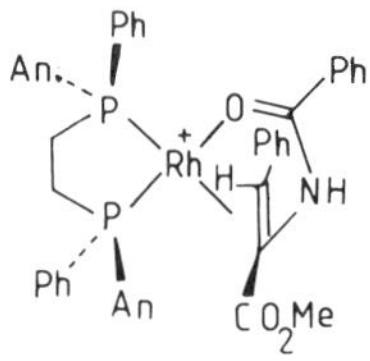

56 major 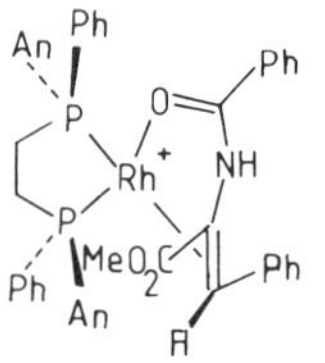    57 minor

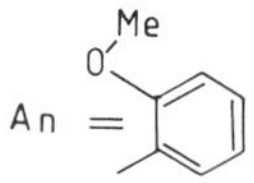

The simplicity of solution equilibria observed here is typical of 5-ring chelate rhodium complexes of z-enamides, where complexation constants are high ($\sim 10^4 M^{-1}$) and the geometry well-defined. Under catalytic conditions, the enamide complex represents the major resting-state, and its bright scarlet color is characteristically observed. In other cases, particularly CHIRAPHOS, only a single diastereomer is observed with most substrates. Although this pattern of complexation is frequently comparable with larger-ring chelate complexes, anomalies are often observed. Under appropriate conditions, z-enamide complexes with bound carboxylate,[87] with 2:1 stoichiometry,[88] with tridentate binding[89] and possible σ-bonding to the benzylidene carbon[27] have all been observed and characterized in solution by $^{13}$C/$^{31}$P nmr spectroscopy, structures being shown as (**58**)–(**61**).

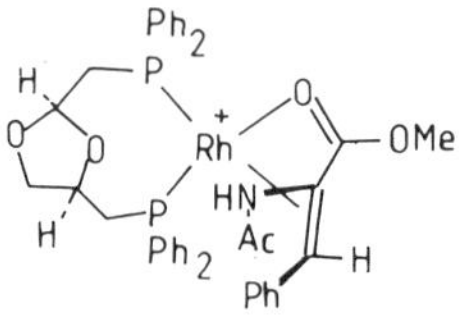

58 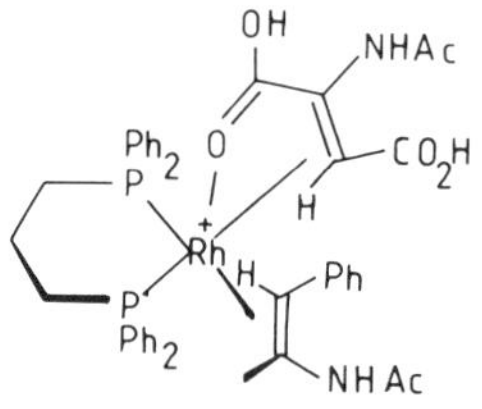    59

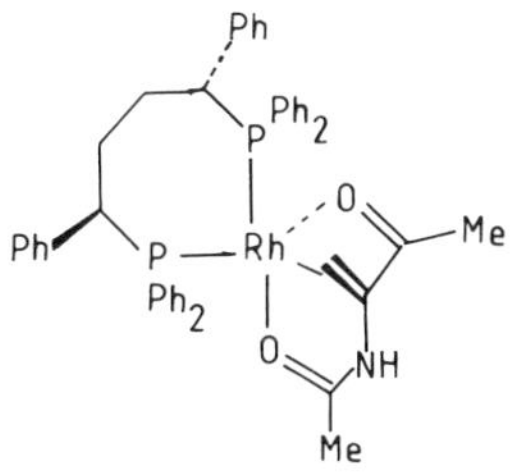

**60**

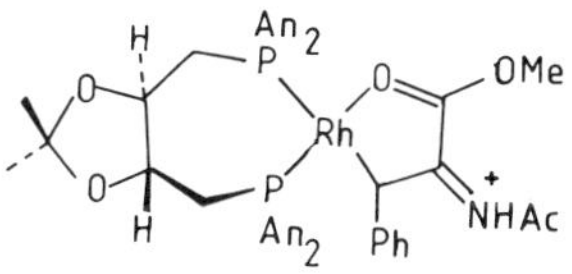

**61**

Complexation of $\alpha\beta$-unsaturated acids and amides has been extensively studied.[58] In such cases, both diastereomers of a species involving coordination of both olefin and carboxylate group are observed, usually in comparable proportions. The extent of complexation is enhanced and the rate of exchange of free and bound substrate reduced when base is added, or a preformed tetraalkylammonium carboxylate is employed. This suggests that structures (**62**) and (**63**) best describe the complex. The spectra observed are almost identical to those derived from E-dehydroamino acid derivatives, implying that the latter have structure (**64**) in accord with labeling studies.[83,88]

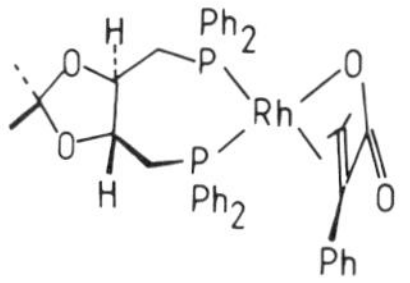

**62**

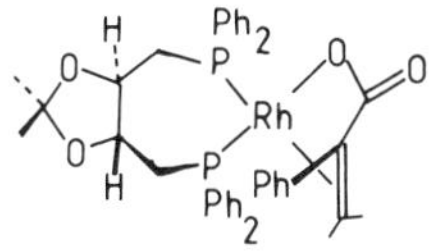

**63**

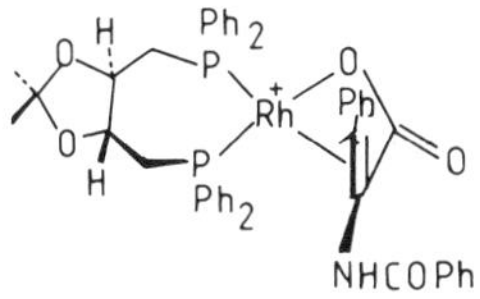

**64** (diastereomers)

### 3.3.3. Rhodium Alkyls

If enamide complexes such as (**46**) are exposed to hydrogen in methanol solution then the color quickly fades, and $^{31}P$ nmr observation suggests direct conversion into solvate by hydrogenation of the bound substrate without the intervention of an intermediate. At low temperatures, (**46**) is converted into (**65**) which is stable at $-80°$ but decomposes on warming to $-50°$.[68] The presence of a solvent molecule *trans* to the rhodium-bound hydride is indicated by a quantitative reaction with one equivalent of $CH_3C^{15}N$ giving (**65a**) with the anticipated changes in nmr spectra. Since

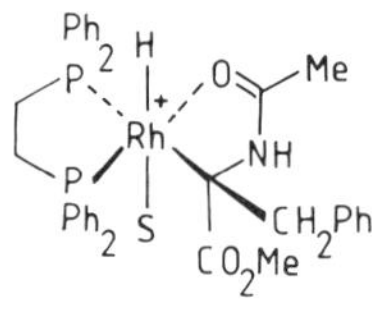

**65** S=MeOH
**65a** S=CH$_3^{15}$CN

addition of hydrogen to a square-planar complex is second-order, but the decomposition of the alkylrhodium hydride, (**65**), is first-order with $\Delta H^{\neq} = 17.1$ kcal mole$^{-1}$ and $\Delta S^{\neq} = +6$ cal mole$^{-1}$ deg$^{-1}$, its stability at low temperature is explicable and reflects a change in the rate-determining stage of reaction.

If the enamide complexes (**56**) and (**57**) are prepared normally and then cooled, the minor diastereomer becomes increasingly disfavored and represents no more than 2% of the species present at 220 K. Under these conditions the solution is unreactive to hydrogen, and warming leads to a reversion to solvate complex with concomitant hydrogenation of the substrate. If the same solvate complex is mixed with a slight excess of methyl z-α-benzamidocinnamate at 200 K and sealed under hydrogen, then the enamide complex mixture which forms under kinetic control is rich in (**57**). Warming this solution to 220 K now leads to a reaction in which the minor diastereomer disappears selectively, and a new set of signals appear.[90] The $^1$H and $^{31}$P nmr of (**66**) suggest that it is closely similar to (**65**).

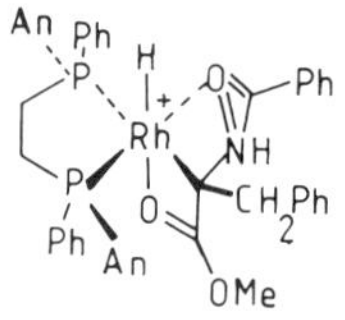

**66**

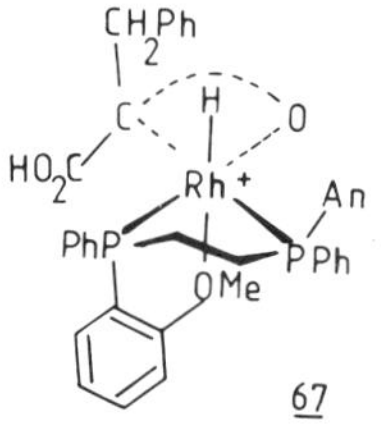

**67**

In the course of characterization of this new alkylrhodium hydride it was demonstrated that the amide remained bound and that the double bond had been reduced because of the loss of *C$_1$*C$_3$ coupling in a doubly labeled sample of enamide on going from (**57**) to (**66**). It is possible that the carboxylate group occupies the coordination site *trans* to hydride, because it experiences a downfield $^{13}$C chemical shift of 12 ppm relative to the free acid or ester. This may be due to the change in its environment combining the effects of saturation and β-rhodium deshielding, and an alternative structure, (**67**), cannot be ruled out on current evidence. Certainly the intermediate is different in detail from the 1,2-*bis*(diphenylphosphino)ethane case; the $^{31}$P chemical shifts and $^{31}$P-Rh, $^{31}$P-$^{31}$P coupling constants are distinct, and it does not react with $CH_3CN$ at low temperatures. The critically important fact is that only the minor diastereomer is reactive to hydrogen, and extrapolating this observation to ambient temperature requires that (**57**) is a true intermediate in catalysis, whereas (**56**) must undergo diastereoisomerism before it can react. Since delivery of hydrogen from rhodium to alkene is known to be *cis*-selective and stereospecific[91] the absolute configuration of (**56**) and (**57**) is defined.†

It is incidentally the case that regiospecific formation of (**65**) and (**66**) requires that the amide-olefin chelate ring remains intact during the catalytic cycle. Formation of the alternative structure is then stereoelectronically precluded as diagram (**68**) indicates.

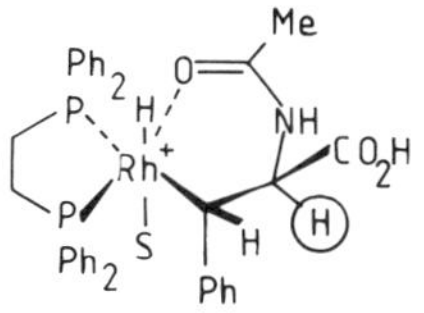

<u>68</u>

## 3.4. Conclusions

The combination of crystallographic and nmr experiments gives a very clear indication of intermediates on the reaction pathway of asymmetric hydrogenation. They leave one major gap in our understanding, since the rate-determining stage involves addition of hydrogen and may generate another intermediate which cannot be observed. There are three possibilities for the rate-determining stage, namely:

(a) concerted reaction which generates the alkylrhodium hydride in a single step;
(b) rate-determining addition of hydrogen followed by collapse of dihydride to alkylrhodium hydride; and
(c) Reversible addition of hydrogen followed by rate-determining formation of alkylrhodium hydride.

No single experiment can discriminate between these three possibilities. Reversible addition of hydrogen is ruled out by the complete

---

† A preliminary experiment in which the CHIRAPHOS enamide complex of Z-α-acetamide-4-methyl-2-pentenoic acid was prepared gave two diastereomeric complexes in the ratio 2:1. When hydrogen was admitted at 220 K, reaction was extremely slow. After 24 hours at 230 K both diastereomers had reacted to some extent to give a mixture of products which could not be well-defined. While this result may not be strictly comparable with the data reported in the literature, it does suggest that conclusions based on comparisons between rate data from different phosphine complexes should be treated with some caution.

lack of *ortho-* ⇌ *para-* hydrogen interconversion during reduction of z-α-benzamidocinnamic acid by DIPAMP, 1,2-*bis*(diphenylphosphino)ethane or 1,4-*bis*(diphenylphosphino)butane rhodium complexes.[70] Distinction between concerted pathway (a) and stepwise pathway (b) is more exacting. The deuterium isotope effect ($H_2$ versus $D_2$) is 1.22 for reduction of the same substrate by the methanol solvate of 1,2-*bis*(diphenylphosphino)ethane under ambient conditions.[69] Addition of HD, although reported to give equipartition of hydrogen and deuterium in reductions by Wilkinson's catalyst[92] and in asymmetric hydrogenation[44] does so in the former but not the latter case. With a range of catalysts, careful integration of the $^1$H nmr of the reduction product (**69**) demonstrates that hydrogen

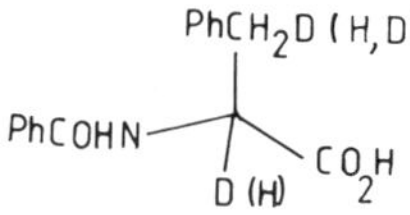

$$PhCOHN-\underset{D(H)}{\overset{PhCH_2D(H,D)}{C}}-CO_2H \qquad \underline{69}$$

prefers the benzylic position and deuterium the α-position by a factor of 1.33:1 to 1.38:1 depending on the catalyst. Thus, the *intra*molecular discrimination (HD) is greater than the *inter*molecular discrimination ($H_2$ vs. $D_2$), which suggests that they cannot arise by a single process. The consequential requirement for an intermediate rules out a concerted mechanism and establishes that rate-determining addition of hydrogen followed by rapid transfer of one rhodium hydride (mechanism (b)) occurs.

Gives the very considerable structural constraints involved in the key intermediates, it becomes apparent why the efficient asymmetric hydrogenation has been limited to enamides and a few related substrates. Both the substrate and the biphosphine must maintain chelation throughout, and the rigidity of the ligand is critical. *Trans-*chelating phosphines seem quite ineffective with any substrate, and perhaps here it will be necessary to introduce groups capable of secondary binding.

The energetics of asymmetric hydrogenation are summarized in Figure 4.10, insofar as they are known. If only the hydrogenation step is stereoselective, then it becomes clear why changing the rate-determining step (e.g., by making the substrate more bulky and inhibiting binding, or by increasing hydrogen pressure) can only lead to a reduction in optical yield.

## 4. THE ORIGIN OF STEREOSELECTIVITY

Given detailed knowledge of *two* intermediates in enamide hydrogenation, both close to the rate-determining transition-state, it should surely

## ASYMMETRIC HYDROGENATION REACTIONS

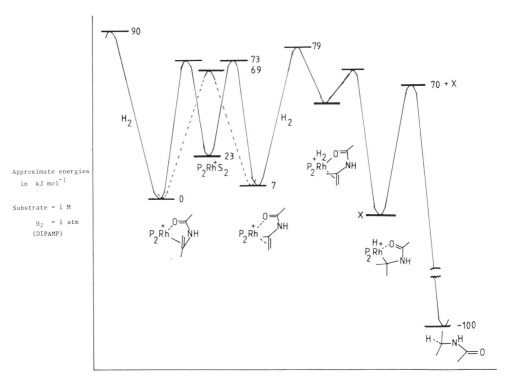

Figure 4.10. Energetics of asymmetric hydrogenation by Rhodium DIPAMP complexes.

be possible to define why the reaction is so stereoselective. In the enamide complex, reference to the X ray structures of Halpern and co-workers helps rationalize the stereoselectivity. The main steric interaction is between an equatorial phenyl-ring and the carbonyl group, which is minimized in the preferred diastereomer (Figure 4.11). It is not at all clear, however, why one diastereomer reacts with hydrogen so much more readily. The steric argument might be that addition of hydrogen in the productive direction causes a change in coordination sphere geometry in which the phosphine *trans* to an amide carbonyl-group rotates to form the octahedral dihydride (Figure 4.11), this being more favorable for the minor isomer of enamide complex. While this *might* be correct, the argument carries a touch of sophistry and cannot be tested by experiment as it stands.

It is puzzling why rigidity in the biphosphine should lead to a reduction in the overall rate of enamide hydrogenation, a point which has been made more than once.[11,25] By itself, this suggests that the path to the rate-determining transition-state requires some distortion of the biphosphine chelate, perhaps a geometrical change brought about by hydrogen addition.

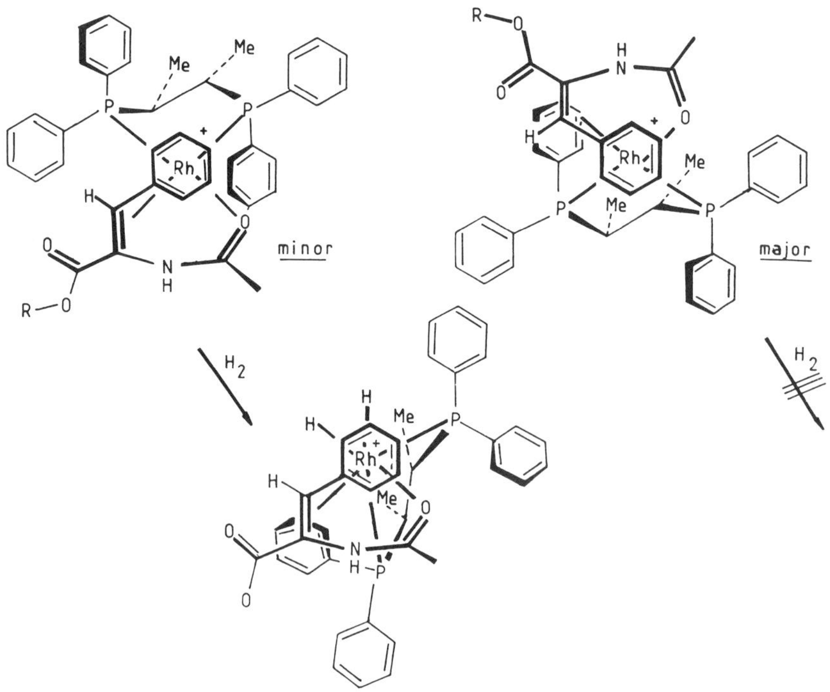

Figure 4.11. Diastereoselection in enamide complexes and the hydrogen-addition process.

Each isomer of enamide complex possesses two diastereotopic faces which give distinct products on hydrogen addition. Only one direction of addition is productive since the Rh-H bond should be *syn*-coplanar with the olefin C=C for maximum overlap in the transfer stage. This means that stereoselectivity is controlled by the relative free-energies of addition along

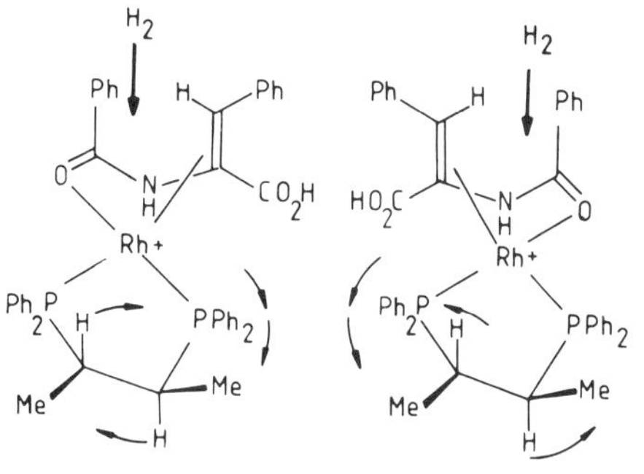

Figure 4.12. The torsional changes involved in hydrogen-addition.

paths A and B in Figure 4.12. These are distinguished by the torsional changes taking place in the rigid chelate ring. Addition B to the minor enamide complex is much softer, since the direction of twist in the chelate is reinforced, while the torsional change in A is harder. If the idea is correct (and it is subject to experimental test!) then it suggests a more general method for stereochemical control in transition-metal catalysts.

## REFERENCES

1. Reviewed by F. H. Jardine, *Prog. Inorg. Chem.* **28**, 63–202 (1981).
2. For example, L. Horner, H. Siegel, and H. Büthe, *Angew. Chem., Int. Ed. Engl.*, **7**, 942 (1968).
3. T. P. Dang and H. B. Kagan, *J. Chem. Soc., Chem. Commun.* **1971**, 481.
4. B. D. Vineyard, W. S. Knowles, M. J. Sabacky, G. L. Bachman, and D. J. Weinkauff, *J. Am. Chem. Soc.* **99**, 5946–5952 (1977).
5. O. Korpiun, R. A. Lewis, J. Chickos, and K. Mislow, *J. Am. Chem. Soc.* **90**, 4842–4846 (1968).
6. M. D. Fryzuk and B. Bosnich, *J. Am. Chem. Soc.* **99**, 6262–6267 (1977).
7. A. Miyashita, A. Yasuda, H. Takaya, K. Toriumi, T. Ito, T. Souchi, and R. Noyori, *J. Am. Chem. Soc.* **102**, 7932–7934 (1980).
8. J. D. Morrison, W. F. Masler, and M. K. Neuberg, *Adv. Catal.* **25**, 81–124 (1976).
9. D. Valentine, Jr. and J. W. Scott, *Synthesis* **1978**, 329–356.
10. V. Čaplar, G. Comisso, and V. Šunjić, *Synthesis* **1981**, 85–116.
11. M. D. Fryzuk and B. Bosnich, *Top. Stereochem.* **12**, 119–154 (1981).
12. L. Markó and J. Bakos, *Aspects Homogeneous Catal.* **4**, 145–202 (1981).
13. J. M. Brown, P. A. Chaloner, B. A. Murrer, and D. Parker, *ACS Symp. Ser.*, **119**, 169–194 (1980).
14. P. A. Chaloner and D. Parker, *Reactions of Coordinated Ligands*, edited by P. S. Braterman, (Plenum Press, New York, in press).
15. W. S. Knowles, M. J. Sabacky, and B. D. Vineyard, *Ann. New York Acad. Sci.* **214**, 119–124 (1973).
16. H. B. Kagan and T. P. Dang, *J. Am. Chem. Soc.* **94**, 6429–6433 (1972).
    H. Kagan and T. P. Dang, *Ger. Offen.* 2,161,200 (1972); *Chem. Abstr.* **77**, 114567k (1972).
17. M. D. Fryzuk and B. Bosnich, *J. Am. Chem. Soc.* **100**, 5491–5494 (1978); **101**, 3043–3049 (1979).
18. R. B. King, J. Bakos, C. D. Hoff, and L. Marko, *J. Org. Chem.* **44**, 1729–1731 (1979).
19. J. M. Brown and B. A. Murrer, *Tetrahedron Lett.* **1979**, 4859–4862.
20. D. P. Riley and R. E. Shumate, *J. Org. Chem.* **45**, 5187–5193 (1980).
21. H. Brunner and W. Pieronczyk, *Angew. Chem. Int. Ed. Engl.* **18**, 620–621 (1979).
22. H. Brunner, W. Pieronczyk, B. Schönhammer, K. Streng, I. Bernal, and J. Korp, *Chem. Ber.* **114**, 1137–1149 (1981).
23. O. Samuel, R. Couffignal, M. Lauer, Z. Y. Zhang, and H. B. Kagan, *Nouv. J. Chim.* **5**, 15–21 (1981).
24. H. B. Kagan, J. C. Fiaud, C. Hoonaert, D. Meyer, and J. C. Poulin, *Bull. Soc. Chim. Belg.* **88**, 923–931 (1979).

25. P. A. MacNeil, N. K. Roberts, and B. Bosnich, *J. Am. Chem. Soc.* **103,** 2273–2280 (1981).
26. T. P. Dang, J-C. Poulin and H. B. Kagan, *J. Organomet. Chem.* **91,** 105–115 (1975).
27. J. M. Brown and B. A. Murrer, *Tetrahedron Lett.* **21,** 581–584 (1980).
28. V. Gramlich and G. Consiglio, *Helv. Chim. Acta* **62,** 1016–1024 (1979).
29. V. Gramlich and C. Salomon, *J. Organomet. Chem.* **73,** C61–C63 (1974).
30. S. Brunie, J. Mazan, N. Langlois and H. B. Kagan, *J. Organomet. Chem.* **114,** 225–232 (1976).
31. J. M. Brown and P. A. Chaloner, *J. Am. Chem. Soc.* **100,** 4307–4309 (1978).
32. G. Balavione, S. Brunie, and H. B. Kagan, *J. Organomet. Chem.* **187,** 125–139 (1980).
33. M. Fiorini and G. M. Giongo, *J. Mol. Catal.* **7,** 411–413 (1980).
34. M. Fiorini, F. Marcati, and G. M. Giongo, *J. Mol. Catal.* **4,** 125–134 (1978); **3,** 385–387 (1977/8).
35. M. Fiorini, G. M. Giongo, F. Marcati, and W. Marconi, *J. Mol. Catal.* **1,** 451–453 (1975/6).
36. K-i. Onuma, T. Ito, and A. Nakamura, *Tetrahedron Lett.* **1979,** 3163–3166.
37. *idem, Bull. Chem. Soc. Jpn.* **53,** 2016–2019 (1980).
38. G. Pracejus and H. Pracejus, *Tetrahedron Lett.* **1977,** 3497–3500.
39. I. Ojima, T. Kogure, and K. Achiwa, *Chem. Lett.* **1978,** 567–568.
40. I. Ojima, T. Kogure, and Y. Yoda, *Chem. Lett.* **1979,** 495–498.
41. D. Lafont, D. Sinou, and G. Descotes, *J. Organomet. Chem.* **169,** 87–95 (1979).
42. J. M. Brown and F. M. Dayrit, to be published.
43. R. Glaser, S. Geresh, J. Blumenfeld, and M. Twaik, *Tetrahedron* **34,** 2405–2408 (1978); R. Glaser, S. Geresh, M. Twaik, and N. L. Benoiton, *Tetrahedron* **34,** 3617–3621 (1978); R. Glaser and S. Geresh, *Tetrahedron* **35,** 2381–2387 (1979). J. M. Brown, P. A. Chaloner, R. Glaser and S. Geresh, *Tetrahedron* **36,** 815–825 (1980); R. Glaser, S. Geresh, and M. Twaik, *Isr. J. Chem.* **20,** 102–107 (1980). D. Lafont, D. Sinou, G. Descotes, R. Glaser, and S. Geresh, *J. Mol. Catal.* **10,** 305–311 (1981) and earlier papers.
44. C. Detellier, G. Gelbard, and H. B. Kagan, *J. Am. Chem. Soc.* **100,** 7556–7561 (1978).
45. K. E. Koenig and W. S. Knowles, *J. Am. Chem. Soc.* **100,** 7561–7564 (1978).
45a. J. W. Scott *et al.*, *J. Org. Chem.* **46,** 5086–5093 (1981).
46. H. B. Kagan, N. Langlois, and T. P. Dang, *J. Organomet. Chem.* **90,** 353–365 (1975); D. Sinou and H. B. Kagan, *J. Organomet. Chem.* **114,** 325–337 (1976).
47. W. C. Cristopfel and B. D. Vineyard, *J. Am. Chem. Soc.* **101,** 4406–4408 (1979).
48. I. Ojima, T. Kogure, and K. Achiwa, *Chem. Lett.* **1978,** 567–568.
49. K. E. Koenig, G. L. Bachman, B. D. Vineyard, *J. Org. Chem.* **45,** 2362–2365 (1980).
50. T. Hayashi, K. Kanehira, and M. Kumada, *Tetrahedron Lett.* **22,** 4417–4420 (1981).
51. O. Pieroni, G. Montagnoli, A. Fissi, S. Merlino, and F. Ciardelli, *J. Am. Chem. Soc.* **97,** 6820–6826 (1975).
52. I. Ojima, T. Kogure, N. Yoda, T. Suzuki, M. Yatabe, and T. Tanake, *J. Org. Chem.* **47,** 1329–1334 (1982).
53. K-i. Onuma, T. Ito, and A. Nakamura, *Chem. Lett.* **1980,** 481–482.
54. D. Meyer, J-C. Poulin, H. B. Kagan, H. Levine-Pinto, J. L. Morgat, and P. Fromageot, *J. Org. Chem.* **45,** 4680–4682 (1980).
55. D. Sinou, D. Lafont, G. Descotes, and A. G. Kent, *J. Organomet. Chem.* **217,** 119–127 (1981).
56. J. D. Morrison, R. E. Burnett, A. M. Aguiar, C. J. Morrow, and C. Phillips, *J. Am. Chem. Soc.* **93,** 1301–1303 (1971). A. M. Aguiar, C. J. Morrow, J. D. Morrison, R. E. Burnett, W. F. Masler, and N. S. Bhacca, *J. Org. Chem.* **41,** 1545–1547 (1976).
57. D. Valentine, Jr., J. F. Blount, and K. Toth, *J. Org. Chem.* **45,** 3691–3698 (1980); D. Valentine, Jr., K. K. Johnson, W. Priester, R. C. Sun, K. Toth, and G. Saucy, *J. Org. Chem.* **45,** 3698–3703 (1980). D. Valentine, Jr., R. C. Sun, and K. Toth, *J. Org. Chem.* **45,** 3703–3707 (1980).

58. J. M. Brown and D. Parker, *J. Chem. Soc., Chem. Commun.* **1980,** 342-344 and *J. Org. Chem.* **47,** 2722-2730 (1982).
59. D. H. G. Crout, M. Lutstorf, P. J. Morgan, R. M. Adlington, J. E. Baldwin, and M. J. Crimmin, *J. Chem. Soc., Chem. Commun.* **1981,** 1175-1176.
60. J. M. Brown and R. G. Naik, *J. Chem. Soc., Chem. Commun.* **1982,** 348-350.
61. I. Ojima, T. Kogure, and N. Yoda, *J. Org. Chem.* **45,** 4728-4739 (1980).
62. D. Sinou, *Tetrahedron Lett.,* **22,** 2987-2990 (1981).
63. A. S. C. Chan, J. J. Pluth, and J. Halpern, *J. Am. Chem. Soc.* **102,** 5952-5954 (1980).
64. B. R. James and D. K. W. Wang, *Can. J. Chem.* **58,** 245-250 (1980).
65. R. H. Crabtree, *Acc. Chem. Res.* **12,** 331-338 (1979).
66. J. Vilím and J. Hetflejš, *Collect. Czech. Chem. Commun.* **43,** 122-133 (1978).
67. A. S. C. Chan, J. J. Pluth, and J. Halpern, *Inorg. Chim. Acta* **37,** L477-L479 (1979).
68. A. S. C. Chan and J. Halpern, *J. Am. Chem. Soc.* **102,** 838-840 (1980).
69. J. M. Brown and D. Parker, *Organometallics* **1,** 950-957 (1982).
70. J. M. Brown, L. R. Canning, A. J. Downs, and A. M. Forster, *J. Organomet. Chem.,* 1983, in press.
71. G. Descotes, D. Lafont, D. Sinou, J. M. Brown, P. A. Chaloner, and D. Parker, *Nouv. J. Chim.* **5,** 167-173 (1981).
72. R. G. Ball and N. C. Payne, *Inorg. Chem.* **16,** 1187-1191 (1977).
73. J. M. Townsend and J. F. Blount, *Inorg. Chem.* **20,** 269-271 (1981).
74. W. S. Knowles, B. D. Vineyard, M. J. Sabacky, and B. R. Stults, *Fundam. Res. Homogeneous Catal.,* edited by M. Tsutsui (Plenum Publishing, 1979), Vol. 3, pp. 531-548.
75. K. Toriumi, T. Ito, H. Takaya, T. Souchi, and R. Noroyi, *Acta Crystallogr.* **B38,** 807-812 (1982).
76. P. C. Chua, N. K. Roberts, B. Bosnich, S. J. Okrasinski, and J. Halpern, *J. Chem. Soc., Chem. Commun.* **1981,** 1278-1280.
77. C. A. Tolman, P. Z. Meakin, D. L. Lindner, and J. P. Jesson, *J. Am. Chem. Soc.* **96,** 2762-2774 (1974).
78. J. Halpern, T. Okamoto, and A. Zakhariev, *J. Mol. Catal.* **2,** 65-68 (1977). M. H. J. M. de Croon, P. F. M. T. van Nisselrooij, H. J. A. M. Kuipers, and J. W. E. Coenen, *J. Mol. Catal.* **4,** 325-335 (1978).
79. J. Halpern, D. P. Riley, A. S. C. Chan, and J. J. Pluth, *J. Am. Chem. Soc.* **99,** 8055-8057 (1977).
80. D. A. Slack, I. Greveling, and M. C. Baird, *Inorg. Chem.* **18,** 3125-3132 (1979).
81. J. M. Brown, P. A. Chaloner, A. G. Kent, B. A. Murrer, P. N. Nicholson, D. Parker, and P. J. Sidebottom, *J. Organomet. Chem.* **216,** 263-276 (1981).
82. B. R. Stults, R. M. Friedman, K. Koenig, and W. Knowles, *J. Am. Chem. Soc.* **103,** 3235-3237 (1981).
83. J. M. Brown and P. A. Chaloner, *J. Am. Chem. Soc.* **102,** 3040-3048 (1980).
84. G. A. Morris and R. A. Freeman, *J. Mag. Res.* **29,** 433-462 (1978).
85. J. M. Brown and P. A. Chaloner, *J. Chem. Soc., Chem. Commun.* **1979,** 613-615.
86. J. M. Brown, P. A. Chaloner, and G. A. Morris, *J. Chem. Soc., Chem. Commun.,* 1983, in press.
87. G. Descotes, D. Lafont, D. Sinou, J. M. Brown, P. A. Chaloner, and R. Glaser, *J. Chem. Soc., Chem. Commun.* **1979,** 611-613.
88. J. M. Brown and P. A. Chaloner, *J. Chem. Soc., Perkin II* **1982,** 711-719.
89. J. M. Brown and B. A. Murrer, *J. Chem. Soc., Perkin II* **1982,** 489-497.
90. J. M. Brown and P. A. Chaloner, *J. Chem. Soc., Chem. Commun.* **1980,** 344-346.
91. G. W. Kirby and J. Michael, *J. Chem. Soc., Perkin I* **1973,** 115-120.
92. F. H. Jardine, J. A. Osborn, and G. Wilkinson, *J. Chem. Soc.* **A 1967,** 1574-1579.

# 5

# Binuclear, Phosphine-Bridged Complexes: Progress and Prospects

## Alan L. Balch

### NOTATION

| | |
|---|---|
| Bu$^t$ | tertiary butyl |
| Cp | ($\eta^5$ = C$_5$H$_5$) |
| dam | bis(diphenylphosphino)arsine, Ph$_2$AsCH$_2$AsPh$_2$ |
| dba | dibenzylideneacetone, PhHC=CHC(O)CH=CHPh |
| dpm | bis(diphenylphosphino)methane, Ph$_2$PCH$_2$PPh$_2$ |
| Me | methyl |
| Ph | phenyl |
| Ph$_2$Ppy, $\overparen{P\ N}$ | 2(diphenylphosphino)pyridine |
| PNP | methylaminobis(difluorophosphine), F$_2$PN(CH$_3$)PF$_2$ |
| P$\frown$P | generalized diphosphine ligand |

## 1. INTRODUCTION

As the existence of this volume attests, transition metal complexes with phosphine ligands have proven to be useful and versatile catalysts for homogeneous reactions. Recently, binuclear metal complexes, particularly those with bridging phosphine ligands, have begun to attract interest because of their potential as catalysts and because of their novel structural

---

*Alan L. Balch* • Department of Chemistry, University of California, Davis, California 95616.

and reactive features. The interest in binuclear complexes arises because of anticipation that they should allow for increased versatility in catalyst design. Complexes with two (or even more) metal atoms can have several advantages over a catalyst containing only a single metal. Small molecules, particularly those like dinitrogen and nitriles which are notoriously difficult to reduce, may be more readily activated by attachment to several metal centers.[1] In catalysts containing two metal centers, one may act to bind the substrate, while the second acts to feed or remove electrons from the first site. The presence of two metal atoms may facilitate multi-electron redox reactions which could not be handled by only a single metal atom.[2]

There already exist examples of significant chemical reactions which demonstrate the unusual reaction characteristic resulting from the presence of two metal centers.[3] Here we examine two, neither of which involves phosphine bridging ligands. The autoxidation of triphenylphosphine catalyzed by four-coordinate iron(II) porphyrins in a noncoordinating solvent like toluene occurs by the mechanism shown in Figure 5.1.[4,5] Two iron(II) porphyrins bind dioxygen sequentially to give the peroxobridged intermediate $P$Fe(III)OOFe(III)$P$ ($P$ is a porphyrin dianion). This is inert to the substrate, triphenylphosphine, but when it spontaneously fragments it forms $P$Fe(IV)O, which is capable of transferring an oxygen atom to

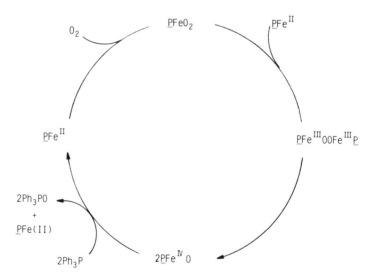

Figure 5.1. The cycle for the iron(II) porphyrin catalyzed oxidation of triphenyl phosphine to triphenylphosphine oxide. $P$ is a porphyrin dianion.

triphenylphosphine. In formal terms, two iron(II) centers are oxidized to iron(IV). This provides the four electrons necessary to break the oxygen–oxygen bond of dioxygen. This is a spontaneous reaction. No bridging ligand (other than the dioxygen) is necessary to facilitate the formation of PFeOOFeP.

A second example of a unique transformation in a bimetallic system is the observation of facile carbon-carbon bond making and breaking reactions in the binuclear complexes shown in reaction 1:[6]

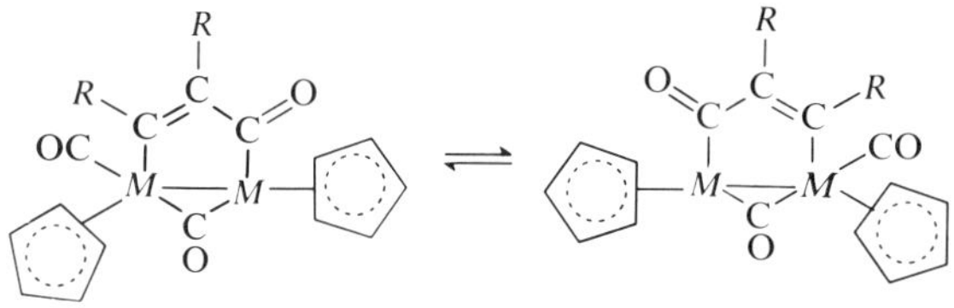

$$\text{(1)}$$

The complex involved is formed by photolysis of $Cp_2M_2(CO)_4$ ($M$ = Ru, Fe) in the presence of an acetylene. The fluxional process shown in reaction 1 occurs so rapidly that the two cyclopentadienyl resonances coalesce at 56°C, and at 90°C the $^{13}C$ resonances due to the ketonic and terminal carbonyl ligand are merged.

Along with the interest in binuclear complexes, there has been considerable development of the chemistry of metal carbonyl clusters. The analogy between metal clusters and metal surfaces as catalytic centers has been drawn.[7,8] A number of clusters have been shown to act as efficient catalysts or catalyst precursors.[9] The preparation of phosphine-bridged, polynuclear complexes can be seen as complementary to the synthesis of cluster catalysts. Whereas metal carbonyl clusters are held together primarily by metal–metal bonds, phosphine-bridged, polynuclear complexes have the phosphine bridges, as well as metal–metal bonds to maintain the integrity of the polynuclear unit. Thus, we find that selective metal–metal bond breaking (as opposed to cluster degradation) is relatively rare for carbonyl clusters. (An example exists in the oxidation of $[Os_{10}C(CO)_{24}]^{2-}$ by iodine to give $[Os_{10}C(CO)_{24}I]^-$ and $Os_{10}C(CO)_{24}I_2$.[10]) However, metal–metal bond breaking and formation is a fairly common reaction for phosphine-bridged binuclear complexes (*vide infra*). Additionally with phosphorus ligand-bridged complexes there is an added dimension of synthetic flexibility that occurs because the bridging ligand can now be tailor-made to meet certain requirements. For example, we shall encounter cases where bridging ligands have been constructed with different sites capable of binding different metal ions.

## 2. STRUCTURAL ASPECTS

A variety of phosphorus containing ligands have been utilized in the preparation of binuclear complexes. Among the most versatile are the diphosphines of the type $Ph_2P(CH_2)_nPPh_2$. These are readily available (at the present time from commercial sources) and are easily handled, air-stable solids.[11,12] The length of the methylene chain can be varied to give different separations between the donor atoms and hence different spacings between metal centers. Many of the diarsine analogs of these ligands are known and form similar complexes. The formation of binuclear complexes with these and other ligand types is not immediately assured, for all of them have the potential for forming monodentate and chelating complexes, as well as diphosphine-bridged complexes. Of particular relevance in this context are the observations of Sanger, who showed that the reaction of $Ph_2P(CH_2)_nPPh_2$ with $[Rh(CO)_2(\mu\text{-}Cl)]_2$ produced binuclear complexes $(\mu\text{-}Ph_2P(CH_2)_nPPh_2)_2Rh_2(CO)_2Cl_2$ for $n = 1, 3, 4$ (and this has later been extended to $n = 5$ and 6), whereas the chelated monomer $Ph_2P(CH_2)_2PPh_2Rh(CO)Cl$ was formed when $n = 2$.[13] While these results are not entirely general for all metal ions,[14] they serve to emphasize the fact that bis(diphenylphosphino)ethane and, of course, cis-bis(diphenylphosphino)ethylene generally have strong preferences to bind as chelating rather than bridging ligands. Additionally, it should be noted that the presence of a diphosphine ligand in a binuclear complex does not imply that the phosphine acts as a bridge. For example, $(Bu^t_2P(CH_2)_3PBu^t_2)_2Pt_2$ is a binuclear species with an unbridged metal–metal bond and a chelating diphosphine bound to each metal.[15,16] Related to these diphenyl phosphino ligands are analogs bearing other groups as the terminal phosphorus substituents. Those in the $t$-butyl groups have been extensively investigated by Shaw and his group.[17]

The aminobis(difluorophosphines), $RN(PF_2)_2$, form another group of bridging ligands that has received considerable attention.[18] These do not have the capacity for variable spacing of the phosphino groups, but other relatives, $F_2PXPF_2$ where $X = O, S, (CH_2)_n$, could offer that possibility. The presence of the fluoro substituents makes this class of ligand a good $\pi$-acceptor. Consequently, these ligands stabilize low-metal oxidation states. Moreover, the small size of the fluoro groups makes such ligands considerably less bulky than the ligands bearing diphenylphosphino groups.

Ligands of the type $R_2OPXPOR_2$ where $X$ can be O, S, NR, $(CH_2)_n$ also are potential candidates for bridging ligands. Of these, tetraethyldiphosphite, $(EtO)_2POP(OEt)_2$,[19–22] has received the most attention, while some work with $(RO)_2PNR'P(OR)_2$ has been reported.[23,24] Tetraethyldiphosphite has been observed to function as either a bridging or a terminal ligand. It is not known to function as a chelating ligand, and it has been speculated that the favored P–O–P bond angle of *ca.* 120–150° precludes chelation.[22]

Phosphide anions, $R_2P^-$ are also known to function as bridging ligands.[25,26] In this regard, they resemble the more familiar bridging halide and chalconide ligands. We have, somewhat arbitrarily, excluded these from this review.

Within this group, the number of bridging phosphorus ligands in a binuclear complex can vary from one to four. However, the most extensive studies have been carried out on species containing two bridging phosphines. Two phosphine ligands are sufficient to hold the two metal centers together while allowing the reactants to have reasonably good access to the metal centers. They also accommodate a significant variation in the metal–metal separation even for a single ligand. This aspect of flexibility appears to be a significant benefit of these diphosphine ligands.

Three and four bridging phosphorus ligands have only been observed with the relatively small ligand $CH_3N(PF_2)_2$, (PNP). With four such ligands, the two metal ions are effectively enclosed in a cage. This can be seen in Figure 5.2 where the structure of $(\mu\text{-PNP})_4Mo_2Cl_2$ is shown.[27] The congestion about the two metal ions would appear to offer a significant impediment

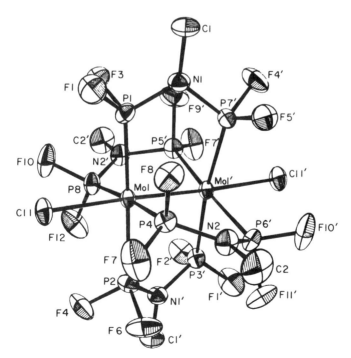

Figure 5.2. The structure of $(\mu\text{-PNP})_4Mo_2Cl_2$ taken from Reference 27 by permission.

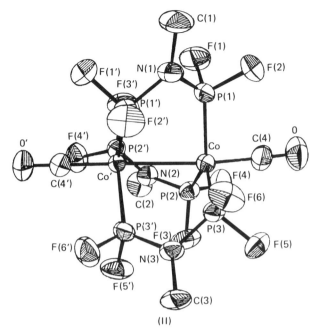

Figure 5.3. The structure of $(\mu\text{-PNP})_3\text{Co}_2(\text{CO})_2$ taken from Reference 28 by permission.

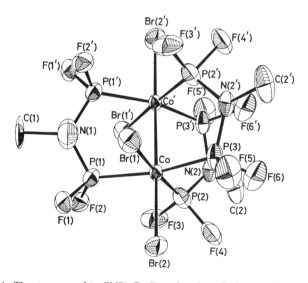

Figure 5.4. The structure of $(\mu\text{-PNP})_3\text{Co}_2\text{Br}_4$ taken from Reference 29 by permission.

to catalytic activity where substrate needs to bind to the metal ions involved, unless bonding only to the axial sites was required. Compounds with three bridging phosphorus ligands suffer from similar although less extensive crowding. The structure of $(\mu\text{-PNP})_3\text{Co}_2(\text{CO})_2$ is shown in Figure 5.3.[28] The related complex $(\mu\text{-PNP})_3\text{Co}_2(\text{PF}_2\text{NHMe})_2$ shows similar structural features.[28] Oxidation of the latter complex with bromine yields $(\mu\text{-PNP})_3\text{Co}_2\text{Br}_4$ which is shown in Figure 5.4.[29] Here, the three phosphorus ligands have moved together to accomodate two additional ligands. In addition to retaining the three PNP ligands, the complex also retains its Co–Co bond.

With two phosphorus ligands acting to support the binuclear compounds, there is an extensive body of data available. Generally, these bridging ligands adopt *trans* arrangements at each metal ion as shown in *A* of Figure 5.5. In most cases, this probably occurs for steric reasons, since it places the phosphorus atoms and their bulky substituents as far apart as possible. However, there are now cases known where the phosphine ligands bridges are both *cis* on each metal (*C* of Figure 5.5) and where the phosphine bridges are *cis* on one metal and *trans* on another (*B* of Figure 5.5). Examples of types *B* and *C* are the platinum methyl complexes $(\mu\text{-dpm})_2\text{Pt}_2(\text{CH}_3)_4$ shown in Figure 5.6[30] and $(\mu\text{-dpm})_2\text{Pt}_2(\text{CH}_3)_3^+$ shown in Figure 5.7.[31]

Within the large group of molecules containing *trans* phosphorus ligands on both metals, several groups of ligand arrangements can be identified. These are shown in Figure 5.8. The face-to-face dimers have the greatest flexibility in that there are no other connections between the metals. With long chain diphosphines, this can lead to the formation of

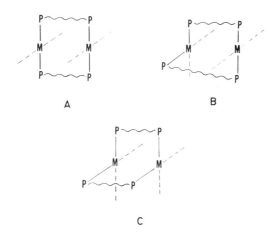

Figure 5.5. Orientation of two diphosphine ligands about two metal ions.

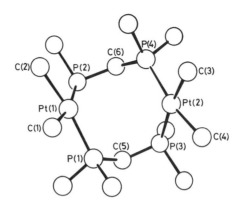

Figure 5.6. The structure of $(\mu\text{-dpm})_2\text{Pt}_2(\text{CH}_3)_4$ taken from Reference 30.

complexes in which the two metal centers are so remote that they are effectively isolated from one another. An example is given by the structure of $(\mu\text{-Ph}_2\text{P}(\text{CH}_2)_2\text{O}(\text{CH}_2)_2\text{PPh}_2)_2\text{Rh}_2(\text{CO})_2\text{Cl}_2$ as shown in Figure 5.9.[32] The two square planar rhodium ions are separated by 7.95 Å. Similar structures with large gaps between metal ions are seen for the

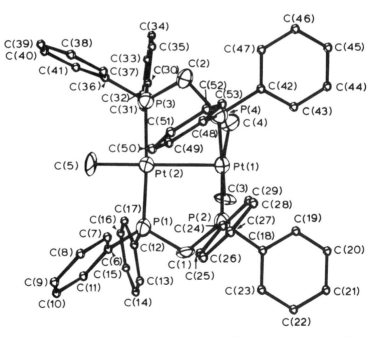

Figure 5.7. The structure of $(\mu\text{-dpm})_2\text{Pt}_2(\text{CH}_3)_3^+$ taken from Reference 31.

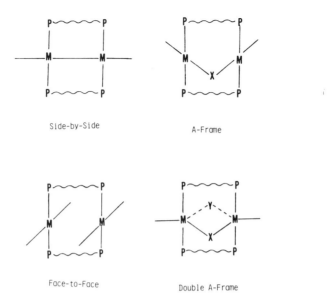

Figure 5.8. Idealized geometries of binuclear-*trans*-diphosphine complexes.

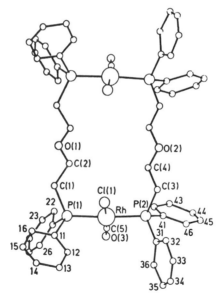

Figure 5.9. The structure of $(\mu\text{-}Ph_2P(CH_2)_2O(CH_2)_2PPh_2)Rh_2(CO)_2Cl_2$ taken from Reference 32 by permission.

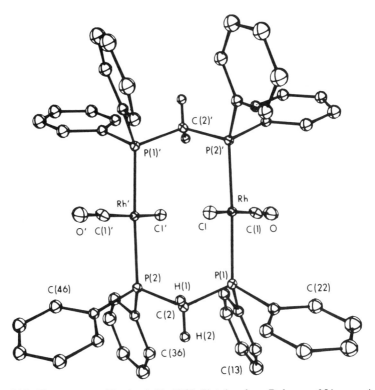

Figure 5.10. The structure of $(\mu\text{-dpm})_2\text{Rh}_2(\text{CO})_2\text{Cl}_2$ taken from Reference 35 by permission.

26-membered rings in $(\mu\text{-Bu}_2^t\text{P}(\text{CH}_2)_{10}\text{PBu}_2^t)_2\text{Rh}_2(\text{CO})_2\text{Cl}_2$ and $(\mu\text{-Bu}_2^t\text{P}(\text{CH}_2)_{10}\text{PBu}_2^t)_2\text{Pd}_2\text{Cl}_4$[33] and in the 16-membered ring in $(\mu\text{-Bu}_2^t\text{P}(\text{CH}_2)_2\text{CMeH}(\text{CH}_2)_2\text{PBu}_2^t)_2\text{Pd}_2\text{Cl}_4$.[34] With fewer methylene (or analogous groups) in the chain, the two metal centers can be brought into very close proximity without necessarily entailing the formation of a classical metal–metal bond. For example, in $(\mu\text{-dpm})_2\text{Rh}_2(\text{CO})_2\text{Cl}_2$ shown in Figure 5.10 the Rh···Rh separation is only 3.24 Å.[35] The structurally similar complex $(\mu\text{-dam})_2\text{Rh}_2(\text{CO})_2\text{Cl}_2$ shows the plasticity of molecules of this sort which are not supported by a metal–metal bond. The structures of two different crystalline forms of this complex show significant variation in the Rh···Rh separation. In the unsolvated form, this separation is 3.396(1),[36] while in the dichloromethane solvate it is 3.236(2).[37]

Other variations of the face-to-face structure can be identified. An additional ligand can be present on one or both metals to give five-coordinate species. The two metals may actually be directly bonded. Such is the case for $(\mu\text{-dpm})_2\text{Mo}_2\text{Cl}_4$ where a Mo–Mo quadruple bond is present, and the Mo–Mo separation is only 2.138 Å.[38]

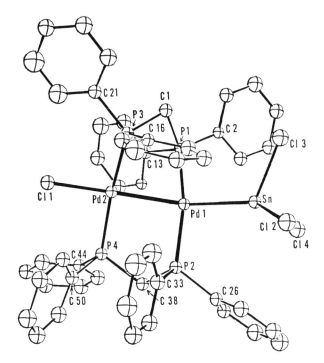

Figure 5.11. The structure of $(\mu\text{-dpm})_2\text{Pd}_2\text{Cl}(\text{SnCl}_3)$ taken from Reference 39 by permission.

The side-by-side dimers have a direct metal linkage and the square planar coordination about each metal is completed by the presence of two terminal axial ligands. A typical example is $(\mu\text{-dpm})_2\text{Pd}_2\text{Cl}(\text{SnCl}_3)$ shown in Figure 5.11.[39]

The dimeric structures can also contain additional bridging ligands. These include the A-frame structure, as well as the doubly-bridged A-frame structure. Figure 5.12 shows the structure of $(\mu\text{-dpm})_2\text{Pd}_2(\mu\text{-SO}_2)\text{Cl}_2$, a typical A-frame,[40] Figure 5.13 shows the structure of a doubly-bridged A-frame.[41] Of some interest is the observation that the structure of these A-frames is amazingly consistent from molecule to molecule. The $CP_2M_2X$ rings generally adopt boat, rather than chair, conformations, and the orientation of phenyl rings appears to be common to many of the complexes.

While complexes of the long-chain diphosphines are largely limited to the face-to-face arrangement of metal centers, the smaller bridging ligands, particularly dpm and dam, have the capacity of stabilizing all four structural types and of allowing for the interconversion between these various forms.[42] The flexibility of these ligands in spanning a variety of metal–metal separations is considerable. These range from cases where

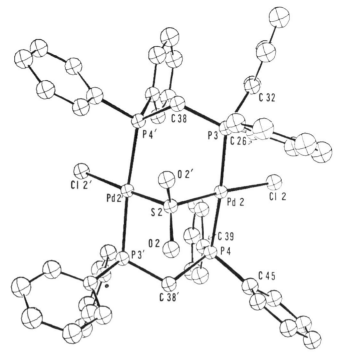

Figure 5.12. The structure of $(\mu\text{-dpm})_2\text{Pd}_2(\mu\text{-SO}_2)\text{Cl}_2$ taken from Refenrence 40 by permission.

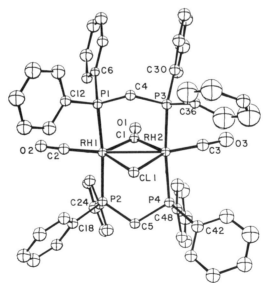

Figure 5.13. The structure of $(\mu\text{-dpm})_2\text{Rh}_2(\mu\text{-CO})(\mu\text{-Cl})(\text{CO})_2^+$ taken from Reference 41 by permission.

Table 1. Structural Parameters for Bis(diphenylphosphino)methane (dpm) Bridged Complexes

| Compound | $M \cdots M$ (Å) | $P \cdots P$ (Å) | P-C-P (°) | Reference |
|---|---|---|---|---|
| $(\mu\text{-dpm})_2\text{Mo}_2\text{Cl}_4$ | 2.138(2) | | 106.2(3) | 38 |
| $(\mu\text{-dpm})_2\text{Mo}_2(\text{NCS})_4$ | 2.167(3) | | 107(1) | 38 |
| $(\mu\text{-dpm})_2\text{Mn}_2(\mu\text{-CO})(\text{CO})_4$ | 2.934(6) | 3.05(1) | 115(2) | 43 |
| | | 3.03(1) | 111(2) | |
| $(\mu\text{-dpm})_2\text{Mn}_2(\mu\text{-CNC}_6\text{H}_4\text{Me-}p)(\text{CO})_4$ | 2.936(2) | 3.129(3) | 115(1) | 44 |
| | | 3.103(3) | 113(1) | |
| $(\mu\text{-dpm})_2\text{Re}_2\text{Cl}_5$ | 2.263(1) | | 107.5(11) | 45 |
| | | | 106.3(10) | |
| $(\mu\text{-dpm})_2\text{Re}_2(\text{CO})_6(\text{H})_2$ | 2.893(2) | | | 46 |
| $(\mu\text{-dpm})\text{Fe}_2(\mu\text{-CO})(\text{CO})_6$ | 2.709(2) | 2.999(4) | 109(1) | 47 |
| $(\mu\text{-dpm})\text{Ru}_3(\text{CO})_7(\mu_3\text{-PPh})(\mu\text{-CHPPh}_2)$ | 2.843 | | | 48 |
| $(\mu\text{-dpm})\text{Co}_2(\mu\text{-C}_2\text{Ph}_2)(\text{CO})_4$ | 2.459(2) | 2.946(4) | 106(1) | 49 |
| $(\mu\text{-dpm})_2\text{Rh}_4(\text{CO})_8$ | 2.696(1) | | 113.8(7) | 50 |
| | 2.671(1) | | 111.0(7) | |
| $[(\mu\text{-dpm})_2\text{Rh}(\mu\text{-H})(\mu\text{-CO})(\text{CO})_2]^+$ | 2.732(2) | | | 51 |
| $(\mu\text{-dpm})_2\text{Rh}_2(\mu\text{-C}_4\text{F}_6)\text{Cl}_2$ | 2.744(9) | 2.959(3) | 108.4(4) | 52 |
| | | 2.981(3) | 110.1(4) | |
| $(\mu\text{-dpm})_2\text{Rh}_2(\mu\text{-CO})\text{Br}_2$ | 2.7566(9) | 3.026 | 111.0(4) | 53 |
| | | 2.992 | 109.9(4) | |
| $(\mu\text{-dpm})_2\text{Rh}_2(\mu\text{-SO}_2)\text{Cl}_2$ | 2.7838(8) | 3.018(3) | 111.6(4) | 54 |
| | | 3.012(3) | 110.5(5) | |
| $(\mu\text{-dpm})_2\text{Rh}_2(\mu\text{-C}_2\text{S}_4)\text{Cl}_2(\text{CO})$ | 2.811(3) | 3.016(9) | 109(1) | 55 |
| | | 3.094(11) | 113(1) | |
| $[(\mu\text{-dpm})_2\text{Rh}_2(\mu\text{-CO})(\mu\text{-Cl})(\text{CO})_2]$ | 2.810(3) | 3.024(10) | 110(1) | 41 |
| | | 3.096(10) | 115(1) | |
| $[(\mu\text{-dpm})_2\text{Rh}_2(\mu\text{-CO})(\mu\text{-Cl})(\text{CO})_2]^+$ | 2.838(1) | 3.060(4) | 114.2(6) | 56 |
| $(\mu\text{-dpm})_2\text{Rh}_2(\mu\text{-S})(\text{CO})_2$ | 3.154(2) | | | 57 |
| $[(\mu\text{-dpm})_2\text{Rh}_2(\mu\text{-Cl})(\text{CO})_2]^+$ | 3.1520(8) | 3.088(2) | 114.7(3) | 58 |
| $(\mu\text{-dpm})_2\text{Rh}_2(\text{CO})_2\text{Cl}_2$ | 3.2386(5) | 3.130(1) | 116.8(2) | 35 |
| $(\mu\text{-dpm})_2\text{Rh}_2(\mu\text{-CO})(\mu\text{-C}_2\{\text{Co}_2\text{Me}\}_2)\text{Cl}_2$ | 3.3542(9) | | | 59 |
| $(\mu\text{-dpm})_2\text{Ir}_2(\mu\text{-S})(\mu\text{-CO})(\text{CO})_2$ | 2.843(2) | 3.045(6) | 11.6(7) | 60 |
| | | 3.028(6) | 111.7(7) | |
| $(\mu\text{-dpm})_2\text{Pd}_2(\text{SnCl}_3)\text{Cl}$ | 2.644(2) | 2880(6) | 102(1) | 39 |
| | | 2.960(6) | 108(1) | |

Table 1. (Continued)

| Compound | $M \cdots M$ (Å) | $P \cdots P$ (Å) | P-C-P (°) | Reference |
|---|---|---|---|---|
| $(\mu\text{-dpm})_2\text{Pd}_2\text{Br}_2$ | 2.699 | | 105(2)<br>103(2) | 61 |
| $[(\mu\text{-dpm})_2\text{Pd}_2(\mu\text{-I})(\text{CH}_3)\text{I}]^+$ | 2.976(6) | 3.15(2)<br>3.12(2) | | 62 |
| $[(\mu\text{-dpm})_2\text{Pd}_2(\mu\text{-CNCH}_3)(\text{CNCH}_3)_2]^+$ | 3.215(2) | 3.066(6) | 114(1) | 63 |
| $(\mu\text{-dpm})_2\text{Pd}_2(\mu\text{-S})\text{Cl}_2$ | 3.258(2) | 3.059(7)<br>3.508(7) | 112.0(10)<br>113.6(9) | 40 |
| $(\mu\text{-dpm})_2\text{Pd}_2(\mu\text{-SO}_2)\text{Cl}_2$ | 3.221(2)<br>3.383(2) | 3.111(5)<br>3.170(5) | 119(1)<br>115(1) | 40 |
| $(\mu\text{-dpm})_2\text{Pd}(\mu\text{-C}_4\text{F}_6)\text{Cl}_2$ | 3.492(1) | 3.115(4)<br>3.113(4) | 114.6(6)<br>115.4(6) | 64 |
| $(\mu\text{-dpm})_2\text{Pt}_2(\text{CO})\text{Cl}^+$ | 2.620 | | 107<br>106 | 65 |
| $(\mu\text{-dpm})_2\text{Pt}_2\text{Cl}_2$ | 2.652 | | | 66 |
| $(\mu\text{-dpm})_2\text{Pt}_2\text{H}(\text{dpm})$ | 2.770(2) | | | 67 |
| $(\mu\text{-dpm})_2\text{Pt}_2(\text{CH}_3)_3^+$ | 2.769(1) | | 115.0<br>111.5 | 31 |
| $(\mu\text{-dpm})_2\text{Pt}_2(\mu\text{-H})(\text{CH}_3)_2^+$ | 2.933(1) | | | 68 |
| $(\mu\text{-dpm})_2\text{Pt}_2(\mu\text{-CS}_2)\text{Cl}_2$ | 3.094 | | | 69 |
| $(\mu\text{-dpm})_2\text{Pt}_2(\text{CH}_3)_4$ | 4.36 | | | 30 |
| $(\mu\text{-dpm})_2\text{Cu}_3\text{Cl}_2$ | 2.678(6) | | | 70 |
| $(\mu\text{-dpm})\text{Cu}_3(\mu\text{-dpm-H})_2$ | 2.836 | | | 71 |
| $(\mu\text{-dpm})_2\text{Cu}_4\text{I}_4$ | 2.908(4) | | 106(1) | 72 |
| $(\mu\text{-dpm})_2\text{Cu}_4\text{Br}_4$ | 2.939(6) | | 112(1) | 72 |
| $(\mu\text{-dpm})_2\text{Cu}_4\text{Cl}_4$ | 3.110(6) | | 114(1) | 73 |
| $(\mu\text{-dpm})_2\text{Cu}_3\text{I}_3$ | 2.916(4) | | | 74 |
| $[(\mu\text{-dpm})_2\text{Ag}_2]^{2+}$ | 3.041 | | 117.5 | 75 |
| $(\mu\text{-dpm})_2\text{Ag}_3\text{Br}_3$ | 3.303 | | 110 | 76 |
| $(\mu\text{-dpm})_2\text{Au}_2\text{Cl}_2$ | 2.962(1) | | 114.2(8) | 77 |
| $(\mu\text{-dpm})\text{Au}_2\text{Cl}_2$ | 3.341 | | 116 | 78 |

Table 2. Structural Parameters for Bis(diphenylarsino)methane (dam) Bridged Complexes

| Compound | $M \cdots M$ (Å) | As$\cdots$As (Å) | As-C-As(°) | Reference |
|---|---|---|---|---|
| $(\mu\text{-dam})_2W_2(\mu\text{-}C_2Me_2)(CO)_5Br$ | 2.937(1) | | | 79 |
| $(\mu\text{-dam})_2Re_2(\mu\text{-Cl})_2(CO)_4$ | 3.806(1) | 3.320(2) | 115(1) | 80 |
| | 3.814(1) | 3.394(2) | 113(1) | |
| $(\mu\text{-dam})_2Ru_3(\mu_3\text{-O})(CO)_6$ | 2.750(2) | | | 81 |
| | 2.723(2) | | | |
| $(\mu\text{-dam})_2Co_2(\mu\text{-}C_2Ph_2)(CO)_2$ | 2.518(4) | 3.165(3) | 105(1) | 82 |
| | | 3.246(3) | 108(1) | |
| $(\mu\text{-dam})_2Rh_2(CO)_2Cl_2$ | 3.396(1) | 3.298(1) | 114(1) | 36 |
| $(\mu\text{-dam})_2Rh_2(CO)_2Cl_2\cdot CH_2Cl_2$ | 3.236(2) | 3.272(2) | 114.4(5) | 37 |
| $(\mu\text{-dam})_2Pd_2(\mu\text{-CO})Cl_2$ | 3.274(8) | 3.22(1) | 111(3) | 83 |
| $(\mu\text{-dam})_2Pt_2(\mu\text{-CO})Cl_2$ | 3.162(4) | 3.22(1) | 107(3) | 84 |

there are no metal–metal bonds but other bridging ligands that require rather large metal–metal separations to cases where there are metal–metal multiple bonds. Some idea of the degree of flexibility can be gained from Tables 1, 2, and 3, which show some structural parameters for a variety of complexes of dpm, dam, and PNP, respectively. Factors responsible for

Table 3. Structural Parameters for Methylaminobis(difluorophosphine) (PNP) Bridged Complexes

| Compound | $M \cdots M$ (Å) | P-N-P (°) | Reference |
|---|---|---|---|
| $(\mu\text{-PNP})_4Mo_2Cl_2$ | 2.457(1) | 111.9(2), 118.4(2) | 27 |
| | | 114.8(2), 109.2(2) | |
| $(\mu\text{-PNP})_2Fe_2(\mu\text{-CO})(CO)_4$ | 2.661(1) | | 85 |
| $(\mu\text{-PNP})_3Fe_2(\mu\text{-PF}_2)(CO)$ $(PF_2NMe)$ | 2.725(2) | | 86 |
| $(\mu\text{-PNP})_2Fe_2(\mu\text{-PF}_2)$ $(\mu\text{-PF}_2NMe)Cp_2$ | 3.646 | | 87 |
| $(\mu\text{-PNP})_2Fe_2(CO)_6$ | 3.90 | | 85 |
| $(\mu\text{-PNP})_3Co_2(CO)_2$ | 2.716(1) | | 28 |
| $(\mu\text{-PNP})_3Co_2Br_4$ | 2.717(5) | | 29 |
| $(\mu\text{-PNP})_3Co_2(PF_2NMeH)_2$ | 2.769(1) | | 28 |

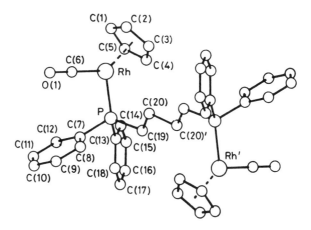

Figure 5.14. The structure of ($\mu$-Ph$_2$P(Ch$_2$)$_4$PPh$_2$)Rh$_2$(CO)$_2$Cp$_2$ taken from Reference 88 by permission.

the flexibility of the dpm ligand include changes in the P–C–P angle,[41] changes in the conformation of the ring formed by the metal(s) and the ligand,[47] and changes in the P–$M$–$M$–P torsional angles.[76]

Compounds with only one phosphorus ligand linking the two metal centers are also known. These have the least constraints placed on the interaction between the two metals. One such molecule, ($\mu$-Ph$_2$P(CH$_2$)$_4$PPh$_2$)Rh$_2$Cp$_2$(CO)$_2$, is shown in Figure 5.14.[88] Other species with a single phosphine linking two essentially independent metal centers include ($\mu$-Ph$_2$P(CH$_2$)$_4$PPh$_2$){Rh(CO)(Ph$_2$P(CH$_2$)$_4$PPh$_2$)}$_2^{2+}$,[89] cis-($\mu$-Ph$_2$PCHCHPPh$_2$)Au$_2$Cl$_2$,[90] and $\mu$-dpm{Ru(bipy)$_2$Cl}$_2^{2+}$.[91]

Recently the problem of constructing bifunctional phosphine ligands with two different binding sites, each capable of bonding a different type of metal ion, has received attention. Two such ligands are 2(diphenylphosphino)pyridine,[92] (Ph$_2$Ppy) **1**, and [dimethyl(diphenylphosphinomethyl)silyl]cyclopentadienyl lithium, **2**.[93]

The behavior of Ph$_2$Ppy as a bridging ligand will be described in a succeeding section.

## 3. NUCLEAR MAGNETIC RESONANCE SPECTROSCOPIC PROBES

NMR spectroscopy has been useful in structural characterization of binuclear phosphine-bridged complexes. The most extensive data available involved complexes involving two *trans* dpm ligands, and this brief section will be devoted to a review of some general features which have been found.

The $^1$H-NMR spectra of dpm ligands are informative in establishing the mode of coordination. In dpm itself the methylene resonance occurs as a triplet with $^3J(PH) = 1.5$ Hz at 2.8 ppm.[94] When dpm acts as a chelating ligand, the methylene resonance is shifted downfield to 4–4.5 ppm, and the resonance appears as a triplet with $J(P, H) \cong 10$ Hz.[94,95]

For bridged complexes, two types of spectra are observed. For complexes like $(\mu$-dpm$)_2$Pd$_2$Cl$_2$, where rapid conformational change in the CP$_2$Pd$_2$ ring can render the two methylene protons equivalent, the methylene resonance appears as a 1:4:6:4:1 quintet with the apparent $J(P, H) = 5$ Hz.[95,40] The quintet pattern arises due to virtual coupling of the phosphorus nuclei, where the *trans* P–Pd–P coupling is expected to be $\sim 300$ Hz. While such a complex belongs to the $AA'XX'X''X'''A''A'''$ ($A = H, X = P$) spin system, the spectra can be more readily analyzed as an $AXX'X''X'''$ system since $J(A, A'')$ and $J(A, A''')$ are expected to be very small. Some spectra simulations are shown in Figure 5.15. The values chosen for $J(2, 3)$, $J(4, 5)$, $J(2, 5)$, and $J(2, 6)$ are in the normally observed range. Once the *trans* P–P coupling constant exceeds 100 Hz, the spectral pattern appears as a symmetrical quintet and the apparent $J(P, H)$ is the average of $^1J(P, H)$ and $^3J(P, H)$.

For molecular A-frames, no motion of the CP$_2$M$_2$ ring can render the two methylene protons of a dpm ligand equivalent. However, for most A-frames the two dpm ligands are generally equivalent by symmetry. Hence, there the methylene group of dpm appears as an AB quartet with $J(H, H)$ in the range of 12–15 Hz. Frequently, the chemical shift difference between the two types of protons reaches 0.5 to 1 ppm. Superimposed on this AB quartet is the phosphorus–proton coupling which further splits each resonance into a quintet. Again virtual coupling is responsible for the splitting pattern. As a result of these considerations, $^1$H-NMR spectra are capable of differentiating between binuclear A-frames and other dimers in which the two methylene protons are equivalent.

Proton-decoupled[31] P-NMR spectroscopy has also been useful in structural characterization. The most frequent use has involved attempts to simulate the observed and often complex spectral patterns on the basis of an assumed structure and reasonable values for the appropriate coupling constants and chemical shifts. It is the coupling constants, more than the chemical shifts, that are structurally useful, and a number of coupling constants are now known that are unique to binuclear complexes. The

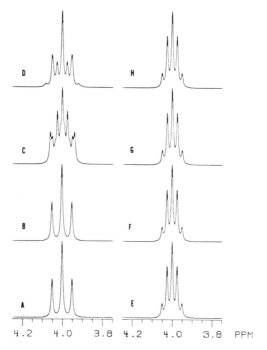

Figure 5.15. Simulations of the $^1$H-NMR spectra of methylene protons of a model ($\mu$-dpm)$_2$M$_2$ complex with trans-phosphine ligands. The model is based on a $AX_2Y_2$ spin system. (Taken from the Ph.D. thesis of C. T. Hunt, University of California Davis 1981.) Sets, A-H, of coupling constants, $J_{ij}$, were entered into a simulation routine and produced spectra A-H. These spectra show the phenomenon of virtual coupling (see text).

| Sets | $J_{ij}$ (Hz) | 12 | 13 | 14 | 15 | 23 | 24 | 25 | 34 | 38 | 45 |
|---|---|---|---|---|---|---|---|---|---|---|---|
| A | | 10 | 10 | 0 | 0 | 0 | 0 | 0 | 0 | 0 | 0 |
| B | | 10 | 10 | 0 | 0 | 30 | 0 | 0 | 0 | 0 | 30 |
| C | | 10 | 10 | 0 | 0 | 30 | 0 | 5 | 5 | 0 | 30 |
| D | | 10 | 10 | 0 | 0 | 30 | 5 | 5 | 5 | 5 | 30 |
| E | | 10 | 10 | 0 | 0 | 30 | 150 | 5 | 5 | 150 | 30 |
| F | | 10 | 10 | 0 | 0 | 30 | 300 | 5 | 5 | 300 | 30 |
| G | | 5 | 5 | 5 | 5 | 30 | 300 | 5 | 5 | 300 | 30 |
| H | | 20 | 20 | −10 | −10 | 30 | 300 | 5 | 5 | 300 | 30 |

question arises as to how much structural information is contained within these parameters.

Most of the available information can be discussed by reference to structure **3**:

$$P_5-M_7 \begin{array}{c} P_1 \\ | \\ | \\ P_2 \end{array} \begin{array}{c} P_3 \\ | \\ | \\ P_4 \end{array} M_8-P_6$$

**3**

which encompasses a $(\mu\text{-dpm})_2M_2$ unit that may contain two additional ligands $P_5$ and $P_6$ (but these need not be present). Coupling constants between *trans* ligands [i.e., $J(1, 2)$] are known to be large (*ca.* 300 Hz) for Group VIII metal ions, while *cis* coupling constants [i.e., $J(1, 5)$] are smaller (*ca.* 10–52 Hz). Major questions to be considered are: do the other coupling constants which are more or less unique to binuclear species reflect the presence and strength of metal–metal bonding, and do they reflect the geometry of the unit?

Pertinent data are given in Table 4. Some attempts at analysis of this data have been made.[101,108] The P–P coupling constant in an unsymmetrical diphosphine $R_2PCH_2PR'_2$ is generally about 110 Hz. (It is 115 Hz for $Ph_2PCH_2PMePh$.)[109] In the complexes, $J(1, 3)$ is always of smaller magnitude but otherwise offers little structural information. It has been proposed that $J(1, 4)$ is sensitive to the presence of a metal–metal bond. Low magnitudes of $J(1, 4)$ correspond to cases where there is no metal–metal bond, and larger magnitudes occur in compounds which are expected to have metal–metal single bonds.[108] Within Table 4, $(\mu\text{-dpm})_2Pt_2(\mu\text{-Cl})H_2^+$ appears to be the exception, since this compound should have no Pt–Pt bond. Although originally proposed for platinum compounds, this relationship also is maintained for all of the palladium and rhodium compounds for which it can be tested. Similarly, it has been proposed that, for platinum, the two-bond coupling $J(1, 8)$ is also effected by metal–metal bonding.[108] For those compounds which have Pt–Pt single bonds, $J(1, 8)$ is negative with values of *ca.*-100 Hz, while with compounds lacking a Pt–Pt bond $J(1, 8)$ is positive. The two-bond P–Pt coupling constant also appears to give geometric data. When the P–Pt–Pt unit is linear, $^2J(Pt, P)$ is positive and falls in the range 150–750 Hz, while for cases where the P–Pt–Pt angle is ~90°, $^2J(Pt, P)$ is negative and smaller in magnitude (85–120 Hz).[107]

Two technical problems in dealing with $^{31}$P-NMR spectra of binuclear complexes are worthy of special notice. Selective population transfer

Table 4. Coupling Constants from the $^{31}P\{^1H\}$ NMR Spectra of $(\mu\text{-}dpm)_2M_2$ Complexes[a]

| Compound | Coupling constants (Hz) | | | | | | | | | | Reference |
| --- | --- | --- | --- | --- | --- | --- | --- | --- | --- | --- | --- |
| | 12 | 13 | 14 | 15 | 16 | 17 | 18 | 56 | 57 | 58 | 78 | |
| $(\mu\text{-}dpm)_2Rh_2(\mu\text{-}Cl)(\mu\text{-}CO)(CO)_2^+$ | ~355 | 68.4 | 21.5 | — | — | 93.4 | 0.8 | — | — | — | 0.1 | 96 |
| $(\mu\text{-}dpm)_2Rh_2(\mu\text{-}Br)(\mu\text{-}CO)(CO)_2^+$ | ~363 | 71.6 | 22.0 | — | — | 92.6 | 0.8 | — | — | — | 0 | 96 |
| $(\mu\text{-}dpm)_2Rh_2(\mu\text{-}I)(\mu\text{-}CO)(CO)_2^+$ | ~548 | 74.3 | 19.9 | — | — | 90.4 | 2.3 | — | — | — | 0.2 | 96 |
| $(\mu\text{-}dpm)_2Rh_2(\mu\text{-}Cl)(\mu SO_2)(CO)_2^+$ | ~350 | 46.9 | 17.5 | — | — | 90.8 | 0.5 | — | — | — | 0 | 96 |
| $(\mu\text{-}dpm)_2Rh_2(\mu\text{-}Cl)(\mu\text{-}CO)(CNBu^t)_2^+$ | ~390 | 82.4 | 21.9 | — | — | 106.9 | 0.8 | — | — | — | 0 | 97 |
| $(\mu\text{-}dpm)_2Rh_2(\mu\text{-}CO)(CNBu^t)_4^{+2}$ | ~327 | 78.2 | 21.9 | — | — | 88.9 | −1.6 | — | — | — | 0 | 97 |
| $(\mu\text{-}dpm)_2Rh_2(\mu\text{-}SO_2)(CNBu^t)_4^{+2}$ | ~329 | 58.6 | 27.2 | — | — | 91.0 | −1.2 | — | — | — | 0 | 97 |
| $(\mu\text{-}dpm)_2Pd_2(\mu\text{-}CHCH_3)I_2$ | ±363.6 | ±104.5 ±44.9 | ±5.5 | — | — | | | — | — | — | | 98 |
| $(\mu\text{-}dpm)_2Pd_2\{\mu\text{-}HC_2(Co_2H)\}Cl_2$ | | 44.7 | 0.8 | — | — | | | — | — | — | | 99 |
| $(\mu\text{-}dppm)_2Pd_2\{\mu\text{-}HC_2Co_2Me\}Cl_2$ | | 48.5 | 1.0 | — | — | | | — | — | — | | 99 |
| $(\mu\text{-}dpm)_2Pd_2ClBr$ | | 43.2 | 35.5 | | | | | | | | | 100 |
| $(\mu\text{-}dpm)_2Pd_2BrI$ | | 39 | 39 | | | | | | | | | 100 |
| $(\mu\text{-}dpm)_2Pd_2ClI$ | | 39 | 39 | | | | | | | | | 100 |
| $(\mu\text{-}dpm)_2Pt_2(\mu\text{-}Cl)Me_2^+$ | | 33 | 3 | | | 3030 | 30 | | | | | 101 |
| $(\mu\text{-}dpm)_2Pt_2(\mu\text{-}Cl)Ph_2^+$ | | 29 | ⩽3 | | | 3037 | 39 | | | | | 102 |
| $(\mu\text{-}dpm)_2Pt_2(\mu\text{-}Cl)(COPh)_2^+$ | | 28 | ⩽3 | | | 3354 | 48 | | | | | 102 |
| $(\mu\text{-}dpm)_2Pt_2(\mu\text{-}CH_2)Cl_2$ | | 23 | 6 | | | 3388 | 73 | | | | | 103 |

| Complex | | | | | | | | | Ref |
|---|---|---|---|---|---|---|---|---|---|
| $(\mu\text{-dpm})_2Pt_2(\mu\text{-S})Cl_2$ | 43 | 0 | | 3588 | 273 | | | | 103 |
| $(\mu\text{-dpm})_2Pt_2(\mu\text{-Cl})H_2^+$ | 52.9 | 15.5 | | 2665 | 6.8 | | | | 104 |
| $(\mu\text{-dpm})_2Pt_2(\mu\text{-CH})Cl_2^+$ | 52.0 | 21.2 | | 2317 | −40.4 | | | | 104 |
| $(\mu\text{-dpm})_2Pt_2(\mu\text{-H})H_2^+$ | 56.0 | 18.2 | | 2769 | 16.6 | | | | 104 |
| $(\mu\text{-dpm})_2Pt_2(CO)Cl^+$ | 46 | 43 | | 2706 | 62 | | | | 105 |
| | | | | 2573 | 92 | | | | |
| $(\mu\text{-dpm})_2Pt_2Cl_2$ | 62.5 | 26.4 | | 2936 | −136 | | | | 103 |
| $(\mu\text{-dpm})_2Pt_2(PPhMe_2)_2^{2+}$ | 48 | 22 | 24 | 2840 | ≈100 | 1938 | 650 | | 106 |
| $(\mu\text{-dpm})_2Pt_2(PPh_2Me)_2^{2+}$ | 50 | 20 | 22 | 2870 | −120 | 1914 | 750 | | 106 |
| $(\mu\text{-dpm})_2Pt_2(C_5H_2N)_2^{2+}$ | 58 | 30 | | 2828 | −113 | | | 4940 | 106 |
| $(\mu\text{-dpm})_2Pt_2(CO)_2^{2+}$ | 46 | 33 | | 2390 | −96 | | | 4810 | 106 |
| $(\mu\text{-dpm})Pt_2(\mu\text{-S})(PPh_3)_2$ | 37.9 | 14.7 | 7.6 | +3537.5 | −108.6 | 174.8 | 3191.6 | 1231.2 | 107 |

[a] The numbering system is:

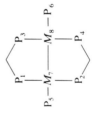

experiments have been shown to be valuable in analyzing the complex spectra that result when several isotopomers (isotope isomers) are present.[107] This is particularly common with platinum where the spin one-half isotope, $^{195}$Pt, is 33.4% naturally abundant. Secondly, the temperature dependence of $^{31}$P chemical shifts can produce unusual effects in the spectra. We have encountered cases in $(\mu\text{-dpm})_2\text{Pd}_2XY$ ($X$ = Cl, $Y$ = Br or I) where the chemical shifts of the two chemically distinct phosphorus atoms approach each other with increasing temperature and, in one case, they actually merge and then move away again.[100]

## 4. REACTIVITY PATTERNS FOR BRIDGED, BINUCLEAR COMPLEXES

Behind the interest manifested in binuclear complexes stands the expectation that they will display fundamentally new modes of reactivity. Naturally, we also expect that they can show the patterns of reactivity known for mononuclear complexes. These include Lewis base association/dissociation, Lewis acid association/dissociation, ligand migration (insertion/deinsertion), oxidative addition/reductive elimination, and oxidative ligand coupling/reductive ligand uncoupling, as well as electron-transfer.[110,111] While these reaction patterns do occur with binuclear complexes, it is the new reactions that are emphasized in this section. Not surprisingly, the bulk of these involve either the formation or breaking of metal–metal bonds or the formation or loss of bridging groups other than the phosphorus ligands. Such reactions generally require that the two metal ions involved get relatively close to one another. As a consequence, ligands which allow the metal to maintain their distance from one another do not induce particularly interesting modifications. Conversely, those ligands which force the metals into close association, also induce the more interesting new chemistry.

The dinuclear rhodium(I) complex $[\mu\text{-Ph}_2\text{P}(\text{CH}_2)_4\text{PPh}_2]\text{Rh}_2(\text{Cp})_2(\text{CO})_2$ offers a good example of a case where the two metal centers are so loosely linked that each acts essentially independently of the other to give reactions which are characterisitc of $\text{R}_3\text{PRh}(\text{Cp})\text{CO}$.[88] For example, $(\text{PhMe}_2\text{P})\text{Rh}(\text{Cp})\text{CO}$ reacts with bromine to form the salt $[(\text{PhMe}_2\text{P})\text{Rh}(\text{Cp})(\text{CO})\text{Br}]\text{Br}$, and $[\mu\text{-Ph}_2\text{P}(\text{CH}_2)_4\text{PPh}_2](\text{Rh}(\text{Cp})(\text{CO}))_2$ reacts similarly to produce $\{[\mu\text{-Ph}_2\text{P}(\text{CH}_2)_4\text{PPh}_2](\text{RhCp}(\text{CO})\text{Br})_2\}\text{Br}_2$.

The doubly bridged complex $[\mu\text{-Ph}_2\text{P}(\text{CH}_2)_3\text{PPh}_2]_2\text{Rh}_2(\text{CO})_2\text{Cl}_2$ shows a different sort of behavior.[112] The mononuclear Rh(I) complex $(\text{Ph}_3\text{P})_2\text{Rh}(\text{CO})\text{Cl}$ adds one molecule of sulfur dioxide or tetracyanoethylene to give the adducts $(\text{Ph}_3\text{P})_2\text{Rh}(\text{CO})\text{Cl}(\text{SO}_2)$ and

$(Ph_3P)_2Rh(CO)Cl(C_6N_4)$, respectively. While both rhodium ions initially reside in identical environments in $[\mu\text{-}Ph_2P(CH_2)_3PPh_2]_2Rh_2(CO)_2Cl_2$, the addition of one mole of substrate to form $[\mu\text{-}Ph_2P(CH_2)_3PPh_2]_2(Rh(CO)Cl)(Rh(CO)Cl(SO_2))$ or $[\mu\text{-}Ph_2P(CH_2)_3PPh_2]_2(Rh(CO)Cl)(Rh(CO)Cl(C_6N_4))$ effectively alters the reactivity of the second (unreacted) site so that further addition of substrate does not occur even when the substrate is in one-hundredfold excess. The details of the origin of this effect are not known. There is no evidence that the substrate has bound to the one rhodium in any usual fashion, and it certainly does not bridge the two sites. However, it does appear probable that there is a steric interaction between the coordination spheres of the two metal ions that prohibits the four-coordinate rhodium from reacting.

In the remainder of this section, we deal with cases in which the two metal centers are in much closer proximity.

## 4.1. Insertions of Small Molecules into Metal–Metal Bonds

The metal–metal bonds in the dimeric Pd(I) and Pt(I) complexes, $(\mu\text{-}dpm)_2M_2X_2$ ($X$ = Cl, Br, I, $N_3$), show an unusual propensity to insert small molecules. The overall reaction converts a side-by-side dimer to an A-frame as shown in Equation 2 and results in a substantial increase in the metal–metal separation:

$$X-M\underset{P\sim\sim\sim P}{\overset{P\sim\sim\sim P}{|\phantom{M}|}}M-X + A \longrightarrow \underset{P\sim\sim\sim P}{\overset{P\sim\sim\sim P}{X\diagdown M\underset{A}{\phantom{M}}M\diagup X}} \qquad (2)$$

The metal–metal separations in $(\mu\text{-}dpm)_2M_2X_2$ are about 2.6 Å. Such distances are comparable to those in unbridged Pd(I) and Pt(I) dimers (e.g., $Pd_2(CNCH_3)_6{}^{2+}$, 2.5310(9);[113] $Pd_3(CNCH_3)_8(PPh_3)_2{}^{2+}$, 2.5921(5);[114] and $Pt_2(CO)_2CL_4{}^{2-}$, 2.584(2).[115] On the other hand, the products of insertion have metal–metal distances in excess of 3.1 Å, which puts them outside the range of full metal–metal bonds. In electron-counting terms, both the reactant complex and the product have 16 electrons per metal, and both exhibit nearly square planar coordination. A theoretical, molecular orbital interpretation of the bonding in these A-frames is available.[116]

The range of substances which insert into the metal–metal bonds of $(\mu\text{-}dpm)_2Pd_2Cl_2$ and $(\mu\text{-}dpm)_2Pt_2Cl_2$ is summarized in Figure 5.16. Note that both a one-atom bridge (CO,[42,63,84] CNR,[42] S,[40,103] $SO_2$,[40,103] $CH_2$,[103] $N_2Ph^+$[117] insertion) and a two-atom bridge (acetylene,[64,99] $CS_2$[69] insertion) can be placed between the two metal atoms. In the case

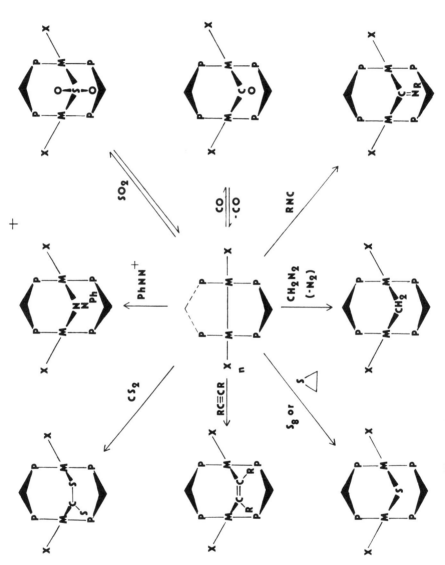

Figure 5.16. Reaction chemistry of $(\mu\text{-dpm})_2Pd_2Cl_2$ and $(\mu\text{-dpm})_2Pt_2Cl_2$.

of acetylene insertion, this two-atom insertion places the two metal ions (Pd) 3.492 Å apart, the largest separation yet found in a dpm bridged complex. A number of other small molecules including dioxygen, dinitrogen, and nitrile are unreactive toward $(\mu\text{-dpm})_2\text{Pd}_2\text{Cl}_2$.

In some cases the insertion reactions are reversible. This is the case with the palladium complexes for carbon monoxide and sulfur dioxide. With acetylenes, preliminary studies indicate that the insertion products can be made to eliminate the acetylene on photolysis. The remaining A-frames shown in Figure 5.16 are formed irreversibly.

Qualitatively, the chemical behavior of palladium and platinum in $(\mu\text{-dpm})_2M_2\text{Cl}_2$ is similar. Both metals appear to undergo similar reactions. Detailed comparisons of equilibrium constants, reaction rates, and structural parameters are generally not possible because relevant data are lacking. The one place where a comparison can be made involves the solid state structures of $(\mu\text{-dam})_2\text{Pd}_2(\mu\text{-CO})\text{Cl}_2$[83] and $(\mu\text{-dam})_2\text{Pt}_2(\mu\text{-CO})\text{Cl}_2$.[84] These molecules have closely similar A-frame structures which are compared in Figure 5.17. Despite the similarities, the metal–metal separation and $M\text{-C(O)-}M$ angle are noticeably lower for $(\mu\text{-dam})_2\text{Pt}_2(\mu\text{-CO})\text{Cl}_2$ than for $(\mu\text{-dam})_2\text{Pd}_2(\mu\text{-CO})\text{Cl}_2$. Since it is known that these parameters are subject to external pressures on the molecule, the differences could be simply due to packing forces. However, the observation of decidedly different carbonyl stretching frequencies for $(\mu\text{-dam})_2\text{Pd}_2(\mu\text{-CO})\text{Cl}_2)$ (1723 cm$^{-1}$) and for $(\mu\text{-dam})_2\text{Pt}_2(\mu\text{-CO})\text{Cl}_2$ (1638 cm$^{-1}$) suggests that the structural variations may have an electronic origin. The difference in carbonyl stretching frequencies of these two molecules originally misled some workers into suggesting that $(\mu\text{-dam})_2\text{Pt}_2(\mu\text{-CO})\text{Cl}_2$ had a four-electron donating carbonyl group similar to that of $(\mu\text{-dpm})_2\text{Mn}_2(\mu\text{-CO})(\text{CO})_4$.[84]

The effects of the terminal ligands $X$ on these insertion reactions have received little attention. In the only systematic study available, it has been shown that the equilibrium for carbon monoxide insertion is dependent on

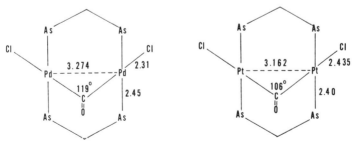

Figure 5.17. A comparison of structural features of $(\mu\text{dam})_2\text{Pd}_2(\mu\text{-Co})\text{Cl}_2$ and $(\mu\text{-dam})_2\text{Pt}_2(\mu\text{-CO})\text{Cl}_2$.

$X$, with the stability of the carbon monoxide adduct increasing with $X$ in the order $I < Br < N_3 \sim Cl$.[42] In the A-frame products themselves, the anionic $X$ ligands can frequently be replaced by neutral ligands such as isocyanides, monodentate phosphines, or pyridine to give cationic derivatives.

While the observations on insertion in metal–metal bonds are most extensive for Pd(I) and Pt(I) dimers, other rather isolated examples of similar behavior are known. Although structural data are lacking $(dpm)_2Rh_2(CO)_2(\nu_{CO}, 1915 \text{ cm}^{-1})$, which is isoelectronic with $(\mu\text{-}dpm)_2Pd_2Cl_2$, adds one equivalent of carbon monoxide to form $(dpm)_2Rh_2(\mu\text{-}CO)(CO)_2(\nu_{CO}, 1940, 1920, 1835 \text{ cm}^{-1})$.[51] The product is valence isoelectronic and, presumably, structurally similar to $(\mu\text{-}dam)_2Pd_2(\mu\text{-}CO)CL_2$. Structural data are available for another example of insertion of carbon monoxide into a Rh–Rh bond. Carbon monoxide reacts with $(\mu\text{-}dpm)_2RH_2(C_2R_2)Cl_2(Rh\text{-}Rh, 2.744(9) \text{ Å})$[52] according to reaction 3:

$$\begin{array}{c}\text{Cl}\underset{\underset{P^R\overset{|}{\underset{}{C}}=\underset{}{\overset{|}{C}}R_P}{\overset{|}{\underset{Rh}{\overbrace{P\qquad P}}}}}{\overset{}{}}\text{Rh}\underset{}{\overset{|}{\underset{}{}}}\text{Cl} + CO \rightarrow \text{Cl}\underset{\underset{P^R\overset{|}{\underset{}{C}}=\underset{}{\overset{|}{C}}R_P}{\overset{|}{\underset{Rh}{\overbrace{P\quad O\quad P}}}}}{\overset{|}{\underset{}{C}}}\text{Rh}\underset{}{\overset{|}{\underset{}{}}}\text{Cl}\end{array} \quad (3)$$

Although an additional bridging group has been added, the Rh–Rh separation has increased to $\sim 3.35$ Å in the product and the Rh–C–Rh angle is $\sim 116$.[59] This reaction type is not limited to phosphine-bridged complexes. Consider reaction 4:

$$\begin{array}{c}\text{Ph}_3P\diagdown\overset{O}{\diagup}\diagdown PPh_3\\ \overset{}{\underset{ON}{\diagup}}\text{Ir}-\text{Ir}\overset{}{\underset{NO}{\diagdown}}\end{array} + PhN_2^+ \rightarrow \begin{array}{c}\text{Ph}_3P\diagdown\overset{O}{\diagup}\diagdown PRh_3^+\\ \overset{}{\underset{ON}{\diagup}}\text{Ir}\underset{\underset{\underset{Ph}{N}}{\overset{\|}{N}}}{}\text{Ir}\overset{}{\underset{NO}{\diagdown}}\end{array} \quad (4)$$

Here, addition of diazonium cation results in an increase in the Ir–Ir separation from 2.555 Å (a reasonable value for a single bond) to 3.063(6) Å.[118,119]

The reactivity of the unique bridging ligand in these A-frames has received some attention, and further exploration in this area may be anticipated. The reactivity of the unique carbonyl groups with wide $M$–$C$–$M$ angles and large $M\cdots M$ separations and a comparison with the more standard bridging carbonyl ligands and organic ketones bear exploration. Such studies with $(\mu\text{-}dpm)_2Pd_2(\mu\text{-}CO)X_2$ have been thwarted because of the ease with which carbon monoxide dissociates. The oxidation of bridging sulfide to bridging sulfur dioxide,[40] the alkylation and protonation of sulfide,[57] and the protonation of a bridging methylene ligand in ($\mu$-dpm)$_2$Pd$_2$($\mu$-CH$_2$)I$_2$ to give a terminal methyl group in {($\mu$-dpm)$_2$Pd$_2$($\mu$-I)(CH$_3$)I}$^+$ [62] are unusual transformations of these bridges which have been documented. Additionally, the catalytic activity of ($\mu$-dpm)$_2$Pd$_2$Cl$_2$ in the cyclotrimerization of dimethylacetylene dicarboxylate (*vide infra*) involves reactivity of the unique bridging site.[64,99]

### 4.2. Introduction of a Bridging Ligand with Contraction of the Metal–Metal Separation

Some 16-electron, four-coordinate dimers add a ligand to a bridging site according to equation 5:

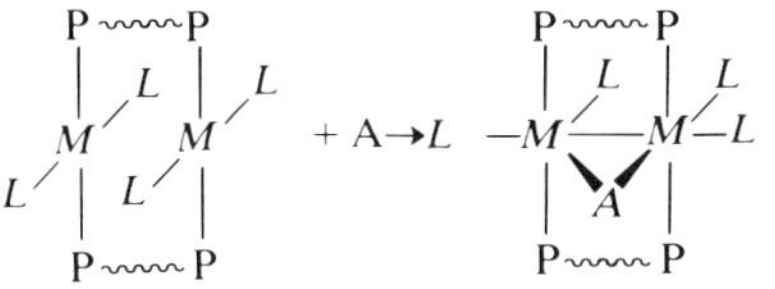

$$\begin{array}{c}\text{P}\sim\sim\text{P}\\|\ \ L\ \ |\ L\\M\ \ \ \ \ M\\L\ |\ \ L\ |\\\text{P}\sim\sim\text{P}\end{array}\ \ +\ \text{A}\rightarrow L\ \ \begin{array}{c}\text{P}\sim\sim\text{P}\\|\ \ L\ \ |\ L\\-M\rule[0.5ex]{1em}{0.4pt}M-L\\|\ \ \vee\ |\\ \ \ A\ \\\text{P}\sim\sim\text{P}\end{array} \quad (5)$$

(In structural changes this reaction is the opposite of the previous class.) The reactant can either be a face-to-face dimer, as shown, or an A-frame. The metal–metal separation is decreased in forming the product to less than 2.9 Å. In electron-counting terms it is convenient to write a metal–metal bond in the product. This gives the product an 18-electron count at each metal and readily accounts for the contraction of the metal–metal distance.

Some examples of the reaction involving the addition of carbon monoxide and sulfur dioxide to rhodium(I) and iridium(I) A-frames are shown in equation 6:[41,60,120,121]

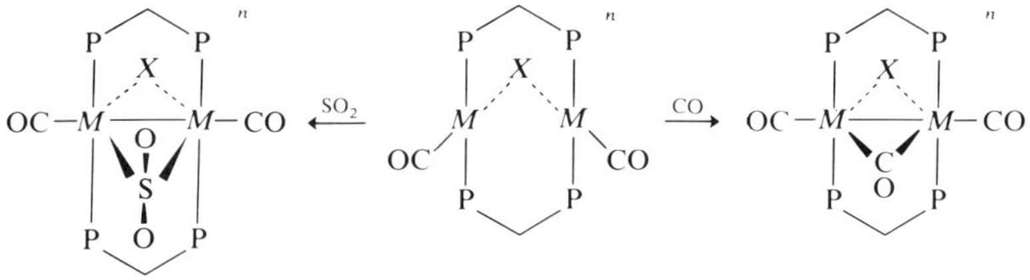

$$\text{(equation 6)}$$

| | | |
|---|---|---|
| $M$ = Rh, $X$ = Cl, $n$ = +1 | $M$ = Rh, $X$ = Cl, $n$ = +1 | $M$ = Rh, $X$ = Cl, $n$ = +1 |
| $M$ = Rh, $X$ = S, $n$ = 0 | $M$ = Rh, $X$ = S, $n$ = 0 | $M$ = Ir, $X$ = S, $n$ = 0 |
| $M$ = Ir, $X$ = S, $n$ = 0 | $M$ = Ir, $X$ = S, $n$ = 0 | |

The structural data for ($\mu$-dpm)$_2$RH$_2$($\mu$-Cl)(CO)$_2$$^+$ (Rh···Rh, 3.1520(8) Å)[58] and for ($\mu$-dpm)$_2$RH$_2$($\mu$-Cl)($\mu$-CO)(CO)$_2$$^+$ (Rh-Rh, 2.838(1) Å)[41] clearly establish the structural changes that are probably typical of the other examples in Equation 6. Another set of related reactions is summarized in Equation 7:[97]

$$I \quad \underset{P}{\overset{P}{\underset{|}{\bigcap}}} \underset{L}{\overset{L}{\underset{|}{Rh}}} \underset{L}{\overset{L}{\underset{|}{Rh}}} \underset{P}{\overset{P}{\bigcap}} \Bigg]^{2+} + CO \rightleftharpoons \Bigg[ \underset{P}{\overset{P}{\underset{|}{\bigcap}}} \underset{L}{\overset{L}{\underset{|}{Rh}}} \underset{\underset{O}{C}}{-} \underset{L}{\overset{L}{\underset{|}{Rh}}} \underset{P}{\overset{P}{\bigcap}} \Bigg]^{2+}$$

$$\Big\updownarrow SO_2 \qquad L = t\text{-}C_4H_9NC$$

$$\Bigg[ \underset{P}{\overset{P}{\underset{|}{\bigcap}}} \underset{L}{\overset{L}{\underset{|}{Rh}}} \underset{\underset{\underset{O}{S}}{O}}{-} \underset{L}{\overset{L}{\underset{|}{Rh}}} \underset{P}{\overset{P}{\bigcap}} \Bigg]^{2+}$$

(7)

These involve addition to face-to-face Rh(I) dimers. Detailed structural characterization of the product is not as yet available, but it should prove interesting. The electronic absorption spectra of the face-to-face reactants show that there is significant steric interaction between the $t$-butyl groups in the reactant.[122,123] Consequently, ligand addition to an already crowded molecule is somewhat surprising. It appears likely that the isocyanide substituents must bend away from one another in the product.

Reactant ligands for this series of reactions appear to be limited so far to carbon monoxide and sulfur dioxide. It would be interesting to learn what other potential ligands (isocyanides, acetylenes, etc.) can also be added to the A-frame and face-to-face reactants in this fashion.

The addition of carbon monoxide to $(\mu\text{-dpm})_2Rh_2(\mu\text{-Cl})(CO)_2^+$ (Equation 6) is one of the few reactions of these binuclear complexes where any mechanistic observations have been made.[120] By following the introduction of $^{13}CO$ into this complex by infrared and $^{13}C$-*NMR* spectroscopy, Mague and Sanger[120] have shown that the entering labeled carbon monoxide initially resides in a terminal position. They propose the mechanism shown in Equation 8 and specifically suggest initial attack of the incoming ligand on a site which is exo to the final bridging site. In support of this mechanism, it is reported that $t$-butyl isocyanide reacts with $(\mu\text{-dpm})_2Rh_2(\mu\text{-Cl})(CO)_2^+$ to form $(\mu\text{-dpm})_2Rh_2(\mu\text{-Cl})(\mu\text{-CO})(CO)$-$(t\text{-BuNC})$. On the other hand, addition of sulfur dioxide in reaction 6, as well as the additions of carbon monoxide and sulfur dioxide to

($\mu$-dpm)$_2$Rh$_2$($t$-BuNC)$_4^{2+}$, appears to place the added molecule in the bridging site with no evidence as yet presented for exo attack.

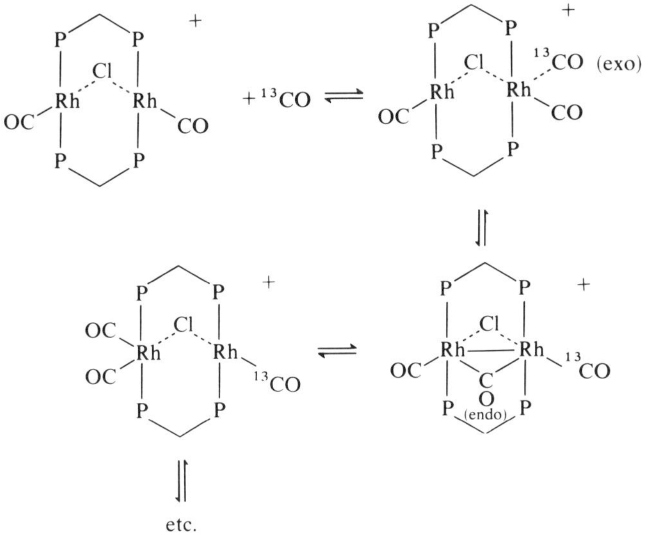

$$\text{(8)}$$

### 4.3. Loss of Carbon Monoxide Accompanied by Carbon Monoxide Bridge Formation

Extrusion of carbon monoxide, as shown in reaction 9, has been seen from a few diphosphine bridged metal complexes. This sort of reaction is also found when simple metal carbonyls condense to form polynuclear clusters (e.g., the conversion of Fe(CO)$_5$ to Fe$_2$(CO)$_9$). In electron-counting terms, all of the examples available involve loss of carbon monoxide from two 18-electron centers. If a metal–metal bond is accepted as part of the

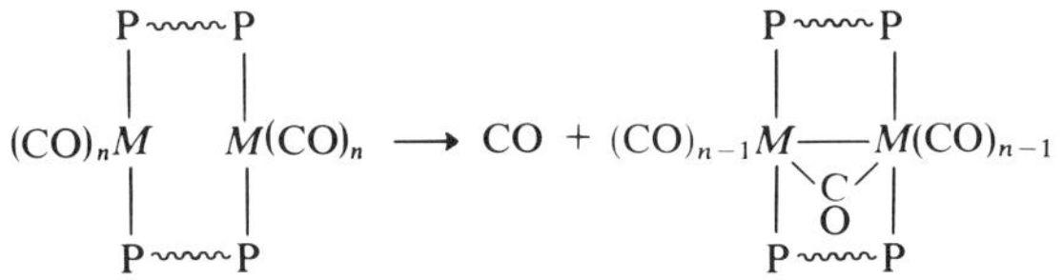

$$\text{(9)}$$

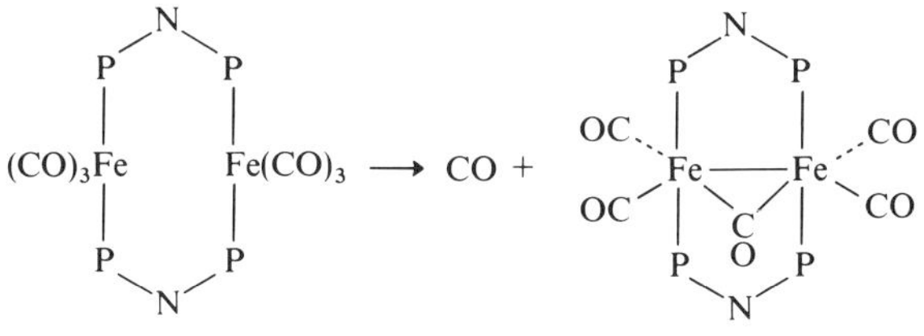

$$\text{(10)}$$

product, then an 18-electron count is readily established there as well. Examples of this behavior are shown in reactions 10,[85] 11,[124] and 12:[121]

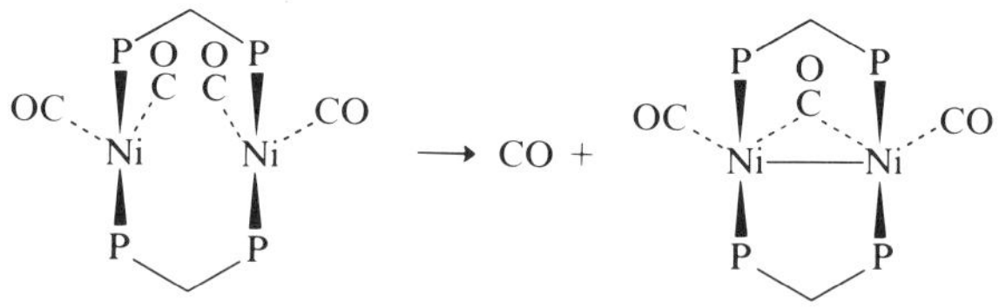

$$(NC)(OC)_2Rh\underset{P\phantom{xxx}P}{\overset{P\phantom{xxx}P}{|\phantom{xxx}|}}Rh(CO)_2CN \rightarrow CO + (NC)(OC)Rh\underset{P\phantom{x}O\phantom{x}P}{\overset{P\phantom{x}\phantom{x}P}{|\phantom{x}C\phantom{x}|}}Rh(CO)(CN) \quad (12)$$

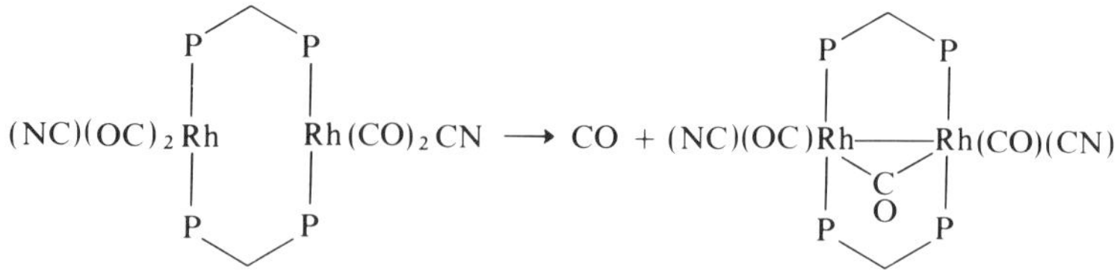

The first example shows the drastic shortening of the iron–iron separation that accompanies carbon monoxide loss. The Fe⋯Fe separation decreases from 3.90 Å in the reactant to 2.661(1) in the product.[85] The product of reaction 11 has a Ni–Ni separation of only 2.508 Å. The geometric details for the other compounds in these reactions are not known. The loss of carbon monoxide is reversible for the nickel complex.

In the reactions in this and the two proceeding sections, a number of compounds with bridging carbonyl ligands are formed. In these, the metal–metal separation can vary significantly. As can be seen from Table 5, there is no simple correlation between the carbonyl stretching frequency of these molecules and the metal–metal distance. Thus, while infrared spectroscopy

Table 5. Properties of Complexes with Bridging Carbonyl Ligands

| Compound | $M \cdots M$ distance (Å) | $\nu(CO)$ cm$^{-1}$ | Reference |
|---|---|---|---|
| $(\mu\text{-dpm})_2Pd_2(\mu\text{-CO})Cl_2$ | — | 1705 | 42 |
| $(\mu\text{-dam})_2Pd_2(\mu\text{-CO})Cl_2$ | 3.274(8) | 1723 | 42, 83 |
| $(\mu\text{-dam})_2Pt_2(\mu\text{-CO})Cl_2$ | 3.162(4) | 1638 | 84 |
| $[(\mu\text{-dpm})_2Rh_2(\mu\text{-CO})(\mu\text{-Cl})(CO)_2]^+$ | 2.838(1) | 1865 | 41, 56 |
| $(\mu\text{-dpm})_2Rh_2(\mu\text{-CO})Br_2$ | 2.7566(8) | 1745 | 53 |
| $(\mu\text{-dpm})_2Rh_2(\mu\text{-CO})(\mu\text{-C}_2\{CO_2Me\}_2)Cl_2$ | 3.3542(9) | ~1700 | 59 |
| $(\mu\text{-dpm})_2Ir_2(\mu\text{-CO})(\mu\text{-S})(CO)_2$ | 2.843(2) | 1760 | 60 |
| $(\mu\text{-Ph}_2Ppy)_2Rh_2(\mu\text{-CO})Cl_2$ | 2.612(1) | 1797 | 92 |

## 4.4. Two-Center, Two-Fragment Oxidative Addition

A number of low-valent, diphosphine bridged dimers undergo oxidative addition of conventional substrates, such as the halogens according to reaction 13.

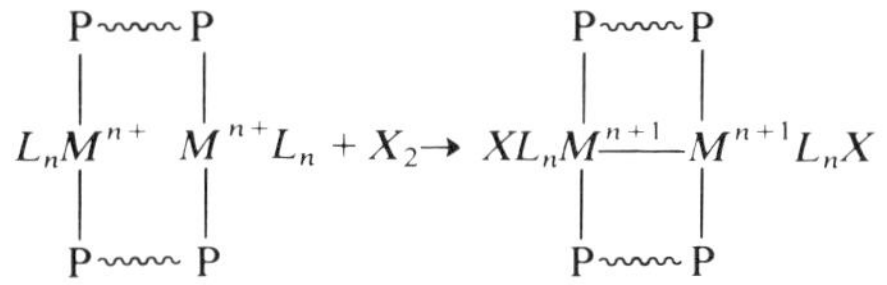

$$L_nM^{n+} \quad M^{n+}L_n + X_2 \rightarrow XL_nM^{n+1}\text{—}M^{n+1}L_nX \quad (13)$$
(with bridging P~~~P groups above and below each metal)

In this case the stoichiometry of oxidant uptake is one-half mole of $X_2$ per metal ion rather than the one mole per metal ion that is characteristic of the monomeric analogs with the same metal ions. As a result, the metal ion is oxidized (formally) by one electron rather than by two electrons, and a metal–metal bond is formed.

This reaction has been characterized for a wide variety of Rh(I) complexes. The behavior of $(\mu\text{-dpm})_2\text{Rh}_2(\text{CN}R)_4^{2+}$ shown in reaction 14 is typical.[122] Oxidants capable of performing this reaction include chlorine, bromine, iodine, and trifluoromethyl disulfide. However, dihydrogen does not add to these Rh(I) compounds. The dimeric Rh(II) products are quite stable. Treating these with excess oxidant does not result in the formation of the more usual Rh(III) complexes. However, some oxidants, including diphenyl disulfide, pentafluorophenyl disulfide, and diphynel diselenide, react with $(\mu\text{-dpm})_2\text{Rh}_2(\text{CN}R)_4^{2+}$ to form Rh(III) products, $(\text{dpm})\text{Rh}(\text{CN}R)_2(X)_2^{2+}$, exclusively.[123] The difference in behavior of these oxidants has not been explained.

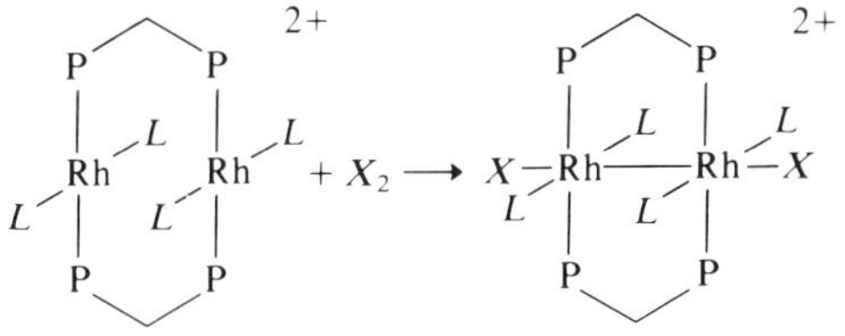

$$\begin{array}{c}[L_2\text{Rh}(\mu\text{-dpm})_2\text{Rh}L_2]^{2+} + X_2 \rightarrow [XL_2\text{Rh}(\mu\text{-dpm})_2\text{Rh}L_2X]^{2+}\end{array} \quad (14)$$

These oxidation reactions are readily monitored by following isocyanide stretching vibrations by infrared spectroscopy. The energy of this vibration increases by about 40 cm$^{-1}$ for oxidation of Rh(I) to Rh(II). For comparison, the oxidation of mononuclear Rh(I) complexes to Rh(III) species results in an 80 cm$^{-1}$ increase in the energy of the isocyanide stretching vibration.

Similar oxidative addition reactions of halogens with the neutral complexes $(\mu\text{-dpm})_2\text{Rh}_2(\text{CO})_2\text{Cl}_2$ have been reported to form $(\mu\text{-dpm})_2\text{Rh}_2(\text{CO})_2\text{Cl}_2X_2$ and $(\mu\text{-dam})_2\text{Rh}_2(\text{CO})_2\text{Cl}_2X_2$.[122]

Other ligands are also capable of sustaining similar oxidative behavior. Among these the bridged isocyanide complexes of Gray and co-workers offer the widest range of examples. The bridging isocyanide is most frequently 1,3-disocyanopropane (bridge) and the reactions of $(\text{bridge})_4\text{Rh}_2^{2+}$ are summarized by reaction 15:[125]

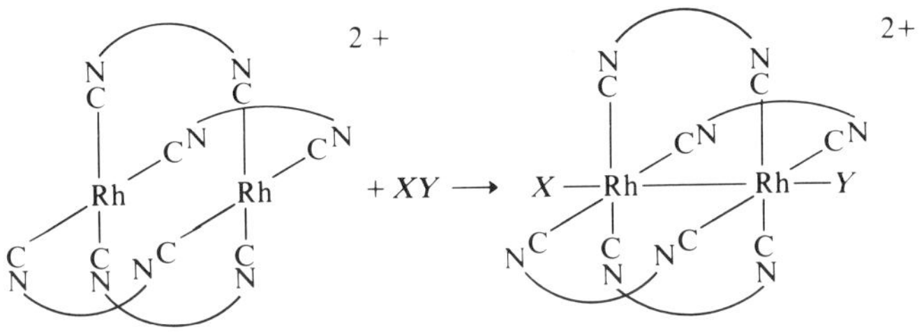

$$(\text{bridge})_4\text{Rh}_2^{2+} + XY \rightarrow X-\text{Rh}-\text{Rh}-Y \quad (15)$$

Here $XY$ can be not only the halogens, chlorine, bromine and iodine, but also methyl iodide and $\text{Mn}_2(\text{CO})_{10}$.[126] The latter forms a linear Mn–Rh–Rh–Mn chain. The Rh···Rh separation in $(\text{Bridge})_4\text{Rh}_2^{2+}$ is 3.262 Å,[127] a value which is simlar to those found for the phosphine- and arsine-bridged analogs $(\mu\text{-dpm})_2\text{Rh}_2(\text{CO})_2\text{Cl}_2$ and $(\mu\text{-dam})_2\text{Rh}_2(\text{CO})_2\text{Cl}_2$. These are nonbonded distances. On oxidation the Rh–Rh distance contracts to give a normal Rh–Rh single bond. In $(\text{bridge})_4\text{Rh}_2\text{Cl}_2^{2+}$ the Rh–Rh distance is 2.837(1) Å.[128] It should be noted that bridging ligands are not absolutely required to form Rh(II) dimers. With simple isocyanides The Rh(II) dimers $\text{Rh}_2(\text{CN}R)_8X_2^{2+}$ are readily formed.[129,130] However, in solution these are subject to fairly facile disproportionation (reaction 16), a reaction which is inhibited when bridging ligands are present.

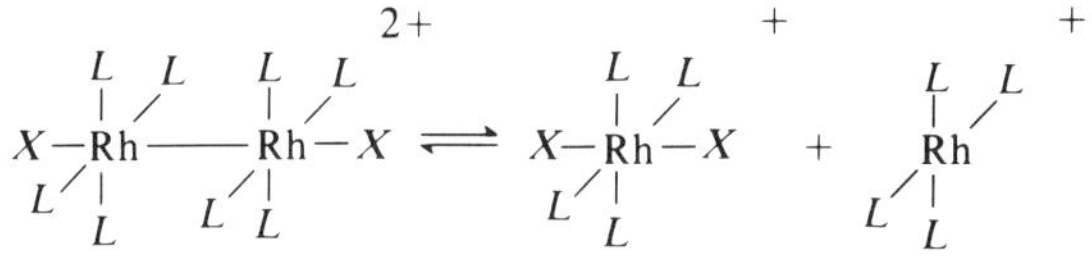

$$X-\text{Rh}-\text{Rh}-X \rightleftharpoons X-\text{Rh}-X + \text{Rh} \quad (16)$$

$L = \text{RNC}$

Oxidative addition to molecular A-frames appears to be little studied. Eisenberg and co-workers[57] have shown that alkyl halides and protons react at the bridging sulfur in $(\mu\text{-dpm})_2M_2(\mu\text{-S})(\text{CO})_2$ ($M$ = Rh, Ir) to give $(\mu\text{-dpm})_2M_2(\mu\text{-SR})(\text{CO})_2^+$ rather than undergoing oxidative addition to the metals. However, $(\mu\text{-dpm})_2\text{Ir}_2(\mu\text{-S})(\text{CO})_2$ does add dihydrogen to yield two hydrido complexes which may have any of the structures **4–6**:[60]

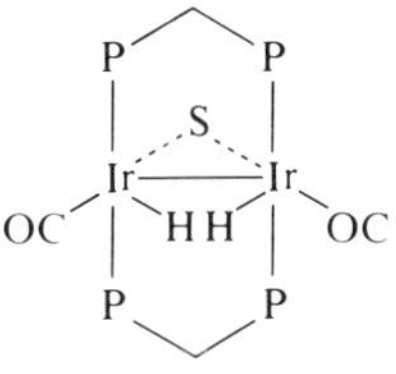

[Structures labeled 4, 5, 6 showing binuclear Ir complexes with P-P bridges, S bridges, H ligands, and CO ligands]

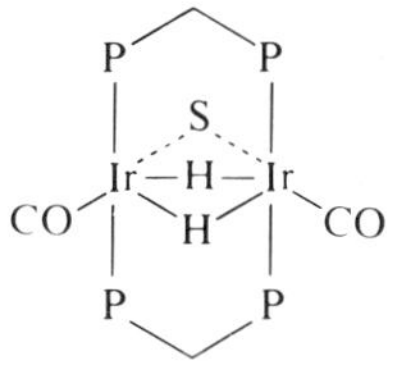

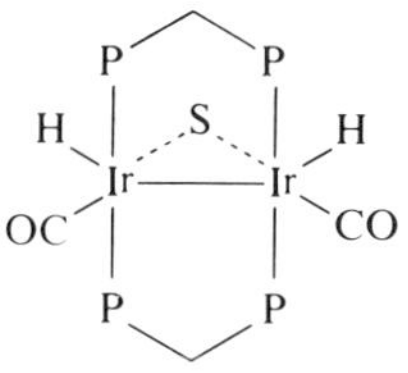

These species are probably involved in the homogeneous hydrogenation activity of $(\mu\text{-dpm})_2\text{Ir}_2(\mu\text{-S})(CO)_2$.

However, the behavior of $\text{Ir}_2(\mu\text{-SR})_2(CO)_2(PR'_3)_2$ shows that molecular A-frames and similarly shaped molecules can undergo two-center, two-fragment oxidative addition. This complex adds diiodine and dihydrogen according to reaction 17:[131,132]

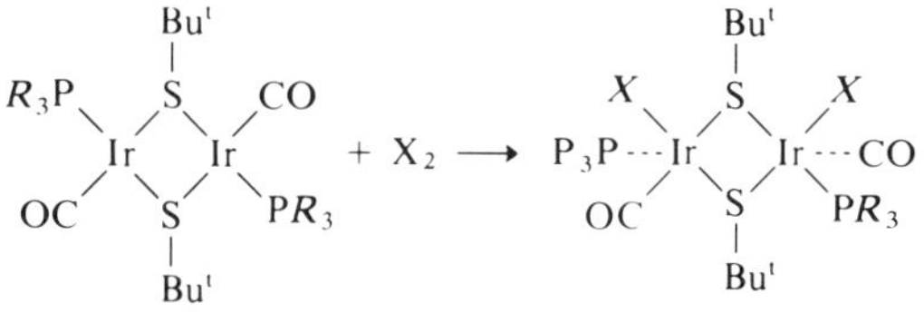

$$R_3P\text{-Ir}(S\text{-Bu}^t)(CO)(S\text{-Bu}^t)\text{-Ir-}PR_3 + X_2 \longrightarrow P_3P\cdots\text{Ir}(X)(S\text{-Bu}^t)(CO)(S\text{-Bu}^t)\text{Ir}(X)\cdots PR_3 \quad (17)$$

In the process the Ir–Ir separation shrinks from 3.216 in $\text{Ir}_2(\mu\text{-SBu}^t)_2(CO)_2\{P(OMe)_3\}_2$ to 2.703 in $\text{Ir}_2(\mu\text{-SBu}^t)_2(CO)_2\{PPhMe_2\}_2I_2$.

All the preceding reactions have involved oxidative additions to $d^8$ metal complexes. Similar reactions are also known for $d^{10}$ metal complexes. The Pd(0) complex, $(dpm)_3Pd_2$, whose detailed structure is unknown, also undergoes oxidative addition according to reaction 18[95]:

$$(dpm)_3Pd_2 + X_2 \xrightarrow{-dpm} (\mu\text{-dpm})_2Pd_2X_2 \xrightarrow{X_2}$$

$$(\mu\text{-dpm})_2Pd_2X_4 \rightarrow 2(dpm)PdX_2 \quad (18)$$

The initial step is a two-center, two-fragment addition which offers a new route to Pd(I) complexes. These, however, unlike the Rh(II) dimers, are subject to further oxidation, which eventually produces the Pd(II) chelates, $(dpm)PdX_2$, by way of intermediates $(\mu\text{-dpm})_2Pd_2X_4$, which have not been isolated. In related chemistry with different bridging ligands, the $d^{10}$ Au(I) complexes are reported to undergo two-center, two-fragment oxidative additions as shown in reaction 19:

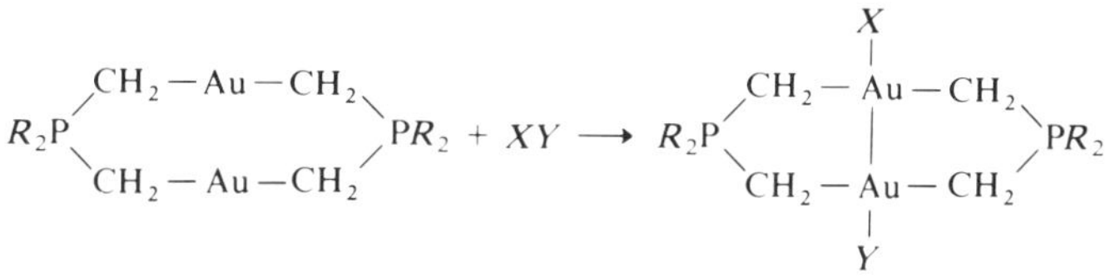

$$R_2P\begin{pmatrix}CH_2-Au-CH_2\\CH_2-Au-CH_2\end{pmatrix}PR_2 + XY \longrightarrow R_2P\begin{pmatrix}CH_2-Au(X)-CH_2\\CH_2-Au(Y)-CH_2\end{pmatrix}PR_2 \quad (19)$$

where $XY$ can be $Cl_2$, $Br_2$, $I_2$ or methyl iodide.[133,134]

## 4.5. Two-Center, Three-Fragment Oxidative Addition

In this reaction a dihalo compound undergoes oxidative addition to two metal centers as represented in reaction 20:

$$\begin{array}{c} P\!\!\sim\!\!\sim\!\!P \\ | \quad | \\ M^{n+} \quad M^{n+} \\ | \quad | \\ P\!\!\sim\!\!\sim\!\!P \end{array} + X_2A \rightarrow \begin{array}{c} P\!\!\sim\!\!\sim\!\!P \\ X\diagdown | \quad |\diagup X \\ M^{n+2} \; M^{n+2} \\ \diagdown_A\diagup \\ P\!\!\sim\!\!\sim\!\!P \end{array} \quad (20)$$

Each of the metal centers undergoes a net two-electron oxidation as in a conventional oxidative addition. However, the two substrate bonds are boken to produce three litigating fragments: two terminal halides, and a bridging ligand. The known reactions of this class all involve a single complex, $(dpm)_3Pd_2$, and one diphosphine ligand is lost from it in the process.[98] The reaction chemistry is summarized in Figure 5.18. The reaction with methylene dihalides offers a new route to methylene bridged complexes. Methylene bridged species have yet to be formed starting with mononuclear metal complexes. In that case, halomethyl complexes are formed.[135-137] The same holds for the addition of $\sigma$-diiodobenzene. With $Pd(PPh_3)_3$ this yields $(Ph_3P)_2Pd(o-C_5H_4I)I$ as the sole product, while the $(dpm)_3Pd_2$, the phenylene bridged complex is formed in high yield. The addition of oxalyl chloride and phenyl isocyanide dichloride to $(dpm)_3Pd_2$ provides alternate routes to molecular A-frames that were originally synthesized by the addition of carbon monoxide and phenyl isocyanide, respectively, to $(\mu\text{-dpm})_2Pd_2Cl_2$. In regard to the behavior of the isocyanide dichloride, it should be noted that three-fragment, one-center oxidative addition has been previously reported. For example, addition of phenyl isocyanide dichloride to $(Ph_3P)_4Pt$ and to $(Ph_3P)_3RhCl$ yields $(Ph_3P)Pt(CNPh)Cl_2$ and $(Ph_3P)_2Rh(CNPh)Cl_3$, respectively.[138]

The methylene bridged complex, $(\mu\text{-dpm})_2Pd_2(\mu\text{-CH}_2)X_2$, and its platinum analog are isoelectronic and probably isostructural with $(\mu\text{-dpm})_2Pd_2(\mu\text{-S})Cl_2$. Consequently, they are expected to have a large metal–metal separation and to differ appreciably from the larger class of dimetallacyclopropanes.[139] Some reaction chemistry of $(\mu\text{-dpm})_2Pd_2(\mu\text{-CH}_2)X_2$ has been explored. While the methylene bridge is resistant to insertion of carbon monoxide, sulfur dioxide, and isocyanides, it is readily protonated.[98] The product is the A-frame $(\mu\text{-dpm})_2Pd_2(\mu\text{-I})\text{-}(CH_3)I^+$, which has been isolated as the fluoroborate salt. It contains a terminal methyl group that resists further protonation.

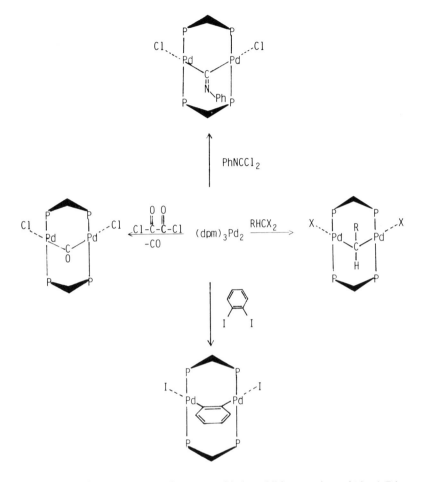

Figure 5.18. Two center, three-fragment oxidative addition reactions of $(dpm)_3Pd_2$.

## 4.6. The Formation or Stabilization of Other Novel Bridging Ligands

This classification involves a catchall for the formation of unusual bridging ligands that have not been accounted for in previous sections. Clearly, some of the preceding sections involve reactions that create unusual bridging ligands. Here we are concerned with some other rather singular reactions that do not fall in the previous categories.

Reaction of dpm with dimanganese decacarboryl produces the substitution product $(\mu\text{-dpm})_2\text{Mn}_2(\text{CO})_6$. On further heating to 140° $(\mu\text{-dpm})_2\text{Mn}_2(\text{CO})_6$ undergoes carbon monoxide loss. Some pertinent reaction chemistry is summarized in Equation 21:[43,44,140,141]

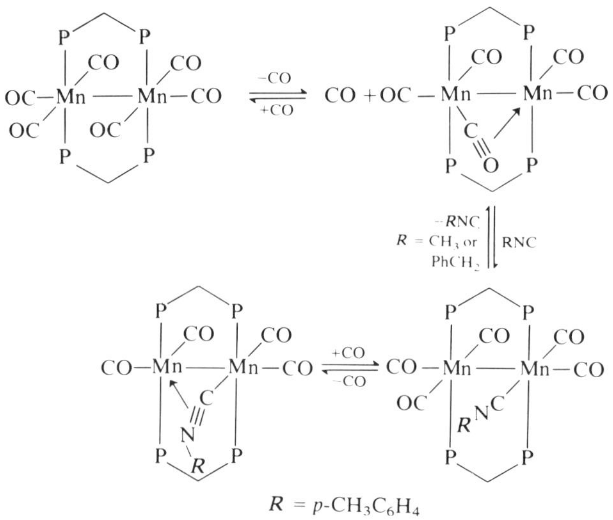

$$\text{(21)}$$

This sequence leads to the formation of the unusual bridging isocyanide and carbon monoxide ligands, which act as normal two-electron donors to one Mn and donate a pair of $\pi$ electrons to the second Mn. A drawing of the in-place portion of the structure of $(\mu\text{-dpm})_2\text{Mn}_2(\mu\text{-}p\text{-CH}_3\text{C}_6\text{H}_4\text{NC})\text{(CO)}_4$ is shown in Figure 5.19.[43] This gives details of the novel mode of isocyanide coordination.

The fluxional behavior of both $(\mu\text{-dpm})_2\text{Mn}_2(\text{CO})_6$ and of $(\mu\text{-dopm})_2\text{Mn}_2(\mu\text{-CO})(\text{CO})_4$ has been examined by both $^{31}\text{P}$ and $^{13}\text{C}$ NMR spectroscopy.[142,143] For $(\mu\text{-dpm})_2\text{Mn}_2(\text{CO})_6$, global scrambling of carbon monoxide ligands over all sites occurs, presumably by pairwise terminal-bridge interchange. In contrast, limited scrambling occurs when the novel bridging carbonyl ligand is present. The proposed mechanism for this intramolecular reaction is shown in reaction 22;

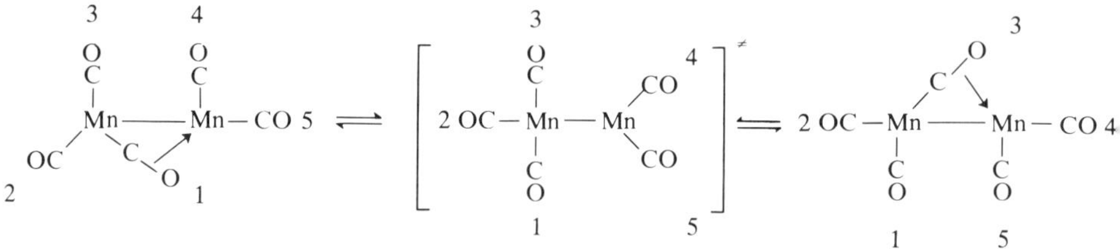

$$\text{(22)}$$

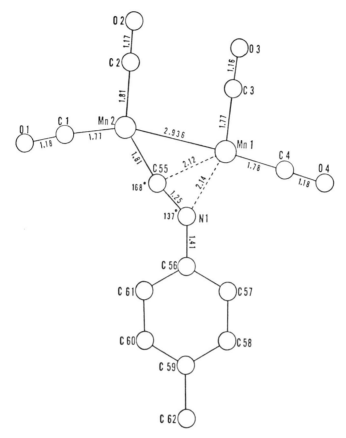

Figure 5.19. A view of a planar portion of $(\mu\text{-dpm})_2\text{Mn}_2(\mu\text{-}p\text{-CH}_3\text{C}_6\text{H}_4\text{NC})(\text{CO})_4$, which contains the bridging isocyanide ligand. The bridging dpm ligands lie above and below this plane and are not shown.

The exact role of the diphosphine ligand in stabilizing the novel 4-electron donor carbonyl and isocyanide ligands is not clear. Other examples of similar carbonyl groups include $\text{Cp}_2\text{NbMo(CO)}_3\text{Cp}$, $\text{Cp}_2\text{Zr(OC)-(OCMe)Mo(CO)}_3\text{Cp}$, and $(\text{C}_5\text{Me}_5)_2\text{ZrCo(CO)}_2\text{Cp}$. None of these requires a bridging phosphine ligand to stabilize the bridging carbonyl group.

In a different vein, the reaction of carbon disulfide with $(\mu\text{-dpm})_2\text{RH}_2(\mu\text{-CO})\text{CL}_2$ produces **7** as the final product.[55] In **7** two molecules of carbon disulfide have condensed to form a tridentate ligand which chelates one metal and also forms a bridge to the second metal.

```
         P       P
    Cl   |       |   CO
     \   |       |   /
    S—Rh————————Rh—Cl
     /    \C—S /
    C—S    \  /
   /      P    P
  S
```

7

## 4.7. Capture of a Second Metal Induced by 2(Diphenylphosphino)pyridine

The bifunctional ligand 2(diphenylphosphino)pyridine can be used to construct binuclear metal complexes in a stepwise fashion.[92] With Group VIII metal ions, this ligand appears to preferentially bind via the phosphorus atom. From this reaction, a host of metal complexes, very similar to those of triphenylphosphine, are available. All have the potential for binding a second metal ion through the uncoordinated pyridine nitrogen. Some of the reactions that result in the formation of binuclear complexes are shown in Figure 5.20.[92,144–146] The structure of one example, ($\mu$-Ph$_2$Ppy)$_2$RhPd(CO)Cl$_3$, is shown in Figure 5.21. In all cases so far examined, the formation of the binuclear complex results in the formation of a metal–metal bond as well. For the crystallographically characterized molecules these are all of similar length: ($\mu$-Ph$_2$Ppy)$_2$Rh$_2$($\mu$-CO)Cl$_2$, 2.612(1) Å,[145] ($\mu$Ph$_2$Ppy)$_2$RhPd(CO)Cl$_3$, 2.594(1) Å,[92] and ($\mu$-Ph$_2$Ppy)$_2$RuPd(CO)$_2$Cl$_2$, 2.660(1) Å.[146] This probably reflects the limited flexibility of Ph$_2$Ppy.

In all but the first example the reactions involve the oxidative addition of a metal–halogen bond to another metal center. Also, in all cases shown the bridging ligands have rearranged themselves so that a head-to-tail orientation of the Ph$_2$Ppy ligands is formed. In other words, one metal–phosphorus bond has broken and that phosphorus has become bound to the second metal atom. This behavior emphasizes the ability of phosphine ligands to dissociate from metal complexes, a fact that is frequently overlooked in viewing the reactions of diphosphine bridged complexes. For example, with the dpm bridged complexes, the diphosphine-dimetal unit generally appears as a sturdy framework about which reactions occur at all of the other coordination sites.

None of the metal–metal bonds shown on the binuclear complexes in Figure 5.20 undergoes facile insertion of small molecules. For example ($\mu$-Ph$_2$Ppy)$_2$Pd$_2$Cl$_2$, which is isoelectronic with ($\mu$-dpm)$_2$Pd$_2$Cl$_2$, does not react with carbon monoxide or sulfur dioxide. This lack of reactivity has been attributed to the limitations that the bridging Ph$_2$Ppy places on the

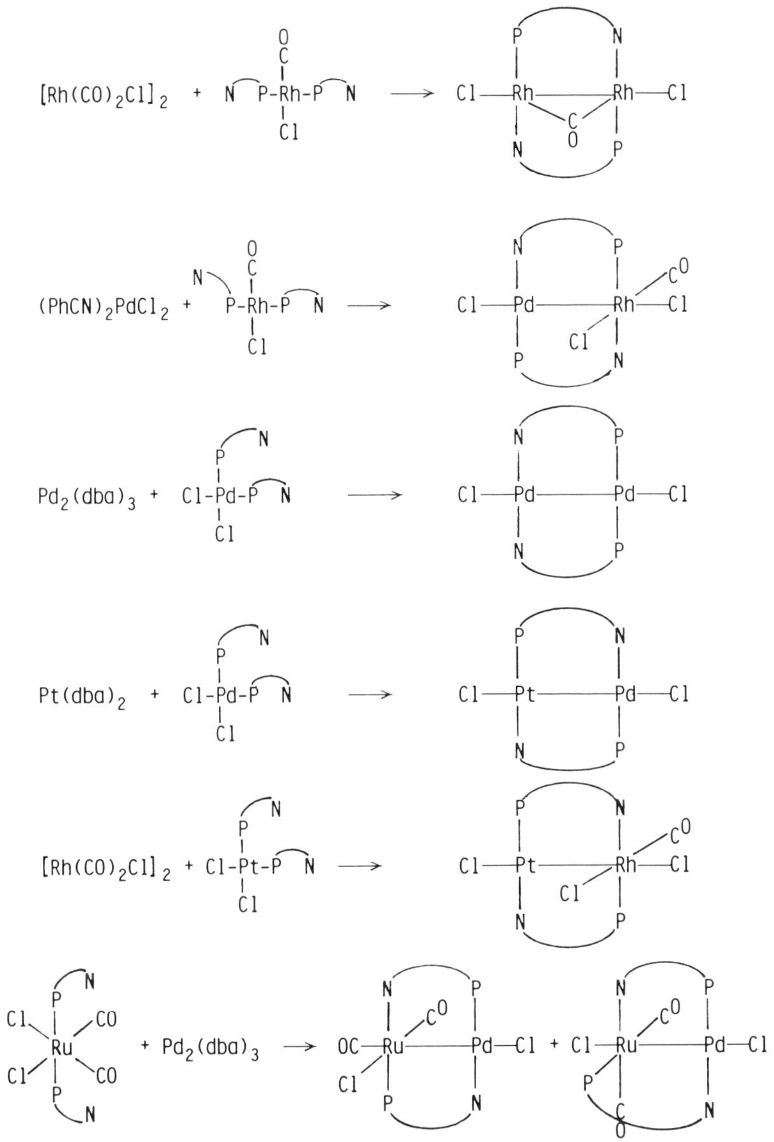

Figure 5.20. Reactions leading to the formation of binuclear complexes through the use of 2-(diphenylphosphino)pyridine as a bridging ligand.

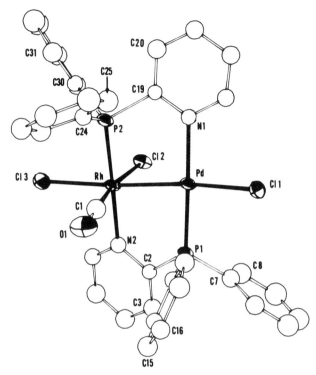

Figure 5.21. The structure of $(\mu\text{-Ph}_2\text{Ppy})_2\text{RhPd(CO)Cl}_3$ taken from Reference 92 by permission.

metal–metal separation and its inability to span metal–metal separations of 3 Å or greater.

## 5. DEMONSTRATED CATALYTIC ACTIVITY OF BINCULEAR COMPOUNDS

A number of binuclear phosphine-ligand bridged complexes have been shown to function as catalysts or catalyst precursors. Here we review these cases briefly. More detailed coverage of hydrogenation and hydroformylation using rhodium catalysts will appear in the next chapter. In all cases of catalytic activity shown by binuclear complexes, there is a serious question about the true identity of the catalytically active species. As with catalysis begun by metal carbonyl clusters, the possibility exists that a small amount of highly active mononuclear compound is the true catalyst.

($\mu$-Dpm)$_2$Pd$_2$Cl$_2$ is a catalyst for the conversion of dimethylacetylene dicarboxylate to hexamethyl mellitate (reaction 23)[99]:

$$3RC\equiv CR \rightarrow \begin{array}{c}\text{(benzene ring with 6 R groups)}\end{array} \quad R = CH_3O\overset{O}{\underset{\|}{C}}- \quad (23)$$

During the process, ($\mu$-dpm)$_2$Pd$_2$($\mu$-C$_2${CO$_2$CH$_3$}$_2$)Cl$_2$ forms. This complex, which has been isolated in crystalline form, can also serve as a catalyst. Other related complexes, dpmPdCl$_2$ and (dpm)$_3$Pd$_2$, are much less effective. The structure of ($\mu$-dpm)$_2$Pd$_2$($\mu$-C$_2${CO$_2$CH$_3$}$_2$)Cl$_2$ shows that, on insertion into the Pd-Pd bond, the acetylene has been reduced to a dimetalated olefin. The central C-C bond distance has elongated to 1.338 Å, and all the bond angles about the central two carbon atoms are near 120°. The catalytic cycle may involve further insertions of the acetylene into the Pd-C bonds of this A-frame, but direct eivdence of the mechanism is lacking. However, an analysis of the molecular orbitals involved indicates that such a process is feasible.[116]

(Dpm)$_3$Pd$_2$ is a catalyst for the hydrogenation and isomerization of olefins and diolefins (both conjugated and nonconjugated).[147,148] It catalyzes the dihydrogen reduction of methyl acetylene to propylene (which is then more slowly reduced to propane) and of propionaldehyde to ethane and ethylene. As a hydrogenation catalyst, it is effected by dioxygen pretreatment and by triphenylphosphine and carbon monoxide, which inhibit activity. After exposure to dioxygen it catalyzes exchange between H$_2$ and D$_2$. There is no information available that bears on the question of what the form of the catalytically active species actually is whether it contains two palladium atoms. So far hydride complexes of palladium with bridging dpm ligands have not been characterized, although a number of dinuclear hydrides of platinum have been isolated and characterized.[67,68]

A brief report of the ability of (dpm)$_3$Pt$_2$ to catalyze reactions 24 and 25 has appeared.[149] (Dpm)$_3$Pt$_2$ is also reported to be less effective

$$2NO + CO \rightarrow N_2O + CO_2 \quad (24)$$

$$O_2 + 2CO \rightarrow 2CO_2 \quad (25)$$

in the hydrogenation of butadiene than is (dpm)$_3$Pd$_2$.[147,148]

($\mu$-Dpm)$_2$Rh$_2$(CO)$_2$ is a catalyst for the reduction of acetylene to ethane by dihydrogen. The rate of acetylene reduction exceeds that of ethylene reduction and no cyclotrimerization of acetylene is observed.[51] Treatment of ($\mu$-dpm)$_2$Rh$_2$(CO)$_2$ with carbon monoxide and NaBH(OMe)$_3$

yields $(\mu\text{-dpm})_2\text{Rh}_2(\mu\text{-CO})(\mu\text{-H})(\text{CO})_2^+$, which is an unusually active catalyst for the water gas shift reaction at 90° and 1 atm pressure.[51]

$(\mu\text{-Dpm})_2\text{Ir}_2(\mu\text{-S})(\text{CO})_2$ is a catalyst for the hydrogenation of acetylene, ethylene, and propylene and can be recovered quantitatively from the reaction.[60]

## 6. CONCLUDING REMARKS

The previous sections have documented the fact that binuclear complexes can be obtained with a variety of novel structural features and that these undergo, in some cases, unprecedented chemical reactions. We can expect that work along these lines will continue to point out new structural types and new classes of reactions. Progress so far can be said to have laid a foundation for the basic coordination chemistry of the most readily obtained binuclear systems. It remains to be seen whether these reaction types can be clearly utilized in the design of new catalysts or whether unusual forms of catalytic activity for binuclear complexes will be discovered serendipitously.

Several areas for future development can be easily identified. Strategies for the incorporation of two different metal ions into the same complex have only recently begun to be considered.[26,92,93,144-146] The range of possibilities for hetero-binuclear complexes is enormous, and we can expect considerable development of new ligands and new combinations of metals. With two different metals, each can perform an individual, separate function, or both could bind to a bridging ligand and induce unusual polarization.

The opportunity afforded by the ability to design phosphine ligands to a variety of geometric specifications allows us to consider templated-based construction of metal clusters with several metal ions, possibly difficult metal ions. This has considerable potential for extending the now popular field of cluster chemistry. With polyphosphine backbones, the cluster need no longer rely solely on metal-metal bonding to determine its shape and stability. Some inroads into this area have begun to appear. The triphosphine, tripod $\text{HC}(\text{PPh}_2)_3$, has been used not only to cap triangular faces of tetrahedral metal clusters[159] but, more importantly, to construct the new triangular array **8**.[151]

Bifunctional ligands such as $\text{Ph}_2\text{Ppy}$, which have the ability to capture a second metal ion, also have the potential to facilitate the transfer of ligands from one metal to another. For example, in Figure 5.20 most of the metal-metal bond formation is accompanied by the transfer of a chloride ligand from one metal to another. It remains to be seen whether the transfer

```
              Ph   H   Ph
               \  C  /
          Ph—P     P—Ph
           O\  |  O|
            \C  \C/
             Ni————Ni—C—O
             |Ph-P-Ph|
            C\\ | //C
           O/  Ni  \O
               |
               C
               ‖
               O
```

**8**

of other ligands, particularly reactive organometallic ligands (alkyl, aryl, vinyl groups, hydrides), can be facilitated by bifunctional ligands.

The design of new ligands to use in constructing binuclear and polynuclear clusters does not have to be restricted to phosphine ligands. 2(Diphenylphosphine) pyridine is an example of a hybrid ligand with two different metal binding sites. Further elaboration of this concept should allow for the development of a variety of ligands capable of binding hard metal ions at one site and soft metal ions at another. The diisocyanide, 1,3-diisocyanopropane, resembles the diphosphine described, here, and we can expect to see the development of further aspects of its chemistry. Some work on the properties of complexes of sulfide ligands, such as $PhSCH_2SPh$, has been reported,[152] but this type of ligand appears to be less promising than others.

Finally, the ability of diphosphine ligands to form stable binuclear complexes as well as chelating complexes raises new interpretations of observations made using these ligands. For example, the activity of platinum(I) chloride/tin(II) chloride mixtures as hydroformylation catalysts depends on the chain length ($n$) of the diphosphine $Ph_2P(CH_2)_nPPh_2$ employed.[153] While this was initially interpreted as an effect of chelate ring size, which may well be the case, it is also just as possible that a binuclear complex is formed and that at $n = 4$ (the phosphine that produces the fastest rate) the optimal spacing in a binuclear catalyst is obtained.

ACKNOWLEDGMENTS

I thank Rich Eisenberg, Alan Sanger, and George Stanely for sharing unpublished data, and the NSF for the support of those aspects of original research done at the University of California, Davis.

## REFERENCES

1. E. L. Muetterties, *Bull. Soc. Chem. Belg.* **85**, 451 (1976).
2. J. P. Collman, P. Denisevich, Y. Konai, M. Marrocco, C. Koval, and F. C. Anson, *J. Am. Chem. Soc.* **102**, 6027 (1980).
3. M. H. Chisholm, ed., "Reactivity of Metal-Metal bonds," *ACS Symposium Series 155*, Washington, D.C., 1981.
4. D.-H. Chin, G. N. La Mar, and A. L. Balch, *J. Am. Chem. Soc.* **102**, 5945 (1980).
5. ———, *J. Am. Chem. Soc.* **102**, 4344 (1980).
6. A. F. Dyke, S. R. Finnimore, S. A. R. Knox, P. J. Naish, A. G. Orpen, G. H. Riding, and G. E. Taylor in "Reactivity of Metal-Metal Bonds," M. H. Chisholm, *ACS Symposium Series 155*, 1981, p.259.
7. E. L. Muetterties, *Bul. Soc. Chem. Belg.* **84**, 959 (1975).
8. ———, *Science* **196**, 839 (1977).
9. R. Whyman, *Transition Metal Clusters*, edited by B. F. G. Johnson (John Wiley and Sons, New York, 1980), p. 545.
10. D. H. Farrar, P. G. Jackson, B. F. G. Johnson, J. Lewis, W. J. H. Nelson, M. D. Vargas, and M. McPartlin, *J. Chem. Soc. Chem. Comm.*, 1009 (1981).
11. J. Chatt and F. A. Hart, *J. Chem. Soc.* 1378 (1960).
12. W. Hewertson and H. R. Watson, *J. Chem. Soc.* 1490 (1962).
13. A. R. Sanger, *J. Chem. Soc. Chem. Comm.* 893 (1975).
14. ———, *J. Chem. Soc., Dalton Trans.* 1971 (1977).
15. T. Yoshida, T. Yamagata, T. H. Tulip, J. A. Ibers, and S. Otsuka, *J. Am. Chem. Soc.* **100**, 2063 (1978).
16. A. Dedieu and R. Hoffmann, *J. Am. Chem. Soc.* **100**, 2074 (1978).
17. C. Crocker, R. J. Errington, R. Markham, G. J. Moulton, K. J. Odell, and B. L. Shaw, *J. Am. Chem. Soc.* **102**, 4373 (1980).
18. R. B. King, *Acc. Chem. Res.* **13**, 243 (1980).
19. A. L. du Preez, I. L. Marais, R. J. Haines, A. Pidcock, and M. Safari, *J. Organomet. Chem.* **141**, C10 (1977).
20. R. J. Haines, A. Pidcock, and M. Safari, *J. Chem. Soc., Dalton Trans.*, 830 (1977).
21. F. A. Cotton, R. J. Haines, B. E. Hanson, and J. C. Sekutowski, *Inorg. Chem.* **17**, 2010 (1979).
22. A. L. du Preez, I. L. Marais, R. J. Haines, A. Pidcock, and M. Safari, *J. Chem. Soc., Dalton Trans.*, 1918 (1981).
23. R. J. Haines, E. Maintjies, and M. Laing, *Inorg. Chim. Acta* **36**, L403 (1979).
24. R. J. Haines, M. Laing, E. Meintjies, and D. Sommerville, *J. Organometal. Chem.* **215**, C17 (1981).
25. E. Keller and H. Vahrenkamp, *Chem. Ber.* **112**, 1626 (1979).
26. R. Müller and H. Vahrenkamp, *Chem. Ber.* **113**, 3517 (1980).
27. F. A. Cotton, W. H. Ilsley, and W. Kaim, *J. Am. Chem. Soc.* **102**, 1918 (1980).
28. M. G. Newton, R. B. King, M. Chang, N. S. Pantaleo, and J. Gimeno, *J. Chem. Soc., Chem. Comm.*, 531 (1977).
29. M. G. Newton, N. S. Pantaleo, R. B. King, and T. J. Lotz, *J. Chem. Soc., Chem. Comm.* 514 (1978).
30. R. J. Puddenphatt, M. A. Thomson, Lj. Manojlovic-Muir, K. W. Muir, A. A. Frew, and M. P. Brown, *J. Chem. Soc., Chem. Comm.* 805 (1981).
31. M. P. Brown, S. J. Cooper, A. A. Frew, Lj. Manojlovic-Muir, K. W. Muir, R. J. Puddephatt, K. R. Seddon, and M. A. Thompson, *Inorg. Chem.* **20**, 1500 (1981).
32. N. W. Alcock, J. M. Brown, and J. C. Jeffery, *J. Chem. Soc., Dalton Trans.*, 888 (1977).

33. F. C. March, R. Mason, K. M. Thomas, and B. L. Shaw, *J. Chem. Soc., Chem. Comm.* 584 (1975).
34. N. A. Al-Salem, W. S. McDonald, R. Markham, M. C. Norton, and B. L. Shaw, *J. Chem. Soc., Dalton Trans.*, 59 (1980).
35. M. Cowie and S. K. Dwight, *Inorg. Chem.* **19,** 2500 (1980).
36. J. T. Mague, *Inorg. Chem.* **8,** 1975 (1969).
37. M. Cowie and S. K. Dwight, *Inorg. Chem.* **20,** 1534 (1981).
38. E. H. Abbott, K. S. Bose, F. A. Cotton, W. T. Hall, and J. C. Sekutowshi, *Inorg. Chem.* **17,** 3240 (1978).
39. M. M. Olmstead, L. S. Benner, H. Hope, and A. L. Balch, *Inorg. Chim. Acta* **32,** 193 (1979).
40. A. L. Balch, L. S. Benner, and M. M. Olmstead, *Inorg. Chem.* **18,** 2996 (1979).
41. M. M. Olmstead, C. H. Lindsay, L. S. Benner, and A. L. Balch, *J. Organometal. Chem.* **179,** 289 (1979).
42. L. S. Benner and A. L. Balch, *J. Am. Chem. Soc.* **100,** 6099 (1978).
43. C. J. Commons and B. F. Hoskins, *Aust. J. Chem.* **28,** 1663 (1975).
44. L. S. Benner, M. M. Olmstead, and A. L. Balch, *J. Organometal. Chem.* **159,** 289 (1978).
45. F. A. Cotton, L. W. Shive, and B. R. Stults, *Inorg. Chem.* **15,** 2239 (1976).
46. M. J. Mays, D. W. Prest, and P. R. Raithby, *J. Chem. Soc., Chem. Comm.*, 171 (1980).
47. F. A. Cotton and J. M. Troup, *J. Am. Chem. Soc.* **96,** 4422 (1974).
48. G. Lavigne and J.-J. Bennet, *Inorg. Chem.* **20,** 2713 (1981).
49. P. H. Bird, A. R. Fraser, and D. N. Hall, *Inorg. Chem.* **16,** 1923 (1977).
50. F. H. Carré, F. A. Cotton and B. A. Frenz, *Inorg. Chem.* **15,** 380 (1976).
51. C. P. Kubiak and R. Eisenberg, *J. Am. Chem. Soc.* **102,** 3637 (1980).
52. M. Cowie and R. S. Dickson, *Inorg. Chem.* **20,** 2682 (1981).
53. M. Cowie and S. K. Dwight, *Inorg. Chem.* **19,** 2508 (1980).
54. ———, *Inorg. Chem.* **19,** 209 (1980).
55. M. Cowie and S. K. Dwight, *J. Organometal. Chem.* **214,** 233 (1981).
56. M. Cowie, *Inorg. Chem.* **18,** 286 (1979).
57. C. Kubiak and R. Eisenberg, *Inorg. Chem.* **19,** 2726 (1980).
58. M. Cowie and S. K. Dwight, *Inorg. Chem.* **18,** 2700 (1979).
59. M. Cowie and T. G. Southern, *J. Organometal. Chem.* **193,** C46 (1980).
60. C. P. Kubiak, C. Woodcock, and R. Eisenberg, *Inorg. Chem.* **19,** 2733 (1980).
61. R. G. Holloway, B. R. Penfold, R. Colton, and M. J. McCormick, *J. Chem. Soc., Chem. Comm.*, 485 (1976).
62. M. M. Olmstead, J. P. Farr, and A. L. Balch, *Inorg. Chim. Acta* **52,** 47 (1981).
63. M. M. Olmstead, H. Hope, L. S. Benner, and A. L. Balch, *J. Am. Chem. Soc.* **99,** 5502 (1977).
64. A. L. Balch, C.-L. Lee, C. H. Lindsay, and M. M. Olmstead, *J. Organometal. Chem.* **177,** C22 (1979).
65. Lj. Manojlovic-Muir, K. W. Muir, and T. Solomun, *J. Orgnometal. Chem.* **179,** 479 (1979).
66. Lj. Manojlovic-Muir, K. M. Muir, and T. Solomun, *Acta Cryst.* **B35,** 1237 (1979).
67. Lj. Manojlovic-Muir and K. W. Muir, *J. Organommetal. Chem.* **219,** 129 (1981).
68. M. P. Brown, S. J. Cooper, A. A. Frew, Lj. Manojlovic-Muir, K. W. Muir, R. J. Puddephatt, and M. A. Thomson, *J. Organometal. Chem.* **198,** C33 (1980).
69. T. S. Cameron, P. A. Gardner, and K. R. Grundy, *J. Organometal. Chem.* **212,** C19 (1981).
70. N. Bresciani, N. Marsich, G. Nardin, and L. Randaccio., *Inorg. Chim. Acta* **10,** L5 (1974).
71. A. Camus, N. Marsich, G. Nardin, and L. Randaccio, *J. Organoetal. Chem.* **60,** C39 (1973).

72. A. Camus, G. Nardin, and L. Randaccio, *Inorg. Chim. Acta* **12,** 23 (1975).
73. G. Nardin and L. Randaccio, *Acta Cryst.* **B30,** 1377 (1974).
74. G. Nardin, L. Randaccio and E. Zangrando, *J. Chem. Soc., Dalton Trans.*, 2566 (1975).
75. H. H. Karsch and V. Schubert, unpublished results quoted in reference 76.
76. U. Schubert, D. Neugebauer, and A. A. M. Aly, *Z. anorg. allg. Chem.* **464,** 217 (1980).
77. H. Schmidbaur, A. Wohlleben, U. Schubert, A. Frank, and G. Hattner, *Chem. Ber.* **110,** 2751 (1977).
78. H. Schmidbaur, A. Wohlleben, F. Wagner, O. Orama, and G. Huttner, *Chem. Ber.* **110,** 1748 (1977).
79. E. O. Fischer, A. Ruhs, P. Friedrich, and G. Huttner, *Angew. Chem. Int. Ed. Eng.* **16,** 465 (1977).
80. C. Commons and B. F. Haskins, *Aust. J. Chem.* **28,** 1201 (1975).
81. G. Lavigne, N. Lugan, and J.-J. Bonnet, *Nov. J. Chim.* **5,** 423 (1981).
82. P. H. Bird, A. R. Fraser, and D. N. Hall, *Inorg. Chem.* **16,** 1923 (1977).
83. R. Colton, M. J. McCormick, and C. D. Pannan. *Aust. J. Chem.* **31,** 1425 (1978).
84. M. P. Brown, A. N. Keith, Lj. Manojlovic-Muir, K. W. Muir, R. J. Puddephatt, and K. R. Seddon, *Inorg. Chim. Acta* **34,** L223 (1979).
85. M. G. Newton, R. B. King, and M. Chang, and J. Gimeno, *J. Am. Chem. Soc.* **99,** 2802 (1977).
86. ———, *J. Am. Chem. Soc.* **100,** 326 (1978).
87. ———, *J. Am. Chem. Soc.* **100,** 1632 (1978).
88. F. Faraone, G. Bruno, G. Tresoldi, G. Faraone, and G. Bombieri, *J. Chem. Soc., Dalton Trans.*, 1651 (1981).
89. L. H. Pignolet, D. H. Doughty, S. C. Nowicki, M. P. Anderson, and A. L. Casalnuovo, *J. Organometal. Chem.* **202,** 211 (1980).
90. P. G. Jones, *Acta Cryst.* **B36,** 2775 (1980).
91. B. P. Sullivan and T. J. Meyer, *Inorg. Chem.* **19,** 752 (1980).
92. J. P. Farr, M. M. Olmstead, and A. L. Balch, *J. Am. Chem. Soc.* **102,** 6654 (1980).
93. N. E. Schore, *J. Am. Chem. Soc.* **101,** 7410 (1979).
94. C. H. Lindsay, L. S. Benner, and A. L. Balch, *Inorg. Chem.* **19,** 3503 (1980).
95. C. T. Hunt and A. L. Balch, *Inorg. Chem.* **20,** 2267 (1981).
96. A. Sanger and J. T. Mague, unpublished results, A. Sanger, personal communication.
97. J. T. Mague and S. H. deVries, *Inorg. Chem.* **19,** 3743 (1980).
98. A. L. Balch, C. T. Hunt, C.-L. Lee, M. M. Olmstead, and J. P. Farr, *J. Am. Chem. Soc.* **103,** 3764 (1981).
99. C.-L. Lee, C. T. Hunt, and A. L. Balch, *Inorg. Chem.* **20,** 2498 (1981).
100. C. T. Hunt and A. L. Balch, *Inorg. Chem.* **21,** 1641 (1982).
101. S. J. Cooper, M. P. Brown, and R. J. Puddephatt, *Inorg. Chem.* **20,** 1374 (1981).
102. G. K. Anderson, H. C . Clark, and J. A. Davies, *J. Organometal. Chem.* **210,** 135 (1981).
103. M. P. Brown, J. R. Fisher, R. J. Puddephatt, and K. R. Seddon, *Inorg. Chem.* **18,** 2808 (1979).
104. M. P. Brown, R. J. Puddephatt, M. Rashidi, and K. R. Seddon, *J. Chem. Soc., Dalton Trans.*, 516 (1978).
105. ———, *J. Chem. Soc., Dalton Trans.*, 1540 (1978).
106. M. P. Brown, S. J. Franklin, R. J. Puddephatt, M. A. Thomson, and K. R. Seddon, *J. Organometal. Chem.* **178,** 281 (1979).
107. C. T. Hunt, G. B. Matson, and A. L. Balch, *Inorg. Chem.* **20,** 2279 (1981).
108. M. P. Brown, J. R. Fisher, S. J. Franklin, and K. R. Seddon, *J. Orgnometal. Chem.* **161,** C46 (1978).
109. S. O. Grim and J. D. Mitchell, *Inorg. Chem.* **16,** 1770 (1977).
110. C. A. Tollman, *Chem. Soc. Rev.* **1,** 337 (1972).

111. J. P. Collman and L. Hegedus, *Principles and Applications of Organotransition Metal Chemistry* (University Science Books, Mill Valley, CA, 1980).
112. A. L. Balch and B. Tulyathan, *Inorg. Chem.* **16,** 2840 (1977).
113. D. J. Doonan, A. L. Balch, S. Z. Goldberg, R. Eisenberg, and J. S. Miller, *J. Am. Chem. Soc.* **97,** 1961 (1975).
114. A. L. Balch, J. R. Boehm, H. Hope, and M. M. Olmstead, *J. Am. Chem. Soc.* **98,** 7431 (1976).
115. A. Modinos and P. Woodward, *J. Chem. Soc., Dalton Trans.*, 1516 (1975).
116. D. M. Hoffman and R. Hoffmann, *Inorg. Chem.* **20,** 3543 (1981).
117. A. D. Rattray and D. Sutton, *Inorg. Chim. Acta* **27,** L85 (1978).
118. G. S. Brownlee, P. Carty, D. N. Cash, and A. Walker, *Inorg. Chem.* **14,** 323 (1975).
119. F. W. B. Einstein, D. Sutton, and P. L. Vogel, *Inorg. Nucl. Chem. Lett.* **12,** 671 (1976).
120. J. T. Mague and A. R. Sanger, *Inorg. Chem.* **18,** 2060 (1979).
121. A. R. Sanger, *J. Chem. Soc., Dalton Trans.*, 228 (1981).
122. A. L. Balch, *J. Am. Chem. Soc.* **98,** 8049 (1976).
123. A. L. Balch, J. W. Labadie, and G. Delker, *Inorg. Chem.* **18,** 1224 (1979).
124. G. G. Stanley and P. H. Bird, unpublished results; G. G. Stanley, personal communication.
125. N. S. Lewis, K. R. Mann, J. G. Gordon, II, and H. B. Gray, *J. Am. Chem. Soc.* **98,** 7461 (1976).
126. D. A. Bohling, T. P. Gill, and K. R. Mann, *Inorg. Chem.* **20,** 194 (1981).
127. K. R. Mann, J. A. Thich, R. A. Bell, C. L. Coyle, and H. B. Gray, *Inorg. Chem.* **19,** 2462 (1980).
128. K. R. Mann, R. A. Bell, and M. B. Gray, *Inorg. Chem.* **18,** 2671 (1979).
129. A. L. Balch and M. M. Olmstead, *J. Am. Chem. Soc.* **98,** 2354 (1976).
130. M. M. Olmstead and A. L. Balch, *J. Organomet. Chem.* **148,** C15 (1978).
131. A. Thorez, A. Maisonnat, and R. Poilblanc, *J. Chem. Soc. Chem. Comm.*, 518 (1977)
132. J. J. Bonnet, P. Kalck, and R. Poilblanc, *Angew. Chem.* **92,** 572 (1980).
133. H. Schmidbaur and R. Franke, *Inorg. Chim. Acta* **13,** 85 (1975).
134. H. Schmidbaur, J. R. Mandl, A. Frank, and G. Huttner, *Chem. Ber.* **109,** 466 (1976).
135. J. R. Moss and J. C. Spiers, *J. Organomet. Chem.* **182,** C20 (1979).
136. N. J. Kermode, M. F. Lappert, B. W. Skelton, A. H. White, and J. Holton, *J. Chem. Soc., Chem. Comm.*, 698 (1981).
137. H. Werner, R. Feser, W. Paul, and L. Hofmann, *J. Organomet. Chem.* **219,** C29 (1981).
138. W. P. Fehlhammer, A. Mayr, and B. Olgemöller, *Angew. Chem. Internal. Ed.* **14,** 369 (1975).
139. W. A. Herrmann, *Adv. Organomet. Chem.* in press.
140. R. Colton and C. J. Commons, *Aust. J. Chem.* **28,** 1673 (1975).
141. A. L. Balch and L. S. Benner, *J. Organomet. Chem.* **135,** 339 (1977).
142. K. G. Caulton and P. Adair, *J. Organomet. Chem.* **114,** C11 (1976).
143. J. A. Marsella and K. G. Caulton, submitted for publication.
144. A. Maisonnat, J. P. Farr, and A. L. Balch, *Inorg. Chim. Acta* **53,** L217 (1981).
145. J. P. Farr, M. M. Olmstead, C. H. Lindsay, and A. L. Balch, *Inorg. Chem.* **20,** 1182 (1981).
146. J. P. Farr, A. Maisonnat, M. M. Olmstead, and A. L. Balch, unpublished observations.
147. E. W. Stern and P. K. Maples, *J. Catalysis* **27,** 120 (1972).
148. ———, *J. Catalysis* **27,** 134 (1972).
149. C.-S. Chin, M. S. Sennett, P. J. Wier, and L. Vaska, *Inorg. Chim. Acta* **31,** L443 (1978).
150. A. A. Arduini, A. A. Bahsoun, J. A. Osborn, and C. Voelker, *Angew. Chem. Int. Ed. Engl.* **19,** 1024 (1980).
151. J. A. Osborn and G. G. Stanely, *Angew. Chem. Int. Ed. Engl.* **19,** 1025 (1980).
152. A. R. Sanger, C. G. Lobe, and J. E. Weiner-Fedorak, *Inorg. Chim. Acta* **53,** L123 (1981).
153. Y. Kawabata, T. Hayashi, and I. Ogata, *J. Chem. Soc., Chem. Comm.*, 462 (1979).

# 6

# *Hydrogenation and Hydroformylation Reactions Using Binuclear Diphosphine-Bridged Complexes of Rhodium*

*Alan R. Sanger*

## NOTATION

| | |
|---|---|
| dppm | $Ph_2PCH_2PPh_2$ |
| dpam | $Ph_2AsCH_2AsPh_2$ |
| dppe | $Ph_2P(CH_2)_2PPh_2$ |
| dppp | $Ph_2P(CH_2)_3PPh_2$ |
| dppb | $Ph_2P(CH_2)_4PPh_2$ |
| tdpme | $(Ph_2PCH_2)_3CCH_3$ |
| triphos | $(Ph_2PCH_2CH_2)_2PPh$ |
| (−)-diop | $(CH_3)_2C\!\!\begin{array}{c}O-\overset{H}{\underset{\phantom{x}}{C}}-CH_2PPh_2\\ |\\ O-\underset{H}{\overset{\phantom{x}}{C}}-CH_2PPh_2\end{array}$ |
| (P) | a tertiary phosphine ligand |

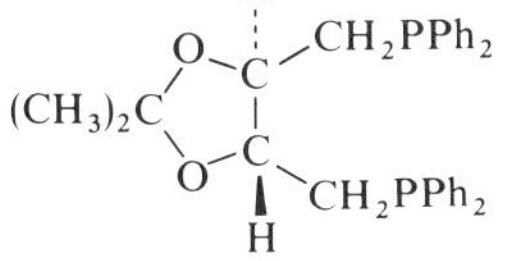

*Dr. Alan R. Sanger* • Alberta Research Council, 11315–87 Avenue, Edmonton, Alberta, Canada T6G 2C2.

(P⌒P)         a di(tertiary phosphine)
(P⌒P)*        a di(tertiary phosphine) coordinated to a metal by one donor atom

⌒
|   |
M   M         a dppm or dpam ligand bridging two metal centers

## 1. INTRODUCTION

Modification of transition-metal complexes by complexation with Lewis-base ligands, especially phosphines or phosphites, has led to major improvements in catalytic reactivity and selectivity.[1,2] Further, by permitting operation of production facilities at more moderate temperatures and pressures, considerable reductions in both capital and operating costs have been realized. For example, this is true of recent installations for the production of butyraldehyde using rhodium hydroformylation catalysts.[3,4] The majority of industrial hydrogenation reactions are performed using heterogeneous catalysts.[5] Nevertheless, the homogeneous hydrogenation catalysts are of importance for laboratory-scale reactions, the synthesis of thermally unstable or otherwise sensitive products (especially those of biological interest), applications of prochiral catalysts for the production of optically-active products, notably L-dopa (see Chapter 4), and mechanistic studies.[1,2,6,7]

The majority of studies involving phosphine-modified homogeneous catalysts have concentrated on the variation of activity with steric effects[8] or the Lewis basicity of the modifying ligands.[1,2] The effect of introducing a bidentate ligand into a catalyst system is frequently considered as merely the sum total of the effects of two separate ligands maintained in a rigid geometry with respect to one another. The assumption that chelation will occur has frequently and sometimes erroneously been made. Chelation is but one possibility; some alternative effects are suggested in Figure 6.1.

The variety of phosphine complexes of the transition metals is extensive,[9] and the variety of complexes of diphosphines is also impressive.[10] However, the *effects* of bridging two or more metal centers with a diphosphine ligand have been relatively neglected. In the case of tri(tertiary phosphine) or higher systems, a greater variety of bridging and chelating bonding modes are available and many have been realized.[11a–11d] Extreme examples of such schemes are "heterogenized homogeneous catalysts": transition metal catalysts bonded to flexible polymer supports, especially those with phosphine anchor-sites.[12] Such systems are discussed in Chapter 14.

In the following sections, binuclear complexes of rhodium bridged by one or two diphosphine ligands, which are active for the hydrogenation or

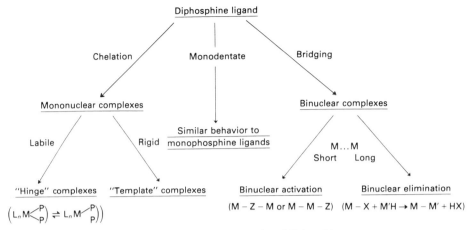

Figure 6.1. Potential bonding modes of diphosphines.

hydroformylation of alkenes or alkynes, will be described. When it is informative to do so, comparison with complexes of other metals, notably cobalt or iridium, will be made.

## 2. HYDROGENATION OF ALKENES OR ALKYNES

The use as hydrogenation catalysts of neutral or cationic complexes of rhodium with one or two chelating diphosphine ligands is well established.[1,2,13–15] The mechanisms of reaction have been intensely studied (see Chapter 2), especially in relation to asymmetric hydrogenation using complexes of chiral ligands (see Chapter 4). In contrast to the above systems, the series of complexes $[RhCl(CO)(P\frown P)]_x$, where $(P\frown P)$ is $Bu^t_2P(CH_2)_nPBu^t_2$ [16] or $Ph_2P(CH_2)_mPPh_2$,[17] unexpectedly were found to be binuclear ($x = 2$) for a variety of values of $n$ or $m$, (**I**), with the sole exception of the dppe complex, (**II**).[17]

**I**   **II**

The structures of binuclear, diphosphine-bridged complexes of various metals, including rhodium, are discussed in more detail in Chapter 5.

## 2.1. Complexes of Ph$_2$PCH$_2$PPh$_2$

Complexes of type (**I**) are of special interest when the ligating phosphorus atoms are separated by one atom as in, for example, a methylene group. The rhodium atoms are sufficiently close to interact in a repulsive manner, as demonstrated by the electronic spectrum of [Rh$_2$Cl$_2$(CO)$_2$(dppm)$_2$][8,9] and the separation between the metal atoms (3.239 Å), which is larger than that between the phosphorus atoms of a bridging dppm ligand (3.130 Å).[20] However, the repulsion between the rhodium centers is not large, as the Rh–Rh separation in the arsine analog is reduced from 3.396 Å[21] to 3.236 Å when the crystal contains dichloromethane of crystallization colinear with the Rh–Rh vector.[22,23] The flexibility of the structure allowed by the dppm (or dpam) ligands is considerable. The Rh–Rh bond length in [Rh$_2$Br$_2$($\mu$-CO)(dppm)$_2$], (**III**), is 2.757 Å.[24] Similarly, the oxidative addition of halogens to dppm complexes of structure (**I**) gives Rh–Rh-bonded complexes of rhodium(II), (**IV**).[18,19] It is, however, noteworthy that no analogous oxidative addition complex was observed when [Rh$_2$Cl$_2$(CO)$_2$(dppm)$_2$] was treated with hydrogen.[19]

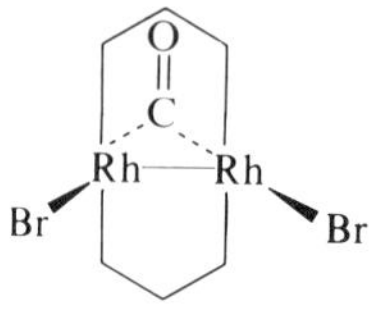

**III** 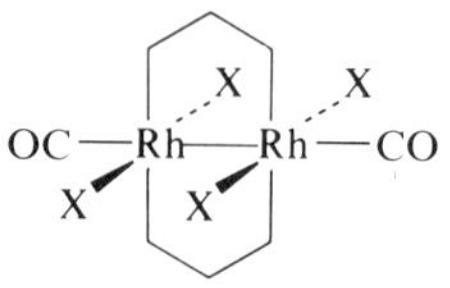    **IV**

Results, using cluster complexes of nickel or iridium, support the proposal that hydrogenation of triple-bonded substrates can be more readily achieved by the use of polynuclear homogeneous catalysts than by use of mononuclear catalysts.[25] The requirement for multinuclear centers to heterogeneously catalyze the hydrogenation of CO has also been proposed.[26] Consequently, complexes of transition metals in which the metal atoms are constrained to be proximate are potentially of interest as catalysts.

However, the only neutral complexes of structure (**V**) that are catalytically active are the cyano-complex and the diarsine analog (see Table I),[27,28] which are formed *in situ* by loss of CO from the tetracarbonyl precursor, (**VI**).[29]

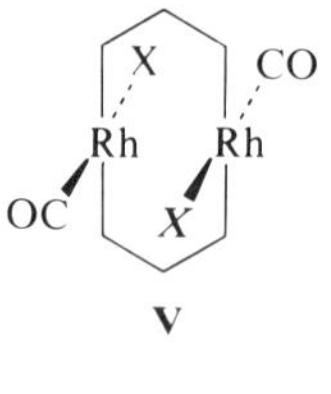

**V** 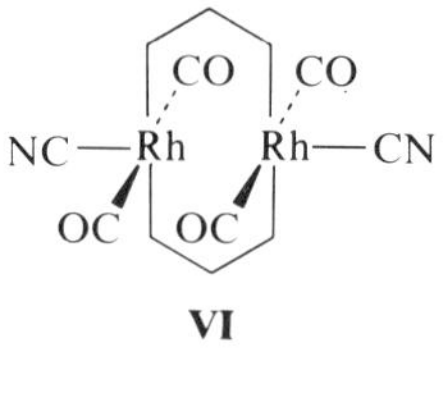    **VI**

Table 1. Hydrogenation Catalytic Activity of A-Frame Complexes[a]

| Catalyst[b] | Concentration ($\times 10^{-3} M$) | Rate[c] |
|---|---|---|
| **For 1-Hexene → $n$-Hexane** | | |
| $[Rh_2Cl(CO)_2(dpm)_2]^+$ | 0.40 | 2.3 |
| $[Rh_2Cl(CO)_2(dam)_2]^+$ | 0.40 | 3.4 |
| $[Rh_2Br(CO)_2(dpm)_2]^+$ | 0.40 | 1.1 |
| $[Rh_2Br(CO)_2(dam)_2]^+$ | 0.40 | 0.17 |
| **For Phenylacetylene → Styrene** | | |
| $[Rh_2Cl(CO)_2(dpm)_2]^+$ | 0.39 | 0.31 |
| $[Rh_2Cl(CO)_2(dam)_2]^+$ | 0.40 | 0.20 |
| $[Rh_2Br(CO)_2(dpm)_2]^+$ | 0.40 | 0.23 |
| $[Rh_2Br(CO)_2(dam)_2]^+$ | 0.42 | 0.45 |
| $[Rh_2(CN)_2(CO)_4(dpm)_2]$ | 0.40 | 3.0 |
| $[Rh_2(CN)_2(CO)_4(dam)_2]$ | 0.40 | 0.31 |
| $[Rh_2(NCO)_2(CO)_2(dpm)_2]$ | 0.40 | 0.03 |
| $[Rh_2(NCO)_2(CO)_2(dam)_2]$ | 0.41 | 0.0001 |
| $[Rh_2(NCS)_2(CO)_4(dpm)_2]$ | 0.38 | 0.002 |
| $[Rh_2(NCS)_2(CO)_4(dam)_2]$ | 0.39 | 0. |
| $[Rh_2Cl(N_3)(CO)_2(dpm)2]$ | 0.42 | 0.12 |

[a] Reference 28. For dpm, read dppm. For dam, read dpam.
[b] The formulation of the catalytically active species may differ from that of the "catalyst" orginally dissolved in methanol under an atmosphere of dihydrogen.
[c] Moles of substrate hydrogenated per minute per mole of catalyst, at 25°C.

The first rhodium complex of the structure now commonly called "A-frame" to be described was the neutral sulphido-bridged complex (**VII**, $X = S$, $n = 2$).[30,31] Unlike the iridium analog,[32] this complex is a very poor catalyst for the hydrogenation of ethylene.[31] In contrast, both the rhodium and iridium cationic complexes (**VII**, $X =$ Cl or Br, $n = 1$)

**VII** ($X$ is an anionic ligand of charge $n$-)

**VIII**

Table 2. Alkenes and Alkynes Hydrogenated[a,b]

| Reagent | Initial product | Subsequent products |
|---|---|---|
| 1-Pentene | $n$-Pentane | |
| 1-Hexene | $n$-Hexane | |
| 2-Hexene | $n$-Hexane | |
| $c$-Hexene | $c$-Hexane | |
| Vinyl-cyclohexane | Ethylcyclohexane | |
| Styrene | Ethylbenzene | |
| cis-Stilbene | 1,2-Diphenylethane | |
| trans-Stilbene | 1,2-Diphenylethane | |
| Phenylacetylene | Styrene | Ethylbenzene |
| 1-Phenylpropyne | cis-1-Phenylpropene | 1-Phenylpropane |
| 1-Phenylpropyne[c] | cis-1-Phenylpropene | trans-1-Phenylpropene, 1-Phenylpropane |
| Diphenylacetylene | cis-Stilbene | 1,2-Diphenylethane |

[a] Catalyst: $[Rh_2(\mu\text{-}Cl)(CO)_2(dppm)_2]^+$ ($2.0 \times 10^{-3}$ M) in methanol (50 cm$^3$), at 22°C and 1 atmosphere total pressure.
[b] References 27, 28.
[c] Catalyst: $[Rh_2(\mu\text{-}Cl)(CO)_2(dpam)_2]^+$; for this catalyst hydrogenation of the alkene was very slow, and concurrent cis to trans isomerization occurred. Reference 35.

are active catalysts for the hydrogenation of a variety of alkenes or alkynes (see Tables 1 and 2).[27,28,33–36]

The rhodium(O) complex, (VIII), reacts with an alkyne to form a 1:1 addition complex which is catalytically active for the hydrogenation of alkynes and (weakly) alkenes.[37] Complexes of types (III), but with chloro-ligands,[38] and (VII, $X = Cl$)[33] also form complexes with acetylenes, which can be subsequently hydrogenated. A series of complexes of structure (VII) with $RO^-$,[31] $RS^-$,[35,39] $RCO_2^-$,[35,39] $HSO_4^-$,[39] or $ClHO^-$ [39] ligands, and $[Rh_2(CH_3C(CH_2)_2)_2(dppm)_2]$[40] have also been prepared, but their catalytic activity has not yet been reported.

Cationic complexes (VII, $X =$ Cl or Br) hydrogenate alkynes successively to the corresponding alkene and alkane (Figures 6.2 and 6.3).[27,33–35] Even though hydrogenation of an alkene may occur at a higher rate than hydrogenation of an alkyne, it does not occur while alkyne is present. In contrast, the corresponding iridium complex, generated in situ ($^{31}$Pnmr) from $[Ir_2Cl(CO)_4(dppm)_2][IrCl_2(CO)_2]$,[36] hydrogenates both the alkyne and alkene simultaneously (Figure 6.4).[28,34]

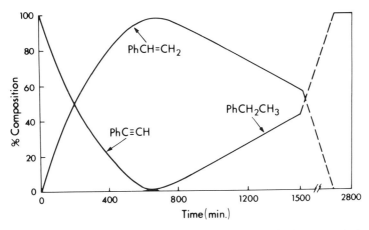

Figure 6.2. Hydrogenation of phenylacetylene catalyzed by a solution of [Rh$_2$Cl(CO)$_3$(dppm)$_2$][RhCl$_2$(CO)$_2$] in methanol (2.00 × 10$^{-3}$ $M$; 22°C).

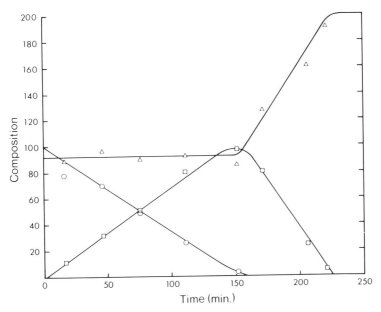

Figure 6.3. Hydrogenation of a second aliquot of phenylacetylene catalyzed by a solution of [Rh$_2$Br(CO)$_3$(dppm)$_2$]BPh$_4$ in methanol (0.70 × 10$^{-3}$ $M$; 25°C). The ethylbenzene present at time = 0 is the product from hydrogenation of the first aliquot of phenylacetylene. ⊙ = PhCCH; □ = PhCHCH$_2$; △ = PhCH$_2$CH$_3$. Reproduced, with permission, from Reference 28.

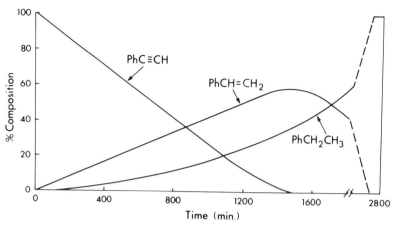

Figure 6.4. Hydrogenation of phenylacetylene catalyzed by a solution of $[Ir_2Cl(CO)_4(dppm)_2][IrCl_2(CO)_2]$ in methanol ($0.27 \times 10^{-3} M$; 22°C). Reproduced, with permission, from Reference 28.

This contrasting behavior may be due to the difference between the abilities of the rhodium and iridium systems to react with hydrogen to form oxidative addition products. Hydrogen reacts with $[Ir_2(\mu\text{-}S)(CO)_2(dppm)_2]$ to form a dihydride,[32] and with $[Ir_2Cl(CO)_3(dppm)_2]BPh_4$ to form an unidentified hydrido-complex.[36] In contrast, no oxidative addition product was observed when complexes (**V**, $X = Cl$),[19] (**VI**; $X = Cl$),[33] or (**VII**, $X = S$)[30,31] were treated with hydrogen.

It is tempting to speculate that only for the dinuclear complexes of iridium is there initial formation of a dihydrido-complex, which subsequently reacts with an alkene to form an alkyl intermediate, or with an alkyne to form an alkenyl intermediate. If such is the case, the activity of dinuclear rhodium complexes must depend on initial formation of an alkyne or alkene complex, which would then react with hydrogen. There exists some evidence for such a scheme. The successive hydrogenation of alkynes and alkenes[28,34] suggests that activation of an alkene is inhibited by an alkyne, probably by preferential coordination of the latter. Further, complexes (**VII**, $X = H$) or (**IX**) do not alone react with hydrogen, but do so after reaction with an alkyne (acetylene or phenylactylene).[37]

**IX**

Table 3. Formation of Carbon Monoxide Bridged Complexes[a]

| Complex | Product |
| --- | --- |
| (i) Association of Carbon Monoxide | |
| $[Rh_2(\mu\text{-}Cl)(CO)_2(dpm)_2]^+$ | $[Rh_2(\mu\text{-}Cl)(CO)_2(\mu\text{-}CO)(dpm)_2]^+$ |
| $[Rh_2(\mu\text{-}Cl)(CO)_2(dam)_2]^+$ | $[Rh_2(\mu\text{-}Cl)(CO)_2(\mu\text{-}CO)(dam)_2]^+$ |
| $[Rh_2(\mu\text{-}Br)(CO)_2(dpm)_2]^+$ | $[Rh_2(\mu\text{-}Br)(CO)_2(\mu\text{-}CO)(dpm)_2]^+$ |
| $[Rh_2(\mu\text{-}Br)(CO)_2(dam)_2]^+$ | $[Rh_2(\mu\text{-}Br)(CO)_2(\mu\text{-}CO)(dam)_s]^+$ |
| $[Rh_2(CN)_2(CO)_2(dpm)_2]$ | $[Rh_2(CN)_2(CO)_2(\mu\text{-}CO)(dpm)_2]$ |
| $[Rh_2(CN)_2(CO)_2(dam)_2]$ | $[Rh_2(CN)_2(CO)_2(\mu\text{-}CO)(dam)_2]$ |
| $[Rh_2(NCO)_2(CO)_2(dpm)_2]$ | none detected |
| $[Rh_2(NCO)_2(CO)_2(dam)_2]$ | none detected |
| $[Rh_2Cl(N_3)(CO)_2(dpm)_2]$ | $(Rh_2(\mu\text{-}N_3)(CO)_2(\mu\text{-}CO)(dpm)_2]^{+\ b}$ |
| (ii) Dissociation of Carbon monoxide | |
| $[Rh_2(CN)_2(CO)_4(dpm)_2]$ | $[Rh_2(CN)_2(CO)_2(\mu\text{-}CO)(dpm)_2]$ |
| $[Rh_2(CN)_2(CO)_4(dam_2]$ | $[Rh_2(CN)_2(CO)_2(\mu\text{-}CO)(dam)_2]$ |
| $[Rh_2(NCS)_2(CO)_4(dpm)_2]$ | none detected |
| $[Rh_2(NCS)_2(CO)_4(dam)_2]$ | none detected |

[a] Reference 29, 33. For dpm, read dppm. For dam, read dpam.
[b] This product is present in low equilibrium amounts in solution; the solid product has not yet been isolated.

It is also of interest to compare the reactivity of both active and inactive dinuclear complexes with CO (see Table 3).[29,33] The only complexes that are of significant activity for the hydrogenation of alkynes are the chloro- or bromo-complexes of structure (VII), and the cyano- or, weakly, azido-complexes of structure (V).[27,28] It is these complexes, and only these complexes, which react readily and reversibly to form carbonyl-bridged complexes.[29,33] Reaction of $^{13}CO$ with complex (VII, $X = Cl$) occurs initially at a terminal site, with subsequent migration of an orginally terminal CO to the bridging position, (X) (Equation 2.1).[33] The neutral, tricarbonyl cyano-complex, (XI), is unique within the series of dinuclear complexes of rhodium(I) described to date (Equation 2.2).[29] The neutral azido complex (V, $X_2 = Cl, N_3$) forms weakly conducting solutions of the cationic complex (X, $X = N_3$) (Equation 2.3).[29]

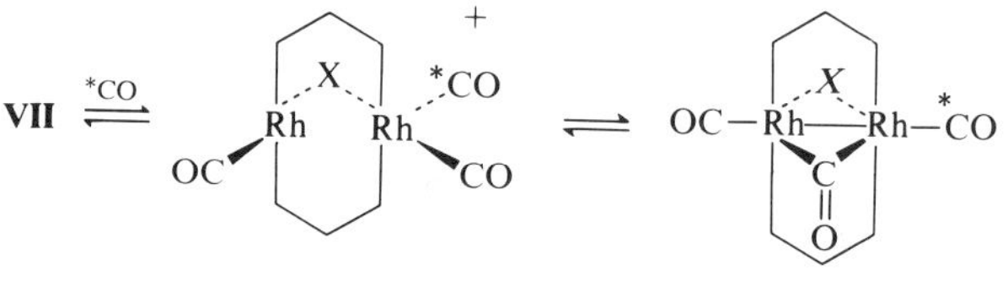

$$\text{VII} \underset{}{\overset{*CO}{\rightleftharpoons}} \left[\begin{array}{c}\text{OC}\diagdown \underset{}{\overset{X}{\underset{}{Rh}}}\diagup \overset{*CO}{\underset{}{Rh}}\diagdown \\ \text{CO}\end{array}\right]^{+} \rightleftharpoons \left[\text{OC}-\overset{X}{\underset{\overset{\|}{C}}{Rh}}-\overset{*}{Rh}-\text{CO}\right]^{+} \quad (2.1)$$

$$\mathbf{X}$$

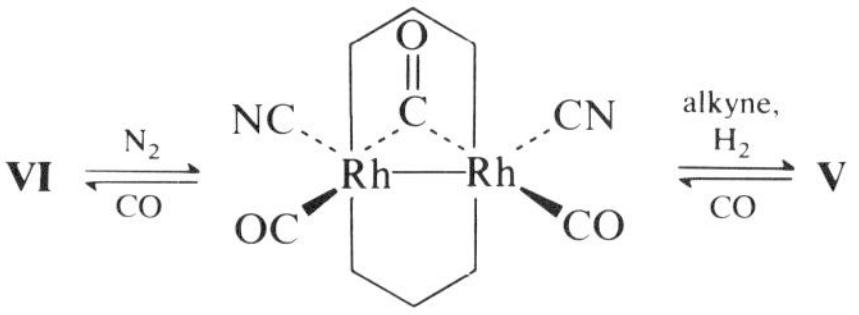

$$\text{VI} \underset{\text{CO}}{\overset{N_2}{\rightleftharpoons}} \left[\text{NC}-\underset{\text{OC}}{Rh}-\underset{}{Rh}-\text{CN} \atop \text{CO}\right] \underset{\text{CO}}{\overset{\text{alkyne, }H_2}{\rightleftharpoons}} \mathbf{V} \quad (2.2)$$

$$\mathbf{XI}$$

$$[\text{Rh}_2(\text{Cl})(\text{N}_3)(\text{CO})_2(\text{dppm})_2] \overset{\text{CO}}{\rightleftharpoons} [\text{Rh}_2(\text{N}_3)(\text{CO})_3(\text{dppm})_2]\text{Cl} \quad (2.3)$$

The cationic complex (**VII**, $X$ = Cl) forms a symmetrical ($^{31}$Pnmr; i.r.) complex with diphenylacetylene, which is very labile.[33] The complexes [Rh$_2$Cl$_2$($\mu$-CO)($\mu$-acet)(dppm)$_2$], where "acet" is hexafluoro-2-butyne or acetylenedicarboxylic acid dimethyl ester, are also symmetrical, and the "acet" ligand is coplanar with the rhodium atoms.[38] The C–C bond length is increased to 1.32 Å, and so the complex is truly described as a *cis*-dimetallated alkene.[38] Comparison with the reactions with CO suggests that the activation of an alkyne (or alkene) to hydrogenation may occur when the alkyne is initially coordinated to the catalyst at a terminal position, e.g., (**XII**),[28] with subsequent oxidative addition and hydrogen transfer steps. A complex of the type (**XII**) may instead be only a minor but active component of a solution of which the alkyne-bridged complex is the major but less active component.[41,42]

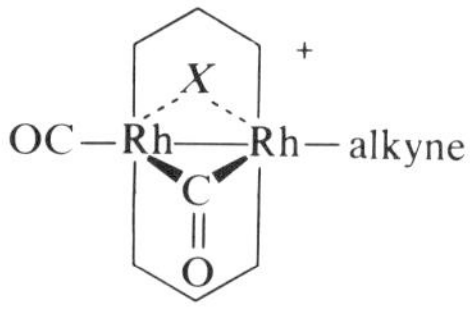

$$\left[\text{OC}-\overset{X}{\underset{\overset{\|}{\underset{O}{C}}}{Rh}}-Rh-\text{alkyne}\right]^{+}$$

**XII**

It must be stressed that the above arguments are conjecture, and that the possibility of the presence of an undetectably small but finite concentration of an active hydrido-complex cannot be excluded on the basis of available data. The above systems are not simple, and the presence of more than one parallel "mechanism" is possible. Comparison of the activities of complexes with different anionic ligands (see Table 1) was made by comparing methanol solutions.[27,28] Acetone and other strong donor solvents form stable, less active complexes with many of the catalysts.[29,32] The activities of both the $BPh_4^-$ and $[RhCl_2(CO)_2]^-$ salts of $[Rh_2Cl(CO)_2(dppm)_2]^+$ were similar at low-to-intermediate (ca. $0.4 \times 10^{-3} M$) concentrations, as is appropriate for fully dissociated complexes of which the cation is the active catalyst. However, at greater concentrations the molar activity of the $[RhCl_2(CO)_2]^-$ salt decreased markedly, indicating that significant changes in the nature of the solutions were occurring. The cause of these phenomena has not yet been determined.

## 2.2. Complexes of Diphosphines with Longer Chain Lengths

The chelated complex $[HRh((-)-diop)_2]$ is active for the hydrogenation of terminal olefins or unsaturated carboxylic acids.[43–45] From kinetic data and spectroscopic evidence an active intermediate containing a monodentate $(-)$-diop ligand has been postulated,[44] analogous to a known carbonyl complex.[17] Polymeric species containing bridging diphosphine ligands were postulated to explain inhibition of catalytic activity in the presence of added $(-)$-diop or dppb, but not added $PPh_3$.[44] However, proof of existence of such bridged complexes was not obtained.

The activity of complexes prepared in situ containing one mole of diphosphine $Ph_2P(CH_2)_nPPh_2$ ($n = 1-6$) per mole of rhodium[46] varied with the diphosphine chain length in a similar manner to the cationic complexes $[Rh(P\frown P)_2]^+$.[45] In each case only structures containing chelating diphosphine ligands were considered.[13–15,43–49] In one instance[49] a number of the complexes formed have independently been shown to be of a bridged structure.[17] Kinetic studies of the different diphosphine complexes gave results so varied as to virtually exclude a common mechanism. For example, whereas no interaction of hydrogen with $[Rh(dppe)_2]^+$ has been observed directly, the dihydrido-complex $[H_2Rh(dppp)_2]^+$ is stable.[45] Bridged complexes analogous to the binuclear carbonyl-complex dictation $[Rh_2(CO)_4(dppb)_2(\mu\text{-}dppb)]^{2+}$,[50] complexes of type (**I**),[17] or the diop-bridged, catalytically active, ruthenium complex $[Ru_2Cl_4(diop)_2(\mu\text{-}diop)]$,[51] have not been considered as possible intermediates. However, the ligands that form the most active complexes are ligands that are known to also form complexes containing a monodentate diphosphine or bridged complexes.

The ligand $p\text{-}C_6H_4[CH_2P(CH_2CH_2PPh_2)_2]_2(=L)$ is able to chelate two separate metal atoms, thereby combining chelation and bridging features.[52] The catalytic activity of the binuclear complex $[RhCl(L)RhCl]$ is said to be due to the independent activity of each metal center, maintained at a distance from one another.[52] Comparison with the mononuclear analog, $[RhCl(triphos)]$, was not attempted.

The mechanisms of hydrogenation by neutral or cationic mononuclear complexes of rhodium are reasonably well understood[1,2,6,41,42,53,54] and are discussed in detail in Chapter 2. The system is, however, sufficiently complex that new information is constantly being obtained. For example, it has recently been shown that a solvated complex is kinetically important when benzene is the solvent.[55] This may have major implications in studies of complexes of aryl-phosphines or diphosphines, such as $Ph_2P(CH_2)_n PPh_2$ or diop.[15] It should be noted that for complexes of chelating chiral diphosphines (see Chapter 4), the enantioselective step involves the least stable of two possible catalytic intermediates; the complex which is stable, isolable, and present in highest concentration is not as important in determining the reaction rate and does not yield the major product.[41,42]

Homogeneous and heterogeneous hydrogenation catalysts have been compared,[56] but multinuclear homogeneous processes have frequently not been considered. This is surprising when one considers the often proposed multicenter mechanism for heterogeneous catalysis[5]:

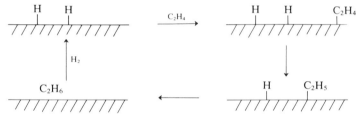

Similar multinuclear processes have been noted for homogeneous hydrogenation by complexes of metals other than rhodium.[7] For example, successive reactions of an alkene with two $[HCo(CN)_5]^{3-}$ anions give the corresponding alkyl complex anion and alkane.[57,58] Similarly, stoichiometric hydrogenation of a terminal alkene by $[HM(CO)_3(C_5H_5)]$ ($M$ = Cr, Mo, W) occurs with formation of $[M_2(CO)_6(C_5H_5)_2]$.[59] For $M$ = Cr, the binuclear complex product reacts with hydrogen to regenerate the hydrido-complex, and the system is cyclic.[59]

Dinuclear elimination reactions of metal alkyls (Equation 2.4, $L$ = all other ligands) have been studied less intensely[60] than, for example, similar reactions of metal amides (Equation 2.5).[61] The following characteristics are common to known dinuclear elimination reactions of metal alkyls: (i) a site of unsaturation must be present *cis* to the alkyl ligand; and (ii) the other reagent must be a hydrido-complex.[60]

$$L_n M(alkyl) + L_m MH \rightarrow H(alkyl) + L_n M \cdot ML_m \quad (2.4)$$

$$L_n MNR + L_m M'H \rightarrow HNR_2 + L_n M \cdot M'L_m \quad (2.5)$$

The existence of a step similar to Reaction 2.4 in the hydrogenation of alkenes or alkynes has not been proven, or even considered, for most systems. If the rate-determining step is either formation of an alkene complex, or hydrogen migration to form an alkyl complex, the reaction will be first-order in hydrido-metal complex, even if a binuclear alkane elimination step occurs as a major or minor reaction. On the available evidence such a binuclear process cannot be totally disregarded. When the metal centers are held in proximity to one another, as with binuclear complexes bridged by a diphosphine ligand, or a polymer with phosphine anchor-sites, such a step may well be significant.

## 3. HYDROFORMYLATION OF ALKENES

Improvements in rate, selectivity, and catalyst lifetime may be achieved when cobalt hydroformylation catalysts are modified by addition of a tertiary phosphine.[62,63] When the phosphine added is dppe, the complex formed, $[HCo(CO)_2(dppe)]$,[64] is a weak catalyst for the hydroformylation of 1-pentene, exhibiting poor selectivity to formation of the unbranched product (Equation 3.1a).[62]

$$HRC=CH_2 + H_2 + CO \begin{cases} \xrightarrow{(a)} RCH_2CH_2CHO \\ \xrightarrow{(b)} RCH(CH_3)CHO \end{cases} \quad (3.1)$$

In contrast, addition of $Ph_2P(CH_2)_nPPh_2$ ($n = 4$ or 5) to solutions of $[HCo(CO)_4]$ gives complexes of unknown structure that are of high activity and good selectivity.[62]

When the substrate to be hydroformylated is methyl acrylate, the above trend is reversed. The most active catalyst is that formed by addition of dppe.[63] The greatest activity was found for a catalyst with dppe:Co = 0.5.[63] However, direct comparisons of the two systems are not reasonable because the conditions of reaction, proportions of diphosphine to cobalt, and the substrate differ markedly. In each case hydrogenation is a competitive reaction.

The mononuclear catalyst $[Ru(CO)_3(dppe)]$ is of lower activity than monophosphine ruthenium complexes, but of higher selectivity.[65] Platinum complexes promoted with $SnCl_2$ are also of low activity if the phosphine ligand is dppe, but are much more active with dppb, or a related bis(phosphinomethyl)cycloalkane ligand,[66] which are known to form bridged structures.[17b]

Reaction of $Ph_2P(CH_2)_nPPh_2$ ($n = 2-5$), or corresponding cyclohexylphosphines, with $[Rh_2Cl_2(CO)_4]$ ($\overparen{P\quad P}$:Rh = 1) gave solutions active for the hydrogenation of $\alpha, \beta$-unsaturated esters, but of low hydroformylation activity.[49] Under these conditions, bridged complexes of type (**I**) (p. 217) are formed, except for the chelate complex of dppe, (**II**).[17] Although chelation of rhodium by the reaction of diop[67] or cyclo-1,2-$(Ph_2PO)_2C_6H_{10}$ with $[Rh_2Cl_2(CO)_4]$ has been assumed, formation of bridged complexes occurs under the conditions reported.[17] At greater $P\frown P$:Rh ratios, the hydroformylation activity and selectivity to formation of the unbranched aldehyde both increased.[49,67] The structures of complexes with ($P\frown P$:Rh > 1) vary with the chain-length of $P\frown P$; dppe forms $[Rh(dppe)_2]^+$, but other diphosphines form carbonyl complexes, some of which contain monodentate or bridging diphosphine ligands.[17,48,50,68] The variation in catalytic activity of chlororhodium complexes with diphosphine ligand chain-length is not, therefore, a simple correlation.

Addition of $Ph_2P(CH_2)_nPPh_2$ ($n = 1-4$),[70-74] diop,[67,69] triphosphines,[71,72] or soluble polymers with phosphine anchor-sites[72] to solutions containing $[HRh(CO)(PPh_3)_3]$ gives complexes active for the hydroformylation of a variety of alkenes under mild conditions. For complexes of diop, asymmetric induction is observed in appropriate cases.[67,69] Complexes of rhodium containing both a monophosphine and a diphosphine are efficient, unbranched product-selective catalysts for hydroformylation of 1-hexene,[73] as are complexes $[\{HRh(CO)(P\frown P)\}_2(\mu\text{-}P\frown P)]$.[74] At higher temperatures (105°C) the maximum activity for the dppb system occurs at the ratio dppb:Rh = 1.5.[74]

The predominant complex in a solution prepared by the dissolution of $[HRh(CO)(PPh_3)_3]$ under an atmosphere of CO and $H_2$ is $[HRh(CO)_2(PPh_3)_2]$ (Equation 3.2)[75-78]:

$$HRh(CO)(PPh_3)_3 + CO \rightleftharpoons HRh(CO)_2(PPh_3)_2 + PPh_3 \tag{3.2}$$

Comparison of spectroscopic data with the catalytic activity of solutions containing various proportions of a polyphosphine and $[HRh(CO)_2(PPh_3)_2]$ indicate both the existence of bridged complexes and their relatively high activity.[70-72]

Addition of a small amount of a diphosphine (Figure 6.5) or triphosphine (Figure 6.6) to a solution containing $[HRh(CO)_2(PPh_3)_2]$ increased the hydroformylation catalytic activity.[70-72] However, addition of equivalent or greater amounts of the same polyphosphines gave solutions of lower activity. For the diphosphine dppe, the following equilibria were proposed (Equations 3.3 and 3.4)[71]:

$$2HRh(CO)_2(PPh_3)_2 + dppe \rightleftharpoons 2PPh_3 + \{HRh(CO)_2(PPh_3)\}_2(dppe) \tag{3.3}$$

(**XIII**)                                (**XIV**)

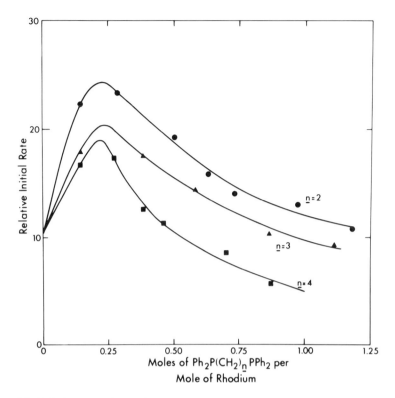

Figure 6.5. Relative rates of hydroformylation of 1-hexene catalyzed by benzene solutions containing [HRh(CO)(PPh$_3$)$_3$] and various amounts of the diphosphine Ph$_2$P(CH$_2$)$_n$PPh$_2$ (n-2-4). Reproduced, with permission, from Reference 72.

$$(\mathbf{XIV}) + \text{dppe} \rightleftharpoons 2\text{PPh}_3 + \{\text{HRh(CO)}_2(\text{dppe})\}_n \quad (3.4)$$
$$(\mathbf{XV})$$

Infrared spectra of solutions of mixtures of [HRh(CO)(PPh$_3$)$_3$] and dppe in molar ratios up to and exceeding 1:2 under hydroformylation conditions show, in each case, bands due to complex (**XIII**) and two substituted species in equilibria (Figure 6.7, Table 4). The bands assigned to (**XIV**) were predominant for solutions of composition dppe/Rh = 0.3–0.5. This ratio is close to but slightly higher than that of the solution of maximum catalytic activity (Figure 6.5).[71,72] The bands assigned to (**XV**) were the most intense for solutions of relatively greater amounts of dppe, which were also solutions of lesser catalytic activity. No other infrared bands (2300–1700 cm$^{-1}$) were observed.[71]

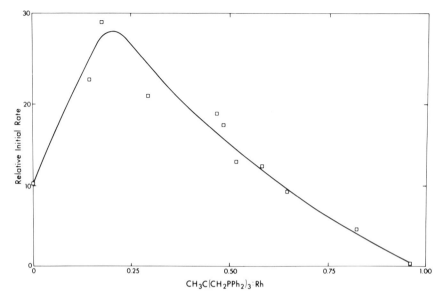

Figure 6.6. Relative rates of hydroformylation of 1-hexene catalyzed by benzene solutions containing [HRh(CO)(PPh$_3$)$_3$] and various amounts of (Ph$_2$PCH$_2$)$_3$CCH$_3$. Reproduced, with permission, from Reference 71.

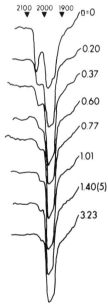

Figure 6.7. Infrared spectra of solutions in benzene of the products of reaction of [HRh(CO)-(PPh$_3$)$_3$] with CO and various amounts of Ph$_2$P(CH$_2$)$_2$PPh$_2$ under hydroformylation conditions.

Table 4. Infrared Spectra (2300–1700 cm$^3$) of Solutions of HRh(CO)$_2$(PPh$_3$)$_2$ and Ph$_2$PCH$_2$CH$_2$PPh$_2$ or (Ph$_2$PCH$_2$)$_3$CCH$_3$

| Complex | $\nu(\text{RhH})^a$ | $\nu(\text{CO})^a$ |
| --- | --- | --- |
| (XIII) | 2040(w) | 1977(vs), 1943(sh) |
| (XIV) | 2031(w) | 1954(vs)$^b$ |
| (XV) | 2015(w) | 1971(vs)$^b$ |
| (XVI) | 2004(w) | 1954(vs)$^b$ |
| (XVII) | 1977(w) | 1909(vs)$^b$ |

$^a$ Values in wave numbers.
$^b$ Nonoverlapping bands; shoulders or weak bands overlapping bands due to other complexes may also be present.

For solutions prepared from [HRh(CO)(PPh$_3$)$_3$] and the tripod triphosphine, tdpme, the following equilibria (Equations 3.5 and 3.6) were postulated[71]:

$$3\text{HRh(CO)}_2(\text{PPh}_3)_2 + \text{tdpme} \rightleftharpoons 3\text{PPh}_3 + \{\text{HRh(CO)}_2(\text{PPh}_3)\}_3(\text{tdpme}) \quad (3.5)$$

(**XVI**)

$$2(\textbf{XVI}) + 4\text{tdpme} \rightleftharpoons 6\text{PPh}_3 + 3\text{CO} + 3\{\text{HRh(CO)(tdpme)}\}_2 \quad (3.6)$$

(**XVII**)

The infrared spectrum of a solution containing equimolar amounts of rhodium and tdpme showed only bands due to the binuclear complex, (**XVII**) (Figure 6.8, Table 4).[71] Solutions containing lesser amounts of tdpme show, in addition to bands due to (**XIII**) or (**XVII**), bands at 2004 and 1954 cm$^{-1}$, assigned to the monosubstituted, tdpme-bridged complex, (**XVI**).[71] These bands were most intense for solutions of composition close to tdpme/Rh = 0.3, which is close to but slightly higher than the ratio for the solution of maximum catalytic activity (Figure 6.6).[71] Further evidence for the nature of (**XVI**) has been obtained from the $^1$Hnmr spectra at different temperatures of a solution of composition tdpme/Rh = 0.33. At 303 K no coupling of phosphorus to the hydrido-ligand was discernable. At 233 K a broad doublet ($\tau$18.0 p.p.m., $^2J$(P–H) ~ 4 Hz) was observed, and at 213 K a doublet of doublets [$^2J$(P–H) ~ $^2J$(P'–H) ~ 4 Hz] was resolved.[72] The resolution of the coupling constants at different temperatures confirms that the two phosphine ligands bonded to rhodium are different. The linear triphosphine, triphos, behaves similarly.[72]

The mechanism of cobalt-catalyzed hydroformylation (see Chapter 2) has been studied intensively. Early postulates that the rate-determining

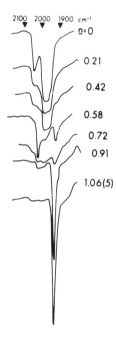

Figure 6.8. Infrared spectra of solutions in benzene of the products of reaction of [HRh(CO)-(PPh$_3$)$_3$] with CO and various amounts of (Ph$_2$PCH$_2$)$_3$CCH$_3$ under hydroformulation conditions.

step is hydrogen addition to an unsaturated, mononuclear, acyl-cobalt complex[79] have been challenged on the basis of *in situ* infrared spectroscopic studies.[80–82] The mechanism has been shown to include a binuclear reaction (Equation 3.7) as the major or only reductive elimination step.[80,81]

$$\text{HCoL}_n + \text{RCO·CoL}_m \rightarrow \text{RCHO} + \text{Co}_2L_{n+m} \qquad (3.7)$$

It has been suggested[80,81] that the acyl complex is [RCO·Co(CO)$_4$], or [RCO·Co(CO)$_3$PBu$_3$] for modified catalysts, but a site of unsaturation *cis* to the acyl ligand may be required (c.f., Reference 60). A more probable formulation is therefore [RCO·Co(CO)$_3$], or [RCO·Co(CO)$_2$PBu$_3$]. A binuclear, free-radical mechanism for the cobalt-catalyzed hydroformylation of styrene or other conjugated substrates has also been proposed.[84] These studies are far-reaching,[83] especially because similar binuclear elimination steps have not received much consideration in studies of rhodium hydroformylation catalysts.[72,83]

Cobalt complexes with diphosphine ligands have shown features consistent with the above proposals. With an equimolar or excess amount of dppe, which forms a mononuclear, chelated complex,[64] poor activity is obtained.[62] However, with $Ph_2P(CH_2)_nPPh_2$ ($n$ = 4,5), which show a greater tendency to form bridged cobalt complexes,[85] good activity and selectivity are obtained.[62] The maximum activity of the dppe-modified cobalt catalyst for hydroformylation of methyl acrylate occurs at a ratio dppe/Rh = 0.5, suggestive of formation of a binuclear system.[63] Similarly, for $PtCl_2/SnCl_2$/diphosphine hydroformylation catalysts it was found that dppe depressed the activity of the platinum complex, but that dppb or related diphosphino-cycloalkane ligands considerably improved the activity.[66] For complexes containing no $SnCl_2$, dppe chelates platinum, but dppb forms bridged complexes.[17b]

The reaction of $[HRh(CO)_2(PPh_3)_2]$ with a diphosphine (P⌒P/Rh = $0.5^{(70-72)}$ or $1.5^{(74)}$) to give a solution containing predominantly a binuclear, diphosphine-bridged complex has been established spectroscopically, as discussed above. A binuclear elimination step in the mechanism of hydroformylation using mononuclear rhodium catalysts cannot be excluded on the basis of data available.[86] The coincidence between the maximum activity of solutions containing rhodium and a diphosphine, and the formation of bridged complexes in maximum concentration under the same conditions suggests that binuclear elimination may indeed be both possible and important. Maintenance of two rhodium, or other metal, centers in close proximity would favor such a reaction (Equation 3.8):

$$HRhL_n \cdot P\frown P \cdot RhL_m COR \rightarrow RCHO + Rh_2L_{n+m}(P\frown P)$$
$$\xrightarrow{H_2,L} \{HRhL_n\}_2(P\frown P) \quad (3.8)$$

Reaction of binuclear complexes of rhodium(O) with hydrogen to regenerate hydridorhodium(I) complexes under hydroformylation conditions has been established.[75,77] The extension of these arguments to complexes with triphosphines or polymers with a high density of phosphine anchor-site has been made for complexes of cobalt[12] and rhodium.[12,72]

## 4. CONCLUDING REMARKS

In Sections 2 and 3, I have attempted to show that when two metal centers are bridged by a diphosphine they do not necessarily react as independent catalytic centers, but may react in a binuclear manner.

The relationship of such systems to polymers,[12] or clusters,[87,89] both of which are important in catalysis, is strong, but comparisons with reactions between neighboring sites at surfaces are more tenuous.[5,87,88,90]

Important developments in the field of bridged binuclear catalysts will undoubtedly come from studies of unsymmetric diphosphines, e.g., $R_2PCH_2PR'_2$,[91] phosphines also containing nitrogen,[92,93] sulphur,[94] or other donor atoms, and other innovations yet to be described.

The importance of bridged, binuclear complexes for decarbonylation reactions (see Chapter 11), the water-gas shift reaction (see Chapter 5), and other catalytic applications is only now beginning to be investigated.

## Notes Added in Proof

The cationic complex reported to be $[Rh_2Cl(CO)_2(dppm)_2(\mu\text{-PhCCPh})]^+$ has been reinvestigated, and the spectroscopic evidence reevaluated. The differences in $^{31}P$ nmr parameters between solutions containing $[Rh_2Cl(CO)_2(dppm)_2]^+$ alone and those with added PhCCPh may be due to differences in conditions alone, such as solvent effects. However, the cationic complex (**XVIII**) has been synthesized.[95]

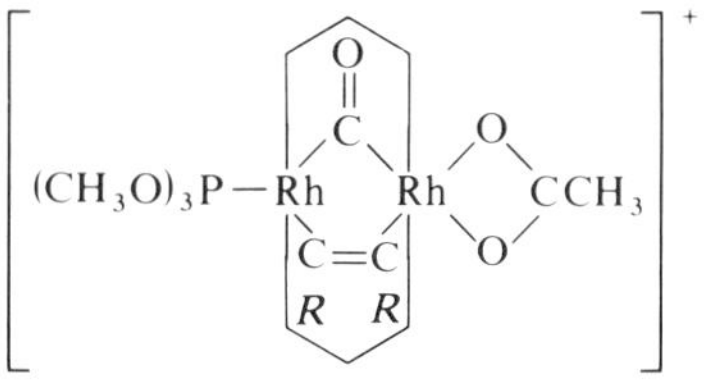

$$\left[ (CH_3O)_3P-Rh\underset{C=C}{\overset{O\underset{\|}{C}}{\underset{R\quad R}{\big|}}}Rh\underset{O}{\overset{O}{\big\backslash}}CCH_3 \right]^+$$

**XVIII**, $R = CH_3CO_2$

The use of 1,1'-bis(diphenylphosphino)ferrocene as a bidentate ligand in rhodium hydroformylation catalysis has been studied in detail.[96] Complexes containing this ligand as a bridge are remarkably stable and effective as catalysts. Although a mechanism involving independent reactions of each metal center as catalyst was proposed, the authors allowed for the possibility of an internuclear interaction.

The diversity of results of studies of hydrogenation and hydroformylation catalysis have prompted a report on the factors governing the different mechanisms of binuclear elimination reactions of complexes leading to carbon–hydrogen bond formation.[97]

## ACKNOWLEDGMENTS

I am grateful to Drs. J. T. Mague, A. L. Balch, O. R. Hughes, M. D. Fryzuk, and K. R. Grundy for communication of results prior to publication, and to these colleagues and Dr. J. P. Collman for helpful and stimulating discussions.

# REFERENCES

1. G. W. Parshall, *Homogeneous Catalysis* (Wiley-Interscience of John Wiley and Sons, New York, 1980).
2. J. P. Collman and L. S. Hegedus, *Principles and Applications of Organotransition Metal Chemistry* (University Science Books, Mill Valley, CA 1980).
3. R. Fowler, H. Connor, and R. A. Baehl, *Hydrocarbon Proc.* **1976,** 247–249.
4. J. Falbe and E. G. Hancock, *Propylene and Its Industrial Derivatives* (Halstead, New York, 1973), Chapt. 9.
5. P. N. Rylander, *Catalytic Hydrogenation in Organic Syntheses* (Academic Press, New York, 1979).
6. F. J. McQuillin, *Homogeneous Hydrogenation in Organic Chemistry* (Reidel, Dordrecht, 1976).
7. R. E. Harmon, S. K. Gupta, and D. J. Brown, *Chem. Rev.* **73,** 21–52 (1973).
8. C. A. Tolman, *Chem. Rev.* **77,** 313–348 (1977).
9. C. A. McAuliffe and W. Levason, *Phosphine, Arsine, and Stibine Complexes of the Transition Metals* (Elsevier, Amsterdam, 1979).
10. W. Levason and C. A. McAuliffe, *Adv. Inorg. Chem. Radiochem.* **14,** 173–253 (1972).
11. a) R. B. King, P. N. Kapoor, and R. N. Kapoor, *Inorg. Chem.* **10,** 1841–1850 (1971); b) R. B. King and J. C. Cloyd, Jr., *Inorg. Chem.* **14,** 1550–1554 (1975); c) R. B. King, J. A. Zinich, and J. C. Cloyd, Jr., *Inorg. Chem.* **14,** 1554–1559 (1975); d) M. M. Taqui Khan and A. E. Martell, *Inorg. Chem.* **13,** 2961–2966 (1974).
12. D. C. Bailey and S. H. Langer, *Chem. Rev.* **81,** 109–148 (1981).
13. a) D. A. Slack and M. C. Baird, *J. Organometal. Chem.* **142,** C69–C72 (1977); b) D. A. Slack, I. Greveling, and M. C. Baird, *Inorg. Chem.* **18,** 3125–3132 (1979).
14. R. R. Schrock and J. A. Osborn, *J. Amer. Chem. Soc.* **93,** 2397–2407 (1971).
15. J. Halpern, D. P. Riley, A. S. C. Chan, and P. L. Pluth, *J. Amer. Chem. Soc.* **99,** 8055–8057 (1977).
16. F. C. March, R. Mason, K. M. Thomas, and B. L. Shaw, *J. C. S., Chem. Comm.* **1975,** 584–585.
17. a) A. R. Sanger, *J. C. S., Dalton Trans.* **1977,** 120–129; b) A. R. Sanger, *J. C. S., Dalton Trans.* **1977,** 1971–1976.
18. A. L. Balch, *J. Amer. Chem. Soc.* **98,** 8049–8054 (1976).
19. A. L. Balch and B. Tulyathan, *Inorg. Chem.* **16,** 2840–2845 (1977).
20. M. Cowie and S. K. Dwight, *Inorg. Chem.* **19,** 2500–2507 (1980).
21. J. T. Mague, *Inorg. Chem.* **8,** 1975–1981 (1969).
22. M. Cowie and S. K. Dwight, *Inorg. Chem.* **20,** 1534–1538 (1981).
23. A. L. Balch, unpublished results.
24. M. Cowie and S. K. Dwight, *Inorg. Chem.* **19,** 2508–2513 (1980).
25. E. L. Muetterties, *Pure Appl. Chem.* **50,** 941–950 (1978).
26. A. Brenner and D. A. Hucul, *J. Amer. Chem. Soc.* **102,** 2484–2487 (1980).
27. A. R. Sanger, *Preprints, 7th Canadian Symposium on Catalysis* (Edmonton, 1980), pp. 67–74.
28. ———, *Canad. J. Chem.* **60,** 1363–1367 (1982).
29. ———, *J. C. S., Dalton Trans.* **1981,** 228–233.
30. C. P. Kubiak and R. Eisenberg, *J. Amer. Chem. Soc.* **99,** 6129–6131 (1977).
31. ———, *Inorg. Chem.* **19,** 2726–2732 (1980).
32. C. P. Kubiak, C. Woodcock, and R. Eisenberg, *Inorg. Chem.* **19,** 2733–2739 (1980).
33. J. T. Mague and A. R. Sanger, *Inorg. Chem.* **18,** 2060–2066 (1979).
34. A. R. Sanger, *Preprints, Sixth Canadian Symposium on Catalysis* (Ottawa, 1979), pp. 34–43.

35. J. T. Mague, unpublished results.
36. A. R. Sanger, unpublished results.
37. C. P. Kubiak and R. Eisenberg, *J. Amer. Chem. Soc.* **102**, 3637–3639 (1980).
38. M. Cowie and T. G. Southern, *J. Organometal. Chem.* **193**, C46–C50 (1980).
39. T. S. Cameron, S. P. Deraniyagala, K. R. Grundy, and K. Jochem, unpublished results.
40. M. D. Fryzuk, *Inorg. Chim. Acta*, in press.
41. J. M. Brown and P. A. Chaloner, *J. C. S., Chem. Comm.* **1980**, 344–346.
42. A. S. C. Chan, J. J. Pluth, and J. Halpern, *J. Amer. Chem. Soc.* **102**, 5952–5954 (1980).
43. W. R. Cullen, A. Fenster, and B. R. James, *Inorg. Nuclear Chem. Letters* **10**, 167–170 (1974).
44. R. G. Ball, B. R. James, D. Mahajan, and J. Trotter, *Inorg. Chem.* **20**, 254–261 (1981).
45. B. R. James and D. Mahajan, *Canad. J. Chem.* **57**, 180–187 (1979).
46. J. C. Poulin, T. P. Dang, and H. B. Kagan, *J. Organometal Chem.* **84**, 87–92 (1975).
47. J. M. Brown and D. Parker, *J. C. S., Chem. Comm.* **1980**, 342–344.
48. M. Tanaka and I. Ogata, *J. C. S., Chem. Comm.* **175**, 735.
49. M. Tanaka, T. Hayashi, and I. Ogata, *Bull. Chem. Soc. Japan* **50**, 2351–2357 (1977).
50. L. H. Pignolet, D. H. Doughty, S. C. Nowicki, M. P. Anderson, and A. L. Casalnuovo, *J. Organometal. Chem.* **202**, 211–223 (1980).
51. a) B. R. James, D. K. W. Wang, and R. F. Voigt, *J. C. S., Chem. Comm.* **1975**, 574–575; b) B. R. James, R. S. McMillan, R. H. Morris, and D. K. W. Wang, *Adv. Chem. Series* **167**, 122–135 (1978).
52. M. M. Taqui Khan, M. Ahmed, and B. Swamy, *Indian J. Chem.* **20A**, 359–362 (1981).
53. J. Halpern, *Inorg. Chim. Acta* **50**, 11–19 (1981).
54. J. Halpern, T. Okamoto, A. Zakhariev, *J. Molecular Catal.* **2**, 65–68 (1976).
55. M. H. J. M. de Croon, P. F. M. T. van Nisselrooij, H. J. A. M. Kuipers, and J. W. E. Coenen, *J. Molecular Catal.* **4**, 325–335 (1978).
56. S. Siegel, *J. Catal.* **30**, 139–145 (1973).
57. J. Kwiatek, *Catal. Rev.* **1**, 37–72 (1968).
58. J. Halpern and L. Y. Wong, *J. Amer. Chem. Soc.* **90**, 6665–6669 (1968).
59. a) A. Miyake and H. Kondo, *Angew. Chem. Internat. Ed. Eng.* **7**, 631–632 (1968); b) A. Miyake and H. Kondo, *Angew. Chem. Internat. Ed. Eng.* **7**, 880–881 (1968).
60. J. Norton, *Accounts Chem. Res.* **12**, 139–145 (1979).
61. M. F. Lappert, P. P. Power, A. R. Sanger, and R. C. Srivastava, *Metal and Metalloid Amides* (Wiley, New York, and Ellis Horwood, Chichester, 1980), Chapt. 12.
62. L. H. Slaugh and R. D. Mullineux, *J. Organometal. Chem.* **13**, 469–477 (1968).
63. K. Murata and A. Matsuda, *Bull. Chem. Soc. Japan* **53**, 214–218 (1980).
64. T. Ikariya and A. Yamamoto, *J. Organometal. Chem.* **116**, 231–237 (1976).
65. R. A. Sanchez-Delgado, J. S. Bradley, and G. Wilkinson, *J. C. S., Dalton Trans.* **1976**, 399–404.
66. Y. Kawabata, T. Hayashi, and I. Ogata, *J. C. S., Chem. Comm.* **1979**, 462–463.
67. C. Salomon, G. Consiglio, C. Botteghi, and P. Pino, *Chimia* **27**, 215–217 (1973).
68. B. R. James and D. Mahajan, *Canad. J. Chem.* **58**, 996–1004 (1980).
69. G. Consiglio, C. Botteghi, C. Salomon, and P. Pino, *Angew. Chem. Internat. Ed. Eng.* **12**, 669–670 (1973).
70. A. R. Sanger, *Preprints, Fifth Canadian Symposium on Catalysis* (Calgary, 1977), pp. 281–287.
71. A. R. Sanger, *J. Molecular Catal.* **3**, 221–226 (1977–1978).
72. A. R. Sanger and L. R. Schallig, *J. Molecular Catal.* **3**, 101–109 (1977–1978).
73. O. R. Hughes, *Hydroformylation Catalyst*, United States Patent No. 4 201 728 (1980).
74. O. R. Hughes and J. D. Unruh, *J. Molecular Catal.* **12**, 71–83 (1981).
75. D. Evans, G. Yagupsky, and G. Wilkinson, *J. Chem. Soc. (A)* **1968**, 2660–2665.
76. G. Yagupsky, C. K. Brown, and G. Wilkinson, *J. Chem. Soc. (A)* **1970**, 1392–1401.

77. C. K. Brown and G. Wilkinson, *J. Chem. Soc. (A)* **1970**, 2753–2764.
78. D. E. Morris and H. B. Tinker, *Chem. Tech.* **2**, 554–559 (1972).
79. R. F. Heck and D. S. Breslow, *J. Amer. Chem. Soc.* **83**, 4023–4027 (1961).
80. M. van Boven, N. H. Alemdaroglu, and J. M. L. Penninger, *Ind. Eng. Chem., Prod. Res. Dev.* **14**, 259–264 (1975).
81. N. H. Alemdaroglu, J. L. M. Penninger, and E. Oltay, *Monatsh. Chem.* **107**, 1153–1165 (1976).
82. R. Whyman, *J. Organometal. Chem.* **66**, C23–C25 (1974).
83. J. P. Collman and L. S. Hegedus, *Principles and Applications of Organotransition Metal Chemistry* (University Science Books, Mill Valley, CA, 1980), pp. 424–426.
84. J. Halpern, *Pure Appl. Chem.* **51**, 2171–2182 (1979).
85. D. J. Thornhill and A. R. Manning, *J. C. S., Dalton Trans.* **1973**, 2086–2090.
86. J. P. Collman and L. S. Hegedus, *Principles and Applications of Organotransition Metal Chemistry* (University Science Books, Mill Valley, CA, 1980), pp. 423, 429.
87. E. L. Muetterties, *Science* **196**, 839–848 (1977).
88. A. K. Smith and J. M. Basset, *J. Molecular Catal.* **2**, 229–241 (1977).
89. E. L. Muetterties, T. N. Rhodin, E. Band, C. F. Brucker, and W. R. Pretzer, *Chem. Rev.* **79**, 91–137 (1979).
90. a) H. Storch, N. Golumbic, and R. Anderson, *The Fischer-Tropsch and Related Syntheses* (John Wiley and Sons, New York, 1951); b) C. N. Satterfield, *Heterogeneous Catalysis in Practice* (McGraw-Hill, New York, 1980).
91. S. O. Grim, P. H. Smith, I. J. Colquhoun, and W. McFarlane, *Inorg. Chem.* **19**, 3195–3198 (1980).
92. J. P. Farr, M. M. Olmstead, C. H. Hunt, and A. L. Balch, *Inorg. Chem.* **20**, 1182–1187 (1981).
93. R. G. Nuzzo, D. Feitler, and G. M. Whitesides, *J. Amer. Chem. Soc.* **101**, 3683–3685 (1979).
94. X. Guo and A. R. Sanger, unpublished results.
95. J. T. Mague, unpublished results.
96. J. D. Unruh and J. R. Christenson, *J. Molecular Catal.* **14**, 19–34 (1982).
97. M. J. Nappa, R. Santi, S. P. Diefenbach, and J. Halpern, *J. Amer. Chem. Soc.* **104**, 619–621 (1982).

# 7

# Functionalized Tertiary Phosphines and Related Ligands in Organometallic Coordination Chemistry and Catalysis

Thomas B. Rauchfuss

## 1. INTRODUCTION

The prospect of predictably manipulating the chemo- and stereoselectivity of metal catalysts is both aesthetically and economically attractive. The most straightforward way to affect the chemical behavior of a given metal ion is through changes in its ligands. It is well recognized that changes within the first coordination sphere of a metal ion can have a relatively dramatic impact on the properties of the entire complex. On the other hand, modifications of the ligand superstructure and substituents have subtler and somewhat more predictable effects. Such ligand modifications as applied to the coordination chemistry of tertiary phosphines have begun to receive increasing attention because of the applicability of this technology to catalysis. In this chapter, I discuss several studies which have concerned the coordination chemistry of tertiary phosphine ligands which bear chemi-

---

*Dr. Thomas B. Rauchfuss* • School of Chemical Sciences, University of Illinois, Urbana-Champaign, Illinois, 61801.

cally active substituents. This review is not comprehensive, rather the ligand systems discussed were selected because of the novel properties they impart to their complexes, the usefulness of their complexes in mechanistic or catalytic studies, or their synthetic versatility. Areas related to the theme of this review which were intentionally de-emphasized are di- and polyphosphines, chiral phosphines, and polymer-bound phosphines. These topics are covered more fully in other chapters of this book.

## 2. ETHER PHOSPHINES

The first reported application of a functionalized phosphine resulted from work done at Monsanto on the use of chiral methoxyaryl phosphines for asymmetric hydrogenation. It was found that the rhodium(I) complexes of the resolved forms of cyclohexylanisylmethylphosphine, $P(C_6H_{11})$-$(o$-$C_6H_4OCH_3)CH_3$, catalyze the hydrogenation of acylaminoacrylic esters with good enantioselectivity.[1,2] Subsequent refinements by the same group culminated in the application of a system based on 1,2-$R$,$R$-bis(phenylanisylphosphino)ethane whose synthesis (Figure 7.1) is itself a significant achievement.[3] In the cases of both of these chiral phosphine ligands, the presence of the methoxy group is crucial to the performance of the catalyst systems since its replacement by sterically comparable substituents results in substantially decreased enantioselectivity.[2] There exist literature precedents which demonstrate that an ether group in similar ligands can either chelate to the metal center or interact with the polar catalytic substrates.

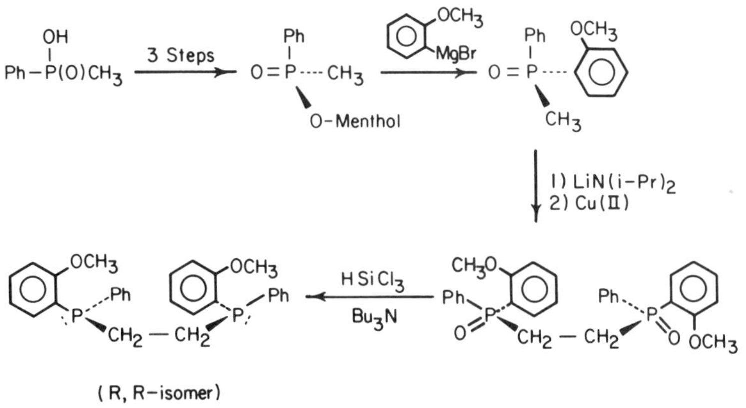

Figure 7.1. The synthesis of 1, 2-$R$, $R$-bis(anisylphenylphosphino)ethane.

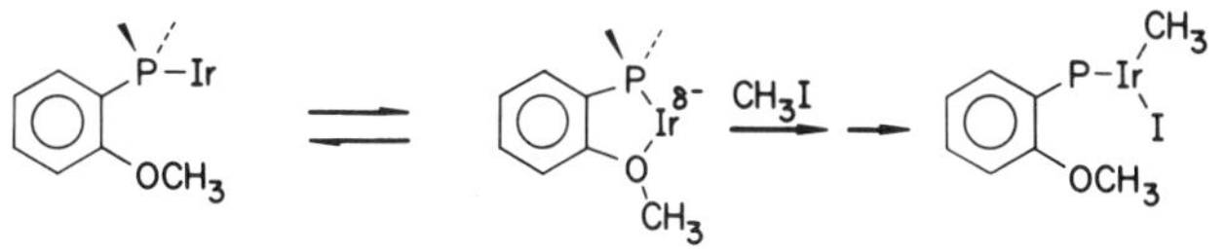

Figure 7.2. Proposed interaction of methoxy group with iridium in the reaction of methyl iodide with trans-IrCl(CO)(o-Me$_2$PC$_6$H$_4$OCH$_3$)$_2$.

Shaw and co-workers have described the influence of o-anisylphosphines vis-à-vis p-anisylphosphines on the reactivity of compounds of the type trans-IrCl(CO)(PMe$_2$C$_6$H$_4$OCH$_3$)$_2$.[4] It was found that these isomeric complexes oxidatively add methyl iodide at very different rates, the o-methoxy derivative reacting ca. 500 times faster than the p-isomer, to give the iridium(III) compounds IrCl(CH$_3$)I(CO)(PMe$_2$C$_6$H$_4$OCH$_3$)$_2$. Since the donor strengths (at phosphorus) for these isomeric ligands are quite similar, it follows that the equilibrium basicities of two iridium(I) complexes are also similar. These findings led these workers to invoke a neighboring-group effect, i.e., anchimeric assistance, to explain the unusual rate enhancement observed for the o-anisylphosphine complex. The methoxy group is proposed to function as a donor ligand which enhances the nucleophilicity of the iridium(I) center and hence the rate of the oxidative addition (Figure 7.2). The methoxy group is in effect a kind of internal catalyst since its influence is almost purely kinetic. This mechanistic rationale would appear to receive support from recent studies which indicate that the rates of oxidative addition of polar substrates to square planar iridium(I) are enhanced by the binding of anions to the 16e$^-$ metal center.[5] The o-anisylphosphines can simulate this situation by coordination of the oxygen donor.

X-ray crystallography has established that the ether oxygens are coordinated in the complex RuCl$_2$(POMe)$_2$,[6] where POMe is o-diphenylphosphinoanisole (Figure 7.3). This remarkable compound is easily prepared in high yield and can be isolated as red, air-stable crystals. The structural study revealed very long Ru···O distances (2.28 ± 0.03 Å), while the Ru···P distances were found to be short and similar to those found in *five-coordinate* ruthenium(II) phosphine complexes. Consistent with the structural results, the ether donors are easily displaced by good $\pi$-acceptor ligands. RuCl$_2$(POMe)$_2$ represents a very unusual example of a complex which reacts readily with carbon monoxide, but is completely resistant to oxygen. This observation is taken to be significant since it demonstrates that this unsymmetrical chelate ligand protects the metal from oxidation while permitting softer substrates to coordinate. Because of the lability associated with one of its two donors, o-diphenylphosphinoanisole has been described as a hemilabile chelating agent. Such ligands are of interest since they

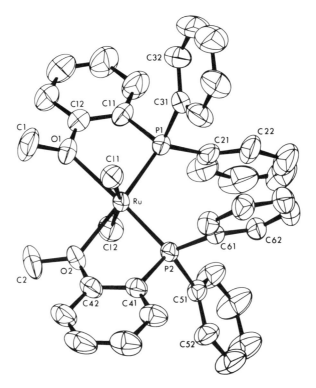

Figure 7.3. ORTEP plot of $RuCl_2(o\text{-}Ph_2PC_6H_4OCH_3)_2$. (Reprinted with permission from *Inorganic Chemistry*.)

represent a means of stabilizing incipiently coordinatively unsaturated complexes. In this regard, it is important to note that although $RuCl_2(POMe)_2$ is not itself catalytically active, it reacts quickly with basic ethanol to afford a yellow, highly active, olefin hydrogenation–isomerization catalyst.[7]

1,5-*Bis*(diphenylphosphino)-3-oxopentane (POP) is a potentially tridentate P$\cdots$O$\cdots$P chelating ligand. This coordination mode has been established crystallographically in [Rh(POP)CO]PF$_6$, although no reactivity studies have been reported for the complex.[8] Also described was 1,11-*bis*(diphenylphosphino)-3,6,9-trioxoundecane(POOOP) and [Rh(POOOP·H$_2$O)CO]PF$_6$. In this complex, the polyetherdiphosphine not only chelates to the rhodium(I) moiety but binds water through hydrogen bonding to the oxygen donors.[8] It is intriguing that the structures of the two rhodium complexes are very similar insofar as both contain *trans* phosphines adjacent to carbonyl and oxygen donors, the latter being derived from an ether or water (Figure 7.4). It is conceivable that compounds of

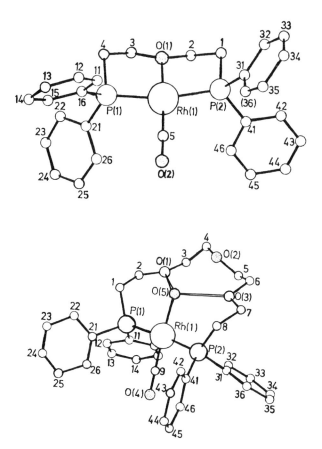

Figure 7.4. Molecular structures of $[Rh(Ph_2PCH_2CH_2OCH_2CH_2PPh_2)\text{-}(CO)]^+$ and $[Rh(Ph_2PCH_2CH_2OCH_2CH_2OCH_2CH_2OCH_2CH_2PPh_2 \cdot H_2O)\text{-}(CO)]^+$. (Reprinted with permission from *J. C. S., Dalton Trans.*)

this type could be useful for reactions which involve hydration or are influenced by inter-ligand hydrogen-bonding effects.

A recent report has described how the ether component of polyether-phosphinite ligands can be used to enhance the reactivity of coordinated carbon monoxide. The polyether diphosphinite complex shown in Figure 7.5 resembles a crown ether, which in turn activates a molybdenum-bound carbonyl towards nucleophilic attack by organolithium compounds.[9] The acyl–lithium interaction of the product is stabilized by complexation of the lithium ion by the polyether chelating agent. The profound influence of Lewis acid–acyl interactions on the facility of the migratory insertion reaction has been previously well-established.[10] In the case of the molybdenum diphosphonite, the effect is clearly a result of the ligand substituents,

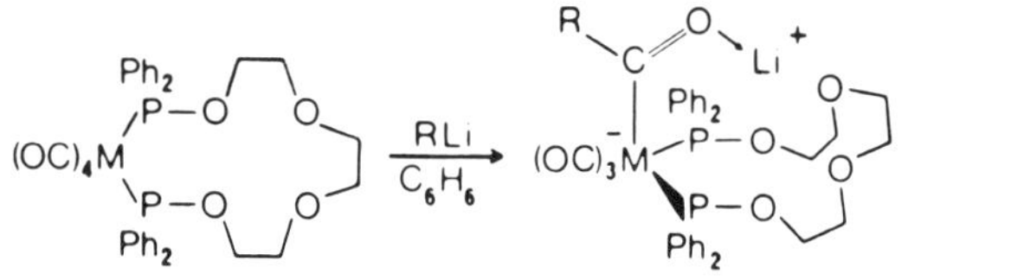

Figure 7.5. Ligand assisted activation of CO in a polyetherdiphosphinite carbonyl complex. (Reprinted with permission from *J. Am. Chem. Soc.*)

since the control compound, cis-Mo(PPh$_2$OCH$_3$)$_2$(CO)$_4$, does *not* react with the same organolithium reagents under mild conditions. These studies establish how remote, but chemically active substituents, can favorably influence the reactivity of the central metal ion.[11]

## 3. AMINOPHOSPHINES

In view of the extensive coordination chemistry of polyamine chelating agents, it is not too surprising that the aminophosphine ligands have become relatively common. Of particular interest are those chelating agents that contain primary or secondary amines whose presence permits further functionalization or promotes metalation.

*Bis*(2-diphenylphosphinoethyl)amine can be prepared on a large scale from sodium diphenylphosphide and the nitrogen mustard, *bis*(2-chloroethyl)amine.[12] The secondary amine function can be acetylated readily, thus providing a reliable means for attaching the diphosphine moiety to a number of molecular and polymeric substrates. These derivatives can then be combined with labile rhodium(I) precursors to afford modified hydrogenation catalysts. One innovative application of this methodology involved the preparation of a protein-bound hydrogenation catalyst. In particular, the P-N-P unit was linked to biotin via amide formation (Figure 7.6).[13] Biotin itself is known to be very tightly bound to the globular protein avidin, and it was found that the rhodium(I) complex of the biotin-derivatized phosphine was also bound by the same protein. The rhodium-protein complex functions efficiently as an asymmetric hydrogenation catalyst, the chirality being transmitted from the chiral protein environment to the coordination sphere of the rhodium ion. While the actual enantioselectivities observed for this particular catalyst are modest

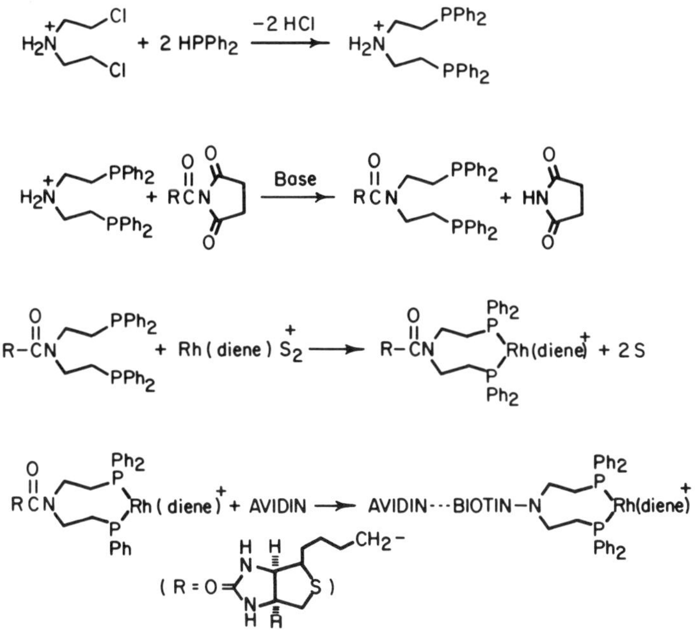

Figure 7.6. Synthetic sequence for the preparation of a protein bound rhodium hydrogenation catalyst.

by 1982 standards, this work demonstrates that appropriately functionalized phosphines can effect the marriage between organometallic and biological homogeneous catalysis.

The aminophosphines depicted in Figure 7.7 have been described.[14,15] Both undergo facile deprotonation in the presence of metal ions to yield anionic P–N or P–N–P chelating agents. Preliminary results are that these hybrid ligands may induce unusual reactivity at the bound metal ion.[14]

Phosphoranes have attracted attention as ligands for reactive metal complexes.[16] Riess and coworkers have focused on the distinctive coordination chemistry of the bicyclophosphorane $Ph(H)P(OCH_2CH_2)_2N$ (phoran) which can, in principle, exist in different tautomeric forms (Figure 7.8).[17] The novelty associated with this ligand system concerns its ability to

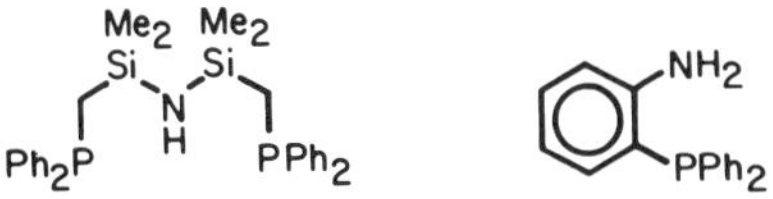

Figure 7.7. 1, 3-*Bis*[(diphenylphosphino)methyl]tetramethyldisilazane and *o*-diphenylphosphinoaniline.

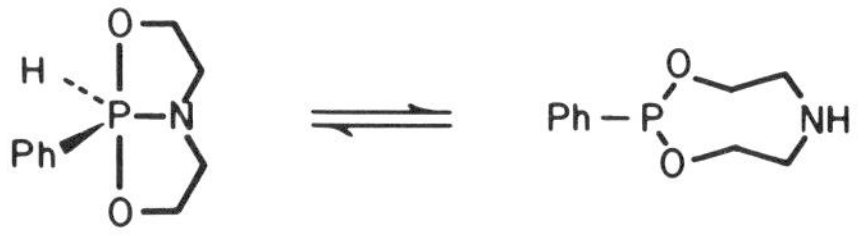

Figure 7.8. The phosphorane–phosphonite tautomerism in phoran.

generate coordinative unsaturation via this tautomerization within the coordination sphere of a metal complex. In this way, phoran represents a particularly innovative version of a hemilabile chelating agent. In the complexes Rh(phoran)$_2$Cl and RhCl(CO)(phoran), the rhodium ions lie in the cradle of the cyclic phosphine–amine chelating agent.[18] These complexes are simply prepared from the phosphorane and Rh$_2$Cl$_2$L$_4$ (L = C$_2$H$_4$, CO) via reactions which illustrate the facility of the ring-chain tautomerization. The lability of the nitrogen donor in the carbonyl complex may explain its reactivity as a hydroformylation catalyst.[19] The enhanced reactivity of other metal complexes of P–N chelates has been previously observed by Roundhill.[20]

The cationic complex CpMo(CO)$_2$(phoran)$^+$ can be reversibly deprotonated with MeLi to generate the $\eta^2$-P-N and $\eta^2$-P-O phosphoranide linkage isomers.[21] Different chemistry has been observed for CpFe(CO)· (phoran)$^+$ where deprotonation of the amine leads to migration of the phenyl group from P to Fe concomitant with the formation of a caged-type phosphite[22] (Figure 7.9). The phenyl migration from P to Fe is all the

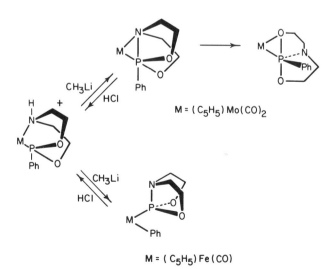

Figure 7.9. Rearrangement pathways observed for organometallic derivatives of phoran.

more remarkable since it is reversible: treatment of the Fe-phenyl derivative with acid promotes the migration (migratory insertion?) of the phenyl group back to phosphorus concomitant with reformation of the bidentate P–N chelate. These interesting interconversions are beautiful illustrations of the novel chemical behavior peculiar to 1,5-diheterocyclooctanes, a theme of recurring interest to chemists.[23]

## 4. CARBONYLPHOSPHINES

The synthetic versatility of the carbonyl functional group virtually guarantees that the carbonylphosphines will become a heavily exploited class of ligands. This view is strengthened by the recognition of the key role that organic carbonyls play in classical coordination chemistry, particularly in the formation of imino- and enolate-type ligands.[24]

The carbonylphosphine particularly worthy of attention is o-diphenylphosphinobenzoic acid.[25–27] This compound is conveniently prepared from sodium diphenylphosphide and the salts of o-chlorobenzoic acid. Workers at Shell have shown that the combination of this ligand with bis(1,5-cyclooctadiene)nickel(0) affords a very active catalyst for ethylene oligomerization.[26,28] This technology is related to the highly successful Shell higher olefin process (S.H.O.P.) which is currently used for preparation of detergent-range $\alpha$-olefins.[29] Understandably, detailed discussions of this catalytic process have not been published by the Shell workers, although insight is provided from other sources.[30] In view of their applicability and their ease of synthesis, it is surprising that the o-diorganophosphinobenzoates have not received greater attention.

Keim and co-workers have found that the treatment of Ni(COD)$_2$ with both triphenylphosphine and the phosphorane, Ph$_3$PCHC(O)Ph, affords a nickel(II) chelate complex formally derived from the enolate of Ph$_2$PCH$_2$C(O)Ph (Figure 7.10).[30] This crystalline compound, which can be conveniently prepared on a large scale, has been characterized by single crystal X-ray diffraction. Much like o-diphenylphosphinobenzoate, the novel enolate ligand functions as an anionic P–O donor. What is particularly intriguing is that its nickel complex also efficiently catalyzes the formation of linear $\alpha$-olefins from ethylene.

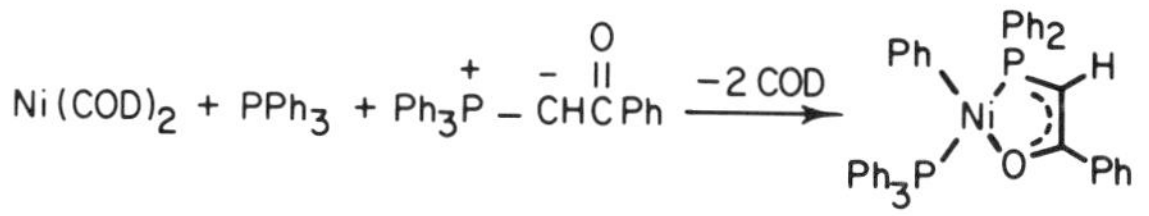

Figure 7.10. Synthesis of a phenylnickel(II) phosphinoenolate from nickel(0) and a phosphorane.

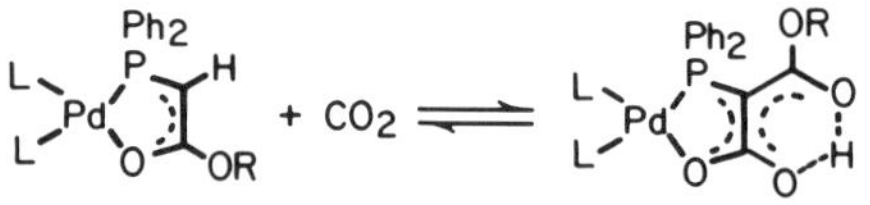

Figure 7.11. Reversible $CO_2$ fixation by a palladium(II) phosphinoenolate complex.

A thorough study of palladium(II) complexes derived from the enolate of ethyl-2-diphenylphosphinoacetate, $Ph_2PCH_2CO_2C_2H_5$, has been published.[31] In many of these complexes, the P–O chelates play a relatively passive role not unlike that found for o-diphenylphosphinophenoxides.[32] The derivatives shown in Figure 7.11, however, distinguish themselves by reversibly binding $CO_2$. The $CO_2$ is captured via addition to the enolate carbon and, once attached, its binding is stabilized by intramolecular hydrogen bonding from the pendant functional groups. This finding is important because of the topical interest in using $CO_2$ as a possible carbon feedstock.

Reaction of methyl o-diphenylphosphinobenzoate, o-$Ph_2PC_6H_4$-$CO_2CH_3$, with the potassium enolate of pinacolone yields the phosphine–diketone ligand shown in Figure 7.12. This chelating agent, called HacacP, is representative of what is known as a compartmentalized ligand, i.e., one which features two or more geometrically and often chemically distinct metal-binding sites.[33] In the case of HacacP, these sites are provided by the O···O or "acac" chelate and the phosphine moiety. A novel homobimetallic coordination compound that can be made from HacacP and copper(I) is $[Cu(acacP)]_2$.[34] This complex possesses an unusual structure wherein the two coordination spheres are planar and mutually cofacial. This species is of particular interest, since not only are the two metal ions adjacent but they are also coordinatively unsaturated.

The use of HacacP for the assembly of heterobimetallic complexes takes full advantage of the compartmentalization of this novel chelating agent. Representative of the synthetic potential of HacacP is the three-step synthesis of $[Pt\{Cu(acacP)\}](BF_4)_2$ (Figure 7.13). The X-ray structure of the intermediate, $PtCl_2\{Cu(acacP)_2\}$ (Figure 7.14), reveals that the platinum and copper entities are square planar and bound to the portion of acacP$^-$ which best suits their soft or hard acid character.[35] The presence of an adjacent, paramagnetic, Lewis acidic $Cu(acac)_2$ moiety could lead

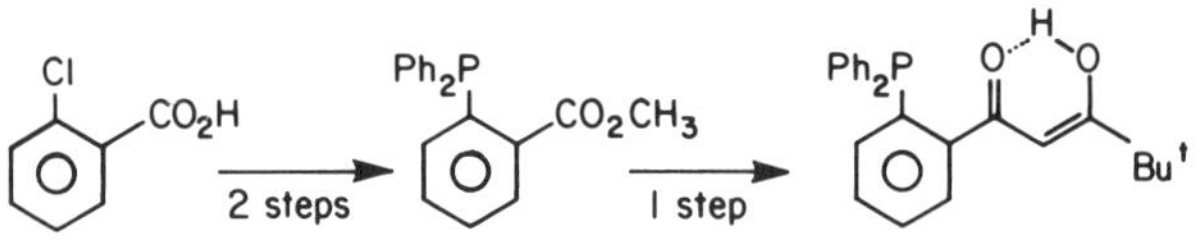

Figure 7.12. Synthesis of o-diphenylphosphinobenzoylpinacolone (HacacP), o-$Ph_2PC_6H_4C(O)CH_2(O)C(CH_3)_3$.

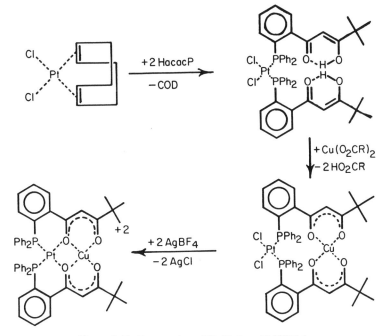

Figure 7.13. Preparation of [Pt{Cu(acacP)$_2$}](BF$_4$)$_2$.

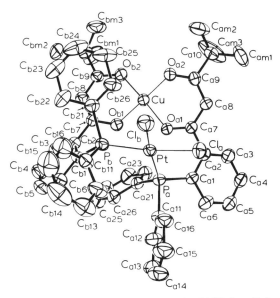

Figure 7.14. The molecular structure of PtCl$_2${Cu(acacP)$_2$}.

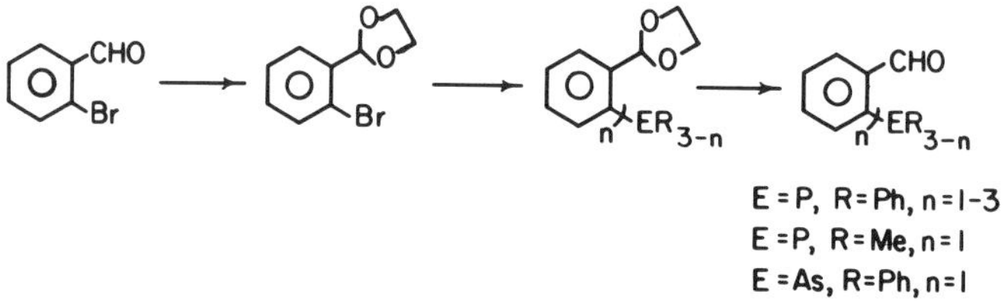

Figure 7.15. The synthetic sequence for the preparation of o-diphenylphosphinobenzaldehyde.

to some interesting perturbations on the chemistry of the platinum center. Another interesting heterobimetallic derivative of HacacP is trans-IrCl(CO){Cu(acacP)$_2$}, wherein a Cu(acac)$_2$ moiety is strapped across the face of a 16$e^-$ iridium(I) center. That species is particularly novel since the iridium(I) component possesses the ability to undergo facile oxidative addition of small electrophilic substrates.[35]

o-Diphenylphosphinobenzaldehyde, or PCHO, can be prepared from o-bromobenzaldehyde (Figure 7.15) and exists as yellow, air-stable crystals.[27,36] Like diphenylphosphinostyrene, PCHO has proven to be a useful mechanistic probe in organotransition metal chemistry.[37-39] The synthetic versatility characteristic of the formyl group suggests that PCHO should represent a useful precursor to a range of new ligand systems. For instance, PCHO condenses easily with a wide variety of aryl and alkyl, mono- and diamines to afford a number of iminophosphine chelating agents (Figure 7.16).[40-42] By varying the steric and electronic characteristics of the

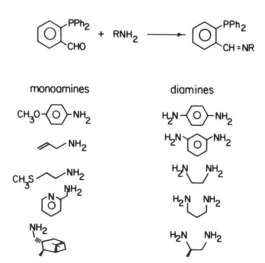

Figure 7.16. Iminophosphines prepared from o-diphenylphosphinobenzaldehyde.

amines, it is possible to modify the properties of the iminophosphine complexes. The resultant iminophosphines form stable, 6-membered P–N chelate rings. The ligand derived from R,R-1,2-diaminocyclohexane and PCHO coordinates stereospecifically to form a chiral tetrahedral copper(I) compound which is geometrically incapable of interconverting between the $\Delta$ and $\Lambda$ forms.[41] The iminophosphines can be easily reduced with $NaBH_4$ to give the corresponding secondary amine ligands. Upon coordination the amine group becomes configurationally stable, and its chirality could, in principle, be exploited for asymmetric synthesis. A number of formyl- and ketophosphine ligands like PCHO have been described although few have received any study as ligands.[37]

## 5. ALKENYLPHOSPHINES

Alkenylphosphine ligands, particularly diphenylphosphinostyrene and the butenylphosphines, have received considerable attention as organometallic chelating agents. Studies involving the reactions of the olefinic component of such ligands have provided a wealth of information relating to metal-olefin reactivity patterns.[43] Those studies have focused on how metal ions can affect the reactivity of the olefinic group. In the present case, we are more concerned with the reverse situation: how the presence of unsaturation in a phosphine ligand can affect the chemistry of the metal ion.

Fenske and co-workers have described a new class of 1,2-diphosphinoalkenes which are unique for their radical anion-forming abilities. This property has led to some very novel chemistry and spectroscopy. The ligands of interest are prepared from the reaction of 2,3-dichloromaleic anhydride (and related derivatives) with trimethylsilylated phosphines[44] (Figure 7.17). The free ligands exist as air-stable, crystalline solids. With classical, electrophilic metal salts, they form "normal" complexes, for instance of the type $NiX_2$(chelate).[44] With electron-rich metal ions, however, these chelating agents react via a curious version of the oxidative

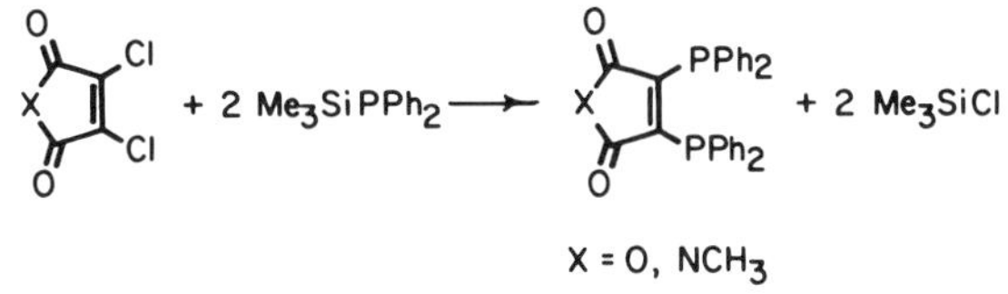

$X = O, NCH_3$

Figure 7.17. The synthesis of 1, 2-*bis*(diphenylphosphino)-maleic anhydride and -*N*-methylmaleimide.

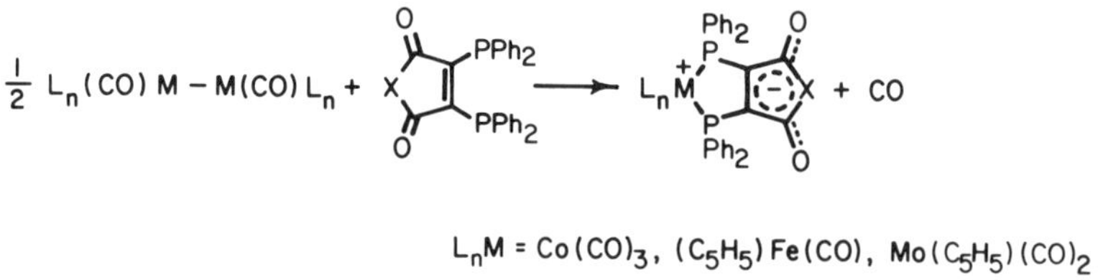

Figure 7.18. The cleavage of metal–metal bonds with 1, 2-*bis*(diphenylphosphino)-*N*-methylmaleimide.

addition reaction. For example, two equivalents of 2,3-*bis*(diphenylphosphino)-*N*-methylmaleimide displace all four PPh$_3$ ligands from yellow Pd(PPh$_3$)$_4$ to form the intensely green, paramagnetic complex, Pd(chelate)$_2$.[45] Electron spin resonance studies indicate the presence of two radical anion ligands attached to a palladium(II) center. A single crystal X-ray diffraction study supports this assignment by revealing a centrosymmetric, square planar palladium complex. Other spectroscopic measurements, especially IR, reinforce the view that these ligands function as one-electron oxidizing agents.

A reaction which is characteristic of Fenske's alkenyldiphosphines is the oxidative cleavage of binuclear metal carbonyls.[46,47] The monometallic products are best described as zwitterions containing radical anionic diphosphine chelating ligands coordinated to cationic (18$e^-$) metal centers (Figure 7.18). These compounds can be subsequently converted to diamagnetic cations via ligand-based oxidations. The ability of ligands to undergo facile reduction/oxidation is consistent with the presence of vacant, low-energy molecular orbitals. Related behavior can be found in the chemistry of metalloporphyrins and the dithiolenes. Since many fundamental organometallic reactions appear to proceed via odd-electron intermediates,[48] complexes which promote or stabilize the formation of paramagnetic species may be expected to display distinctive chemistry. A demonstration of such ligand-based reactivity is provided by the ability of the (C$_5$H$_5$)Mo(CO)$_2$ derivative of Fenske's diphosphine to abstract a hydrogen atom from toluene. The products of this reaction are the diamagnetic molybdenum cation, C$_5$H$_5$Mo(CO)$_2$(diphosphineH)$^+$, and biphenyl[46] (Figure 7.19).

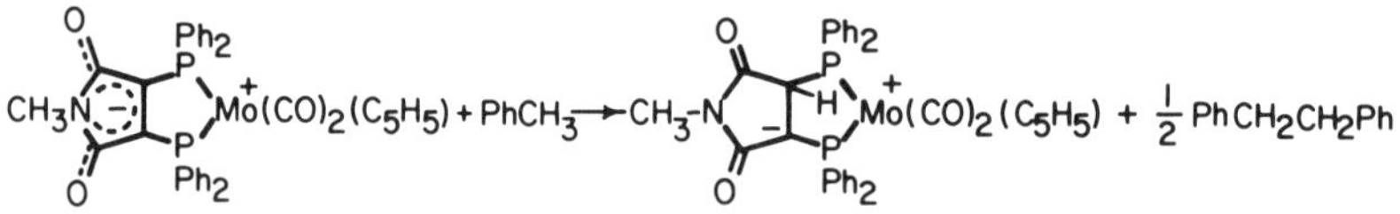

Figure 7.19. Hydrogen atom abstraction from toluene by a 1, 2-*bis*-(diphenylphosphino)-*N*-methyl maleimide complex.

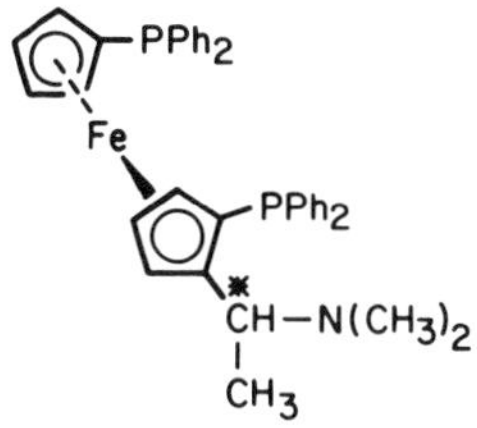

Figure 7.20. Kumada's chiral ferrocenyldiphosphine ligand.

## 6. CYCLOPENTADIENYLPHOSPHINES

Cyclopentadienylphosphines have received considerable attention as they represent the fusion of two very popular ligand types. Depending on the nature of the linkage between the cyclopentadienyl moiety and the phosphine, such systems can either function as net 7e$^-$ chelating agents or as binucleating ligands. In the former capacity, it is not exactly clear what new chemistry will ensue from these monometallic complexes, although they are stereochemically unusual.[49]

Considerable effort has been expended on using cyclopentadienylphosphines as templates for the assembly of bimetallic complexes. The central problem here is that effective strategies have yet to be devised for connecting two different *reactive* metals. For this reason, one metallic component of cyclopentadienylphosphine complexes is generally chemically dormant. This gives rise to the all too familiar situation of structurally novel compounds that are chemically inert. Clearly the challenge of the coming years will be the preparation of reactive heterobimetallic complexes which are constructed in such a way that both metals can interact in a concerted manner with the substrates.

Ferrocenylphosphines have become common, and while the iron of the ferrocene moiety is relatively inert, these ligands are stereochemically novel. The rhodium(I) derivative of the ferrocenyl*di*phosphine shown in Figure 7.20 has been shown to function as a highly selective asymmetric hydrogenation catalyst.[50]

Perhaps the most versatile approach to substituted cyclopentadienyl phosphines has been described in a series of reports by Kauffmann *et al.* concerning the chemistry of a spiro-cycloheptadiene.[51,52] This derivative, which apparently can be easily prepared from cyclopentadiene and ethylenedibromide, reacts with a range of lithiated ligand precursors. This route provides access to a wide range of relatively complex hybrid ligands.

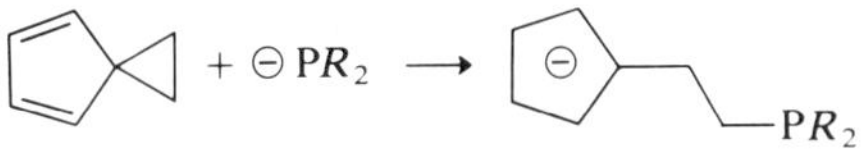

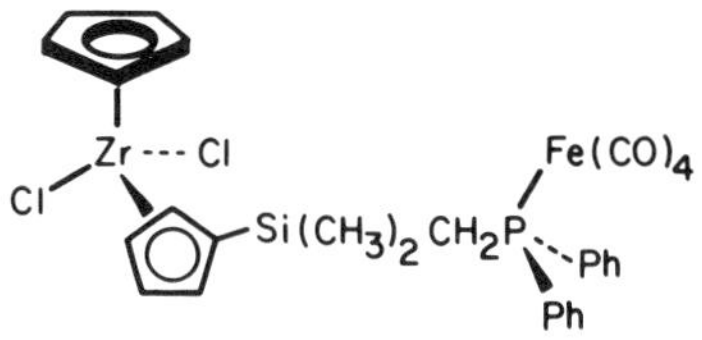

Figure 7.21. Schore's iron(0)–zirconium(IV) complex prepared using a phosphinocyclopentadiene ligand system.

The preparation of a phosphinocyclopentadiene containing an intervening silane bridge has been reported by Schore.[53] This work has involved not only the preparation of the usual ferrocenyl derivatives and their subsequent complexation, but a novel zirconium(IV)-iron(0) complex (Figure 7.21). Since cyclopentadienyl zirconium halides represent precursors to useful synthetic reagents, this work represents a significant step towards the assembly of heterobimetallic complexes which contain *two* reactive metal ions.

## 7. CONCLUDING REMARKS

In the coming years one can expect to see greater use of functionalized phosphines in synthesis and catalysis. One reason for this trend is the increasing infusion of organic synthetic methodology into the field of organometallic coordination chemistry.† Furthermore, the recent applications of homogeneous platinum metal catalysts justifies the modest expense associated with the use of somewhat elaborate phosphines. In this regard, it is important to remind ourselves that aside from considerations of convenience, there is no reason for assuming that the perennial favorite, triphenylphosphine, is the optimum phosphine ligand for *any* catalytic application.

ACKNOWLEDGMENTS

Our own work on phosphines has been funded by the Petroleum Research Fund and by Dow Chemical Company. I thank John Hoots, Edith Landvatter, and Debra Wrobleski for their assistance in the preparation of this review.

---

† For an impressive illustration of this trend, see Reference 54.

## REFERENCES

1. W. S. Knowles and M. J. Sabacky, *Ger. Offen.* **2**, 123, 063 (1971).
2. W. S. Knowles, M. J. Sabacky, and B. D. Vineyard, *Adv. Chem. Ser.* **132**, 274–282 (1972).
3. B. D. Vineyard, W. S. Knowles, M. J. Sabacky, G. C. Bachman, and D. J. Weinkauff, *J. Am. Chem. Soc.* **99**, 5946–5952 (1977).
4. E. M. Miller and B. L. Shaw, *J. C. S., Dalton Trans.*, 480–485 (1974).
5. W. J. Louw, D. J. A. de Waal, and J. E. Chapman, *J. C. S., Chem. Comm.*, 845–846 (1977).
6. J. C. Jeffery and T. B. Rauchfuss, *Inorg. Chem.* **18**, 2658–2666 (1979).
7. T. B. Rauchfuss, *Plat. Met. Rev.* **24**, 95–99 (1980).
8. N. W. Alcock, J. M. Brown, and J. C. Jeffery, *J. C. S., Dalton Trans.* 583–588 (1976).
9. J. Powell, A. Kuksis, C. J. May, S. C. Nyberg, and S. J. Smith, *J. Am. Chem. Soc.* **103**, 5941–5943 (1981).
10. J. P. Collman, R. G. Finke, J. N. Cawse, and J. I. Brauman, *J. Am. Chem. Soc.* **100**, 4766–4772 (1978).
11. For an example of true crown ether-phosphine, see E. M. Hyde, B. L. Shaw, and I. Shepherd, *J. C. S., Dalton Trans.*, 1693–1705 (1978).
12. M. E. Wilson, R. G. Nuzzo, and G. M. Whitesides, *J. Am. Chem. Soc.* **100**, 2269–2270 (1978).
13. M. E. Wilson and G. M. Whitesides, *J. Am. Chem. Soc.* **100**, 306–307 (1978).
14. M. K. Cooper and J. M. Downes, *Inorg. Chem.* **17**, 880–884 (1978); *J. C. S., Chem. Comm.*, 381 (1981).
15. M. D. Fryzuk and P. A. MacNeil, *J. Am. Chem. Soc.* **103**, 3592–3593 (1981).
16. W. Keim, R. Appel, A. Storeck, C. Kruger, and R. Goddard, *Angew. Chem. Int. Ed. Eng.* **20**, 116–117 (1981).
17. D. Houalla, J. F. Brazier, J. Sanchez, and R. Wolf, *Tetrahedron Lett.*, 2969–2970 (1972).
18. C. Pradat, J. G. Riess, D. Bondoux, B. F. Mentzen, I. Tkatchenko, and D. Houalla, *J. Am. Chem. Soc.* **101**, 2234–2235 (1979).
    D. Bondoux, B. F. Mentzen, and I. Tkatchenko, *Inorg. Chem.* **20**, 839–848 (1981).
19. D. Bondoux, D. Houalla, C. Pradat, J. G. Riess, I. Tkatchenko, and R. Wolf, *Fundamental Research in Homogeneous Catalysis*, edited by M. Tsutsui (Pergamon Press, New York, 1979), Vol. 3, pp. 969–981.
20. D. M. Roundhill, R. A. Bechtold, and S. G. Roundhill, *Inorg. Chem.* **19**, 284–289 (1980).
21. F. Jeanneaux, A. Grand, and J. G. Riess, *J. Am. Chem. Soc.* **103**, 4272–4273 (1981).
22. P. Vierling, J. G. Riess, and A. Grand, *J. Am. Chem. Soc.* **103**, 2466–2467 (1981).
23. W. K. Musker, *Acc. Chem. Res.* **13**, 200–208 (1980).
24. D. St. C. Black and A. J. Hartshorn, *Coord. Chem. Rev.* **9**, 219–274 (1972–1973).
25. K. Issleib and H. Zimmerman, *Z. anorg. allg. Chem.* **353**, 197–206 (1967).
26. D. M. Singleton, P. W. Glockner, and W. Keim, British Patent 1 364 870 (1974).
27. J. E. Hoots, T. B. Rauchfuss, and D. A. Wrobleski, *Inorg. Syn.* **21**, 175–179 (1982).
28. For related patents, see also R. F. Mason, United States Patent 3 686 351 (1972) and E. F. Lutz, United States Patent 3 825 615 (1974).
29. E. R. Freitas and C. R. Gum, *Chem. Eng. Progs.* **75(1)**, 73–75 (1979).
30. W. Keim, F. H. Kowaldt, R. Goddard, and C. Krüger, *Angew. Chem. Int. Ed. Eng.* **17**, 466–467 (1978); M. Peuckert and W. Keim, *Organometal.* **2**, 594–597 (1983).
31. P. Braunstein, D. Matt, Y. Dusausoy, J. Fischer, A. Mitschler, and L. Ricard, *J. Am. Chem. Soc.* **103**, 5115–5125 (1981).
32. T. B. Rauchfuss, *Inorg. Chem.* **16**, 2966–2968 (1977).
33. O. Casellato, P. A. Vigato, D. E. Fenton, and M. Vidali, *Chem. Soc. Rev.* **8**, 199–220 (1979).

34. T. B. Rauchfuss, S. R. Wilson, and D. A. Wrobleski, *J. Am. Chem. Soc.* **103,** 6769–6770 (1981).
35. D. A. Wrobleski and T. B. Rauchfuss, *J. Am. Chem. Soc.* **104,** 2314–2315 (1982).
36. G. P. Shiemenz and H. Kaack, *Liebigs Ann. Chem.*, 1480–1495 (1973).
37. T. B. Rauchfuss, *J. Am. Chem. Soc.* **101,** 1045–1047 (1979).
    E. F. Landvatter and T. B. Rauchfuss, *Organometallics* **1,** 506 (1982).
38. T. B. Rauchfuss, *Fundamental Research in Homogeneous Catalysis*, edited by M. Tsutsui, (Pergamon Press, New York, 1979), Vol. 3, pp. 1021–1032.
39. G. D. Vaughn and J. A. Gladysz, *J. Am. Chem. Soc.* **103,** 5608–5609 (1981).
40. T. B. Rauchfuss, *J. Organometal. Chem.* **162,** C19–C21 (1978).
41. J. E. Hoots, J. C. Jeffery, T. B. Rauchfuss, S. P. Schmidt, and P. A. Tucker, *Adv. in Chem. Ser.* **196,** 303–311 (1982).
42. J. C. Jeffery, T. B. Rauchfuss, and P. A. Tucker, *Inorg. Chem.* **19,** 3306 (1980).
43. M. A. Bennett, R. N. Johnson, G. B. Robertson, I. B. Tomkins, and P. O. Whimp, *J. Am. Chem. Soc.* **98,** 3514–3523 (1976) and references therein.
44. D. Fenske and H. J. Becher, *Chem. Ber.* **108,** 2115–2123 (1975).
    H. J. Becher, W. Bensmann, and D. Fenske, *Chem. Ber.* **110,** 315–321 (1977).
45. W. Bensmann and D. Fenske, *Angew. Chem. Int. Ed. Eng.* **18,** 677–678 (1979).
46. D. Fenske and A. Christidis, *Angew. Chem. Int. Ed. Eng.* **20,** 129–131 (1981).
47. D. Fenske, *Angew. Chem. Int. Ed. Eng.* **15,** 381–382 (1976).
48. J. P. Collman and L. S. Hegedus, *Principles and Applications of Organotransition Metal Chemistry* (University Science Books, Mill Valley, CA 1980).
49. C. Charrier and F. Mathey, *J. Organometal. Chem.* **170,** C41–C43 (1979).
50. T. Hayashi, T. Mise, S. Mitachi, K. Yamamoto, and M. Kumada, *Tetrahedron Lett.*, 1133–1137 (1976).
51. Th. Kauffmann, J. Ennen, H. Lhotak, A. Rensing, F. Steinseifer, and A. Woltermann, *Angew. Chem. Int. Ed. Eng.* **19,** 328–329 (1980).
52. K. Berghus, A. Hamsen, A. Rensing, A. Woltermann, and Th. Kauffmann, *Angew. Chem. Int. Ed. Eng.* **20,** 117–118 (1981).
53. N. E. Schore, *J. Am. Chem. Soc.* **101,** 7410–7412 (1979).
54. E. P. Kyba, A. M. John, S. B. Brown, C. W. Hudson, M. J. McPhaul, A. Harding, K. Larsen, S. Niedzwiecki, and R. E. Davis, *J. Am. Chem. Soc.* **102,** 139–147 (1980).

## Note Added in Proof

The following papers are related to the topics discussed in Sections 7.2 and 7.3, respectively: J. Powell, M. Gregg, A. Kuksis, and P. Meindl, *J. Am. Chem. Soc.* **105,** 1064–1065 (1983), and M. D. Fryzuk and P. A. MacNeil, *Organometallics* **2,** 355–356 and 682–684 (1983).

# 8

# Polydentate Ligands and Their Effects on Catalysis

Devon W. Meek

## 1. INTRODUCTION

In recent years, tremendous interest and research activity has been expended on transition metal complexes of tertiary phosphine ligands.[1,2] Metal-phosphine complexes display a tremendous range of coordination numbers, stereochemistries, catalytic properties, and selectivities. The effects can also be "fine-tuned" by purposeful variation of the steric and electronic effects of the substituent groups on phosphorus and from synthesis of polydentate phosphines.[2]

It is expected that the use of polyphosphine ligands will become increasingly important in future studies, especially in the areas of catalysis, asymmetric synthesis, and organometallic stereochemistry, since the special properties of phosphine ligands can be accentuated by a chelating polyphosphine.[2,3] This chapter will concentrate on the unique aspects that can be achieved with polyphosphine ligands, since other chapters in this book will discuss monodentate phosphine complexes. In fact, this chapter will emphasize tridentate ligands, since Balch and Sanger are discussing binuclear and phosphine-bridged complexes with ligands like $Ph_2PCH_2PPh_2$ (Chapters 5 and 6) and Brown and Chaloner are discussing chiral diphosphine complexes (Chapter 4). However, selected examples of studies with diphosphine ligands will be presented in areas other than asymmetric catalysis and

---

Dr. Devon W. Meek • Department of Chemistry, The Ohio State University, Columbus, Ohio, 43210.

dinuclear complexes, where the studies show particular advantages or implications for future work with polydentate ligands.

## 2. ADVANTAGES OF CHELATING POLYDENTATE LIGANDS

If one wishes to control the number of phosphine groups attached to a metal, clearly a chelating polyphosphine ligand will provide more control than a monodentate phosphine.[4] In addition, one can better predict the basicity (or nucleophilicity) of the metal and the stereochemistry of a complex resulting from a polydentate ligand. Also, slow intra- and intermolecular exchange processes and detailed structural and bonding information in the form of metal–phosphorus and phosphorus–phosphorus coupling constants can be observed, particularly when the metal has a magnetic nucleus, e.g., $^{103}$Rh, $^{195}$Pt, or $^{199}$Hg.[5]

Polydentate phosphine ligands may be used to change the usual magnetic states, coordination geometries, and reactivity of complexes by judiciously selecting parameters such as (i) sets of donor atoms, (ii) 'chelate bite angle," and (iii) sterically demanding substituent groups.[4–6] For example (1) Venanzi has used the rigid ligand, **1**, to form "transspanning" complexes of the type, **2**,[7]; (2) Otsuka et al.[8] found that the

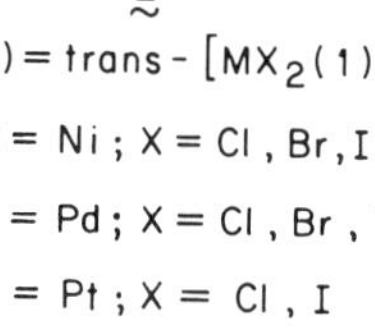

$$(2) = trans - [MX_2(1)]:$$
$$M = Ni; X = Cl, Br, I, NCS$$
$$M = Pd; X = Cl, Br, I$$
$$M = Pt; X = Cl, I$$

bulky phosphine $(t\text{-Bu})_3$P made the crowded 14-electron species [Pd{P$(t\text{-Bu})_3$}$_2$] unreactive toward O$_2$; and (3) Shaw[9] has used long-chain diphosphines of the type $(t\text{-Bu})_2$P(CH$_2$)$_n$P$(t\text{-Bu})_2$ ($n = 9, 10, 12$) to produce the mono-, di-, and tri-nuclear complexes [$M$Cl$_2${$(t\text{-Bu})_2$P(CH$_2$)$_n$P$(t\text{-Bu})_2$}]$_x$ ($M$ = Pd, Pt) with 12- to 45-membered rings, whereas the same type of ligand with a (CH$_2$)$_{5 \text{ or } 6}$ chain produces dimeric [$M_2$Cl$_4${$(t\text{-Bu})_2$P(CH$_2$)$_n$P$(t\text{-Bu})_2$}$_2$] complexes, **3**, and cyclometallated complexes of

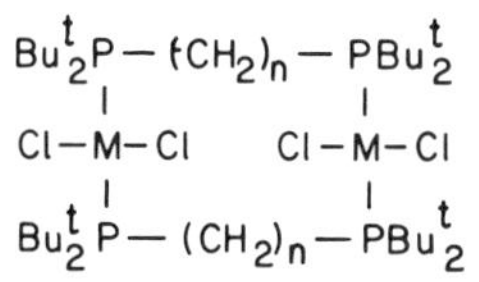

$$Bu^t_2P-(CH_2)_n-PBu^t_2$$
$$|\qquad\qquad\qquad|$$
$$Cl-M-Cl\qquad Cl-M-Cl$$
$$|\qquad\qquad\qquad|$$
$$Bu^t_2P-(CH_2)_n-PBu^t_2$$

**3**

M = Pd, Pt

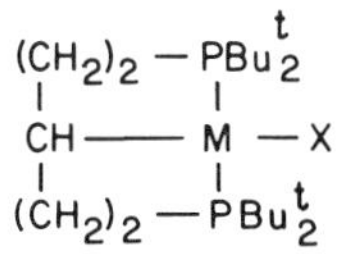

$$(CH_2)_2-PBu^t_2$$
$$|\qquad\qquad|$$
$$CH\longrightarrow M\;-X$$
$$|\qquad\qquad|$$
$$(CH_2)_2-PBu^t_2$$

**4a**

M = Pt, Pd

the type **4a**. In the case of the pentamethylene chain, the cyclometallation reaction occurs under remarkably mild conditions. Treatment of rhodium trichloride with the pentamethylenediphosphine $(t\text{-}Bu)_2P(CH_2)_5P(t\text{-}Bu)_2$ gives two principal products (the binuclear, square-pyramidal rhodium(III) hydride **4b**, with a 16-atom ring and the cyclometallated rhodium(III) hydride **4c**) and a small amount of an olefinic diphosphine rhodium(I) complex, **4d**.

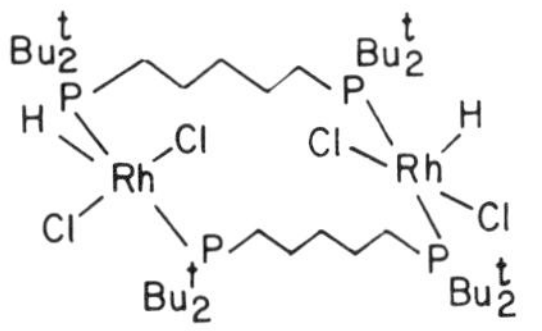

**4b** 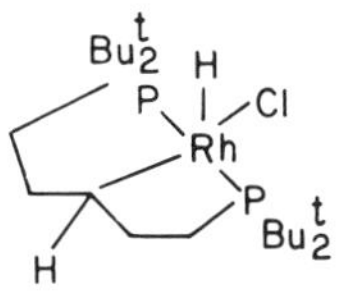     **4c** 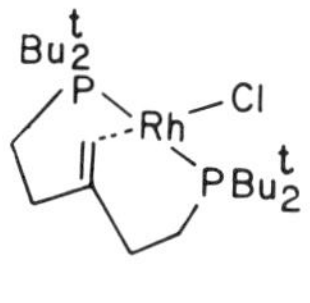     **4d**

Flexible polyphosphine ligands (e.g., **5–9**), which contain either ethylene or trimethylene linkages, have been used to form mononuclear

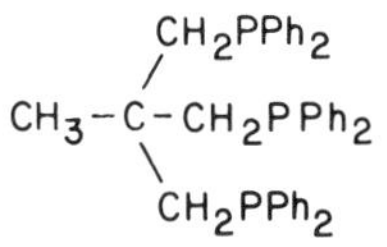

$$CH_3-C\begin{array}{l}CH_2PPh_2\\CH_2PPh_2\\CH_2PPh_2\end{array}$$

**5**, tripod

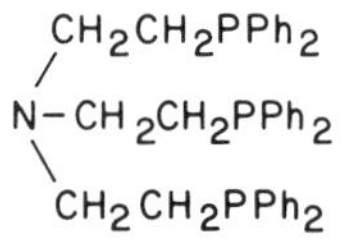

$$N\begin{array}{l}CH_2CH_2PPh_2\\CH_2CH_2PPh_2\\CH_2CH_2PPh_2\end{array}$$

**6**, np$_3$

$$P\begin{array}{l}CH_2CH_2PPh_2\\CH_2CH_2PPh_2\\CH_2CH_2PPh_2\end{array}$$

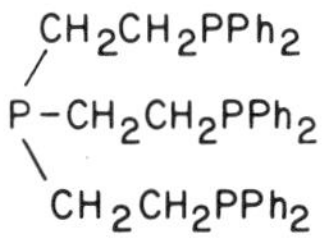

**7**, PP$_3$

$$PhP(CH_2CH_2PPh_2)_2$$

**8**, etp

$$PhP(CH_2CH_2CH_2PPh_2)_2$$

**9**, ttp

complexes with different coordination numbers (four, five, and six) and different structures (distorted tetrahedral, planar, square pyramidal, trigonal bipyramidal and octahedral).[10–12] A properly designed polyphosphine ligand can fix the relative positions of the phosphorus atoms while allowing the remaining ligands to be varied; thus, the stereochemical positions and the nature of the variable ligands can now be monitored by $^{31}$P nmr spectroscopy by observing the chemical shifts and coupling constants of the phosphorus atoms in the polyphosphine ligand.[5]

One can modify the coordination stereochemistry for a given donor set (e.g., $P_3X_2$) around a metal ion simply by changing the number of methylene groups in the connecting chains of a polyphosphine ligand. Such a study with the flexible ligands **8** (etp) and **9** (ttp) permitted the effect of the chelate ring size on the preferred structure of some 5-coordinate cobalt(I) complexes to be determined.[13] The details of the stereochemistry and exchange processes for the 5-coordinate complexes $[Co(CO)_2(etp)]X$, $[Co(CO)_2(ttp)]X$, $[Co(CO)L(etp)]X$, and $[Co(CO)L(ttp)]X$ ($X = BF_4$ or $PF_6$; $L = P(OMe)_3$, $PPh_2H$, $PPh_2Me$, $PEt_3$, or $PPh_3$) depend on both the monodentate ligand and on the triphosphine ligand. For example, both the mono- and di-carbonyl complexes $[Co(CO)(etp)L]^+$ and $[Co(CO)_2(etp)]^+$ are trigonal-bipyramidal with the equatorial positions occupied by a carbonyl ligand and the two terminal phosphorus atoms ($P^2$) of the triphosphine

$Ph_2PCH_2PPh_2$          $HC(PR_2)_3$          $Me_3Si(PBu_2)_3$
**10**                    **11**                **12**
                          R = Me, Ph

ligand; the axial positions are occupied by the central phosphorus atom ($P^1$) of the triphosphine and either a carbonyl, **13**, or a monophosphine ligand ($P^3$), **14**. The $[Co(CO)L(ttp)]^+$ complexes have square-pyramidal structures with $L$ at the axial position. The change from square-pyramidal coordination geometry in $[Co(CO)\{P(OMe)_3\}(ttp)]^+$, **15**, to a trigonal-bipyramidal geometry for the etp complexes can be attributed to a decrease

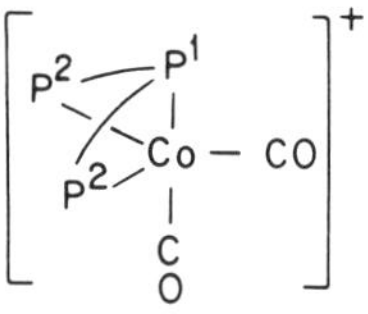

**13** 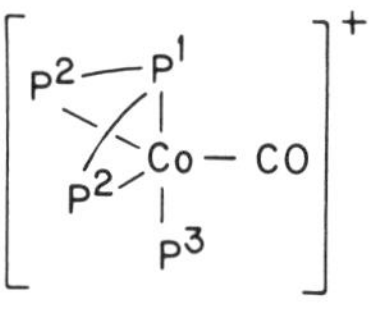          **14** 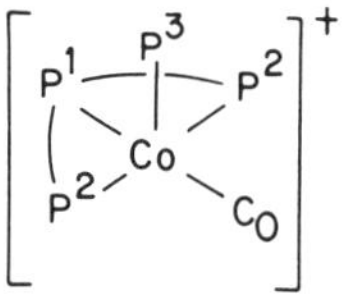          **15**

in chelate bite angle of the etp ligand. As a consequence of the smaller bite angle in 5-membered chelate rings, the central phosphorus atom ($P^1$) either would be pulled in abnormally close to the metal in a square-pyramidal geometry or the ligand must adopt a different coordination geometry where the central phosphorus atom is more equivalent to those of the terminal phosphorus atoms ($P^2$), e.g., as in the trigonal-bipyramidal geometries **13** and **14**.

If the connecting chain between phosphorus atoms is only one atom (e.g., dppm, **10**, $HC(PMe_2)_3$, and $HC(PPh_2)_3$, **11**, or $CH_3Si(PPh_2)_3$, **12**), the phosphino units tend to bridge across two metals (see Chapters 4 and 5 for dppm examples). In fact, this bridging tendency has been used to advantage by Osborn[14,15] and Masters[16] in attaching the tridentate ligands **11** and **12** to a face of a metal cluster, i.e., complexes of type **16**.

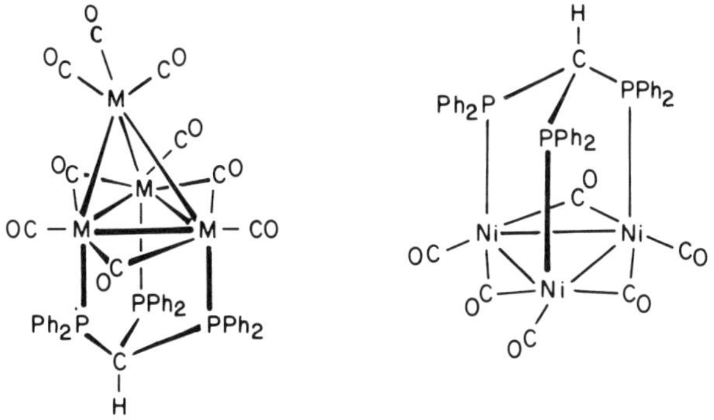

(a)
$M_4 = Co_4, Rh_4, Co_2Rh_2, HFeCo_3$

**16**

(b)

## 3. METHODS FOR SYNTHESES OF POLYDENTATE LIGANDS

The complexes cited above illustrate that the steric and electronic properties of ligands can be varied relatively easily by changing the organic substituents on phosphorus, provided viable synthetic schemes are available. Until 1971, most polytertiary phosphines were prepared by reactions of organic polyhalides with alkali metal dialkyl- or diaryl-phosphides (e.g., the relatively easy preparation of $Ph_2PCH_2CH_2PPh_2$) by treating $ClCH_2CH_2Cl$ with excess $LiPPh_2$, which is most economically prepared by lithium cleavage of $Ph_3P$).[17]

### 3.1. Grignard and Alkali-Metal Reagents

Most of the tetradentate "tripod-like" ligands of the type 

$$\begin{matrix} & B \\ A & -C \\ & D \end{matrix}$$

that were prepared by the research groups of Venanzi,[12] Meek,[10] and Sacconi[11] in the 1960s and early 1970s utilized the Grignard or alkali-metal organophosphide methods (Equations 1–4). However, synthesis of more complicated tri- and tetra-phosphines by these methods is severely limited by the unavailability of the appropriate phosphorus or organic polyhalide intermediates.

$$N(CH_2CH_2Cl)_3 + 3MPPh_2 \rightarrow N(CH_2CH_2PPh_2)_3 + 3MCl \quad (1)$$

$$Me_2AsCH_2CH_2CH_2Cl \xrightarrow[2)\ PCl_3]{1)\ Mg\ in\ Et_2O} P(CH_2CH_2CH_2AsMe_2)_3 \quad (2)$$

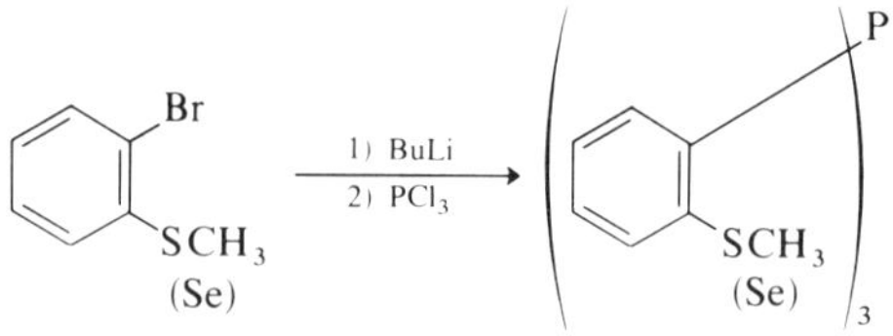

$$\underset{\underset{(Se)}{SCH_3}}{\text{o-Br-C}_6H_4} \xrightarrow[2)\ PCl_3]{1)\ BuLi} \left( \underset{\underset{(Se)}{SCH_3}}{\text{o-C}_6H_4} \right)_3 P \quad (3)$$

$$CH_3C(CH_2Cl)_3 + 3NaPR_2 \rightarrow CH_3C(CH_2PR_2)_3 + 3NaCl \quad (4)$$
$$R = Ph,\ Et$$

Although there are limitations with synthetic routes that use the classical Grignard and organolithium reagents, these methods are still widely used. Some recent applications to somewhat unusual ligands are illustrated below. Capka et al.[18,19] treated a series of alkoxylsilyl-substituted alkyl halides with $LiPPh_2$ to produce alkyldiphenylphosphine chains that could be attached covalently to silica (Equations 5–6).

$$(EtO)_3Si(CH_2)_nBr + LiPPh_2 \rightarrow (EtO)_3Si(CH_2)_nPPh_2 + LiBr \quad (5)$$
$$n = 1\text{–}6$$

$$(EtO)_m(CH_3)_{3-m}SiCH_2Cl + LiPPh_2 \rightarrow (EtO)_m(CH_3)_{3-m}SiCH_2PPh_2 + LiCl \quad (6)$$
$$m = 1\text{–}3$$

The interesting tripod molecule $MeSi(PBu_2)_3$ was prepared in 70% yield by treating $MeSiCl_3$ with $LiPBu_2$ in diethyl ether at $-80°C$.[16,20] The corresponding carbon ligand $CH_3C(PPh_2)_3$ was synthesized similarly,[21,22] whereas the tris(phosphino)methane ligands $HC(PPh_2)_3$ and $HC(PMe_2)_3$ are best prepared by the route shown in Equation 7.

$$R_2PCH_2PP_2 \xrightarrow[-BuH]{t\text{-BuLi, hexane}} Li[HC(PR_2)_2] \xrightarrow[Et_2O]{R_2PCl} HC(PR_2)_3 \quad (7)$$
$$R = Me,\ Ph \qquad\qquad R = Me,\ Ph$$

Syntheses of derivatives of tris(diphenylphosphino)methane have been achieved by a more circuitous route,[23] which is required because of C–P bond cleavage when the elements (e.g., $O_2$, $S_8$, Se) are combined directly with the triphosphine. Thus, lithiation of the methylene group of a *bis*-

phosphorus compound followed by reaction of a phosphinous chloride (Equation 8) or reaction of lithiomethyldiphenylphosphine sulfide (in excess) with a phosphinous chloride (Equation 9) produces the triphosphine derivatives in reasonable yields.

$$R_2P(X)CH_2P(X)R_2 \xrightarrow{BuLi} R_2P(X)CHLiP(X)R_2 \xrightarrow{R'_2PCl} [R_2P(X)]_2CHPR'_2 \quad (8)$$

$$X = O, S, \text{lone pair}$$

$$Ph_3PS \xrightarrow{MeLi} Ph_2P(S)CH_2Li \xrightarrow{Ph_2PCl} Ph_2P(S)CH_2PPh_2 \xrightarrow[2)\ Ph_2PCl]{1)\ BuLi} Ph_2P(S)CH(PPh_2)_2 \quad (9)$$

Stille has used the alkali-metal phosphide route to prepare polymer-attached optically-active diphosphine ligands. Either a monomer (e.g., **17**) was copolymerized and treated with an excess of $PPh_2^-$ or the monomer-diphosphine unit (e.g., **18**) was synthesized before polymerization.[24] In both cases, the polymeric materials were treated with a rhodium complex

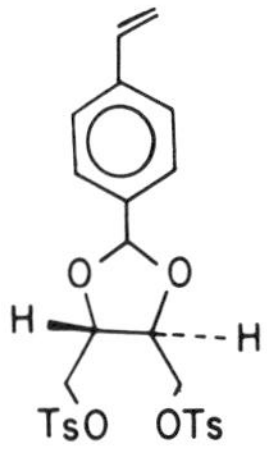

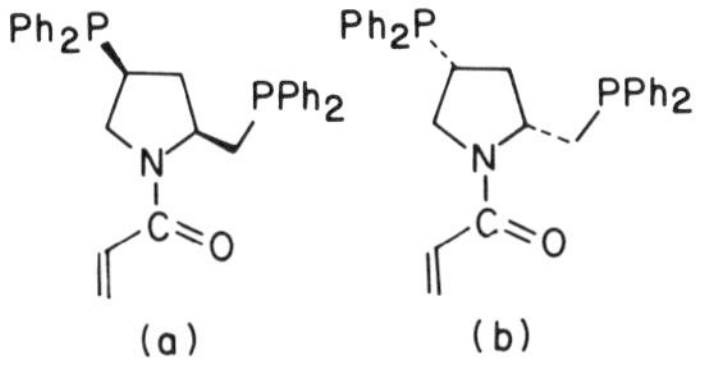

and used as supported catalysts (Schemes 1 and 2; see also Section 5.4). In an important application of the organophosphide synthetic method, Whitesides et al.[25] synthesized the chelating diphosphinoamine, **19**, and then attached it to several different types of organic molecules by coupling the amine with carbonyl halides, anhydrides, active esters, and isocyanides (Scheme 3); they even attached the $NP_2$ unit to a protein as an asymmetric support. To quote from Whitesides's paper, "the usefulness of this procedure for diphosphine synthesis rests on four features. First, the formation of amides by acylation of amines is one of the best understood and most general coupling methods in organic chemistry. The fact that it is possible to acylate the secondary amine of **19** without interference by the diphenylphosphine groups makes it possible to utilize this reaction for the

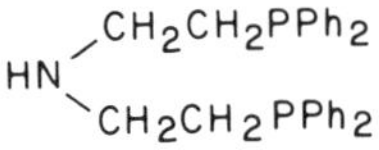

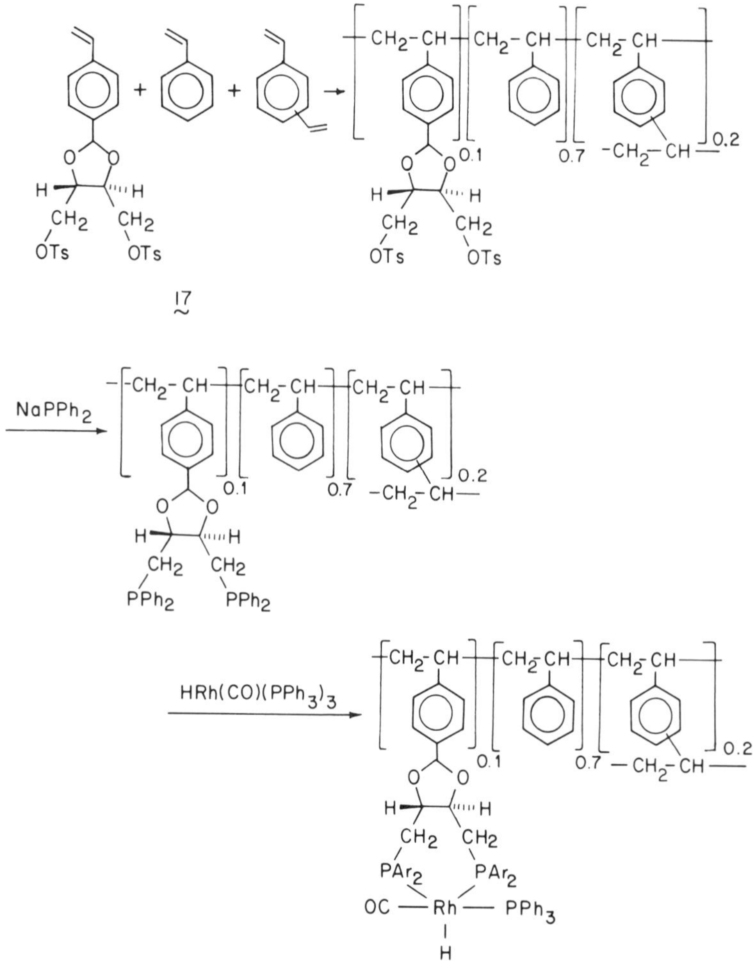

Scheme 1[24e]

preparation of a wide variety of diphosphines. Second, since the preformed diphosphine moiety is introduced as a unit, yields are relatively high. Third, the coupling reaction is compatible with a range of functionalities. Fourth, carboxylic acids and their derived acylating agents are readily available in great variety."[25c]

Scheme 2[24d]

Scheme 3[25a]

## 3.2. Other Synthetic Methods

During the past ten years, two valuable alternatives to the Grignard and alkali-metal-organophosphide syntheses of tertiary phosphines have been developed. By choosing the appropriate route and reagents, one should now be able to "tailor-make" polydentate ligands to incorporate just about any electronic and/or steric effects desired.

### 3.2.1. Base-Catalyzed P–H Addition to Vinyl Compounds

King et al.[26] developed a route to poly(tertiary phosphines) by use of a base-catalyzed addition of a phosphorus–hydrogen bond to the carbon–carbon double bond of various vinylphosphine derivatives; the first extensive series of methylated polyphosphines were prepared this way. The synthesis of unsymmetrical diphosphines of the type $R_2PCH_2CH_2P(CH_3)_2$ involves addition of a P–H bond to $CH_2{=}CHP(S)(CH_3)_2$ via potassium tert-butoxide catalysis, followed by desulfurization with $LiAlH_4$ in boiling dioxane (i.e., Equations 10, 11).

$$R_2\text{P-H} + CH_2{=}CH-\underset{\underset{S}{\|}}{P}(CH_3)_2 \xrightarrow{\text{base}} R_2PCH_2CH_2\underset{\underset{S}{\|}}{P}(CH_3)_2 \qquad (10)$$

$$R_2PCH_2CH_2\underset{\underset{S}{\|}}{P}(CH_3)_2 \xrightarrow{LiAlH_4} R_2PCH_2CH_2P(CH_3)_2 \qquad (11)$$

Variations of the base-catalyzed method were used to prepare the three tritertiary phosphines $R'P(CH_2CH_2PR_2)_2$ ($R' = CH_3$ or $C_6H_5$, $R = C_6H_5$; $R' = C_6H_5$, $R = CH_3$), the mixed aliphatic-aromatic, open-chained tetratertiary phosphine, **20**, the branched pentatertiary phosphine, **21**,

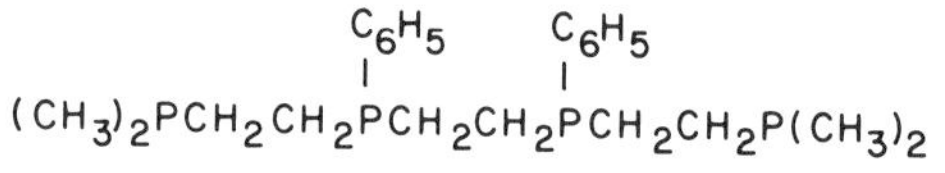

$$(CH_3)_2PCH_2CH_2\overset{\overset{C_6H_5}{|}}{P}CH_2CH_2\overset{\overset{C_6H_5}{|}}{P}CH_2CH_2P(CH_3)_2$$

**20**

$$\begin{array}{c}(CH_3)_2PCH_2CH_2\\ \phantom{(CH_3)_2PCH_2CH_2}\diagdown\\ \phantom{(CH_3)_2PCH_2CH_2}\phantom{xx}PCH_2CH_2\overset{\overset{C_6H_5}{|}}{P}CH_2CH_2P(CH_3)_2\\ \phantom{(CH_3)_2PCH_2CH_2}\diagup\\ (CH_3)_2PCH_2CH_2\end{array}$$

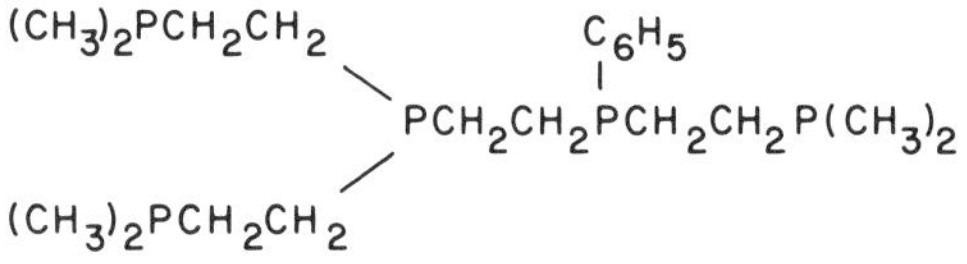

**21**

and the optically active di- and tri-phosphines, **22** and **23**. Compound **23** is an example of a chiral triphosphine, a rarity.[26c]

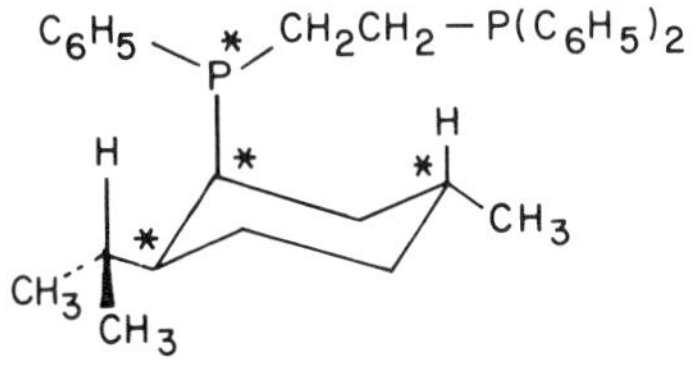

### 3.2.2. Radical-Catalyzed P–H Addition to Vinyl and Allyl Compounds

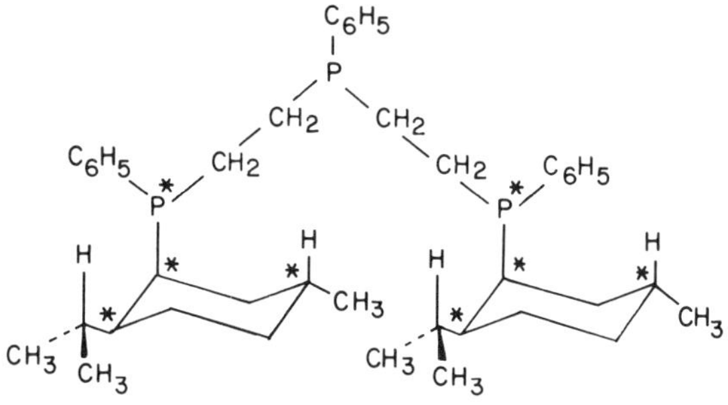

Although the base-catalyzed process is valuable, it is limited to $R_2PCH_2CH_2PR_2$ units and to available vinyl phosphines. A more general method for the preparation of polyphosphines, "mixed" phosphorus-sulfur, "mixed" phosphorus–arsenic, and "mixed" phosphorus–nitrogen compounds is the radical-catalyzed addition of phosphorus–hydrogen, arsenic–hydrogen, or sulfur–hydrogen bonds across carbon–carbon double bonds in either vinyl or allyl compounds.[27] Even in those cases where reagents are available for the base-catalyzed method, the radical method appears to give higher yields and purer compounds, especially if UV radiation is used.[27c] The reactions are fast, high-yield, and general for vinyl compounds and selected allyl (i.e., electronegative substituents such as Cl, OH, and $NH_2$; but not Br, I, SR, or $PPh_2$); they are summarized by Equations 12–15.

$$R_2\text{P-H} + CH_2=CH\text{-}PR_2 \rightarrow R_2PCH_2CH_2PR_2 \quad (12)$$

$$R_2\text{As-H} + CH_2=CH\text{-}PR_2 \rightarrow R_2AsCH_2CH_2PR_2 \quad (13)$$

$$R\text{S-H} + CH_2=CH\text{-}PR_2 \rightarrow RSCH_2CH_2PR_2 \quad (14)$$

$$R_2\text{P-H} + CH_2=CH\text{-}C(O)NH_2 \rightarrow R_2PCH_2CH_2C(O)NH_2 \quad (15)$$

The radical-catalyzed method for preparation of "mixed" PP, PAs, PS, and PN combinations in polydentate ligands is noteworthy, as it represents an easy route to many potentially useful combinations of donor units in ligands. Compounds of the type $\underset{i\text{-PrO}}{\overset{R}{\diagdown}}\!\!\!\diagup\!\text{PCH}=\text{CH}_2$ may be exploited for the syntheses of more complicated "mixed" ligands. For example, reduction of the $R_2\underset{\underset{O}{\|}}{P}\text{-}OR$ group to the secondary phosphine, $R_2$PH, will then make additional reactions possible, *via* either the base-catalyzed or the radical-catalyzed method. One can envisage using several different types of vinyl compounds for addition of $R_2$P-H, $R$S-H, or $R_2$As-H groups. For example, treatment of diphenylphosphine with vinyltriethoxysilane produced Ph$_2$PCH$_2$CH$_2$Si(OE$t$)$_3$ in 90% yield (Equation 16).[27c]

$$\text{Ph}_2\text{PH} + \text{CH}_2 = \text{CH--Si(OE}t)_3 \rightarrow \text{Ph}_2\text{PCH}_2\text{CH}_2\text{Si(OE}t)_3 \qquad (16)$$

This phosphine can easily be attached to a solid support (e.g., silica gel) by a reaction of the triethoxysilane end of the ligand.

Mono- or di-substitution on a primary phosphine (e.g., PhPH$_2$) can be controlled by the stoichiometric ratio of the reagents. For example, Uriarte[27c] prepared PhP(CH$_2$CH$_2$CH$_2$NH$_2$)$_2$ in 85% yield using excess allylamine, whereas the monosubstituted product $\underset{H}{\overset{Ph}{\diagdown}}\!\!\!\diagup\!\text{PCH}_2\text{CH}_2\text{CH}_2\text{NH}_2$ is the predominant product when a 1:1 ratio of PhPH$_2$ and allylamine is used, particularly when both UV and AIBN are used. The molecule $\underset{H}{\overset{Ph}{\diagdown}}\!\!\!\diagup\!\text{PCH}_2\text{CH}_2\text{CH}_2\text{NH}_2$ is a valuable intermediate for additional condensation reactions that would produce "mixed" tridentate ligands (e.g., P⌒P⌒NH$_2$ or As⌒P⌒NH$_2$ combinations).

The radical-chain addition reactions appear generally applicable for vinyl compounds; in the case of allyl derivatives, the clean, high-yield reactions appear to be limited to allylamine, allylalcohol, and allylchloride. However, the alcohol function is easily converted to a chloride, bromide, or tosylate derivative, which can be used to add another ligand donor group (e.g., P$R_2$, As$R_2$, or S$R$). The principal advantages of the radical-catalyzed method are (i) the flexibility of designing many different related polydentate ligands that contain a variety of donor groups with either -CH$_2$CH$_2$- or -CH$_2$CH$_2$CH$_2$-connecting units by simply choosing the appropriate $R_2$P-H, $R_2$As-H, $R$S-H, and vinyl derivatives; (ii) the experimental simplicity of the homogeneous solutions and the one-pot reaction; (iii) the faster reaction times compared to the base-catalyzed addition to vinyl phosphines; (iv) the ease with which the reaction impurities or by-products are removed *in vacuo*, leaving a crude material that is most often sufficiently pure to use as a ligand; and (v) the selective monosubstitution of a $RPH_2$ molecule, especially when the radical is generated by UV at room temperature or below.[27]

### 3.2.3. Phosphonium Route to Unsymmetrical Bisphosphines

Although symmetrical bisphosphine ligands of the type $Ph_2P(CH_2)_n-PPh_2$ ($n = 1-6$) can be prepared relatively easily, attempts to synthesize unsymmetrical bisphosphines, in which the two phosphorus atoms have different substituents, have met with little success. However, recently Briggs and Dyer[28] have developed a relatively rapid and efficient synthesis of the unsymmetrical alkane-based bisphosphines $RPhP(CH_2)_nPPh_2$ ($n = 3$, $R =$ ethyl; $n = 4$, $R =$ methyl or ethyl; $n = 6$, $R =$ methyl, ethyl or cyclohexyl) by the general method shown in Scheme 4, which involves new unsymmetrical biphosphonium salts and bisphosphine dioxides as intermediates.

$$(n \geq 3\,;\; Ph = phenyl,\; R = alkyl)$$

$$Br(CH_2)_nBr \xrightarrow{Ph_3P} [Br(CH_2)_nPPh_3]Br \xrightarrow{RPh_2P} [RPh_2P(CH_2)_nPPh_3]Br_2$$

$$\xrightarrow{2NaOH} RPhP(O)(CH_2)_nP(O)Ph_2 \xrightarrow{2HSiCl_3} RPhP(CH_2)_nPPh_2$$

*Scheme 4*

The synthesis route (Scheme 4) fails for $n = 1$ or 2, due to difficulties at the hydrolysis stage, but it is expected to be general for unsymmetrical bisphosphines of types $RPhP(CH_2)_nPPh_2$, $R'_2P(CH_2)_nPPh_2$, $RPhP(CH_2)_nPPhR$, and other permutations, where $R =$ alkyl, $R' =$ alkyl or aryl, and $n \geq 3$.[28] This method should be particularly attractive for synthesis of chiral bisphosphines of the type $RPhP^*(CH_2)_nPPh_2$ ($R =$ Me, Et, $C_6H_{11}$, $t$-$C_4H_9$, etc.), all of which contain a chiral center at $P^*$.

### 3.2.4. High-Dilution Syntheses of Phosphorus Macrocycles

During the past four years, Kyba has achieved an important breakthrough in the syntheses of phosphorus-containing macrocycles.[29] He has

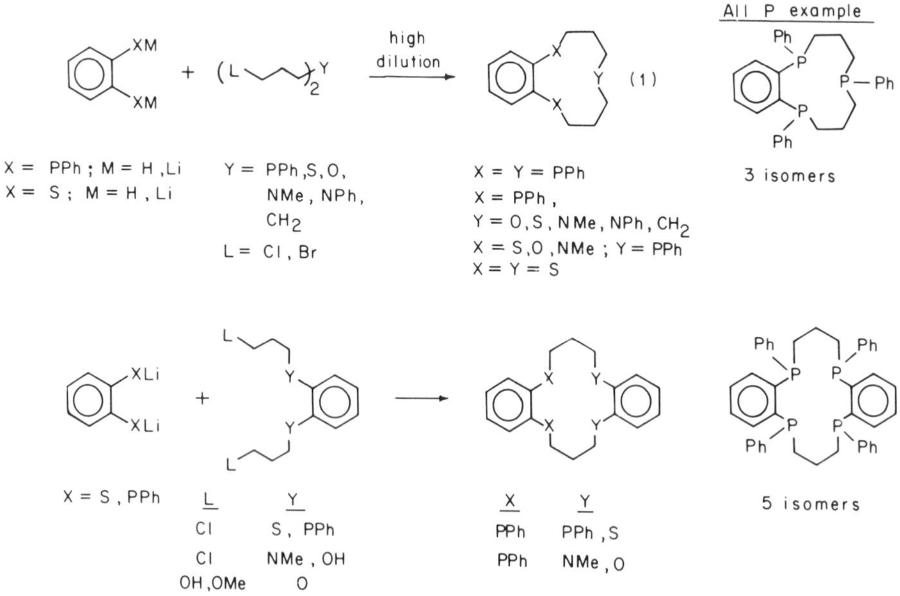

*Scheme 5*

designed a high-dilution apparatus that facilitates condensation of phosphorus or sulfur nucleophiles with halide or tosylate ligand units to produce 11-membered $P_3$ and mixed $P_2S$, $PS_2$, and $P_2N$ cycles, as well as 14-membered $P_4$, $P_2S_2$, $P_2N_2$, and $P_2O_2$ macrocyclic ligands (Scheme 5) in 18 to 45% yields.[29d] Owing to the tetrahedral stereochemistry of the phosphorus atoms, a large number of isomers is produced with this synthetic method, however; the five possible isomers of a $P_4$ macrocycle are illustrated below.

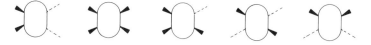

Although no catalysis results have been reported with these macrocycles to date, the $P_3$ macrocycles may provide interesting catalytic effects, as they are constrained to occupy the face of a polyhedron. Open-chain $P_3$ ligands, which can chelate around an edge of a polyhedron, are proving to be useful for homogeneous catalysis (see Sections 4 and 5).

Ciampolini *et al.* very recently reported the synthesis of the first crown-ether-type phosphorus-containing macrocycle 4,7,13,16-tetraphenyl-1,10-dioxa-4,7,13,16-tetraphosphacyclooctadecane ([18]ane $P_4O_2$) in 18% yield in one step by treating the dilithioderivative of

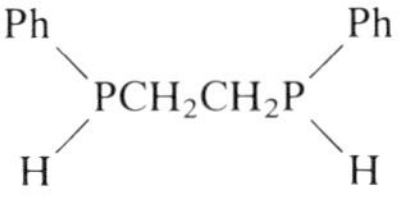

$$\text{Ph}_{\phantom{,}}\!\!\diagdown_{\phantom{,}}\text{PCH}_2\text{CH}_2\text{P}\diagup^{\phantom{,}}\!\!\text{Ph}$$

with *bis*(2-chloroethyl)ether (Equation 17).[30]

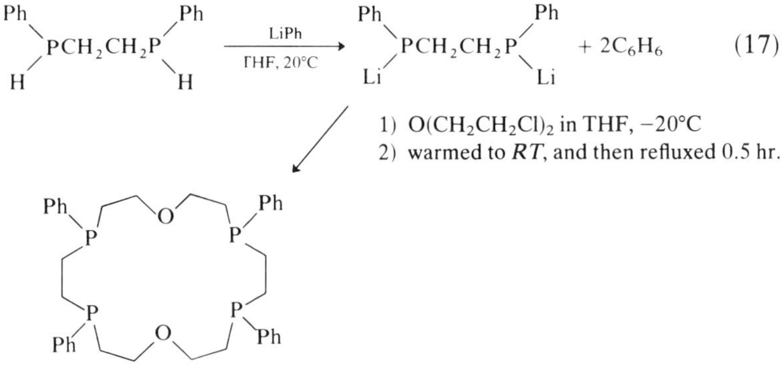

$$\begin{array}{c}\text{Ph}\diagdown\phantom{,}\diagup\text{Ph}\\\text{PCH}_2\text{CH}_2\text{P}\\\text{H}\diagup\phantom{,}\diagdown\text{H}\end{array}\xrightarrow[\text{THF, 20°C}]{\text{LiPh}}\begin{array}{c}\text{Ph}\diagdown\phantom{,}\diagup\text{Ph}\\\text{PCH}_2\text{CH}_2\text{P}\\\text{Li}\diagup\phantom{,}\diagdown\text{Li}\end{array}+2\text{C}_6\text{H}_6 \qquad (17)$$

1) $O(CH_2CH_2Cl)_2$ in THF, $-20°C$
2) warmed to *RT*, and then refluxed 0.5 hr.

**24**

Although the synthesis of **24** was not done under high dilution conditions, the 18% yield could probably be increased significantly by using Kyba's apparatus.[29d] Synthesis of other complicated crown-ether-type polyphosphine molecules can be expected in the near future.

All of the five possible diastereoisomers of **24** were isolated and characterized. Ciampolini *et al.* found that any given diastereoisomer racemizes at 140°C through inversion at the phosphorus atoms, thus giving an isomeric mixture whose composition is close to the statistically expected one. The two most easily available isomers were used to form nickel(II) and cobalt(II) complexes of general formula $[M([18]\text{ane}P_4O_2)]X_2$, ($X = BF_4^-$, $BPh_4^-$). The nickel(II) and cobalt(II) complexes of **24** are low-spin, with the hexa- and penta-coordinate geometries of the cobalt(II)-([18]ane$P_4O_2$) complexes being due to the configurations of the two ligand isomers. The four phosphorus atoms occupy the equatorial positions of both the octahedral and square-pyramidal Co(II) structures; for the α-isomer only one oxygen atom is located in a proper stereochemical position for coordination to the metal.[30]

### 3.2.5. Template Synthesis of Polyphosphine Ligands

Whereas many tri- and tetra-dentate combinations of nitrogen, oxygen, and sulfur donors have been accomplished *via* the "template effect" for two decades, syntheses of polyphosphine ligands *via* a template synthesis have been accomplished only recently.[31-33] As extensions of the base-catalyzed and radical-catalyzed additions of P–H bonds to vinyl groups (discussed in Sections 3.2.1 and 3.2.2), diphosphine[31,32] and triphosphine[33] ligands have been obtained in high yields by performing the condensations on coordinated phosphines (Schemes 6 and 7). This method

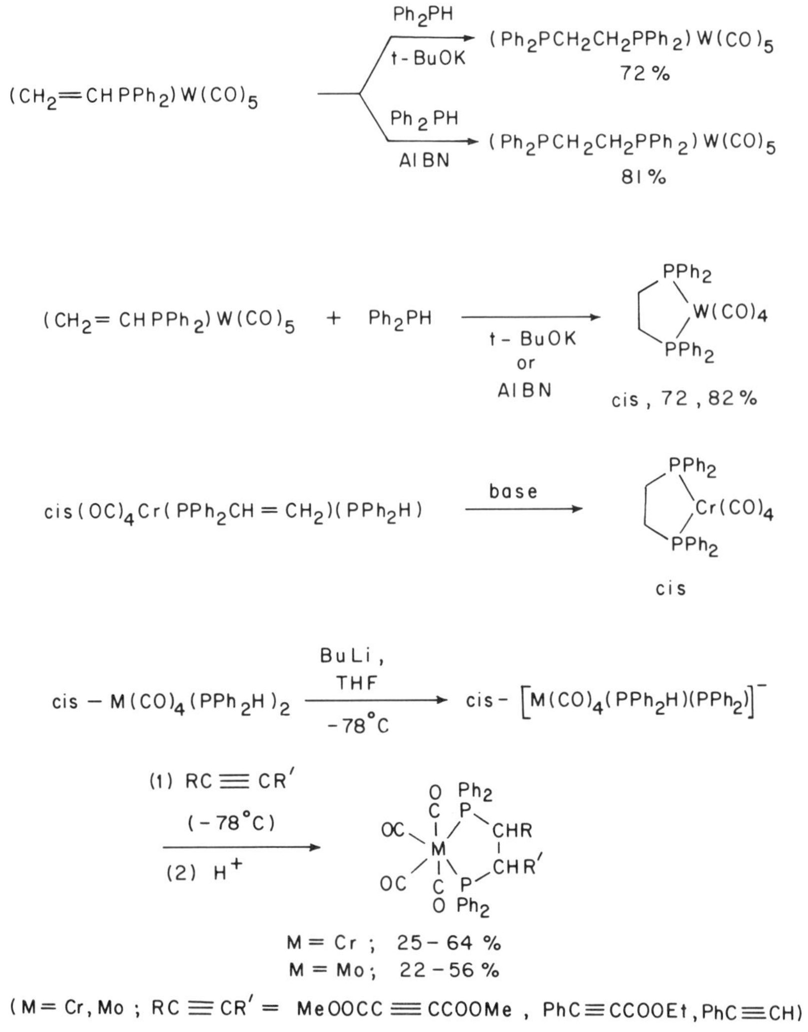

$$(CH_2=CHPPh_2)W(CO)_5 \xrightarrow{\begin{array}{c}Ph_2PH\\t\text{-BuOK}\end{array}} (Ph_2PCH_2CH_2PPh_2)W(CO)_5 \quad 72\%$$

$$(CH_2=CHPPh_2)W(CO)_5 \xrightarrow{\begin{array}{c}Ph_2PH\\AIBN\end{array}} (Ph_2PCH_2CH_2PPh_2)W(CO)_5 \quad 81\%$$

$$(CH_2=CHPPh_2)W(CO)_5 + Ph_2PH \xrightarrow{\begin{array}{c}t\text{-BuOK}\\\text{or}\\AIBN\end{array}} \begin{array}{c}PPh_2\\ \diagup\quad\diagdown\\ \quad\quad W(CO)_4\\ \diagdown\quad\diagup\\ PPh_2\end{array} \quad \text{cis, 72, 82\%}$$

$$cis(OC)_4Cr(PPh_2CH=CH_2)(PPh_2H) \xrightarrow{base} \begin{array}{c}PPh_2\\ \diagup\quad\diagdown\\ \quad\quad Cr(CO)_4\\ \diagdown\quad\diagup\\ PPh_2\end{array} \quad cis$$

$$cis\text{-}M(CO)_4(PPh_2H)_2 \xrightarrow[-78°C]{BuLi, THF} cis\text{-}[M(CO)_4(PPh_2H)(PPh_2)]^-$$

$$\xrightarrow[\begin{array}{c}(1)\ RC\equiv CR'\\(-78°C)\\(2)\ H^+\end{array}]{} \begin{array}{c}\quad O\ Ph_2\\\quad \|\ \ \diagup\\OC\diagdown\ C\ P\diagdown CHR\\\quad\ M\\OC\diagup\ C\ P\diagup CHR'\\\quad \|\ \ \diagdown\\\quad O\ Ph_2\end{array}$$

$$M = Cr;\ 25\text{-}64\%$$
$$M = Mo;\ 22\text{-}56\%$$

(M = Cr, Mo; RC≡CR' = MeOOCC≡CCOOMe, PhC≡CCOOEt, PhC≡CH)

Scheme 6[31,32]

Addition Reactions to Coordinated Secondary Phosphines

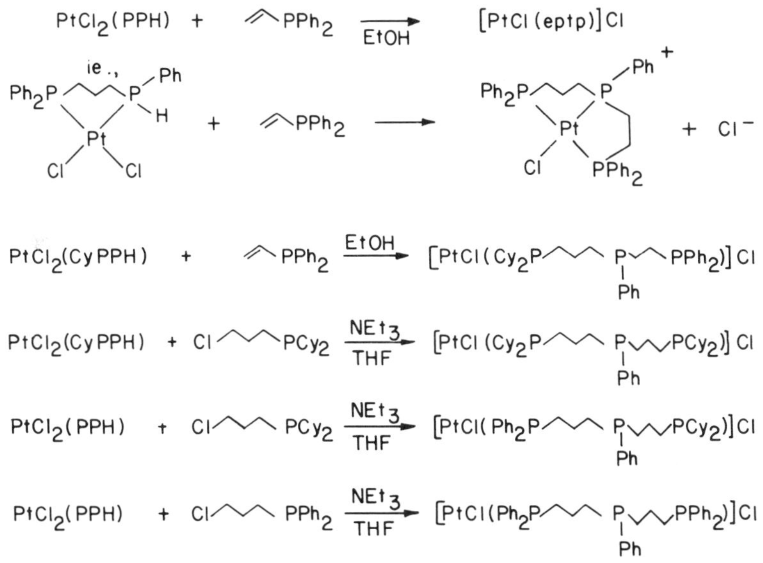

$$PtCl_2(PPH) + \diagup\!\!\!\diagdown PPh_2 \xrightarrow{EtOH} [PtCl(eptp)]Cl$$

i.e.,

$$\begin{array}{c}Ph_2P\!\!\diagup\!\!\diagdown\!\!P(Ph)(H) \\ | \quad\quad | \\ Cl-Pt-Cl \end{array} + \diagup\!\!\!\diagdown PPh_2 \longrightarrow \begin{array}{c} Ph_2P\!\!\diagup\!\!\diagdown\!\!P(Ph) \\ | \quad\quad\quad | \\ Cl-Pt-PPh_2 \end{array}^+ + Cl^-$$

$$PtCl_2(CyPPH) + \diagup\!\!\!\diagdown PPh_2 \xrightarrow{EtOH} [PtCl(Cy_2P\!\diagup\!\!\diagdown\! P(Ph)\!\diagup\!\!\diagdown\! PPh_2)]Cl$$

$$PtCl_2(CyPPH) + Cl\!\diagup\!\!\diagdown\! PCy_2 \xrightarrow[THF]{NEt_3} [PtCl(Cy_2P\!\diagup\!\!\diagdown\! P(Ph)\!\diagup\!\!\diagdown\! PCy_2)]Cl$$

$$PtCl_2(PPH) + Cl\!\diagup\!\!\diagdown\! PCy_2 \xrightarrow[THF]{NEt_3} [PtCl(Ph_2P\!\diagup\!\!\diagdown\! P(Ph)\!\diagup\!\!\diagdown\! PCy_2)]Cl$$

$$PtCl_2(PPH) + Cl\!\diagup\!\!\diagdown\! PPh_2 \xrightarrow[THF]{NEt_3} [PtCl(Ph_2P\!\diagup\!\!\diagdown\! P(Ph)\!\diagup\!\!\diagdown\! PPh_2)]Cl$$

*Scheme 7*[33]

appears extremely attractive for making unsymmetrical ligands like $Ph_2PCH_2CH_2\overset{Ph}{\underset{|}{P}}CH_2CH_2PR_2$ and $R_2PCH_2CH_2\overset{Ph}{\underset{|}{P}}CH_2CH_2CH_2PR'_2$.[33] In addition, the "template control" may make certain isomers of $P_3$ and $P_4$ ligands much more favored, as compared to the high-dilution method.

Base- or radical-catalyzed additions of P—H bonds (also As—H and S—H) to vinyl groups should be advantageous for building certain desired coordination geometries or mixed polymetallic complexes. Keiter has used this method to prepare all six mixed-metal complexes of the type $(OC)_5M\widehat{P\quad P}M'(CO)_5$ (where $M$ = Cr, Mo, W and $\widehat{P\quad P}$ = $Ph_2PCH_2CH_2PPh_2$) by the condensation of the appropriate $(OC)_5M(PPh_2CH=CH_2)$ and $(OC)_5M'(PPh_2H)$ complexes (Equation 18).[34]

$$(OC)_5M(PPh_2CH=CH_2) + (OC)_5M'(PPh_2H) \xrightarrow{KO(t-Bu)}$$
$$(OC)_5MPh_2PCH_2CH_2PPh_2M'(CO)_5 \quad (18)$$
$$M = Cr, Mo, W \text{ and } M' = Cr, Mo, W$$

Keiter's data suggest that the best results are obtained by building the polyphosphine from a vinylphosphine ligand bonded to the transition metal complex of interest. This approach should be limited only by the lability of the complex and the susceptibility of other attached ligands to reaction under the experimental conditions employed. It should also lead to syntheses of new cyclic and acyclic tri- and tetra-phosphines coordinated to the metal (or cluster) used for the template.

## 4. POLYPHOSPHINE HOMOGENEOUS CATALYSTS

As illustrated in the first three sections, chelating di-, tri-, and tetradentate ligands (particularly those incorporating more than one phosphorus donor) have become an important class of ligands in inorganic and organometallic chemistry. The author strongly believes that polyphosphine ligands will become increasingly important, particularly in the areas of asymmetric synthesis, supported catalysts, and clusters. The much lower tendency for M–P bond dissociation in such ligands allows the chemist unique possibilities for "fine tuning" electronic and/or structural parameters, while maintaining a fixed stoichiometry and stereochemistry in the complex.

For metal-phosphine catalysts of the types $RhCl(PPh_3)_3$, $RuHCl(PPh_3)_3$, and $RhH(CO)(PPh_3)_3$, it has become accepted that dissociation of triphenylphosphine is a necessary condition for hydrogenation or hydroformylation catalysis by such complexes. In the sections to follow, examples have been selected to show that three phosphino groups are bonded to the metal centers of the "active species;" yet rapid hydrogenation and selective hydroformylation of terminal olefins still occur.

For a metal catalyst that oscillates between a sequence of 16- and 18-electron complexes, the catalyst cycle *does not require* dissociation of one of the donor groups of a tridentate ligand. The primary requirement for hydrogenation and hydroformylation catalysis is that three sites be available on the metal for bonding two hydrogen atoms and the substrate molecule (olefin or CO, respectively). Thus, a tightly bound tridentate ligand can occupy the other three sites of an octahedral metal. Even a 4-coordinate $d^8$ metal [e.g., Co(I), Rh(I), Ir(I), Pd(II), Pt(II)] complex of a chelating tridentate ligand can function as an effective catalyst, *if the fourth position is a participating ligand in the catalysis cycle* (e.g., hydride or CO for hydrogenation and hydroformylation reactions, respectively). Thus, a complex such as HCo(ttp) or HRh(ttp) (ttp = $PhP(CH_2CH_2CH_2PPh_2)_2$) should undergo the accepted sequence of steps, i.e., for hydrogenation: coordination of an olefin, hydride transfer to produce the metal-alkyl, oxidative-addition of $H_2$, a second hydride transfer, elimination of the alkane, and regeneration of the HM(ttp) catalyst.

Subtle metal-ligand and metal-substrate interactions are critical for understanding catalytic mechanisms involving organometallic and coordination compounds. Phosphorus-31 chemical shifts and coupling constants are uniquely sensitive to these interactions, since a phosphorus atom is bonded directly to the metal center in most cases. Consequently, phosphorus-31 nmr is rapidly becoming the analytical method of choice for rapid characterization of organometallic and coordination compounds that contain phosphorus ligands. It is expected that future $^{31}$P studies will provide definitive identification of "intermediates" in reaction mechanisms and in catalytic systems, since $^{31}$P nmr spectra provide direct characterization of the phosphorus-containing species in solution.

### 4.1. Hydrogenation Catalysis with Chelating Triphosphine Ligands

DuBois and Meek[36] found that the metal-hydride complexes H$_3$Co(ttp), HCo(ttp), and HRh(ttp) (where ttp = the chelating triphosphine ligand PhP(CH$_2$CH$_2$CH$_2$PPh$_2$)$_2$) rapidly catalyzed the hydrogenation of terminal olefins at room temperature and 1 atm of H$_2$ pressure. Also, RhCl(ttp), which can be handled in air as a solid, is an effective homogeneous catalyst in ethanol in the presence of NaBH$_4$, which generates RhH(ttp) *in situ*. The most rapid hydrogenation rates were with "HCo(etp)" (etp = PhP(CH$_2$CH$_2$PPh$_2$)$_2$), which was generated *in situ*, and with an isolated sample of HRh(ttp). Under ambient conditions, these two hydrides convert 1-octene quantitatively to *n*-octane at rates comparable to the Wilkinson catalyst. However, the relative stabilities of the HRh(ttp) and HCo(etp) catalysts differ markedly in solution. The HRh(ttp) complex retains high catalytic activity through several batch operations in benzene solution, whereas the activity of "HCo(etp)" ceases when the olefin is depleted. DuBois also observed that initially H$_3$Co(ttp) rapidly hydrogenated 1-octene to *n*-octane;[36] however, the rate decreases rapidly due to isomerization of 1-octene to *cis* and *trans*-2-octene in the presence of the cobalt catalyst.

These HRh(ttp), RhCl(ttp), HCo(ttp), and H$_3$Co(ttp) complexes differ significantly from Wilkinson's catalyst, RhCl(PPh$_3$)$_3$, in that the triphosphine donor groups remain bonded through the catalyst cycle, whereas some PPh$_3$ dissociation occurs when RhCl(PPh$_3$)$_3$ is used.[37] The most important distinction to be gained from the catalysis experiments using chelating triphosphine ligands is that dissociation from the metal of one of three phosphino groups of the triphosphine ligand is not required for catalytic hydrogenation rates comparable to (or faster than) those obtained with RhCl(PPh$_3$)$_3$.[36]

Subsequently, Niewahner demonstrated that the complex RhCl(ttp) in the presence of either triethylaluminum or diethylaluminum chloride is an effective homogeneous catalyst for hydrogenation of terminal olefins and 1-octyne.[38] Proton and phosphorus-31 nmr spectra were used to identify several different chemical species [including RhH(ttp)] in the catalytically active RhCl(ttp)-AlEt$_2$X solutions. The observed rate of hydrogenation of 1-octene to $n$-octane at $20 \pm 0.3°C$ and under a constant H$_2$ pressure of 750 torr is $6.4 \times 10^4$ M$^{-1}$ min$^{-1}$, i.e., 25 times more rapid than the Wilkinson catalyst, RhCl(PPh$_3$)$_3$, under comparable conditions. In the presence of excess olefin and >2:1 Al:Rh ratio, the observed rate law for the catalytic hydrogenation of 1-octene is $-d[H_2]/dt = k_{obsd}[Rh]_T[H_2]$.

Proton and $^{31}$P{$^1$H} NMR spectra of toluene solutions of RhCl(ttp) and AlEt$_3$ (or AlEt$_2$Cl) indicate that RhH(ttp) is formed rapidly and cleanly in the solutions and that addition of terminal olefins results in very rapid conversion of RhH(ttp) to RhR(ttp). The nmr data also indicate that one ethylaluminum species is bonded directly to the rhodium and another is bridged via the Rh–H linkage.[38] The exact nature of the ethylaluminum species is not known because of exchange between coordinated and free ethylaluminum units and the likelihood that AlEt$_2$Cl is the product from the first step of the reaction between RhCl(ttp) and AlEt$_3$. Hence, the ethylaluminum species are indicated by EtAl̬ in Equations 19 and 20.

$$\text{RhCl(ttp)} + xs\text{AlEt}_3 \rightleftharpoons \overset{\overset{\text{EtAl}<}{|}}{\text{(ttp)Rh–H–AlEt}} + C_2H_4 + \text{EtAl}< \qquad (19)$$

$$\overset{\overset{\text{EtAl}<}{|}}{\text{(ttp)Rh–H–AlEt}} + \text{olefin} \rightleftharpoons \overset{\overset{\text{EtAl}<}{|}}{\text{(ttp)Rh–R}} + \text{EtAl}< \qquad (20)$$

Formation of the hydride complex RhH(ttp) in toluene results from β-hydride elimination of ethylene from RhEt(ttp), which is formed by an initial alkylation of RhCl(ttp) by either AlEt$_3$ or AlEt$_2$Cl.

On the basis of the $^{31}$P{$^1$H} and the $^1$H nmr spectra, the kinetic data, and the fact that solutions of RhCl(ttp) and either AlEt$_3$ or AlEt$_2$Cl rapidly catalyze the hydrogenation of olefins at ambient conditions, the following mechanism involving RhR(ttp), RhH(ttp) associated with an ethylaluminum species, and H$_2$ was proposed.[38] (Association of the ethylaluminum species with the rhodium complexes is omitted for clarity.)

$$\text{RhCl(ttp)} + \text{AlEt}_2\text{Cl} \underset{}{\overset{K_1}{\rightleftharpoons}} \text{RhEt(ttp)} + \text{AlEtCl}_2$$

$$\text{RhEt(ttp)} \underset{}{\overset{K_2}{\rightleftharpoons}} \text{RhH(ttp)} + C_2H_4$$

$$\text{RhH(ttp)} + \text{alkene} \underset{}{\overset{K_3}{\rightleftharpoons}} \text{RhR(ttp)}$$

$$\text{RhR(ttp)} + H_2 \overset{k}{\longrightarrow} \text{RhH(ttp)} + \text{alkane}$$

The analogous system containing $Et_2AlCl$ and the rhodium complex of $PhP(CH_2CH_2CH_2PCy_2)_2$, where $Cy$ = the cyclohexyl group, also catalyzes 1-octene at a rate comparable to that of $RhCl(ttp) + Et_2AlCl$; however, in this case, the rate depends on the concentration of olefin.

In another recent study with the triphosphine ligand ttp, Mazanec, Letts, and Meek discovered that excess $NaBH_4$ in refluxing THF converts $[RuCl_2(ttp)]_2$ into the mixed hydride-borohydride complex $[RuH(\eta^2\text{-}BH_4)(ttp)]$, **26**.[39] Compound **26** is unique in that: (1) the $\eta^2\text{-}BH_4$ group is static on the nmr time scale at room temperature; and (2) the $BH_4^-$ protons undergo a distinct two-step exchange process above room temperature, with the $H_a$ protons being scrambled before the $H_b$ protons (Figure 8.1). Owing to the presence of both Ru–H and doubly-bridged Ru–$BH_4$ linkages in **26**, addition of an acid (e.g., $HBF_4$) in the presence

<chemical structure of compound 26>

of neutral ligands $L$ produces the cationic complexes $[Ru(H)(L)_2(ttp)]^+$ ($L$ = CO, $CH_3CN$, and $P(OMe)_3$), whereas addition of a base produces a different series of complexes, i.e., the neutral complexes $RuH_2(L)(ttp)$ ($L$ = CO, $PPh_3$, and $P(OMe)_3$).[39]

Several of the $[RuH(L)_2(ttp)]^+$ and $RuH_2(L)(ttp)$ complexes may prove to be effective hydrogenation and hydroformylation catalysts. For example, preliminary studies on the hydrogenation of 1-octene showed that the parent compound $RuH(\eta^2\text{-}BH_4)(ttp)$, **26**, is inactive as a catalyst at 25°C and 1 atm of $H_2$ pressure. However, after addition of one equivalent of $HBF_4 \cdot Et_2O$, catalytic hydrogenation commenced immediately at a rate ~0.75 that of $RhCl(PPh_3)_3$ in THF; the hydrogenation gave only $n$-octane as a product.[39] In a separate set of hydrogenations with $NEt_3$ as a cocatalyst, an induction period of about 10 minutes was needed before the maximum rate was obtained; the slow step appears to be cleavage of the $Ru{<}^H_H{>}B$ bridge. From the spectral data, it may be inferred that $RuH_2(ttp)$ is the active catalyst in the presence of base (Figure 8.2). In the presence of acid, the catalytic species is probably cationic, e.g., $[RuH(\text{solvent})_x(ttp)]^+$. The isolated, stable, 18-electron cationic complexes $[RuH(CH_3CN)_2(ttp)]BF_4$ and $[RuH(P(OMe)_3)_2(ttp)]BF_4$ are not effective catalysts; this difference in the catalysis rates may simply reflect the need for one of the nonphosphine ligands to dissociate to generate a vacant

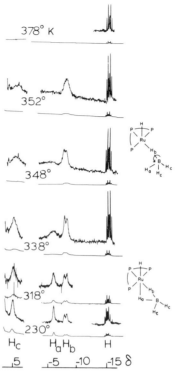

Figure 8.1. Proton ($^{11}$B decoupled) nmr spectra of RuH($\eta^2$-BH$_4$)(ttp), **26**, in $d_8$-toluene for the temperature range 230–378°K. Note that the resonance for H$_a$ begins to collapse in the temperature range 338–348°K, whereas the resonance of H$_b$ remains. Both H$_a$ and H$_b$ resonances collapse as the temperature is raised to 378°K. Reprinted with permission from *J. Am. Chem. Soc.* **104**, 3898–3905 (1982). Copyright (1982) American Chemical Society.

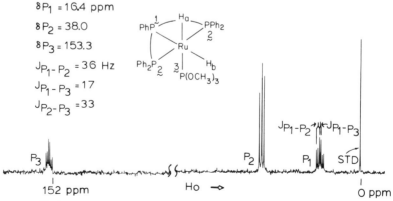

Figure 8.2. $^{31}$P{$^1$H} nmr spectrum of *cis*-[RuH$_2$(P(OMe)$_3$)(ttp)] in C$_6$D$_6$, note that all of the $J_{P-P}$ couplings are relatively small, in contrast to the $^2J_{P-P}$ values of 300–500 Hz observed for *trans* $^2J_{P-P}$ couplings in similar complexes. Reprinted with permission from *J. Am. Chem. Soc.* **104**, 3898–3905 (1982). Copyright (1982) American Chemical Society.

coordination site before the ruthenium-ttp complex can function as a hydrogenation catalyst under mild conditions.[39]

In both the RhH(ttp)- and RuH($\eta^2$-BH$_4$)(ttp)-catalyzed hydrogenation studies discussed above, the $^{31}$P{$^1$H} and $^1$H NMR spectra show that the ttp ligand does not dissociate from the metal over the temperature range studied, i.e., 203–353°. Thus, rapid hydrogenation catalysis occurs at ambient conditions, even when three phosphino groups are bound to the metal center. In the future, we should see particularly selective catalytic results with chelating tridentate ligands, as a result of the steric and electronic control that can be incorporated into such a metal complex.

### 4.2. Hydroformylation Catalysis with Chelating Diphosphine Ligands

Slaugh and Mullineaux of Shell[40] and Eisenman[41] of Diamond Alkali were the first investigators to report that rhodium-phosphine (or -phosphite) complexes are efficient catalysts for hydroformylation of 1-alkenes (Equation 21), using 1-hexene for illustration. Their work was followed

$$\diagup\!\!\diagup\!\!\diagdown\!\!\diagup\!\!\diagdown\!\!= + H_2 + CO \longrightarrow \diagup\!\!\diagdown\!\!\diagup\!\!\diagdown\!\!\diagup\!\!\diagdown CHO + \diagup\!\!\diagdown\!\!\diagup\!\!\diagdown\!\!\!\underset{CHO}{|} \quad (21)$$

$$+ \diagup\!\!\diagdown\!\!\diagup\!\!= + \diagup\!\!\diagdown\!\!\diagup\!\!\diagdown$$

closely by extensive Rh–PPh$_3$-catalyzed studies by Wilkinson and coworkers between 1965 and 1970.[42–46] They found that at 1 atm of 1:1 H$_2$:CO, 25°C, and 50 mM HRh(CO)(PPh$_3$)$_3$, a 20:1 linear-to-branched aldehyde (l:b) ratio was obtained by hydroformylation of 1-pentene.[44] However, at the higher temperatures and pressures and low-rhodium concentrations anticipated for commercial operation, l:b ratios of only 2–3 were reported.[47] This poor selectivity was improved substantially by increasing the PPh$_3$ concentration[46,48] or by decreasing the partial pressure of carbon monoxide.[47b] This requirement for excess PPh$_3$ ligand was attributed to a need to maintain at least two phosphine ligands bonded to rhodium during the selectivity step of the reaction.[46,48]

Oliver and Booth[49] and Pruett and Smith[48] showed qualitatively that both rate and selectivity of hydroformylation increase as the phosphine ligands are changed from the more basic alkyl phosphines to the less basic aryl phosphines. These substituent effects, as well as the need for a large concentration of monodentate phosphines, suggest that increased l:b selectivity could be achieved by chelating diphosphine or polyphosphine ligands, since dissociation of a chelated ligand from the metal should be more difficult than dissociation of a monodentate ligand. In fact, Unruh, Hughes,

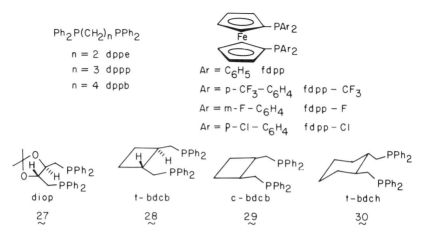

Figure 8.3. Examples of bidentate diphosphine ligands used in the Celanese hydroformylation studies (**27–30**).

and Christenson at Celanese[50] recently reported hydroformylation studies with the chelating ligands shown in Figure 8.3.

The Celanese investigators[50] found that $t$-bdcb, DIOP, and ferrocene-type ligands led to highest selectivities and when the ligand:rhodium mole ratio was $\geq 1.5:1$. When substituted phenyl groups were attached to phosphorus, good correlations were obtained between the Hammett $\sigma$ parameters and the l:b aldehyde ratios, as well as the rate of formation of linear aldehyde (Figure 8.4). Electron-withdrawing substituents on the benzene rings (less basic ligands) improved the hydroformylation rates and selectivities to linear aldehyde. The 3:2 diphosphine–rhodium complex that has been proposed[50a,b] to account for these observations is shown in Figure 8.5; it has one bulky diphosphine ligand (e.g., $t$-bdcb, DIOP, or the ferrocene-based $P_2$ ligands) chelated to each rhodium atom and a third diphosphine ligand bridging the two rhodium atoms. The 3:2 diphosphine–rhodium ratio is based on the catalysis results, and the distorted trigonal-bipyramidal structure is based on the $^{31}P\{^1H\}$ nmr spectral data.[50c]

Consistent with the 3:2 diphosphine:Rh stoichiometry, a mixed ligand effect was observed, wherein 1:1 molar combinations of a diphosphine and $HRh(CO)(PPh_3)_3$ require a monophosphine of optimum size (i.e., phosphine ligands approximately the size of $Ph_2PEt$ and $Ph_2PCH_2-$, where $\theta \cong 140°$) to produce linear aldehyde selectivities comparable with stoichiometric (i.e., 1.5:1 ratio) use of diphosphines alone.

In the studies on the Rh–fdpp–$CF_3$ catalyst system,[50d] the effects of $H_2$ and CO partial pressures and total reaction pressure (1:1 $H_2/CO$) indicate that CO, rather than the bridging diphosphine ligand, dissociates

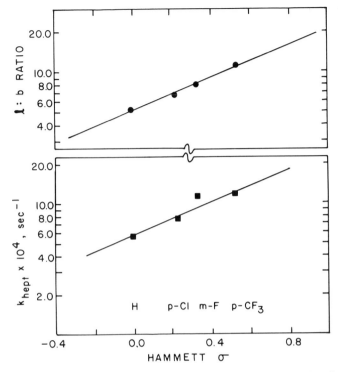

Figure 8.4. Hammett $\sigma\rho$ plot for fdpp and derivatives. $k_{hept} = (\text{Eff}_{hept})(k_{obs})/100\,[\text{Rh}]$, where [Rh] = catalyst concentration in millimoles/liter. Conditions: 790 kPa, 110°C, 1:1 $H_2:CO$.

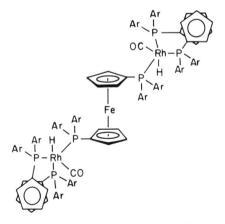

Figure 8.5. Proposed structure of the 3:2 diphosphine:rhodium selective catalyst species.

Table 1. 1-Hexene Hydroformylation[a] with Rigid Diphosphines

| Added P~P | Ratio P~P:Rh | 1:b | Percent selectivity to | | | |
|---|---|---|---|---|---|---|
| | | | Hexane | 2-Hexene | 2-Methyl-hexanal | Heptanal |
| (+)-diop[b] | 1.0 | 2.1 | 1.0 | 53.0 | 15 | 31 |
| | 1.5 | 5.5 | 0.1 | 1.2 | 15 | 83 |
| | 2.0 | 5.5 | 0.8 | 0.2 | 15 | 84 |
| t-bdcb[c] | 1.0 | 3.2 | 0.4 | 23.0 | 18 | 58 |
| | 1.5 | 7.2 | 0.5 | 1.4 | 12 | 86 |
| | 2.0 | 7.2 | 0.5 | 1.2 | 12 | 87 |
| | 3.0 | 7.2 | 0.5 | 1.3 | 12 | 86 |
| | 10.0 | 7.9 | 0.5 | 1.0 | 11 | 87 |
| c-bdcb[d] | 1.5 | 3.2 | 0.4 | 24.0 | 18 | 58 |
| | 2.0 | 3.4 | 0.4 | 12.0 | 20 | 68 |
| | 5.0 | 3.7 | 0.6 | 0.6 | 21 | 78 |
| | 55.0 | 3.7 | 0.7 | 0.7 | 21 | 78 |
| t-bdch[e] | 1.5 | 2.4 | 0.0 | 29.0 | 21 | 50 |
| | 5.0 | 1.1 | 0.0 | 1.0 | 46 | 52 |

[a] All runs with 790 kPa of 1:1 $H_2$:CO.
[b] 106 ± 3°C.
[c] 106 ± 2°C.
[d] 100 ± 1°C.
[e] 103 ± 1°C.

in order to provide a 16-electron complex capable of binding the olefin. In contrast to the results with t-bdcb, DIOP, and the ferrocene diphosphines, the "nonrigid" diphosphines, dppe, dppp, and dppb did not lead to high selectivities for linear aldehyde. Also, they did not maximize total aldehyde efficiencies at a ligand-to-rhodium ratio of 1.5:1, as did the more rigid ligands. The t-bdch ligand (Table 1) behaved very much like the nonrigid ligands, while c-bdcb (Table 1) was intermediate between the high selectivities of the ferrocene-based ligands and the nonrigid ligands.

The mechanism suggested recently by the Celanese investigators to account for the fdpp, t-bdcb, and DIOP results is shown in Figure 8.6. The scheme does not specify the geometry of the 3:2 complex, and it assumes that the two rhodium atoms within the complex act independently. The Celanese mechanism depicted in Figure 8.6 differs in two important aspects

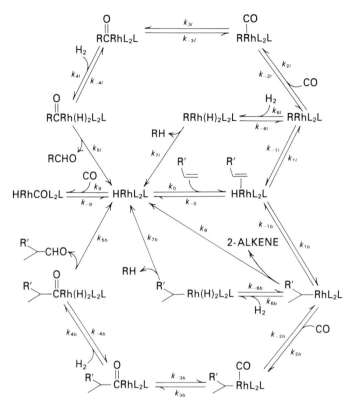

Figure 8.6. Proposed Celanese mechanism for the rhodium-phosphine catalyzed hydroformylation. $L_2$ = chelating ligand; $L = 1/2$ of bridging ligand; $R = CH_2CH_2R'$.

from the Wilkinson mechanism,[44–46] which has been the mechanism generally accepted for rhodium-phosphine catalyzed hydroformylation reactions[51,52]:

1. Wilkinson suggested that only two phosphine ligands were bound to rhodium in the "active-catalyst," i.e., dissociation of one $PPh_3$ ligand from $HRh(CO)(PPh_3)_3$ is a prerequisite for formation of an active catalyst. According to mechanistic studies that are summarized by Figure 8.7, $HRh(CO)_2(PR_3)_2$ can be formed readily from $HRh(CO)(PR_3)_3$ and CO, $HRh(CO)_2(PR_3)_2$ being the active species capable of reacting with the olefin. In contrast, the Celanese catalysis and $^{31}P\{^1H\}$ results suggest that under hydroformylation conditions, three phosphorus atoms must be bound to rhodium at the instant that straight-chain selectivity is determined.[50d] The high 1:b selectivity obtained with rigid

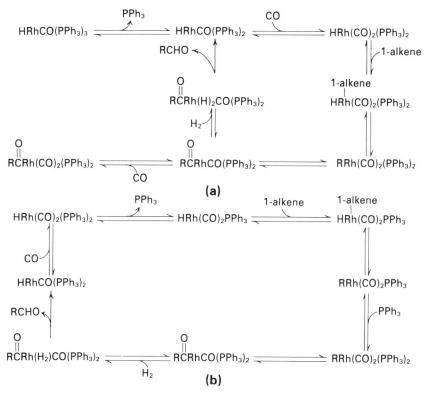

Figure 8.7. Wilkinson mechanism for rhodium-catalyzed hydroformylation: (a) associative path; (b) dissociative path. $R = R'CH_2CH_2-$ or $R'CH(CH_3)-$.

diphosphine ligands indicates that complexes of the type $HRh(CO)L_2L$ (where $L_2$ = a chelating phosphine, $L$ = a monophosphine or a diphosphine acting as a monodentate to that rhodium atom) influence the important step in the catalytic cycle at which Rh–H adds to coordinated olefin.

2. Variations of the $H_2:CO$ ratio show that the rate-controlling steps for linear and branched aldehydes are different, and that CO may not be bound to rhodium when the olefin is first coordinated to Rh. These results suggest: (a) either the CO ligand dissociates to form the 16-electron complex $HRh(L_2)(L)$, which then coordinates an olefin molecule to form $HRh(L_2)(L)(olefin)$; or (b) $L$ dissociates to allow formation of $HRh(CO)(L_2)(olefin)$, but it quickly reassociates, driving the Rh–H addition step preferentially to the linear alkyl intermediate ($n$-alkyl)$Rh(CO)(L)_2(L)$. In either case, the stereochemistry imposed

by the three phosphine groups provides the influence for linear aldehyde selectivity.

The important differences between the Celanese and Wilkinson mechanisms (i.e., composition and coordination geometry of *tris*-phosphine vs. *bis*-phosphine active species) merit further careful study. This may prove to be an area where chelating specially designed triphosphine ligands can provide definitive choices between the two mechanisms.

## 5. SOLID-SUPPORTED POLYPHOSPHINE CATALYSTS

Applications of di-, tri-, and tetra-dentate ligands in the areas of asymmetric synthesis and catalysis by polymer-bound metals have begun to require incorporation of polyphosphine units into complicated organic structures. Introduction of multiple phosphine units onto a solid support can be accomplished (1) either by attachment of the entire polydentate ligand *via* functionality on one of the substituent chains, or (2) by introduction of multiple individual phosphine units into a precursor, ordinarily by nucleophilic displacement of halides or tosylates by a $R_2P^-$ ion (most commonly the anion has been $Ph_2P^-$) or by addition of $R_2PH$ molecules to activated olefins.[53] The latter routes suffer from unpredictable yields and incompatibility with many functional groups. Three studies, which illustrate the special advantages that can be gained by covalent attachment of chelating polydentate ligands, are presented below.

### 5.1. Coordination of Multiple Phosphino Groups from Supported Monophosphines

In contrast to organic polymers such as polystyrene, rigid nonswellable inorganic supports such as silica or alumina should increase the efficiency of anchored catalysts by minimizing undesirable deactivation reactions, such as intermolecular aggregation of active species or multiple-ligand chelation of the metal center. The latter factor is a problem with most of the supported monophosphine catalysts that have been studied. For example, spacing of the $-PR_2$ groups uniformly on the support surface is very difficult to control during the synthesis. Consequently, one metal may be bonded to only one $-PR_2$ group, whereas another metal may be bonded to two $-PR_2$ groups (Figure 8.8). Thus, the electronic and stereochemical properties of the two metals will differ significantly, causing the catalysis specificity and rate to differ at the two metal sites.

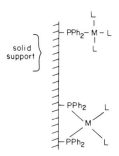

Figure 8.8. Illustration of two types of metal-phosphine bonding that can occur with supported monophosphine ligands.

In an interesting demonstration of the variable coordination environments created by anchored monodentate diphenylphosphino groups, Capka et al. compared the kinetics of liquid-phase hydrogenation of alkenes catalyzed by **homogeneous** rhodium(I) complexes prepared *in situ* from $\mu,\mu'$-dichloro-*bis*-[di(alkene) rhodium] and tertiary phosphines of the type $R\text{PPh}_2$ ($R = -(CH_2)_n Si(OEt)_3$; $n = 1-6$ and $R = -CH_2SiMe_{3-m}(OEt)_m$; $m = 1-3$) and by their **heterogenized analogues** anchored to silica.[18,19] The hydrogenation with both types of catalysts at 1.1 atm hydrogen pressure and 37–67°C were first order in the alkenes. With one exception, the rate constants of the homogeneous catalysts do not depend on the length of the phosphine-alkylene chain, $-(CH_2)_n PPh_2$. Except for the $(CH_2)_3$ chain, all the rates for the homogeneous catalysts are in the range of 4.5–5.5 liter mol$^{-1}$ min$^{-1}$. The length of the $-CH_2CH_2CH_2-$ chain apparently allows the ethoxysilyl group to interact—either sterically or electronically—with the rhodium dimer, and this rate is one-half the other values.

The ligands $(EtO)_3Si(CH_2)_n PPh_2$ and $(EtO)_2MeSi(CH_2)_n PPh_2$ were anchored onto silica and treated with $[RhCl(ethylene)_2]_2$ to produce the supported catalysts. The resulting heterogenized complexes were highly active for both hydrogenation and hydrosilylation of alkenes under mild conditions. The observed variation in the catalytic activity of $(CH_2)_n PPh_2$–RhClL$_n$ complexes clearly shows that two different active sites are formed on the silica surface. The activity is significantly higher (a factor of 10 for hydrosilylation and a factor of 3–7 for hydrogenation) when the rhodium atom is linked through a short chain, i.e., the $-CH_2PPh_2$ group, compared to linkage *via* alkylene chains containing 2–6 methylene groups. Variations in the activity were attributed to differences in the nature of the catalytic species formed on the surface. From calculations of the surface area and the phosphine concentrations, it was concluded that the short $-CH_2PPh_2$ chain forms monophosphine, monorhodium complexes (i.e.,

**31**), whereas the longer chains lead to dimeric structures (**32**) for 2–4 carbons and (**33**) for 5, 6 carbons, respectively; such types of dimeric complexes are known to be less active hydrogenation and hydrosilylation catalysts for olefins.[35]

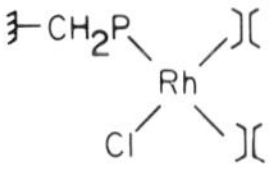

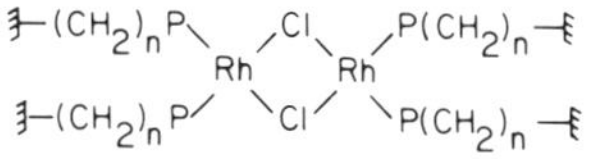

### 5.2. A Triphosphine Ligand Attached to Glass Beads

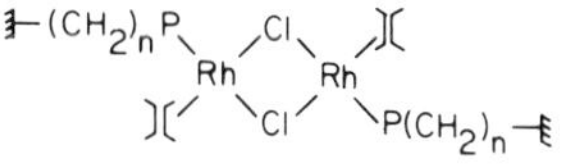

The structure and stoichiometry problems that are inherent with supported monophosphines, which were illustrated above with Capka's studies, can be avoided by choosing a polydentate ligand to contain the requisite number of phosphine donors. Uriarte and Meek recently demonstrated the utility of this approach by: (1) attaching the first examples of tridentate ligands onto a controlled-pore glass (C.P.G.) solid support; (2) characterizing the $CoCl_2$ complex of the attached triphosphine ligand; and (3) demonstrating efficient hydrogenation catalysis of terminal olefins at 25°C and 1 atm $H_2$ pressure by this bound $CoCl_2$·triphosphine complex in the presence of a hydride source, OMH-1. The attachment sequence of reactions is presented below.[27d]

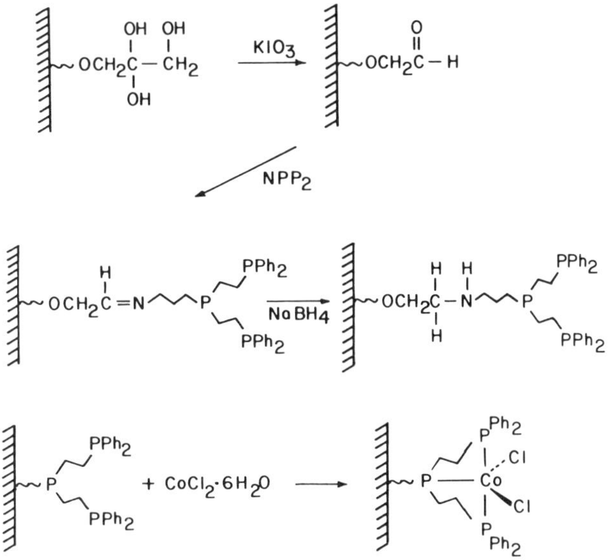

### 5.3. A NP₂ Ligand Attached to Organic Compounds

Whitesides et al. introduced a chelating diphosphine unit into several different organic molecules by taking advantage of the selective reactivity of the amino group compared to the phosphorus atoms of $HN(CH_2CH_2PPh_2)_2$ toward carbonyl halides, anhydrides, active esters, and isocyanates (see Section 3.1 for the attachment reactions).[25] They demonstrated that the attached amino-diphosphine ligand (NP₂) leads to complexes of transition metals having a wide range of structures and physical properties. Particularly, the water-soluble complexes offer tremendous potential for homogeneous catalysis in aqueous solutions. For ease of synthesis and isolation, ligand **34** provides the most practical rhodium catalyst for effective homogeneous hydrogenation in aqueous solutions.[25d]

$$\text{HOOC} \underset{\text{HOOC}}{\diagdown} \!\!\!\! \bigcirc \!\!\!\! - \overset{O}{\overset{\|}{C}} - N \underset{CH_2CH_2PPh_2}{\overset{CH_2CH_2PPh_2}{\diagup}}$$

**34**

The NP₂ unit and the resultant achiral $[Rh(NP_2)(NBD)]^+$ moiety can also be attached easily at a specific site in a protein. The protein structure then provides the chirality required for enantioselective hydrogenation. Thus, hydrogenation of α-acetamidoacrylic acid to N-acetylalanine catalyzed by $[Rh(NP_2)(NBD)]^+$ bound to avidin at $RT$ and 1.5 atm of $H_2$ showed ~40% S enantiomeric excess. Although these hydrogenation results with avidin are modest, it does demonstrate that asymmetric synthesis is accomplished by the bis-phosphine rhodium catalyst attached covalently to a protein.[25b]

### 5.4. Chelating Diphosphines Attached to Organic Polymers

Stille et al. have prepared a series of rhodium-bound catalysts containing optically-active diphosphine ligands by two different routes. In some cases, they first prepared monomers containing the diphosphine unit and then performed the polymerization.[24] In other cases, the polymer was treated with excess diphenylphosphide to replace tosylate groups (Schemes 1 and 2). For example, the optically-active monomer 2-p-styryl-4,5-bis[(tosyloxy)methyl]-1,3-dioxolane was polymerized with a variety of co-monomers to provide polymer-attached optically active ligands for rhodium-catalyzed hydroformylation[24e] and hydrogenation[24a,b] reactions that produced optically active aldehydes and amino acids with the same enantioselectivity as their homogeneous counterparts. More recently, two optically active bis(phosphino)pyrrolidine monomers were prepared and copolymerized with hydrophilic co-monomers and a divinyl monomer to provide a cross-linked insoluble polymer that will swell in polar solvents. Exchange of rhodium(I) onto the polymer gave catalysts that were active for the asymmetric hydrogenation of N-acyl α-amino acids in high optical yields, where the phosphine that was derived from the enantiomer of the naturally occurring 4-hydroxy-proline gave (s)-amino acids. The catalysts could be recycled by simple filtration with no loss in selectivity.[24c,d]

## 6. CONCLUDING REMARKS

It is anticipated that polydentate ligands will become increasingly important in studies of catalysis, particularly for definition of the stoichiometry and stereochemical specificity of "active species." Increasing use of polyphosphorus ligands can also be expected, since $^{31}$P nmr spectra are now readily available on FT nmr instruments, and the $^{31}$P spectra provide direct information on the phosphorus-containing species in solution.[5] As illustrations of the special situations one can control with polydentate ligands, a few recent examples are cited below.

In a particularly interesting application of the radical- and base-catalyzed synthesis of polyphosphine ligands (Section 3.2.2), Keiter et al.[54] have prepared all five possible nonchelated triphosphine complexes of $W(CO)_5$ (**35–39**).

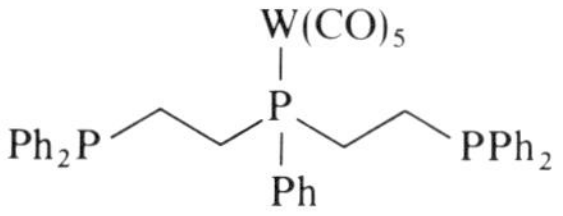

$Ph_2P\frown\overset{W(CO)_5}{\underset{Ph}{P}}\frown PPh_2$    $(OC)_5WPPh_2\frown\overset{P}{\underset{Ph}{|}}\frown PPh_2$

         **35** 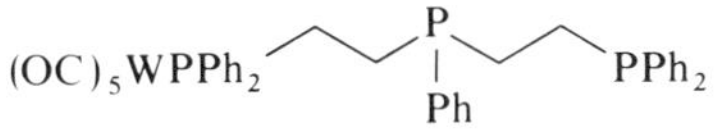                               **36**

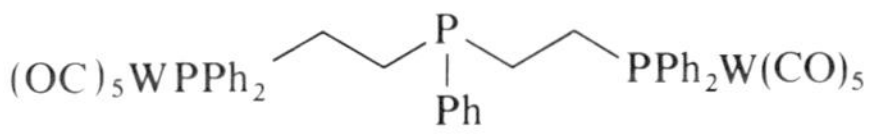

$(OC)_5WPPh_2\frown\overset{P}{\underset{Ph}{|}}\frown PPh_2W(CO)_5$

**37**

$(OC)_5WPPh_2\frown\overset{W(CO)_5}{\underset{Ph}{P}}\frown PPh_2$    $(OC)_5WPPh_2\frown\overset{W(CO)_5}{\underset{Ph}{P}}\frown PPh_2W(CO)_5$

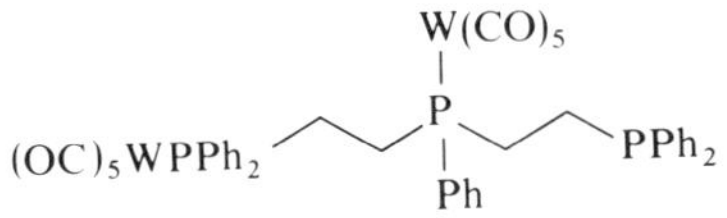

        **38** 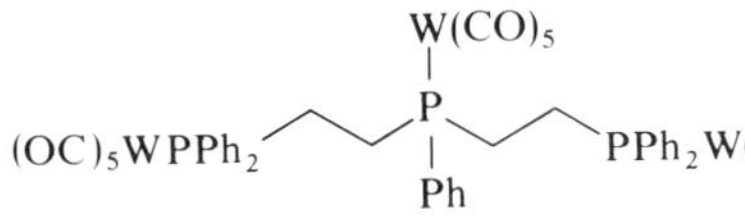                               **39**

The synthesis strategy involved building each specific complex from judiciously selected coordinated fragments (Equations 22–24), similar to the method used by Waid and Meek for platinum complexes of new unsymmetrical triphosphines (Scheme 7, p. 273).[33]

$$(OC)_5WPPh(CH=CH_2)_2 + 2PPh_2H \xrightarrow[75°C]{AIBN} Ph_2PCH_2CH_2\underset{\underset{W(CO)_5}{|}}{\overset{\overset{Ph}{|}}{P}}CH_2CH_2PPh_2 \quad (22)$$
$$\mathbf{35}$$

Complex **36**, a structural isomer of **35**, is synthetically difficult and best prepared by a two-step reaction involving $Ph_2PCH_2CH_2PPh(CH=CH_2)$ (Equations 23 and 24):

$$PhP(CH=CH_2)_2 + PPh_2H \xrightarrow{AIBN} PhP(CH=CH_2)(CH_2CH_2PPh_2) \quad (23)$$

$$PhP(CH=CH_2)(CH_2CH_2PPh_2) + (OC)_5WPPh_2H \xrightarrow{AIBN}$$

$$(OC)_5WPPh_2CH_2CH_2\overset{\overset{Ph}{|}}{P}CH_2CH_2PPh_2 \quad (24)$$
$$\mathbf{36}$$

Complex **37** was prepared also by the free radical reaction, whereas complexes **38** and **39** were prepared by the base-catalyzed route (see p. 289).[54]

Tridentate tripod ligands, such as $CH_3C(CH_2PPh_2)_3$, triphos, $CH_3C(CH_2PEt_2)_3$, etriphos, and $CH_3C(CH_2AsPh_2)_3$, triars, have been used to stabilize a remarkable series of $P_3$, $As_3$, and $P_4$ fragments in triple-decker sandwich dinuclear cations of the type $[(tripod)M(\mu-(\eta^3-P_3)M(tripod)]^{2+}$.[55] More recently, these tripod ligands have led to hydride- and halide-bridged dinuclear complexes, e.g., $[(triphos)Fe(\mu-H)_2Fe(triphos)]PF_6$, $[(triars)Co(\mu-H)_2Co(triars)]BPh_4$, and $[(etriphos)Fe(\mu-Cl)_2Fe(etriphos)]PF_6$.[56] The hydride-bridged complexes should be examined for potential hydrogenation catalysis.

Tripod ligands, such as $HC(PPh_2)_3$ and $CH_3Si(PPh_2)_3$, have been suggested[14–16] as good candidates for coordinating to three different metal centers in a cluster complex, and thus holding those centers together. However, complications can also result. For example, treatment of $Ru_3(CO)_{12}$ with $HC(PPh_2)_3$ in THF solution leads to formation of at least eight products, each in <10% yield.[57] One of the products is the desired axial symmetrically capped $Ru_3(CO)_9[HC(PPh_2)_3]$, **40**, analogous to the previously reported $Ru_3(CO)_9[CH_3Si(PPh_2)_3]$.[16] Spectroscopic data indicate that two other products contain bridging CO ligands (as for **40**); however, at least three products contain only terminal CO ligands, as well as two equivalent and one unique phosphine group. One such product was characterized crystallographically as the *ortho*-disubstituted complex $Ru_3(CO)_9\{HC(PPh_2)(PhPC_6H_4PPh)\}$, **41**.[57] In the latter complex, the

## POLYDENTATE LIGANDS AND THEIR EFFECTS ON CATALYSIS

triphosphine ligand has been transformed so that a phenyl ring of one phosphine moiety has undergone *ortho*-substitution by the phosphorus atom of a second phosphine moiety, together with the loss of benzene.

Other perturbations sometimes arise from polydentate ligands, i.e., either not all the donor atoms are bound to metal atoms, or each donor atom may bond to a different metal atom, leading to polynuclear complexes. Examples of these two situations are **42**[58] and **43**,[59] respectively.

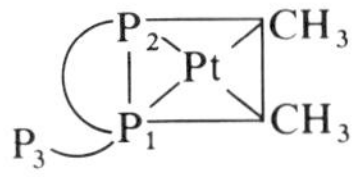

**42**

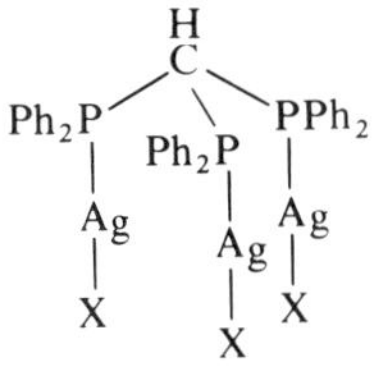

**43**

In the case of the potential tetradentate ligand *tris*(2-pyridyl)phosphine, $P(py)_3$, Wilkinson et al. have shown that it functions preferentially as a monodentate phosphorus ligand.[60] Complexes can also be isolated in which $P(py)_3$ acts as a bidentate, chelating P, N ligand, e.g., $RuHCl[P(py)_3]_3$ and $RhCl[P(py)_3]_2$. In the presence of excess $P(py)_3$ and at low $CO + H_2$ (1:1) pressures, the complex $RhH(CO)[P(py)_3]_3$ is a catalyst for the selective (l:b ratio = 13:1) hydroformylation of 1-hexene to $n$-heptanal.

Nickel(II) complexes of the tridentate ligand 2,6-*bis*(diphenylphosphinomethyl)pyridine, *pnp*, react readily with CO in $H_2O$-EtOH at room temperature and pressure to give a Ni(O)-carbonyl complex.[61] This Ni(II)-Ni(O)-*pnp* system is an effective homogeneous catalyst for the water-gas shift reaction in water-alcohol (ethanol, propanol, or butanol) at low temperatures and pressures.

The triphosphine ligand $Ph_2PCH_2CH_2\overset{|}{\underset{}{P}}CH_2CH_2PPh_2$, *etp*, has been used recently to prepare the complexes **44** (see Scheme 8)[62] and **45**,[63]

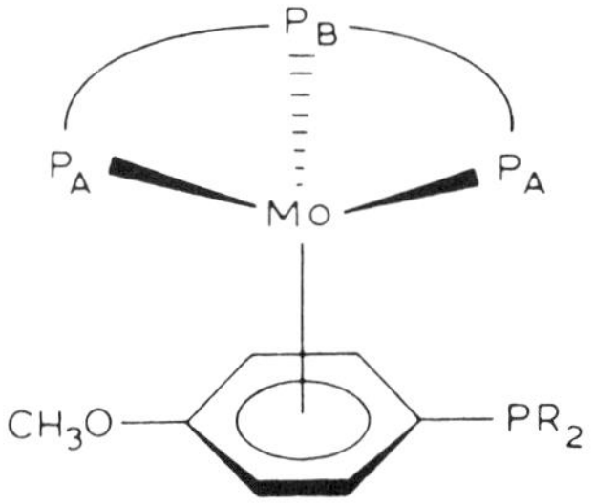

**45**

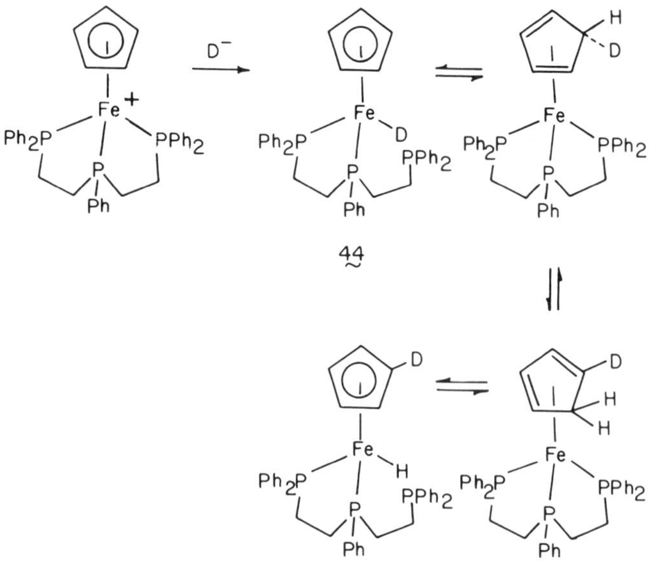

*Scheme 8*

which display unique properties. Reduction of the cation $[(\eta^5\text{-}C_5H_5)\text{Fe}(etp)]^+$ with LiAlD$_4$ gives **44**, which slowly undergoes H–D scrambling at 20°C. The mechanism for H–D scrambling is shown in Scheme 8; this scrambling process appears to be the first example of an *exo*-hydrogen sigmatropic shift around a hydrocarbon ligand bonded to a transition metal.

Complex **45** was first isolated as a co-product during the reduction of MoCl$_3$(*etp*) (Equation 25).

$$\text{MoCl}_3(etp) + PR_3 \xrightarrow[\text{THF}]{\text{Na/Hg; N}_2} trans\text{-Mo}(N_2)_2(etp)PR_3 \qquad (25)$$
$$\mathbf{46}$$

Acid treatment of **46** converts a Mo–N$_2$ group into NH$_3$ and permits isolation of MoBr$_3$(*etp*) in >90% yield (Equation 26).[64]

$$2\text{Mo}(N_2)_2(etp)(PPh_3) + 8HBr \rightarrow 2NH_4Br + 2MoBr_3(etp) + 3N_2 + 2PPh_3 \qquad (26)$$

Use of the triphosphine ligand permitted, for the first time, characterization of the metal-containing product (MoBr$_3$(*etp*) in this case) after ammonia formation from a metal-dinitrogen complex.[64]

Compared to the number of papers on chelating polyphosphine ligands, there is a paucity of papers on chelating phosphites, even though such ligands should make interesting comparisons by exhibiting stronger $\pi$-acceptor properties than a similar polyphosphine. It is expected that the number of studies with chelating phosphites will increase rapidly. In fact, King reported an initial study on the bidentate $(CH_3O)_2PCH_2CH_2P(OCH_3)_2$ in 1978;[65] recently Fryzuk[66] showed that this ligand and the corresponding isopropyl ligand formed the interesting polynuclear rhodium bridged-hydride complexes

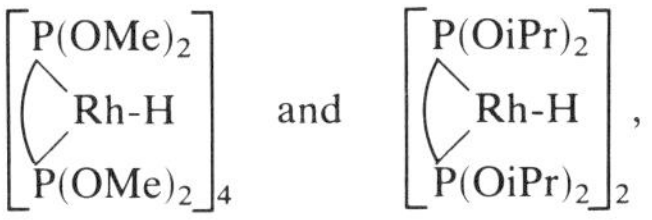

$$\begin{bmatrix} P(OMe)_2 \\ \diagdown \\ Rh\text{-}H \\ \diagup \\ P(OMe)_2 \end{bmatrix}_4 \text{ and } \begin{bmatrix} P(OiPr)_2 \\ \diagdown \\ Rh\text{-}H \\ \diagup \\ P(OiPr)_2 \end{bmatrix}_2 \text{, respectively.}$$

      **47**               **48**

Complexes **47** and **48** are extremely efficient homogeneous hydrogenation catalysts for simple olefins; however, it is not yet known whether the integrity of the hydride-bridged cluster is maintained during the hydrogenation cycle. These compounds should lead to some fascinating results in the future.

King has recently studied the first examples of chelating tridentate phosphine–phosphite ligands, i.e., $R\text{P}[CH_2CH_2P(OCH_3)_2]_2$ ($R = CH_3$ and $C_6H_5$).[67] These ligands react with metal(II) chlorides of Fe, Co, and Ni in methanol solution to form the $[M\text{Cl}(P_3 \text{ ligand})]^+$ cations, which are planar for nickel and tetrahedral for cobalt and iron.[67]

*Acknowledgments*

The author is grateful to the Guggenheim Foundation for a 1981–1982 Fellowship, to The Ohio State University for a Professional Leave, and to Professor E. Muetterties and the Chemistry Department at the University of California, Berkeley, for their hospitality during the period this chapter was written.

*References*

1. (a) J. Chatt, *Advances in Organometallic Chemistry*, edited by F. G. A. Stone and R. West (Academic Press, New York, 1974), Vol. 12, pp. 1–29; (b) G. Booth, *Advances in Inorganic Chemistry and Radiochemistry*, edited by H. J. Emeleus and A. G. Sharpe

(Academic Press, New York, 1964), Vol. 6, pp. 1–69; (c) A. Pidcock, *Transition Metal Complexes of Phosphorus, Arsenic and Antimony Ligands*, edited by C. A. McAuliffe (Wiley and Sons, New York, 1973), pp. 1–31.
2. R. Mason and D. W. Meek, *Angew. Chem. Int. Ed. Engl.* **17**, 183–194 (1978).
3. D. W. Meek, *Strem Chemiker* **5**, 3–11 (1977).
4. For reviews on the spectral and magnetic data of transition metal complexes of polydentate ligands, see: (a) C. Furlani, *Coord. Chem. Revs.* **3**, 141–167 (1968); (b) M. Ciampolini, *Structure and Bonding (Berlin)* **6**, 52–93 (1969); (c) J. S. Wood, *Progress in Inorganic Chemistry*, edited by S. J. Lippard (Wiley-Interscience of John Wiley and Sons, New York, 1972), Vol. 16, pp. 227–486.
5. D. W. Meek and T. J. Mazanec, *Acct. Chem. Res.* **14**, 266–274 (1981).
6. L. Sacconi, *Transition Metal Chemistry*, edited by R. L. Carlin (Marcel Dekker, New York, 1968), Vol. 4, pp. 199–298.
7. N. J. DeStefano, D. K. Johnson, and L. M. Venanzi, *Angew. Chem. Int. Ed. Engl.* **13**, 133–134 (1974).
8. M. Matsumoto, H. Yoshiska, K. Nakatsu, T. Yoshida, and J. Otsuka, *J. Am. Chem. Soc.* **96**, 3322–3324 (1974).
9. (a) N. A. Al-Salem, W. S. McDonald, R. Markham, M. C. Norton, and B. L. Shaw, *J. C. S. Dalton*, 59–63 (1980); (b) N. A. Al-Salem, H. D. Empsall, R. Markham, B. L. Shaw, and B. Weeks, *J. C. S. Dalton*, 1972–1982 (1979); (c) B. L. Shaw, *Catalytic Aspects of Metal Phosphine Complexes*, edited by E. C. Alyea and D. W. Meek, *Advances in Chemistry Series* (Am. Chem. Soc., Washington, 1982), Vol. 196, pp. 101–115.
10. (a) G. S. Benner, W. E. Hatfield, and D. W. Meek, *Inorg. Chem.* **3**, 1544–1549 (1964); (b) G. Dyer and D. W. Meek, *Inorg. Chem.* **4**, 1398–1402 (1965); (c) C. A. McAuliffe and D. W. Meek, *Inorg. Chim. Acta* **5**, 270–272 (1971).
11. (a) L. Sacconi, *Coord. Chem. Revs.* **8**, 351–367 (1972); (b) L. Sacconi, *J. Chem. Soc. A*. **1970**, 248–256; (c) R. Morrassi, I. Bertini, and L. Sacconi, *Coord. Chem. Revs.* **11**, 343–402 (1973).
12. L. M. Venanzi, *Angew. Chem.* **76**, 621–628 (1964); *Angew. Chem. Int. Ed. Engl.* **3**, 453–460 (1964).
13. D. L. DuBois and D. W. Meek, *Inorg. Chem.* **15**, 3076–3083 (1976).
14. A. A. Arduini, A. A. Bahsoun, J. A. Osborn, and C. Voelker, *Angew. Chem. Int. Ed. Engl.* **19**, 1024–1025 (1980).
15. J. A. Osborn and G. S. Stanley, *ibid.*, **19**, 1025–1026 (1980).
16. J. J. deBoer, J. A. van Doorn, and C. Masters, *J. C. S., Chem. Comm.* **1978**, 1005–1006.
17. (a) J. Chatt and F. A. Hart, *J. Chem. Soc.* **1960**, 1378–1389; (b) W. Hewertson and H. R. Watson, *J. Chem. Soc.* **1962**, 1490–1494.
18. M. Czakova and M. Capka, *J. Mol. Catal.* **11**, 313–322 (1981).
19. Z. M. Michalska, M. Capka, and J. Stoch, *J. Mol. Catal.* **11**, 323–330 (1981).
20. G. Fritz, G. Becher, and D. Kummer, *Z. anorg. allgem. Chem.* **372**, 171–179 (1970).
21. K. Issleib and H. P. Abicht, *J. Prakt. Chem.* **312**, 456 (1970).
22. H. H. Karsch, U. Schubert, and D. Neugelbauer, *Angew. Chem. Int. Ed. Engl.* **18**, 484 (1979).
23. S. O. Grim and E. D. Walton, *Phosphorus and Sulfur* **9**, 123–126 (1980).
24. (a) N. Takaishi, H. Imai, C. A. Bertelo, and J. K. Stille, *J. Am. Chem. Soc.* **100**, 264–268 (1978); (b) T. Masuda and J. K. Stille, *J. Am. Chem. Soc.* **100**, 268–272 (1978); (c) G. L. Baker, S. J. Fritschel, J. R. Stille, and J. K. Stille, *J. Org. Chem.* **46**, 2954–2960 (1981); (d) G. L. Baker, S. J. Fritschel, and J. K. Stille, *J. Org. Chem.* **46**, 2960–2965 (1981); (e) S. J. Fritschel, J. J. H. Ackerman, T. Keyser, and J. K. Stille, *J. Org. Chem.* **44**, 3152–3157 (1979).

25. (a) R. G. Nuzzo, S. L. Haynie, M. E. Wilson, and G. M. Whitesides, *J. Org. Chem.* **46**, 2861–2867 (1981); (b) M. E. Wilson and G. M. Whitesides, *J. Am. Chem. Soc.* **100**, 306–307 (1978); (c) M. E. Wilson, R. G. Nuzzo, and G. M. Whitesides, *J. Am. Chem. Soc.* **100**, 2269–2270 (1978); (d) R. G. Nuzzo, D. Feitler, and G. M. Whitesides, *J. Am. Chem. Soc.* **101**, 3683–3685 (1979).
26. (a) R. B. King, *Acct. Chem. Res.* **5**, 177–185 (1972); (b) R. B. King and J. C. Cloyd, Jr., *J. Am. Chem. Soc.* **97**, 53–60 (1975); (c) R. B. King, J. Bakos, C. D. Hoff, and L. Markó, *J. Org. Chem.* **44**, 3095–3100 (1979).
27. (a) D. L. DuBois, W. H. Myers, and D. W. Meek, *J. C. S. Dalton*, 1011–1015 (1975); (b) R. Uriarte, T. J. Mazanec, K. D. Tau, and D. W. Meek, *Inorg. Chem.* **19**, 79–85 (1980); (c) R. Uriarte, Ph.D. Dissertation, The Ohio State University, Columbus, Ohio, 1978; (d) R. J. Uriarte and D. W. Meek, *Inorg. Chim. Acta* **44**, L283–L284 (1980).
28. J. C. Briggs and G. Dyer, *Chem. Ind.*, 163–165 (1982).
29. (a) E. P. Kyba, R. E. Davis, C. W. Hudson, A. M. John, S. B. Brown, M. J. McPhaul, L.-K. Liu, and A. C. Glover, *J. Am. Chem. Soc.* **103**, 3868–3875 (1981); R. E. Davis, E. P. Kyba, A. M. John, and J. M. Yep, *Inorg. Chem.* **19**, 2540–2544 (1980); (c) E. P. Kyba and S. B. Brown, *Inorg. Chem.* **19**, 2159–2162 (1980); (d) E. P. Kyba and S.-S. P. Chou, *J. Org. Chem.* **46**, 860–863 (1981).
30. M. Ciampolini, P. Dapporto, A. Dei, N. Nardi, and F. Zanobini, *Inorg. Chem.* **21**, 489–495 (1982).
31. (a) R. L. Keiter, R. D. Borger, J. J. Hamerski, S. J. Garbis, and G. S. Leotsakis, *J. Am. Chem. Soc.* **99**, 5224–5225 (1977); (b) R. L. Keiter, Y. Y. Sun, J. W. Brodack, and L. W. Cary, *J. Am. Chem. Soc.* **101**, 2638–2641 (1979).
32. P. M. Treichel and W. K. Wong, *J. Organomet. Chem.* **157**, C5–C9 (1978).
33. R. L. Waid and D. W. Meek, unpublished data (1981); R. L. Waid, Ph.D. Dissertation, The Ohio State University, Columbus, Ohio, June, 1982.
34. R. L. Keiter, S. L. Kaiser, N. P. Hansen, J. W. Brodack, and L. W. Cary, *Inorg. Chem.* **20**, 283–284 (1981).
35. (a) B. R. James, *Homogeneous Hydrogenation* (John Wiley and Sons, New York, 1973); (b) E. L. Muetterties, ed., *Transition Metal Hydrides* (Marcel Dekker, New York, 1971).
36. D. L. DuBois and D. W. Meek, *Inorg. Chim. Acta* **19**, L29–L30 (1976).
37. (a) J. A. Osborn, F. H. Jardine, J. F. Young, and G. Wilkinson, *J. Chem. Soc. A.* **1966**, 1711–1732; (b) C. A. Tolman, P. Meakin, D. L. Lindner, and J. P. Jesson, *J. Am. Chem. Soc.* **96**, 2762–2774 (1974).
38. (a) J. Niewahner and D. W. Meek, *Inorg. Chim. Acta* **64**, L123–L125 (1982); (b) J. Niewahner and D. W. Meek, *A. C. S. Advances in Chemistry Series* **196**, 257–272 (1982).
39. (a) T. J. Mazanec, J. B. Letts, and D. W. Meek, *J. Chem. Soc., Chem. Commun.* 356 (1982); (b) J. B. Letts, T. J. Mazanec, and D. W. Meek, *J. Am. Chem. Soc.* **104**, 3898–3905 (1982).
40. L. H. Slaugh and R. D. Mullineaux, United States Patent No. 3 239 566 (1966).
41. J. L. Eiseman, United States Patent No. 3 290 379 (1966).
42. J. A. Osborn, G. Wilkinson, and J. F. Young, *J. Chem. Soc., Chem. Comm.* **1965**, 17–18.
43. D. Evans, G. Yagupsky, and G. Wilkinson, *J. Chem. Soc. A.* **1968**, 2660–2665.
44. D. Evans, J. A. Osborn, and G. Wilkinson, *J. Chem. Soc. A.* **1968**, 3133–3142.
45. M. C. Baird, J. T. Mague, J. A. Osborne, and G. Wilkinson, *J. Chem. Soc.* **1967**, 1347–1360.
46. C. K. Brown and G. Wilkinson, *J. Chem. Soc. A.* **1970**, 2753–2764.
47. (a) F. B. Booth, United States Patent No. 3 511 880 (1970); (b) K. L. Oliver and F. B. Booth, *Hydrocarbon Processing* **49** (4), 112–114 (1970); (c) A. Hershman, K. K. Robinson, J. H. Craddock, and J. F. Roth, *Ind. Eng. Chem. Prod. Res. Dev.* **8**, 372–375 (1969); (d) J. H. Craddock, A. Hershman, F. E. Paulik, and J. F. Roth, *ibid.* **8**, 291–297 (1969).

48. R. L. Pruett and J. A. Smith, *J. Org. Chem.* **34,** 327–330 (1969).
49. K. L. Oliver and F. B. Booth, *Am. Chem. Soc. Pet. Div. Prepr., Gen. Pap.* **14** (3), A7 (1969).
50. (a) J. D. Unruh and J. R. Christenson, *J. Mol. Catal.* **14,** 19–34 (1982); (b) O. R. Hughes and J. D. Unruh, *J. Mol. Catal.* **12,** 71–83 (1981); (c) J. D. Unruh, O. R. Hughes, J. R. Christenson, and D. Young, in *Proceedings of Eighth Conference on Catalysis in Organic Synthesis*, edited by W. R. Moser, in press; (d) J. D. Unruh and J. R. Christenson, *Organometallic Reactions and Syntheses*, edited by E. I. Becker and M. T. Tsutsui, Vol. 7 (1982); (e) O. R. Hughes and D. A. Young, *J. Am. Chem. Soc.* **103,** 6636–6642 (1982).
51. B. Cornils, "Hydroformylation, Oxo Synthesis, Roelen Reaction," *New Syntheses with Carbon Monoxide*, edited by J. Falbe (Springer-Verlag, Berlin, 1980), pp. 1–225.
52. R. L. Pruett, "Hydroformylation," *Advances in Organometallic Chemistry*, edited by F. G. A. Stone and R. West (Academic Press, New York, 1979), pp. 1–60.
53. G. M. Kosolapoff and L. Maier, eds., *Organic Phosphorus Compounds* (Wiley-Interscience of John Wiley and Sons, New York, 1972), Vol. 1; (b) B. W. Bangerter, R. P. Beatty, J. K. Kouba, and S. S. Wreford, *J. Org. Chem.* **42,** 3247–3251 (1977), and references therein.
54. R. L. Keiter, J. W. Brodack, R. D. Borger, and L. W. Cary, *Inorg. Chem.* **21,** 1256–1259 (1982).
55. (a) C. A. Ghilardi, S. Midollini, A. Orlandini, and L. Sacconi, *Inorg. Chem.* **19,** 301–306 (1980); (b) M. Di Vaira, S. Midollini, and L. Sacconi, *J. Am. Chem. Soc.* **101,** 1757–1763 (1979); (c) M. Di Vaira, C. A. Ghilardi, S. Midollini, and L. Sacconi, *J. Am. Chem. Soc.* **100,** 2550–2551 (1978).
56. (a) P. Dapporto, S. Midollini, and L. Sacconi, *Inorg. Chem.* **14,** 1643–1650 (1975); (b) C. Bianchini, P. Dapporto, C. Mealli, and A. Meli, *Inorg. Chem.* **21,** 612–615 (1982).
57. M. M. Harding, B. S. Nicholls, and A. K. Smith, *J. Organomet. Chem.* **226,** C17–C20 (1982).
58. (a) K. D. Tau and D. W. Meek, *J. Organomet. Chem.* **139,** C83–C86 (1977); (b) K. D. Tau, R. Uriarte, T. J. Mazanec, and D. W. Meek, *J. Am. Chem. Soc.* **101,** 6614–6619 (1979).
59. A. F. M. J. Van Der Ploeg and G. Van Koten, *Inorg. Chim. Acta* **51,** 225–239 (1981).
60. K. Kurtev, D. Ribola, R. A. Jones, D. J. Cole-Hamilton, and G. Wilkinson, *J. Chem. Soc., Dalton* 55–58 (1980).
61. P. Giannoccaro, G. Vasapollo, and A. Sacco, *J. C. S., Chem. Commun.* 1136–1137 (1980).
62. S. G. Davies, H. Felkin, and O. Watts, *J. C. S., Chem. Commun.* 159–160 (1980).
63. M. C. Davies and T. A. George, *J. Organomet. Chem.* **224,** C25–C27 (1982).
64. J. A. Baumann and T. A. George, *J. Am. Chem. Soc.* **102,** 6153–6154 (1980).
65. R. B. King and W. M. Rhee, *Inorg. Chem.* **17,** 2961–2963 (1978).
66. M. D. Fryzuk, *Organometallics* **1,** 408–409 (1982).
67. R. B. King and J. W. Bibber, *Inorg. Chim. Acta* **59,** 197–201 (1982).

# 9

# Cationic Rhodium and Iridium Complexes in Catalysis

## Robert H. Crabtree

### 1. INTRODUCTION

Among the factors that affect the reactivity of transition metal compounds, the net charge carried by the complex has up to now tended to be neglected. In this review, we will first survey the effects of net charge † on the reactivity of metal complexes in general, before going on to the specific case of cationic complexes of rhodium and iridium in catalysis. We will not discuss work bearing on asymmetric hydrogenation, since this is covered in Chapter 3.

While we will emphasize the effect of overall charge, particularly of positive charge, we cannot ignore such well-recognized effects as the oxidation state of the metal or the nature of the coligands. Overall charge is just one factor among these. We will see, however, that it can in some cases be the major factor determining reactivity.

---

† By net charge, we refer to the charge carried by the complex as a whole. $RhClL_3$ has no net charge, $RhL_4{}^+$ a net charge of +1.

---

Dr. Robert H. Crabtree • Sterling Chemistry Laboratory, Yale University, 225 Prospect Street, New Haven, Ct. 06511.

## 2. THE EFFECTS OF NET IONIC CHARGE

### 2.1. Effects on the Reactivity of Bound Ligands

The charge on a complex cation or anion tends to be delocalized on to the ligands in accordance with Pauling's electroneutrality principle.[1] Hydrogens in the immediate coordination sphere of a cation, for example, tend to become positively charged. This is one reason for the striking ability of $H_2O$ or $NH_3$ to stabilize the simple cations of the transition metals (e.g., $Cu(H_2O)_4^{2+}$), compared to the far poorer abilities of even relatively unhindered alcohols, ethers, and primary, secondary, and tertiary amines. One of the roles a metal can play in catalysis is to change the normal reactivity pattern of the substrate on going from the free state to the bound state. Clearly one would expect electrophilic behavior of a substrate to be encouraged by binding to a cation, and nucleophilic behavior by binding to an anion.

The activation of the nitrile group, RCN, by metals is a good example. Free nitriles are very resistant to nucleophilic attack by both $BH_4^-$ and $H_2O$. Coordination to ruthenium in a complex with an overall charge of 2+ has a dramatic effect. The nitrile can now be very rapidly reduced by $NaBH_4/EtOH$ as follows:

$$[Ru(RCN)_3L_3]^{2+} \xrightarrow[EtOH]{NaBH_4} RuH(BH_4)L_3 + RCH_2NH_2 + 2RCN$$

In this case, the $BH_4^-$ attacks the nitrile, transferring two $H^-$ groups to carbon. The nitrogen atom is then protonated in the ethanolic medium, releasing the free amine.[2] Catalysis of RCN reduction by $NaBH_4/EtOH$ has been observed for a variety of simple metal ions. The same mechanism probably applies.[3a] In the catalyzed hydrolysis of nitriles the metal ion plays the same role, but the nucleophile is now $H_2O$.[3b]

The same type of behavior can be found for nucleophilic attack of coordinated olefins and arenes. The C=C group of $(cod)PdCl_2$ is attacked only by basic MeOH, while $[(cod)Pd(PR_3)Cl]^+$ is rapidly attacked even in neutral MeOH.[4] Hydride addition to coordinated pi-ligands tends to be controlled by the charge on the metal, for example[5]:

$$Re(C_6H_6)_2^+ \xrightarrow{H} Re(C_6H_6)(C_6H_7) \xrightarrow{H} \times$$
$$Fe(C_6H_6)_2^{2+} \xrightarrow{H} Fe(C_6H_6)(C_6H_7)^+ \xrightarrow{H} Fe(C_6H_7)_2$$

Overall positive charge also tends to increase the acid dissociation constant of a bound ligand. The best known example is that of water. Many metal ions are so polarizing that the $H_2O$ ligand immediately loses protons on binding.

$$U^{6+} + 2H_2O \rightarrow UO_2^{2+} + 4H^+$$

Transition metal ions having fewer positive charges are far less polarizing, but the charge effect can still be seen.[6]

$$[(C_5Me_5)IrCl_2]_2 \xrightarrow[H_2O]{Ag^+} [(C_5Me_5)Ir(H_2O)_3]^{2+} \rightarrow [((C_5Me_5)Ir)_2(\mu\text{-}OH)_3]^+$$
$$\text{not isolated}$$

A single positive charge seems to be insufficient to cause ionization of the coordinated water.[7]

$$IrH_2(H_2O)_2L_2^+ \not\rightleftharpoons IrH_2(OH)(H_2O)L_2 + H^+$$

## 2.2. Effects on the Types of Ligand Bound

An overall positive charge on a complex most directly affects the binding of charged ligands. In particular, anionic ligands, or ligands which can dissociate a proton to give anions, are bound very strongly. This is one reason why the tetra-anion of ethylenediaminetetracetic acid is so effective at sequestering metal ions of high-positive charge.

The same effect can make $BPh_4^-$ a coordinating anion, to the extent that it can even displace tertiary phosphine[7]:

$$[IrH_2(Me_2CO)_2(PCy_3)_2]^+ \xrightarrow[Me_2CO]{BPh_4} [(\eta^6\text{-}C_6H_5BPh_3)Ir(PCy_3)H_2]$$

Its use as a counter ion in catalytic studies can lead to catalyst deactivation[8] for this reason.

Strong effects are also seen in the binding of neutral ligands. As mentioned above, ligands with several H-substituents are very effective at delocalizing positive charge. This feature increases the binding constant of $H_2O$, for example, to the extent of giving relatively stable aquo-complexes even with soft metal ions, which on hard-soft considerations would not be expected to bind well. Some examples of hard-ligand complexes of soft metals follow:

$$[(C_6H_6)Os(H_2O)_3]^{2+[9]}, Pddpe(thf)_2^{2+[10]}, Rh(C_5Me_5)(Me_2CO)_3^{2+[6]},$$
$$Pd(MeCN)_4^{2+[11]}, IrH_2(PPh_3)_2(H_2O)_2^{+[7]}, Rhdpe(MeOH)_2^{+[12]}.$$

These solvent ligands can be quite labile, and, for this reason, the solvento-complexes themselves are often useful catalyst precursors, as discussed in a recent review by Davies and Hartley.[10]

Among complexes of harder ions, such as those in the first row, the absolute tendency to bind hard ligands such as $H_2O$ is greater, yet the same trends mentioned above seem to apply. The vast majority of isolable aquo complexes, for example, bear a net positive charge.

The binding of *dmso* to metals by sulfur or oxygen is often thought to be dictated by the hardness or softness of the metal. Yet Rh(I), Pd(II), and Pt(II), normally soft, often bind *via* O. This often occurs in complexes of net positive charge, e.g., $[Rh(cod)(O-Me_2SO)_2]^+$, $[Pd(O-Me_2SO)_2(S-Me_2SO)_2]^{2+}$, $[Pd(dpe)(O-Me_2SO)_2]^+$ or $[Pd(O-R_2SO)_4]^{2+}$ ($R$ = isoamyl), while in cationic or neutral complexes $S$-binding often occurs, e.g.,

$$Pt(S-Me_2SO)_2Cl, K[Pt(S-Me_2SO)Cl_3],$$

$$[Rh(S-Me_2SO)_2(CO)Cl].^{(13)}$$

The arguments we have presented suggest that one reason for $O$-binding in the cationic systems may be the ability to delocalize positive charge in the product as illustrated below. Other solvents can delocalize positive charge; for example, *dmf* probably binds via an oxygen lone pair, in which case charge can be delocalized on to nitrogen.

$$M^+ \leftarrow O{=}SR_2 \leftrightarrow M{-}O{-}SR_2^+$$
$$M^+ \leftarrow O{=}CHNR_2 \leftrightarrow M{-}O{-}CH{=}NR_2^+$$

Clearly, a continuous range of ligand-binding ability exists. This will change with the metal and coligands. So one might ask when a solvent is coordinating, and when it is not. We will refer to a solvent as noncoordinating when the solvent does not appear to occupy a vacant site in the inner coordination sphere, as weakly coordinating when solvento-complexes can be detected or isolated but the substrate(s) for the catalytic reaction can displace the solvent, and as strongly bound when no detectable solvent dissociation in favor of the substrate takes place.

Anomalous displacement effects can sometimes be seen among classical ligands in ionic complexs. For example, $H_2$ would not normally be expected to displace CO, but rather the reverse. In a cationic system, however, the positive charge on the metal seems to weaken the $M$-CO bond more than the $MH_2$ system so that $H_2$ does become able to displace CO in a series of cationic iridium complexes.[14]

$$Ir(CO)_3L_2^+ + H_2 \rightarrow IrH_2(CO)_2L_2^+ + CO$$

Probably, CO is a better overall acceptor than is $H_2$. Indeed we have argued that $H_2$ addition to some cationic iridium species can result in net electron transfer to the metal (metal reduction) rather than away from it (oxidation), as implied by the classical nomenclature "oxidative addition." This is consistent with the idea that part of the net positive charge can be delocalized over the H-ligands, which become $\delta+$ rather than $\delta-$ in character.

In the case of alkene binding, ethylene normally binds better than, say, butene, so that an ethylene dimerization catalyst would not usually give $C_6$ or higher products because $C_2H_4$ would exclude butenes from the metal site. Sen[15] has decribed interesting catalytic reactions of $[Pd(MeCN)_4]^{2+}$ in which the reverse seems to be true. The electrophilic character of the metal encourages the binding of substituted alkenes rather than $C_2H_4$, and so $C_6$, $C_8$, and even $C_{10}$, species are formed directly from $C_2H_4$.

One might imagine that a charged mononuclear complex might be incapable of agglomeration to give polynuclear species, because of coulombic repulsion. This does not seem to be always the case. Cluster formation may be slowed but is not prevented in 1+ species. This process can lead to the deactivation of cationic catalysts.[16] The presence of potentially bridging groups (RS, Cl, H) can encourage oligomerization, especially of systems also containing weakly bound ligands, which can be displaced by the bridging groups.

### 2.3. Effects on Redox Properties

The main point of interest for our purposes is the greater resistance to oxidation shown by cationic complexes. They can be less oxidation- and air-sensitive than their neutral analogues. As extreme examples, $[CpIrHL_2]BF_4$ ($L = PPh_3$) can be refluxed in $CCl_4/CHCl_3$ for 3 hr in a vessel open to the air without change, and $[Ir(cod)L_2]BF_4$ ($L = PMe_2Ph$) seems to be unique among hydrogenation catalysts in being catalytically fully active even in $H_2/O_2$ mixtures.[7] The iso-electronic $CpIrL_2$ and $Ir(cod)LCl$, in contrast, are air-sensitive.

### 2.4. Effects on Solubility and Catalyst Separation

Since all the complexes under discussion are ionic, it is not surprising to find that most are soluble only in polar solvents. This can be very useful in separating the desired product from the spent catalyst. For example, Professor Bill Suggs has shown that our catalyst, $[Ir(cod)PCy_3(py)]PF_6$, is useful in certain steroid reductions. Here, the reaction solvent is $CH_2Cl_2$. To isolate the steroidal product, the solvent is removed and the residue extracted with hexane. The extract contains only the steroid and not the insoluble catalyst.[17]

An ion exchange resin has been used to immobilize a cationic catalyst.[18] Although in this case the cationic center was formed by protonation of a remote nitrogen group in a P-bound P(C$_6$H$_4$NMe$_2$)$_3$ ligand, the same effect could probably be obtained with complex having an overall change.

Anionic catalysts should have the same advantages mentioned above. Organometallic species are not normally water-soluble, but net charge encourages solubility in H$_2$O, particularly if some hydrogen-bonding functionality is present. Water-soluble catalysts may well prove to be most useful in two-phase systems, e.g., hexane/water, because catalyst separation then becomes much easier than in classical systems. Examples of water-soluble systems are [RhL$_4$]BF$_4$ ($L$ = 1-phospha-3,5,7-tri-azaadamantane)[19] and [Rh(cod)(PPh$_2$C$_6$H$_4$SO$_3$)$_2$], which is a zwitterionic monomer in MeCN, but a P,O-bridged dimer in the solid state.[20]

## 2.5. Counter Ion Effects

Some anions are more coordinating than others. The rough order of increasing binding ability Rh(I) and Ir(I) 1+ systems seems to be S$b$F$_6$, PF$_6$, BF$_4$, CF$_3$SO$_3$, ClO$_4$, CF$_3$CO$_2$, CH$_3$CO$_2$, BPh$_4$. Some transition metal systems show substantial anion effects even between such apparently noncoordinating anions as BF$_4$ and PF$_6$. In our alkane activation experiments[21] (see Section 4.3), for example, BF$_4$ salts were found to be three times more effective than PF$_6$ salts. Curiously, in alkene hydrogenation with a closely related system, the PF$_6$ salts are slightly superior to the BF$_4$ salts.[7] Such effects are relatively rare, however, and are to be expected only where the substrate is a poor ligand.

## 2.6. Mechanistic Effects

The overall charge on a complex ion can dictate the mechanism of an ionic reaction. For example, oxidative addition of HCl normally occurs by protonation, followed by anation, as exemplified by[22]

$$\text{PtMeCl}L_2 \xrightarrow{\text{H}^+} [\text{PtMeHCl}L_2]^+ \xrightarrow{\text{Cl}^-} \text{PtMeHCl}_2L_2$$

In some cationic iridium complexes, the reaction occurs by the inverse mechanism[23]:

$$[\text{Ir(cod)}L_2]^+ \xrightarrow{\text{Cl}^-} \text{IrCl(cod)}L_2 \xrightarrow{\text{H}^+} [\text{IrHCl(cod)}L_2]^+$$

More than just charge effects are involved, since some related neutral complexes also add HCl in this way, but in the neutral cases the intermediates are not stable enough to be isolated, in contrast to the cationic system shown above.

By increasing the affinity of the catalyst for substrate functional groups such as C=O, $-NH_2$, and $-OH$, a positive charge may assist catalytic reactions such as asymmetric hydrogenation where such chelate-formation has been shown to be important.[24] Solvolysis of $M$–Cl bonds to give cationic species may account for the increase in optical yields sometimes encountered with $RhClL_n$ systems on going to more ionizing solvents.[25]

Normally, $20e$ species are so unstable that they are not found. In one case where a transition state of this type has been postulated, axial-equatorial site-exchange in $[NiBrL_3]Br$ ($L = PMe_3$), attack of the $Br^-$ ion on the organometallic cation takes place.[26] This attack may well be favored by the neutralization of charge in the transition state.

## 2.7. Conclusion

Just as one can vary the chemical behavior of a metal by changing the coligands, one can also do so by changing the overall charge. Increasing positive charge makes the metal more polarizing, oxidizing, and harder, and makes the complexes much less soluble in nonpolar solvents. This is a variable that has perhaps received less study than it merits.

## 3. SYNTHETIC METHODS

### 3.1. Halide Abstraction

Chloride abstraction is perhaps the most generally useful method of forming cationic organometallic complexes. Where good ligands (e.g., $PR_3$, CO, $C_2F_4$) are to replace the halide in a labile system, often all that is needed is the ligand, a polar solvent, and a simple salt of the counter-ion[27] (e.g., $NH_4PF_6$ or $NaBF_4$).

When a poor ligand (e.g., $H_2O$, $Me_2CO$, or $C_2H_4$) is to replace the halide, or the metal is substitution inert, this method is less likely to work. $AgPF_6$ in a polar solvent is often useful[28] but suffers from the disadvantage that $Ag^+$ is a good oxidizing agent. The formation of metallic Ag indicates that an alternative method must be used. $TlPF_6$ is suitable,[29] since it is far less oxidizing. $[Me_3O]PF_6$ has also been used;[30] this has the advantage that the products, MeCl and $Me_2O$, can be pumped away. This is simpler than the removal of the fine precipitates of AgCl and TlCl that are obtained in the above methods.

Chloride abstraction by such reagents as $SbCl_5$ and $AlCl_3$ is relatively unusual.[31]

## 3.2. Protonation

Protonation is an excellent method of producing cationic complexes. Quite a large number of complexes $MX_xL_y$ will protonate, especially if a lone pair is available (i.e., where $x + y < 9$ and $x < max$ valency of metal).

In the particular case of cationic rhodium and iridium systems, examples[7] are:

$$IrH_5L_2 \xrightarrow[S]{H^+} IrH_2S_2L_2^+ \quad (S = \text{coordinating solvent})$$

$$RhHL_4 \xrightarrow{H^+} RhL_4^+ + H_2$$

Sometimes an excess of acid can be used, but where an equivalent is required, it has been found[32] that hydrolysis of an equivalent of $[Ph_3C]PF_6$ is a convenient acid source, since the organic salt can be weighed out precisely. Where water must be avoided, the etherate $HPF_6 \cdot Et_2O$ can be used. Methylation with $Me_3O^+$ should be completely analogous, although it has not been widely used.

## 3.3. Electrophilic Attack at a Ligand

Electrophilic attack at a coordinated ligand is also useful. Some examples[33,34] follow:

$$Ru(CO_2Me)_2L_2 + HA \xrightarrow{S} [RuL_2S_x]^{2+} + MeCOOH$$

$$(S = \text{solvent}; A = PF_6, BF_4; L = PPh_3)$$

$$Rh(cod)(acac) \xrightarrow[Ph_3C^+ \text{ or } H^+]{L} [Rh(cod)L_2]^+$$

## 3.4. Redox

Oxidation of organometallic species electrochemically, or by chemical oxidants such as $FeCl_3$, $Br_2$, or $Ph_3C^+$, is still rare but will probably grow more important in the future. No catalytic applications have yet been described.

Oxidation of the bulk metal itself by a noncoordinating oxidant has been used[11] but, among the platinum metals, is probably limited to Pd.

$$Pd \xrightarrow[MeCN]{NOPF_6} [Pd(MeCN)_4](PF_6)_2 + 2NO$$

## 4. CATIONIC RHODIUM AND IRIDIUM COMPLEXES IN CATALYSIS

### 4.1. Hydrogenation of C=C Groups

During the development of $RhCl(PPh_3)_3$ as the first efficient alkene hydrogenation catalyst by Wilkinson et al.,[35] it was realized that only two of the three $PPh_3$ groups remained permanently bound to the metal during the catalytic cycle. An early inference from this idea was that if the P:Rh ratio ($n$) could be adjusted to 2, rather than 3 as in $RhCl(PPh_3)_3$ itself, a more efficient catalyst might be obtained. The effect of a variable P:Rh ratio was studied using mixtures of $PR_3$ and $[Rh(cod)Cl]_2$ as catalyst precursors. Not only was $n = 2.0$ found to be the most effective ratio, but those phosphines, such as $PEt_3$, which were totally inactive at $n = 3$, became active at $n = 2$.[36]

Osborn,[37] who had contributed to this work at Imperial College, subsequently took up the problem at Harvard. He made the important discovery of the existence and catalytic activity of the first examples of the cationic rhodium and iridium complexes which have been used the most extensively in catalytic studies. For example, the readily available $[Rh(diene)Cl]_2$ reacts with excess $PPh_3$ in EtOH to give $[Rh(diene)(PPh_3)_2]^+$, which can be isolated as the perchlorate, tetrafluoroborate, or hexafluorophosphate. These are air-stable, crystalline complexes, and very convenient for use as catalyst precursors. Under $H_2$, the diene is hydrogenated, generating the reactive $RhL_2^+$ fragment.

As expected, this is an excellent catalyst and it has several unusual properties that depend in part on its ionic character. For example, intermediates can be isolated relatively easily from the coordinating solvents that Osborn used. These complexes $[RhH_2S_2L_2]A$ ($S$ = solvent ($Me_2CO$, EtOH); A = noncoordinating anion ($ClO_4$, $PF_6$, $BF_4$)) have two solvent ligands bound through their oxygen lone pairs in the immediate coordination sphere of the complex. This relatively firm binding of hard ligands by a soft organometallic center seems to be particularly favored by a positive charge on the complex, as mentioned above, and by the presence *trans* to $S$ of the strong *trans*-influence hydride ligands (the antisymbiotic effect).[10]

Interesting selectivity effects were found for various dienes and alkynes, the reduction of both of which could be stopped at the monoene stage. Almost exclusive *cis*-addition of $H_2$ to the alkyne was observed. Ketones can be reduced, although this is not a result of the cationic character of the system, since the catalysts can be deprotonated with $NEt_3$ to give neutral species that also effect this reduction equally well. This illustrates a possible ambiguity in studying cationic catalysts if dissociation of a positive species, usually $H^+$, can occur. Fortunately, in this particular case, each system can be separately prepared and studied.[37,38]

As might be expected in a system that binds solvents well, rates of reduction of various substrates are very solvent dependent,[37] the substrate being in competition with the solvent for the metal.

In contrast to the relatively high solvent dependence, the Osborn system shows a relatively small dependence on $L$. Wilkinson's catalyst requires $PPh_3$ or ligand of very similar cone angle, because the steric compression of these relatively bulky phosphines encourages ligand dissociation, which is needed to generate a site at the metal. The smaller $PEt_3$, as mentioned above, is inactive at $n = 3$ but active for $n = 2$. It seems likely that, as in the Wilkinson system, two $L$s remain bound to the metal through the catalytic cycle in the Osborn system, too. No dissociation of $L$ is required, so the steric requirement is relaxed and $L$ can be varied rather widely. Simple amines and nitriles can be used in certain circumstances, for example.[38]

Even $[Rh(diene)L_3]A$ salts were shown to be active,[37] probably because alkene hydrogenation requires no more than three sites at the metal, two for hydrogen and one for the substrate. Induction periods are sometimes observed in these cases, no doubt because $L$(or cod) does not have to dissociate before $H_2$ can add. Compared to the Wilkinson system, the side-reaction leading to isomerization of the alkene (e.g., 1-butene 2-butene) is a more serious competitor with hydrogenation itself. In applications requiring specific deuteration or where the isomerization product is nonhydrogenable, this can be a problem.[7]

$$\text{3-cyclohexenal} \xrightarrow[\text{acetone}]{[Rh(nbd)(PPh_3)_2]^+} \text{1-cyclohexenal}$$

$$\text{3-cyclohexenal} \xrightarrow[\text{ethanol}]{RhCl(PPh_3)_3} \text{cyclohexanal}$$

Osborn et al.[37] showed that the active isomerization catalysts were indeed the dihydrides $[RhH_2S_2L_2]^+$ and not $[RhS_2L_2]^+$, which were prepared by an independent route.

$$[L_2RhCl]_2 \xrightarrow[S]{Ag^+} L_2RhS_2^+$$

However, the addition of acid to the catalyst system (70% aq $HClO_4$) partially suppresses isomerization, while the addition of base ($NEt_3$) encourages it. This suggests that neutral monohydrides are formed from $RhH_2S_2L_2^+$ with base and that these are the more active isomerization catalysts. It also suggests that the untreated normal Osborn catalyst is really a mixed system containing solvated $RhH_2L_2^+$ and $RhHL_2$. Possibly the presence of the extra equivalent of dissociated $PPh_3$ in the Wilkinson system helps prevent reversal of the first hydrogen transfer (leading to isomerization) by coordination of $L$ to the metal (see Figure 9.1). In the Osborn system, step (a) can be followed by step (b), leading to isomerization. In the Wilkinson system, the extra equivalent of $L$, which dissociates during the catalytic cycle, may reassociate [step (c)]. This may help prevent $\beta$-elimination in the alkylrhodium intermediate. The Osborn system was found to catalyze H/D exchange between $D_2$ and $H_2O$, suggesting that the same deprotonation process in the $RhH_2S_2L_2^+$ cations is responsible.

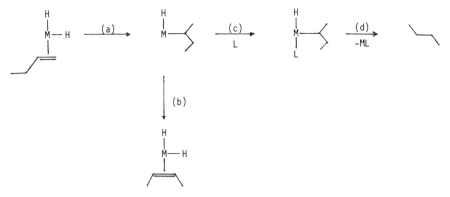

Figure 9.1. A proposed scheme for olefin isomerization.

Exchange between $D_2$ and the *ortho*-hydrogens of $P(OPh)_3$ was also observed, suggesting that reversible *ortho*-metallation occurs.

Deprotonation tended to be suppressed in systems containing the more basic phosphines, such as $[Rh(diene)(PMe_2Ph)_3]A$. Using a combination of acidic conditions and a basic phosphine, isomerization was substantially reduced (e.g., for 1-hexene $k$ isom./$k$ hydrog. = 0.03 for $L = PMe_2Ph$, $HClO_4$: 1.4 mol-equiv, compared to $k_i/k_h = 1$ for $L = PMePh_2$, no $HClO_4$ added).

Our own work[16] took these results as a starting point. We were particularly interested to discover what would be the effect of using a solvent that would be less coordinating than those used by Osborn, and, in particular, less coordinating than the substrate olefins. After many failures, George Morris, then a graduate student in Paris, at last found the successful solvent: $CH_2Cl_2$. It seems that the solvent must be polar so that it can dissolve the ionic catalyst and may be slightly coordinating (see below), though much less so than the substrate. For the rhodium system, the results were encouraging but not dramatically different on changing solvent. The reduction rates increased by six- to eightfold on going from acetone to $CH_2Cl_2$. Deprotonation could be effected with $NEt_3$, as in coordinating solvents, in which case reaction rates tended to fall threefold or so, and isomerization rates tended to rise.

It was only when we extended our studies to the analogous iridium complexes $[Ir(cod)L_2]A$ that exceptional rate effects were observed. Osborn had found these to be poor hydrogenation catalysts in his coordinating solvents and only briefly studied them, concentrating instead on the rhodium system.

We found that in $CH_2Cl_2$, a rate acceleration of $10^3$–$10^6$ was observed over the rates in acetone. Not only were monosubstituted alkenes reduced very rapidly, but the much more hindered tri- and tetra-substituted alkenes were reduced at similar rates.[39]

The ability to rapidly reduce hindered substrates makes the iridium system unique among homogeneous catalysts. The rate acceleration observed on going from $Me_2CO$ to $CH_2Cl_2$ seems to be the result of a changeover of active species (or more accurately, principal species present), from $IrH_2(Me_2CO)_2L_2^+$ in acetone to $IrH_2(olefin)_2L_2^+$ in $CH_2Cl_2$ (olefin = substrate). In acetone, the olefin has to displace $Me_2CO$ to be reduced, so only unhindered olefins that are good ligands are reduced. In $CH_2Cl_2$, that restriction is lifted.

An interesting feature of the iridium system is that these olefin dihydride intermediates can be observed directly by nmr at $-80°$. Two isomers of $(cod)IrH_2L_2^+$, for example, have been isolated. Only the isomer containing a coplanar $M(C=C)H$ grouping is active in hydrogenation; in the other, the $M(C=C)$ groups are orthogonal to the cis-$M$-H bonds, and insertion is not possible. In this system, hydrogenation is observed, but no alkyl hydride intermediate can be trapped. In contrast, no olefin dihydride complexes have yet been observed for the analogous rhodium system, but instead Halpern et al.[40] have successfully trapped the first alkyl hydride intermediate.

A disadvantage of the iridium systems is that they deactivate under certain conditions to give the unusual complexes: $[L_2HIr(\mu\text{-}H)_3IrHL_2]A$ and $[(IrHLL')_3(\mu_3\text{-}H)]A_2$ ($L = PR_3$; $L$ = amine, or $RCN$). Certain solvents (e.g., PhCl) seem to prevent deactivation without affecting the rate of reaction very much, at least for $[Ir(cod)(PMePh_2)_2]PF_6$.[41] Presumably these solvents bind less well than the olefin but better than $CH_2Cl_2$. We suggest that even $CH_2Cl_2$ itself can act as a chelate to iridium via the halogens. We have recently shown that the softer iodine can bind much more strongly to these iridium systems, for example, in the novel complex $[IrH_2(o\text{-diiodobenzene})L_2]A$,[42a] in which a crystallographic study shows that the aromatic ligand is indeed chelated via the halogens. Such species may be important intermediates in oxidative addition of $RX$ to metals.

The catalysts have the activity of a heterogeneous catalyst in their ability to reduce hindered olefins, but they differ sharply from these in retaining the selectivity of homogeneous catalysts. Heterogeneous catalysts tend to reduce 1 to 3 (cis and trans); classical homogeneous catalysts tend to give 2 and only very slowly 3. Only the trans-isomer of 3 is formed in the latter case. $[Ir(cod)(PCy_3)py]PF_6$ catalyzes the reduction of 1 to 2 by one equivalent of $H_2$, and of 2 to trans-3 by an excess.[17] These substrates bind to the iridium to form complexes of the type $[IrH_2(\text{substrate})_2L_2]^+$, in which they appear to be O- not C=C bound. The order of binding is $1 > 2 > 3$. This helps account for the selectivity and suggests that unsaturation promotes Ir-O binding. Indeed $PhCH=CHCO_2Et$ binds so strongly that the C=C function cannot be reduced, at least by [Ir(cod)-(PMePh$_2$)$_2$]PF$_6$.

The catalyst is also insensitive to O$_2$ and oxidizing or other sensitive functionality. For example, 4 is reduced cleanly to 5, while other types of catalyst tend to cleave C–Br or cyclopropane C–C bonds, or both.

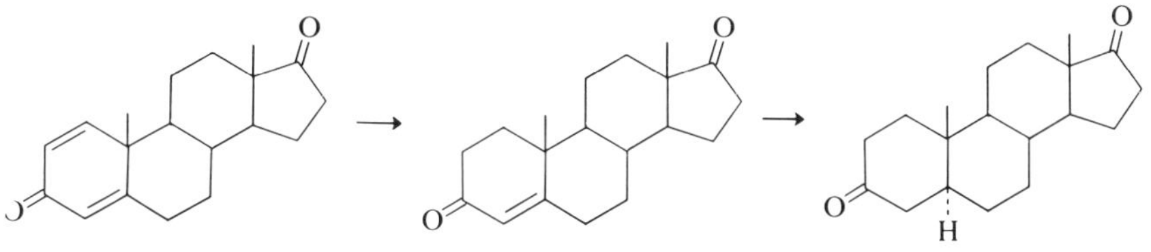

[IrH$_2$(thf)$_2$L$_2$]A systems, prepared *in situ*, are effective in the isomerization of allylic alcohols to carbonyl compounds,[43] and isomerization is always a side reaction in hydrogenation reactions with the iridium catalyst.[42b]

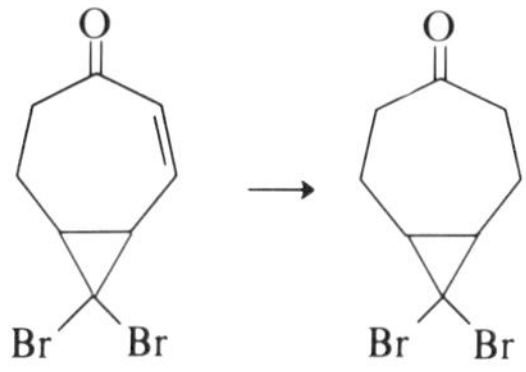

Halpern[44] has made some significant observations in the chemistry of the cationic rhodium catalysts. Because much of this relates to asymmetric induction, it is treated in Chapter 4. It is appropriate to mention here that the catalyst formed from [Rh(cod)(*dpe*)]A (*dpe* = 1,2-diphenylphosphinoethane) has been shown to dimerize reversibly in the absence of substrate. An aromatic ring from one of the *dpe* ligands of the first metal binds in an $\eta^6$-fashion to the second metal and *vice versa*.[44] The system is labile, in contrast to the nonlability of simple arene complexes of Ir(I).

Other cationic systems have been studied as hydrogenation catalysts. For example, [Rh(NH$_3$)$_5$H]$^{2+}$ was found to reduce a variety of aromatic acids in aqueous solution.[45] Since metal precipitated in the absence of substrate, it remains a possibility, however, that the organic substrates stabilized the formation of colloidal Rh, itself an excellent catalyst.

For Maitlis's[46] interesting hydrogenation catalysts based on [(C$_5$Me$_5$)MX$_2$]$_2$ (M = Rh or Ir), no evidence exists for the involvement of cationic species, although this is by no means out of the question in isopropanol, the ionizing, coordinating reaction solvent used.

Interestingly, $[((C_5Me_5)M)_2 HX_2][HX_2]$ catalyzes the air-oxidation of isopropanol to acetone over $3d$. at 20° ($M$ = Rh, Ir).[46b]

Since supported homogeneous catalysts are fully described in Chapter 14, we only mention two particularly imaginative examples here, in both of which the positive charge of the catalyst system has a role to play. Whitesides et al.[47] have bound a $RhL_2^+$ species to avidin, a globular protein. Biotin, which binds strongly to avidin, was covalently attached to a diphosphine fragment and allowed to bind to the protein. A $[(nbd)RhS_2]^+$ fragment was then introduced; this presumably binds to the diphosphine. The resulting catalyst gave ca 45% e.e., in the reduction of $\alpha$-acetamido-acrylic acid. The success of this system requires the rhodium fragment to be compatible in the aqueous environment; the cationic character of the metal probably encourages this.

Catalysts of the $RhL_2^+$ type have also been incorporated into the layer silica mineral hectorite. Once again, the positive charge on the metal strongly encourages binding. The resulting catalyst has unusual selectivity patterns, no doubt due to the requirement that the substrate has to intercalate before it can be reduced.[48]

### 4.2. Hydrogenation of C=O Groups

The reduction of the C=O function is relatively unusual among homogeneous hydrogenation catalysts. $[Rh(cod)(PPh_3)]A$-based systems reduce various ketones, both in the native cationic form, where $H_2O$ is a cocatalyst,[28,37] and in the deprotonated form.[38] This does not necessarily occur via the enol, since $Ph_2CO$ can be reduced by the neutral system in this way.

So far, esters have only been reduced by anionic systems, such as $K[RuH_2(PPh_3)_2(PPh_2o\text{-}C_6H_4)]$, where the hydrogen ligands presumably take on a more strongly hydridic character by delocalization of the net negative charge.[49] The role of the $K^+$ counter-ion in polarizing the C=O group by $O$-binding is probably also important. This interesting problem has had less attention than it merits.

Ketones can be very efficiently reduced by H-transfer from $i$-PrOH in presence of $[Ir(cod)(phen)]Cl$[50] (phen = 1,10-phenanthroline). The positive charge on the metal probably assists coordination of $i$-PrOH, the first step of the reduction.

$$M^+ + i\text{PrOH} \xrightarrow{-H^+} MOi\text{Pr} \rightarrow M\text{H} + Me_2CO$$

### 4.3. Alkane Activation

It is well known that C–H bonds in coordinated ligands can be cleaved by metals. This cyclometalation reaction has been studied for many years;[51] a classic example[52] is:

$$(Np)HRu\{(Me_2PCH_2)_2\}_2 \xrightarrow{-NpH} [(Me_2PCH_2)_2RuH(Me_2PCH_2CH_2PMeCH_2)]_2$$

(Np = 2-napthyl)

in which a relatively unactivated $sp_3$ CH bond is cleaved by the metal.

A question that has not been answered until very recently is how this reaction could be effected intermolecularly to cleave CH bonds in alkanes. An early report by Shilov[53] et al. that $PtCl_4^{2-}$ in $CH_3COOD$ homogeneously effected H/D exchange attracted much attention. The reaction may go via electrophilic attack of the metal on the alkane, but the details are not yet clear.[54]

In a recent example[21] of an alkane dehydrogenation by a metal complex, $[IrH_2S_2L_2]BF_4$ was shown to react with cyclopentane to give $[CpIrHL_2]BF_4$. Yields were increased by the addition of $Bu^tCH=CH_2$, which accepts the hydrogens stripped off the metal. This olefin is unique among those tried in taking up $H_2$ from a metal, but in itself binding only weakly to it. $C_2H_4$, in contrast, is bound to the metal so strongly as to exclude the alkane almost completely, and only a trace of alkane dehydrogenation was observed. Olefins containing allylic hydrogens are themselves dehydrogenated in preference to the alkane, and various styrene derivatives give the unusual deactivation reaction shown below instead of alkane activation.[55]

$$PhCH=CH_2 + [IrH_2S_2L_2]^+ \rightarrow [(\eta^6-PhEt)IrL_2]^+$$

Cycloheptane and cyclooctane gave $\eta^5$-cycloheptadienyl and cyclooctadiene complexes, respectively, but other alkanes, e.g., methylcyclopentane, adamantane, or bicyclooctane, give as yet uncharacterized products which probably contain the dehydrogenated alkane.

Mechanistic experiments tend to exclude heterogeneous, radical, or carbonium ion mechanisms. For example, dynamic light scattering measurements fail to detect particulates in the reaction mixture, although a 1 nanomolar Ir colloid can be easily detected. The reaction mixtures also fail to catalyze the reduction of $PhNO_2$ to $PhNH_2$ by $H_2$, a reaction that does occur with authentic Ir colloids. Gas chromatography experiments show that neither the carbonium ion rearrangements expected for $t$-butyl-methylcarbonium ion nor the chlorine atom abstraction reaction expected for cyclopentyl radical occur.

Cyclometalation might be expected to have occurred in preference to alkane activation in these $PPh_3$ complexes. Recent work on other systems has shown, however, that alkane activation can occur even in $PPh_3$ complexes.[56]

It is to be expected that some ligands will prove to be more metalation-resistant than others. Possibly, for example, the order of increasing metalation-resistance for the following ligands might be:

$$P(OPh)_3 < PPh_3 < PMe_3 < P(C_6F_5)_3$$

This is supported by some important results recently obtained by Bergman in which a variety of alkanes have been activated photochemically. Bergman[57] has shown that $[(C_5Me_5)IrLH_2]$ on photolysis in cyclohexane gives $[(C_5Me_5)IrL(Cy)H]$, but that the cyclohexyl hydride formation is competitive with cyclometalation for $PPh_3$, but the major product for $PMe_3$.

It is possible that the positive charge on the metal in our cationic iridium system gives it an electrophilic character, which helps C–H activation to occur. The Bergman system does not, at first sight, look electrophilic, but this could still be the case if a photoexcited state were the active species. That photoexcited states can be active in C–H chemistry is shown by Ozin's discovery that Cu atoms photoexcited in a $CH_4$ matrix give $CH_3CuH$ as the first-formed product even at 12 K.[58]

Ion cyclotron resonance experiments[59] have also shown that positively charged ions are more active than neutral atoms, e.g.,

$$Fe + alkane \xrightarrow{\times}$$

$$Fe^{\delta+} + alkane \rightarrow products$$

An electrophilic character is also found for low-coordination number metal atoms at kinks and steps in metal surfaces.[60] These sites are also known to be much more active for alkane reactions than the flat metal surfaces.[61]

These results suggest that, at least in electrophilic complexes, electron withdrawal from the C–H bond of the alkane may be more important than electron donation into the C–H$\sigma^*$ orbital. The most likely transition state in which this can be achieved would be T-shaped:

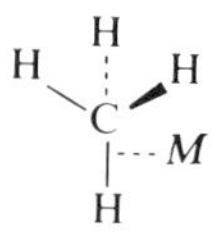

$$\begin{array}{c} H \\ H - \underset{|}{\overset{|}{C}} - H \cdots M \\ H \end{array}$$

Slightly bent CH $\cdots$ M systems have been observed in several studies[62]: intermediates of this type may lie on the route of approach to the transition state.

There is currently much interest in cluster complexes as models for surface chemistry. These studies have shown the unusual bonding modes that are possible on metal arrays. Catalysis, however, requires the availability of sites. The types of clusters usually studied, metal carbonyls, have tended not to be catalytically active. One important aspect of the metal surface, its lack of masking ligands, has not been modeled in the vast majority of cluster studies. In contrast, ligand-deficient clusters containing open sites or very labile ligands may prove to have useful catalytic activity. Bifunctional catalysts in which two chemically different metals are close together but not directly bonded may have more interesting properties.

Another feature of the metal surface that may be important for certain reactions is its electrophilic character. The interesting catalytic behavior of the $[M(\text{diene})L_2]A$ systems, mentioned in this review, may in part depend on the ligand-deficiency and electrophilicity of these systems. In these two properties, they may parallel the properties of a metal surface. It remains to be seen whether those cooperative effects that may occur are incidental or essential to catalytic action in heterogeneous, as well as homogeneous, systems.

### 4.4. Hydroformylation

A number of reports show that $[Rh(\text{cod})L_2]A$ complexes can be precursors for alkene hydroformylation.[63] Both the cationic and deprotonated neutral systems are active. The deprotonated system appears to be indistinguishable from that derived from $RhH(CO)(PPh_3)_3$; the cationic system is distinct, though broadly comparable.

A study of the rates and product ratios in the neutral system suggested that dinuclear species may play a role in the catalytic cycle.

### 4.5. Decarbonylation

Pignolet has shown that the trinuclear iridium compounds $[(IrL_2H_2)_3(\mu_3\text{-H})]A_2$ are catalysts for aldehyde decarbonylation.[64] The use of cationic complexes in this connection is interesting. Decarbonylation with $RhCl(PPh_3)_3$ (6) gives $RhCl(CO)(PPh_3)_2$, and subsequent reaction is very slow. CO loss is probably the slow step that limits the activity of the usual catalysts. Indeed, 6 is usually used as a stoichiometric decarbonylation reagent for this reason. Since an overall position charge should weaken $M$–CO bonding, cationic complexes may prove to be catalytically active under milder conditions than are the classical ones.

### 4.6. Water Gas Shift (WGS)

In the WGS reaction, CO is activated to nucleophilic attack by water. Clearly cationic complexes are well suited to this task, and it is not surprising to find that $[Ir(cod)L_2]A$ is an effective catalyst.[65]

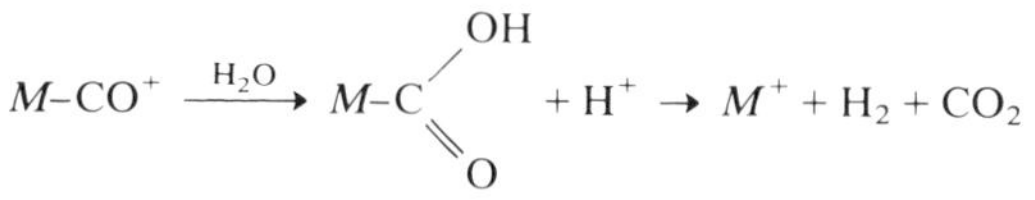

$$M-CO^+ \xrightarrow{H_2O} M-C\begin{matrix}OH\\ \\O\end{matrix} + H^+ \rightarrow M^+ + H_2 + CO_2$$

It is not yet clear if the system remains cationic throughout the catalytic cycle; added base does not suppress the reaction.

### 4.7. Polymerization

$[Rh(NO)(MeCN)_4]A_2$[66] is an interesting alkene oligomerization catalyst, apparently involving allylic intermediates; it may resemble Sen's[15] $[Pd(MeCN)_4]A_2$.

## 5. CONCLUDING REMARKS

The effects of net charge on catalysis have not been given the emphasis due them. In this review, I have tried to outline some of these effects, and hope the ideas presented will provoke further studies in the area.

#### ACKNOWLEDGMENTS

I thank Gregory Hlatky for discussions, and Yale University for a Junior Faculty Fellowship, which allowed me to write this review.

## REFERENCES

1. L. Pauling, *J. Chem. Soc.* **150,** 1461 (1948).
2. R. H. Crabree and A. J. Pearman, *J. Organometal. Chem.* **157,** 335 (1978).
3. a) T. Satoh, S. Suzuki, Y. Suzuki, Y. Miyaji, and Z. Imai, *Tet. Lett.* **1969,** 4555. b) S. E. Diamond, B. Grant, G. M. Tom, and H. Taube, *Tet. Lett.* **1774,** 4025.
4. J. Chatt, L. M. Vallarino, and L. M. Venanzi, *J. Chem. Soc.* **1957,** 2496 and 3413; C. Eaborn, N. Farrell, and A. Pidcock, *J. C. S., Dalton* 289 (1976).
5. S. G. Davies, M. L. H. Green, and D. M. P. Mingos, *Tetrahedron* **34,** 3047 (1977).
6. C. White, S. J. Thompson, and P. M. Maitlis, *J. Chem. Soc.* **1977,** 1654.
7. R. H. Crabtree and J. M. Mihelcic, recent observations (1981).
8. L. M. Haines, *Inorg. Chem.* **19,** 1685 (1971); G. Mestroni, G. Zassinovich, and A. Camus, *J. Organometal Chem.* **140,** 63 (1977).

9. Y. Hung, W-J. King, and H. Taube, *Inorg. Chem.* **20**, 157 (1981).
10. J. A. Davies, F. R. Hartley, and S. G. Murray, *J. C. S., Dalton* **1980**, 2246.
11. R. F. Schramm and B. Wayland, *Chem. Comm.* **1968**, 898.
12. J. M. Brown, P. A. Chaloner, and P. N. Nicholson, *Chem. Comm.* **1978**, 646.
13. J. A. Davies, *Adv. Inorg. Radiochem.* **24**, 116 (1981).
14. M. J. Church, M. J. Mays, R. N. F. Simpson, and F. P. Stefanini, *J. Chem. Soc., (A)* **1970**, 2909 and 3000.
15. A. Sen and T.-W. Lai, *J. Amer. Chem. Soc.* **103**, 4627 (1981).
16. R. H. Crabtree, *Accounts Chem. Res.* **12**, 331 (1979).
17. J. W. Suggs, S. D. Cox, R. H. Crabtree, and J. M. Quirk, *Tet. Lett.* **22**, 303 (1981).
18. S. C. Tang, T. E. Paxson, and L. Kim, *J. Mol. Cat.* **9**, 313 (1980).
19. R. Uriarte and R. H. Crabtree, recent observations (1981).
20. A. F. Bozowski, D. J. Cole-Hamilton, and G. Wilkinson, *Nouveau Journal de Chimie* **2**, 137 (1978).
21. R. H. Crabtree, J. M. Mihelcic, and J. M. Quirk, *J. Amer. Chem. Soc.* **101**, 7738; (1979); R. H. Crabtree, M. F. Mellea, J. M. Mihelcic, and J. M. Quirk, *J. Amer. Chem. Soc.* **104**, 107 (1982).
22. U. Belluco, M. Giustiniani, and M. Graziani, *J. Amer. Chem. Soc.* **89**, 6494 (1967).
23. W. J. Louw, D. J. A. de Waal, and J. E. Chapman, *Chem. Comm.* **1977**, 845; R. H. Crabtree, J. M. Quirk, T. Khan-Fillebeen, and G. E. Morris, *J. Organometal. Chem.* **157**, C13 (1978).
24. A. S. C. Chan, J. J. Pluth, and J. Halpern, *J. Amer. Chem. Soc.* **102**, 5952 (1980).
25. T. Hayashi, T. Mise, S. Mitachi, K. Yamamoto, and M. Kumada, *Tet. Lett.* **1976**, 49.
26. P. Meier, A. E. Merback, M. Dartiguenave, and Y. Dartiguenave, *Chem. Comm.* **1979**, 49.
27. J. R. Shapley, R. R. Schrock, and J. A. Osborn, *J. Amer. Chem. Soc.* **91**, 2816 (1969).
28. R. R. Schrock and J. A. Osborn, *J. Amer. Chem. Soc.* **98**, 2134 (1976).
29. F. W. S. Benfield, M. L. H. Green, and B. R. Francis, *J. Organometal. Chem.* **44**, C13 (1972).
30. C. Eaborn, N. Farrell, and A. Pidcock, *J. C. S., Dalton* **1976**, 289.
31. N. G. Gaylord and H. F. Mark, *Adv. Chem. Ser.* **34**, 127 (1962).
32. O. W. Howarth, C. H. McAteer, P. Moore, G. E. Morris, *J. Chem. Soc., Dalton* **1981**, 1481.
33. R. W. Mitchell, A. Spencer, and G. Wilkinson, *J. Chem. Soc., Dalton* **1973**, 846.
34. M. Green, T. A. Kuc, and S. H. Taylor, *Chem. Comm.* **1970**, 1553.
35. J. A. Osborn, F. H. Jardine, J. F. Young, and G. Wilkinson, *J. Chem. Soc., (A)* **1964**, 1711; F. H. Jardine, J. A. Osborn, and G. Wilkinson, *ibid* **1967**, 1574.
36. S. Montelatici, A. van der Ent, J. A. Osborn, and G. Wilkinson, *J. Chem. Soc., (A)* **1968**, 1054.
37. R. R. Schrock and J. A. Osborn, *Chem. Comm.* **1970**, 567, and *J. Amer. Chem. Soc.* **13**, 2397 and 3089 (1971); J. R. Shapley, R. R. Schrock, and J. A. Osborn, *J. Amer. Chem. Soc.* **91**, 2816 (1969); R. R. Schrock and J. A. Osborn, *J. Amer. Chem. Soc.* **98**, 2134 and 2134 (1976).
38. R. H. Crabtree, A. Gautier, G. Giordano, and T. Khan, *J. Organometal Chem.* **141**, 113 (1977).
39. R. H. Crabtree, H. Felkin, and G. E. Morris, *J. Organometal Chem.* **141**, 205 (1977).
40. A. S. C. Chan and J. Halpern, *J. Amer. Chem. Soc.* **102**, 838 (1980).
41. E. Martin and K. B. Wiberg, unpublished results (1980)
42. a) R. H. Crabtree, J. W. Faller, and M. F. Mellea, manuscript in preparation, (1981); b) R. H. Crabtree, H. Felkin, T. Khan and G. E. Morris, *J. Organomet. Chem.* **168**, 183 (1979).

43. D. Baudry, M. Ephritikine, and H. Felkin, *Nouv. J. Chim.* **2,** 355 (1978).
44. J. Halpern, D. P. Riley, A. S. C. Chan, and J. J. Pluth, *J. Amer. Chem. Soc.* **99,** 8055 (1977).
45. A. R. Powell, *Platinum Metal. Revs* **11,** 58; (1967); G. C. Bond, British Patent no. 197 723 (*Chem. Abs.* **73,** 98401 (1970)).
46. a) P. M. Maitlis, *Accts. Chem. Res.* **11,** 301; b) C. White, A. J. Oliver, and P. M. Maitlis, *J. Chem. Soc., Dalton* **1973,** 1901.
47. M. E. Wilson and G. M. Whitesides, *J. Amer. Chem. Soc.* **100,** 306 (1978).
48. T. J. Pinnavia, R. Raythatha, J. G-S. Lee, L. J. Halloran, J. F. Hoffman, *J. Amer. Chem. Soc.* **101,** 6891 (1979).
49. G. Pez, *Chem. Comm.* **1980,** 783.
50. A. Camus, *J. Mol. Cat.* **6,** 231 (1979).
51. M. I. Bruce, *Ang. Chem. (Int. Ed.)* **16,** 73 (1977).
52. J. Chatt and J. M. Davidson, *J. Chem. Soc.* **1965,** 843.
53. N. F. Gol'dschleger, M. D. Tyabin, A. E. Shilov, and A. A. S'hteinman, *Zh. Fiz. Khim.* **43,** 2174 (1969).
54. A. E. Shilov, *Sov. Sci. Revs B* **4,** 71 (1982).
55. R. H. Crabtree, M. F. Mellea, and J. M. Quirk, *Chem. Comm.* **1981,** 1217.
56. D. Baudry, M. Ephritikine, and H. Felkin, *Chem. Comm.* **1980,** 1243; M. A. Green, J. C. Huffman, and K. G. Caulton, *J. Organometal. Chem.* **218,** C39 (1981).
57. R. G. Bergman, personal communication (1981).
58. G. A. Ozin, D. F. McIntosh, and S. A. Mitchell, *J. Amer. Chem. Soc.* **103,** 1574 (1981).
59. J. Allison, R. B. Feas and D. R. Ridge, *J. Amer. Chem. Soc.* **101,** 1332, 4998 (1979).
60. T. H. Upton, W. A. Goddard, and C. F. Melius, *J. Vac. Sci. Technol.* **16,** 531 (1979); B. Krahl-Urban, E. A. Niekisch, and H. Wagner, *Surf. Sci.* **64,** 52 (1977); J. Holzl and F. K. Schulte, *Springer Tracts Mod. Phys.* **85,** 1 (1979).
61. H. Wagner, *Springer, Tracts Mod. Phys.* **85,** 151 (1979).
62. R. K. Brown, T. M. Williams, A. J. Schultz, G. D. Stucky, S. D. Ittel, and R. L. Harlow, *J. Amer. Chem. Soc.* **102,** 981 (1980).
63. R. H. Crabtree and H. Felkin, *J. Mol. Catal.* **5,** 75 (1979).
64. H. H. Wang and L. H. Pignolet, *Inorg. Chem.* **19,** 1470 (1980).
65. J. Kaspar, R. Spogliarich, G. Mestroni, and M. Graziani, *J. Organometal. Chem.* **208,** C15 (1981).
66. N. G. Connelly, P. T. Draggett, and M. Green, *J. Organometal Chem.* **140,** C10 (1977).

# 10

# Hydrogenation Reactions of CO and CN Functions Using Rhodium Complexes

*Bálint Heil, László Markó, and Szilárd Tőrös*

## 1. INTRODUCTION

Stimulating results achieved in hydrogenation of C=C double bonds with phosphinerhodium complexes led to the application of these compounds for the catalytic reduction of C=O and C=N double bonds. The present review summarizes the results in this area reported up until now. To give a more complete picture of the subject, some reports on iridium and ruthenium phosphine complex catalysts have been included as well.

The catalytic reduction of C=O and C=N double bonds of a hydrogen acceptor (A) molecule may be carried out by "direct" addition of hydrogen (1), by transfer hydrogenation (2), or through hydrosilylation (3). In the case of transfer hydrogenation, the hydrogen necessary for reduction is abstracted from donor (D) organic compounds like secondary alcohols. Hydrosilylation with mono- or dihydrosilanes leads to alkoxi- or alkylaminosilanes, which subsequently have to be hydrolized.

$$A + H_2 \overset{cat.}{\rightleftharpoons} AH_2 \tag{1}$$

$$A + DH_2 \overset{cat.}{\rightleftharpoons} AH_2 + D \tag{2}$$

---

*Drs. Bálint Heil and László Markó* • University of Chemical Engineering, Institute of Organic Chemistry, Veszprém, Schönherz Zoltán u. 12, Hungary.
*Dr. Szilárd Tőrös* • Research Group for Petrochemistry, Hungarian Academy of Sciences, Veszprém, Schönherz Zoltán u. 12, Hungary.

$$A + H_xSiR_{4-x} \xrightleftharpoons{\text{cat.}} R_{4-x}SiH_{x-1}AH \xrightarrow{H_2O} AH_2 \qquad (3)$$

Catalytic reduction of a carbon–oxygen double bond is generally more difficult than that of an olefinic one. This experience is probably related to the following two factors, among others:

a. the low stability—and accordingly small number—of complexes between $H_2$ activating metals and organic carbonyl compounds (as compared to the vast number of metal olefin complexes), and
b. the ability of alcohols, the products of the reaction, to act as ligands, in contrast to saturated hydrocarbons, which are inert against the catalytically active complexes.

The efficiency of ketone reduction may be diminished also by transfer hydrogenation, the secondary alcohol serving as a hydrogen source and, consequently, the hydrogenation ending up in a reversible reaction determined by thermodynamic parameters. This is supported by papers describing dehydrogenation reactions with catalysts successfully used in ketone reduction at lower temperatures. For example, Ohkubo[1] dehydrogenated racemic 1-phenylethanol with a $RhCl(NMDPP)_3$ catalyst prepared *in situ*, and according to Fragale,[2] 1-phenylethanol and other secondary alcohols are partly transformed to the corresponding ketones in the presence of $RhCl(C_8H_{12})PPh_3$ + NaOH under nitrogen at 80–150°.

## 2. HYDROGENATION OF ALDEHYDES

Wilkinson-type rhodium-phosphine catalysts are reported to be not suitable for aldehyde reduction[3] because of decarbonylation of the substrate under the reaction conditions and formation of $RhCl(CO)(PPh_3)_2$, which is catalytically inactive. In contrast to this, Fujitsu[4] achieved the hydrogenation of phenylacetaldehyde with $[Rh(NBD)(PR_3)_n]^+ClO_4^-$ ($n$ = 2 or 3) catalysts under mild conditions. The catalytic activities decreased in the order of $PEt_3 > PPh_3 \sim PMe_3 \gg DPE$ complexes. The total turnover number of the $PEt_3$ complex was greater than 77.5 with this aldehyde.

A very active aldehyde hydrogenation catalyst is formed by treating $IrH_3(PPh_3)_3$ with acetic acid.[4,5] While Coffey[5] hydrogenated *n*-butyraldehyde under mild reaction conditions, Strohmeier[6] obtained turnover numbers up to 8000 in the reduction of several saturated and unsaturated aldehydes with the same catalyst at 80–110° and 1 bar working without solvent. The acetato complex $IrH_2(OOCCH_3)(PPh_3)_3$ was suggested as the active intermediate.

Ruthenium catalysts have also been successfully used[7-10] for the hydrogenation of aliphatic and aromatic aldehydes. Although decarbonylation of the substrate may not be avoided using $RuCl_2(PPh_3)_3$ or $RuHCl(PPh_3)_3$, the carbonyl complex $RuHCl(CO)(PPh_3)_3$ formed is, in contrast to the corresponding rhodium derivative, also active in hydrogenation.[9] Strohmeier reported turnover numbers up to 95,000 with a $RuCl_2(CO)_2(PPh_3)_2$ catalyst.[10]

In hydrogen transfer from different organic compounds to aldehydes, excellent catalytic activity of $RuH_2(PPh_3)_4$ was observed by Japanese scientists[11] to override that of $RhH(PPh_3)_4$.

Selective hydrogenation of a C=O bond in the presence of a C=C bond was studied by Meguro[12] using cinnamaldehyde and crotonaldehyde as model compounds. With $RhCl_3 \cdot 3H_2O$ or $Rh_2Cl_2(CO)_4$ and tertiary amines in benzene, unsaturated alcohols were produced with high (up to 85%) selectivity. Addition of $PPh_3$, however, retarded the reduction of the formyl group, and saturated aldehydes were the products.

## 3. HYDROGENATION OF KETONES NOT CONTAINING OTHER FUNCTIONAL GROUPS

### 3.1. Catalytic Activity and Stereoselectivity

Homogeneous hydrogenation of ketones with rhodiumphosphine catalysts was first carried out by Schrock and Osborn[13] in 1970. The authors reduced acetone, acetophenone, cyclohexanone, and ethyl methyl ketone with a $[RhH_2P_2S_2]^+A^-$ type ionic complex (P = basic phosphine, S = solvent) at 25° and normal pressure. They also found that the initial rate of the reaction is markedly enhanced (up to 30 times) by adding 1% of water, while reduction of olefins is inhibited in aqueous solution. The reaction mechanism suggested by the authors is presented below (see Figure 10.1). Crabtree reported on the catalytic activity of $[Rh(COD)(PPh_3)_2]^+PF_6^-$ in benzene adding $Et_3N$.[14,15]

Preparing the active catalytic system *in situ* from $[Rh(diene)Cl]_2$ and the corresponding tertiary phosphines,[16] mainly covalent rhodium-phosphine complexes are formed. Their activity is highly influenced by the quality of the ligand used. Best results have been achieved with phosphines of high basicity and low steric requirements.[17] $RhCl(PPh_3)_3$ is completely inactive under such conditions.

Gargano reduced ketones with $RhCl(C_8H_{12})PPh_3$ and $Rh_2H_2Cl_2(C_8H_{12})(PPh_3)_2$ in presence of strong alkali. Pretreatment of these precursors with $NaBH_4$ increased the catalytic activity of the system further.[18]

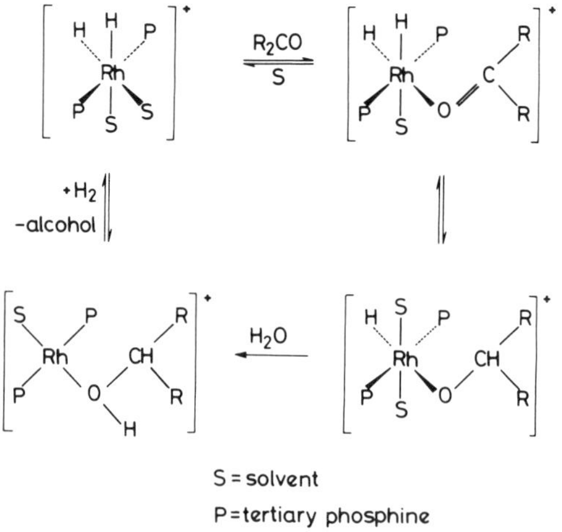

Figure 10.1. Mechanism of ketone hydrogenation as proposed by Schrock and Osborn.[13]

Rhodium hydroxo complexes have been considered as active intermediates (see Figure 10.2).

Addition of small quantities of $Et_3N$ transformed the inactive $RhCl(PPh_3)_3$ into an active catalyst for ketone reduction.[19] Catalytic systems of similar activity were prepared *in situ* from rhodium diene derivatives and also other aryl phosphines in presence of bases like $Et_3N$.

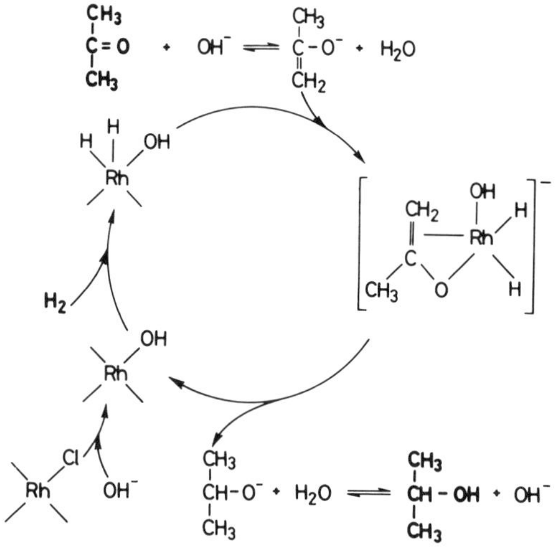

Figure 10.2. Mechanism of ketone reduction in strongly basic media as proposed by Gargano.[18]

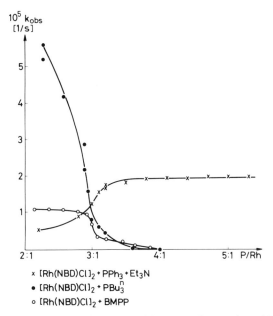

Figure 10.3. Effect of the P:Rh ratio on rate of ketone hydrogenation with different phosphinerhodium catalysts.

These latter transformed the catalyst into a hydridorhodium(I)phosphine complex and enabled the catalytic reduction of different dialkyl-, alkylaryl-, diaryl-, and cyclic ketones with acceptable reaction rates under mild conditions (50°, 1 bar).

It was shown that by changing the phosphine/rhodium ratio in catalytic systems prepared *in situ* several active species are formed,[20] but while increasing the P/Rh ratio in systems prepared from $[Rh(NBD)Cl]_2 + PR_3$ ($PR_3$ = BMPP, $PBu_3^n$, etc.), the rate of hydrogenation decreases, this tendency is reversed in the case of the $[Rh(NBD)Cl]_2 + PPh_3 + Et_3N$ catalysts (see Figure 10.3).[21]

Polymer-bound heterogenized homogeneous rhodium catalysts have also been successfully used in ketone reduction. Italian scientists studied ionic rhodium complexes[22] supported on a Merrifield resin (4), but the Wilkinson-type analog proved to be active only in the presence of $Et_3N$.[23]

$R' = Ph; R'' = Me, menthyl$ \hfill (4)

A few authors also reported on the catalytic activity of ruthenium complexes in hydrogenation of ketones.[8,24,25]

Transfer hydrogenation is another method widely used in ketone reduction. As a source of hydrogen, secondary alcohols like iPrOH are preferred, and the reaction is carried out in presence of KOH at the boiling point of the solvent. Reactions catalyzed with ruthenium complexes working at higher reaction temperatures were often performed in cyclohexanol or benzylalcohol,[26] but, as a consequence of hydrogen transfer from the secondary alcohols produced, the reaction becomes reversible.[11]

Spogliarich reported on hydrogen transfer studies with $[Rh(diene)P_2]^+$ (P = $PPh_3$, $PMePh_2$, $PMe_2Ph$, $PBzPh_2$, DPM, DPE, DPP, DPB, DPET, DIOP; diene = COD, NBD) and Wilkinson-type complexes.[27,28] Reducing cyclohexanone, the best result (90% conversion in 15 min; cyclohexanone/cat. mole ratio = 1900) was achieved with the $[Rh(COD)DPE]^+$ catalyst.[28] The activity of the complexes depends on the nature of the phosphine used and follows the order DPE > DPP ≈ $PPh_3$ > DPB ≈ DIOP ≈ DPET > $PMe_2Ph$ > $PBzPh_2$ > $PMePh_2$. This means that, in general, higher activities were obtained with bidentate ligands. The catalytic activity is also dependent on the number of carbon atoms ($n$) in $Ph_2P(CH_2)_nPPh_2$, decreasing with increasing $n$. Similar ionic complexes were used by Uson to reduce acetophenone.[29]

Apart from the phosphine complexes described, the hydrogen transfer activity of several other derivatives has also been investigated. Among them rhodium,[30] and iridium[31] bipyridyl, and phenanthroline compounds displayed the highest catalytic activity with turnover numbers up to 900 cycles/min. Reducing unsaturated ketones, first the C=C double bond is saturated, but the successive reduction of the C=O group is comparable in rate with that of the olefinic bond.

From many ketones, stereoisomeric *cis* and *trans* alcohols may be formed. The stereoselectivity of the catalysts was mainly studied with 4-*t*-butylcyclohexanone as a model compound. Some of the representative results are summarized in Table 1. The stereoselectivity with catalysts prepared *in situ* is determined by the structure of the phosphine ligand. The *cis* alcohol is preferred, using phosphines like $PPh_3$ and NMDPP in presence of $Et_3N$ where probably $HRhP_3$ is formed as the active intermediate. This suggestion is supported by the fact that the reaction rate is enhanced by increasing the P/Rh ratio from 2.2 to 3. Using basic phosphines like $PBu_3^n$, BMPP, IMPP, etc., the key intermediate formed is apparently $H_2RhP_2$ (highest rates are achieved at P:Rh = 2.2:1, and the *trans* isomer is preferred). Addition of $Et_3N$ is ineffective in this case or slightly decreases both the reaction rate and the stereoselectivity.[21]

The reduction of 4-methyl- and 3-ethylcyclohexanone by hydrogen transfer with $PPh_3$ and $PPh_2NR_2$ type aminophosphine-rhodium(I) com-

Table 1. Stereoselective Hydrogenation of 4-t-Bu-cyclohexanone

| Catalyst | Trans Alcohol, % | Hydrogen donor | Solvent | Temp. | Reference |
|---|---|---|---|---|---|
| $[RhH_2(PPhMe_2)_2S_2]^+ClO_4^-$ | 86 | $H_2$ | | 25° | 13 |
| $[Rh(COD)(PPh_3)_2]^+PF_6]^- + Et_3N$ | 63 | $H_2$ | $C_6H_6$ | 20° | 15 |
| $[Rh(NBD)(DPE)]^+ClO_4^- + KOH$ | 67 | $iPrOH$ | $iPrOH$ | 82° | 28 |
| $[Rh(COD)(DPE)]^+ClO_4^- + KOH$ | 85 | $iPrOH$ | $iPrOH$ | 82° | 28 |
| $[Rh(4,7-Me_2Phen)_2Cl_2]^+Cl^- + KOH$ | 20 | $iPrOH$ | $iPrOH$ | 82° | 30 |
| $[Ir(3,4,7,8-Me_4Phen)(COD)]^+Cl^- + KOH$ | 86 | $iPrOH$ | $iPrOH$ | 82° | 31 |
| $[Rh(NBD)Cl]_2 + PPh_3 + Et_3N$ | 27 | $H_2$ | $C_6H_6/MeOH$ | 50° | 21 |
| $[Rh(NBD)Cl]_2 + PBu_3^n + Et_3N$ | 88 | $H_2$ | $C_6H_6/MeOH$ | 50° | 21 |
| $[Rh(NBD)Cl]_2 + PBu_3^n$ | 90 | $H_2$ | $C_6H_6/MeOH$ | 50° | 21 |

plexes prepared *in situ* has been investigated by Svoboda.[32] By hydrogenating the 4-methyl-derivative with a PPh$_3$ containing catalyst, mainly the *cis* isomer was formed, while employing aminophosphines, the thermodynamically more stable *trans* isomer was the major product.

Japanese scientists succeeded in stereoselective hydrosilylation of substituted cyclohexanones catalyzed by RhCl(PPh$_3$)$_3$ and a silica-linked rhodium(I) complex ($\equiv$Si-O-$\overset{|}{\underset{|}{Si}}$CH$_2$CH$_2$PPh$_2$)$_3$RhCl. They found that the reduction of 2-substituted derivatives leads with each catalyst preferentially to the thermodynamically less stable alcohol isomer, but the influence of the bulkiness of silanes was also significant: increasing steric hindrance favored the more stable stereoisomer.[33]

### 3.2. Enantioselective Hydrogenation of Ketones

Homogeneous enantioselective hydrogenation of ketones has been first carried out with the ionic complexes of Schrock and Osborn using chiral phosphines like BMPP, EMPP, DIOP, etc., but the optical yields were rather low.[34-36] By preparing the catalyst with the same chiral phosphines *in situ*, the optical selectivity (which is determined in this case by the covalent and ionic complexes present simultaneously) was improved considerably.[16] RhClP$_2$ and RhP$_3^+$ type intermediates are regarded to be responsible for this effect.[20]

Many of the chiral bidentate phosphines synthesized in the last years have also been tested for enantioselective ketone reduction. Some of the results achieved are compiled in Table 2. The influence of phosphine structure on optical selectivity and catalytic activity is considerable, but a reliable correlation could not yet be found. It seems that chiral bidentate *bis*(diphenyl)phosphines, like *prophos* forming 5-membered chelate rings with the rhodium atom and used with great success for the hydrogenation of dehydroaminoacids, are not suitable for ketone reduction because of very low reaction rates.

Apart from the structure of the chiral phosphine, optical selectivity of homogeneous asymmetric hydrogenation of ketones is strongly influenced by other reaction parameters, such as temperature, pressure, the P/Rh mole ratio with *in situ* prepared systems,[20,42] the quality and quantity of other additives, and the solvent, etc. While the effect of the latter is shown in Table 3, the influence of Et$_3$N on enantioselectivity in different solvents[19,43] is demonstrated in Figure 10.4. By increasing the Et$_3$N/Rh mole ratio in methanol, there is a maximum in enantioselectivity (ascribed to a HRh(P-P) type catalyst), but further amounts of the base decrease the optical purity of the secondary alcohol considerably. This may be explained

Table 2. Enantioselective Hydrogenation of Acetophenone

| Catalyst | Solvent | Temp. °C | Pressure, bar | Reaction time, h | Chemical yield, % | Optical yield, % | Reference |
|---|---|---|---|---|---|---|---|
| [Rh(NBD)((R)-BMPP)$_2$]$^+$ClO$_4^-$ | EtOH | 20 | 1 | 24–80 | 20–40[a] | 8.6 | 34 |
| [Rh(NBD)((R)-EMPP)$_2$]$^+$PF$_6^-$ | THF | 50 | 1 | 72 | 77 | 0.24 | 35 |
| [Rh(NBD)Cl]$_2$ + IMPP | EtOH | 60 | 3.1 | 43 | ? | 11.4 | 37 |
| [Rh(NBD)(−)-DIOP]$^+$ClO$_4^-$ | iPrOH | 30 | 1 | — | ? | 8.1 | 36 |
| [Rh(HD)Cl]$_2$ + (+)-DIOP | MeOH/C$_6$H$_6$ | 30 | 1 | 173 | 10 | 51 | 16 |
| [Rh(COD)(BPPFOH)]$^+$ClO$_4^-$ | MeOH | 0 | 50 | 8 | 96 | 43 | 39 |
| [Rh(HD)Cl]$_2$ + BDPCP | ? | 80 | 200 | — | ? | 22 | 40 |
| [Rh(COD)(PNNP)]$^+$ClO$_4^-$ | MeOH/C$_6$H$_6$ | 20 | 12 | 36 | 72 | 16.7 | 41 |
| [Rh(C$_2$H$_4$)$_2$Cl]$_2$ + prophos | MeOH | 50 | 1 | 100 | 4 | 44.2 | 42 |
| [Rh(NBD)Cl]$_2$ + (+)-DIOP + Et$_3$N | C$_6$H$_6$ | 50 | 70 | 6 | 64 | 80 | 43 |
| [Rh(NBD)(cycphos)]$^+$PF$_6^-$ + Et$_3$N | MeOH | 25 | 100 | 140 | 71 | 6 | 44 |
| RhCl-(−)-DIOP-PSt[b] | C$_6$H$_6$ | 25 | 4 | 27 | 11 | 7.7 | 45 |
| [Rh(COD)Cl]$_2$ + DPPEA + Et$_3$N | C$_6$H$_6$/H$_2$O | 50 | 1 | — | ? | 22 | 46 |
| [H$_4$Ru$_4$(CO)$_8$[(−)-DIOP]$_2$ | — | 130 | 100 | 5 | 40 | 8.1 | 38 |

[a] moles / mole Rh.
[b] PSt = Styrene–divinylbenzene copolymer.

Table 3. Enantioselective Hydrogenation of Ketones in Different Solvents

| Ketone | Solvent | Optical yield, % | |
|---|---|---|---|
| 2-Octanone[a] | EtOH | (S) | 1.6 |
|  | THF | (S) | 0.3 |
|  | EtOAc | (R) | 0.9 |
|  | DMF | (R) | 5.1 |
|  | $(CH_3)_2CHCOOH$ | (R) | 11.2 |
|  | $CH_3COOH$ | (R) | 12.0 |
| Acetophenone[b] | $C_6H_6$ | (S) | 19 |
|  | $C_6H_6$ + 2% MeOH | (S) | 24 |
|  | $C_6H_6$ + 2% $H_2O$ | (S) | 27 |
|  | MeOH | (S) | 37 |
|  | MeOH + 2% $CH_3COOH$ | (S) | 43 |
|  | $CH_3COOH$ | (S) | 56 |

[a] Catalyst: $[Rh(COD)(ACMP)_2]^+BF_4^-$ [47]
[b] Catalyst: $1/2[Rh(NBD)Cl]_2$ + (S)-BMPP; reaction conditions: 50°, 1 bar $H_2$ [21]

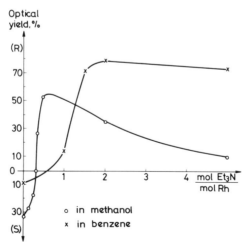

Figure 10.4. Effect of $Et_3N$ on enantioselectivity of acetophenone hydrogenation with a $[Rh(NBD)Cl]_2$ + (−)-DIOP catalyst.

by the effect of excess $OMe^-$ ions which transform the catalyst into a $[Rh_3(diphos)_3(OMe)_2]^+$ type complex described by Halpern.[48] In benzene, this latter effect is small, and the enantioselectivity of the system is mainly determined by the HRh(P-P) derivative.

Optical selectivity in these reactions is obviously strongly depending also on the structure of the substrate. It has been generally observed that enantioselectivity is higher in the case of alkyl-aryl ketones than with dialkyl ketones.[43]

First results with polymer-supported rhodium(I) chiral diphosphine complexes were published by Ohkubo,[45] but the optical selectivity reported was rather moderate (7.7% e.e.). Chiral ruthenium catalysts have also been successfully used, and the results were compiled in a review published in 1981.[49]

Only little information is available on asymmetric ketone reduction using hydrogen transfer. With secondary alcohols or indoline as hydrogen donors, optical yields up to 9.9% were obtained in the presence of $H_4Ru_4(CO)_8[(-)DIOP]_2$ as a catalyst.[50] Iridium compounds like $[Ir(COD)(PPEI)]^+ClO_4^-$ proved to be more enantioselective in the presence of KOH.[51] Reducing propiophenone 30% e.e. was observed at 50% conversion, but it decreased with increasing conversion. The activity of this iridium catalyst was rather high: 98% conversion was obtained in 4 hours with high substrate/catalyst ratios (1000–2000) under mild conditions.

As mentioned earlier, hydrosilylation can be used as an alternative route of ketone reduction. By performing the addition of hydrosilanes in the presence of suitable chiral transition metal catalysts, prochiral ketones are transformed in the subsequent hydrolysis step into optically active alcohols (5):

$$\begin{array}{c} R^1 \\ \diagdown \\ R^2 \diagup \end{array}SiH_2 + \begin{array}{c} R \\ \diagdown \\ R' \diagup \end{array}C=O \xrightarrow{cat.^*} \begin{array}{c} R^1 \\ \diagdown \\ R^2 \diagup \end{array}Si \begin{array}{c} H \\ \diagdown \\ \diagup \\ O\overset{*}{C}HRR' \end{array} \xrightarrow{H_2O} R\overset{*}{C}HOHR' \quad (5)$$

This method of ketone reduction has been widely used all the more, since there are relatively few catalysts suitable for the homogeneous direct hydrogenation of ketones and the enantioselectivity of hydrosilylation may be rather high. Because of a comprehensive review published in 1977,[52] only a few general remarks will be made and some of the latest results presented (Table 4). Both optical purity and configuration of the secondary alcohols formed are strongly dependent on the structure of the hydrosilane. Rather low enantioselectivities were observed with monohydrosilanes using a rhodium-DIOP catalytic system, but activity and enantioselectivity are enhanced if dihydrosilanes are used. Best results have been achieved with $\alpha$-NpPhSiH$_2$. Reasonable selectivities could be obtained also with other chiral diphosphine derivatives using dihdyrosilanes, but in the presence of

Table 4. Asymmetric Hydrosilylation of Ketones

| Ketone | Silane | Phosphine | Optical yield, % | Reference |
|---|---|---|---|---|
| PhCOMe | α-NpPhSiH$_2$ | (+)-DIOP | (S) 58 | 53 |
| PhCOMe | Ph$_2$SiH$_2$ | MPFA | (R) 49 | 52 |
| PhCOMe | α-NpPhSiH$_2$ | naphos | (S) 18 | 54 |
| PhCOMe | Ph$_2$SiH$_2$ | glup | (S) 47 | 55 |
| PhCOMe | α-NpPhSiH$_2$ | PPPM | (R) 60 | 56 |
| PhCOMe | α-NpPhSiH$_2$ | glucophinite | — 65 | 57 |
| PhCOCH$_2$Cl | α-NpPhSiH$_2$ | (+)-DIOP | (S) 63 | 58 |
| PhCOEt | α-NpPhSiH$_2$ | (+)-DIOP | (S) 56 | 59 |
| PhCOEt | Ph$_2$SiH$_2$ | cycDIOP | (R) 52 | 60 |
| PhCOEt | α-NpPhSiH$_2$ | camphinite | — 52 | 57 |
| PhCO(iPr) | PhMe$_2$SiH | (S)-BMPP | (R) 56 | 61 |
| PhCO(tBu) | EtMe$_2$SiH | (R)-BMPP | (R) 56 | 61 |
| PhCO(tBu) | PhMe$_2$SiH | (R)-BMPP | (S) 61.8 | 62 |
| PhCO(cHex) | PhMe$_2$SiH | (R)-BMPP | (S) 58 | 63 |
| CH$_3$(CH$_2$)$_5$COCH$_3$ | Ph(cHex)SiH$_2$ | (−)-DIOP | (R) 44 | 58 |

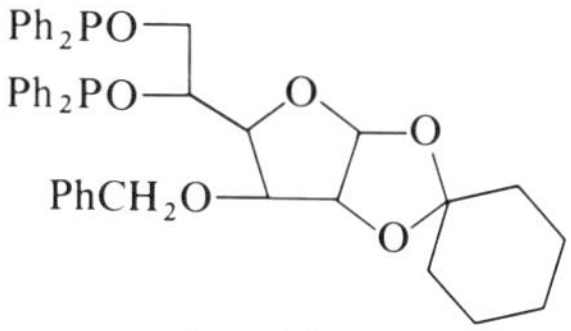

Ph$_2$PO
Ph$_2$PO
PhCH$_2$O

glucophinite

CH$_2$OPPh$_2$
CH$_2$OPPh$_2$

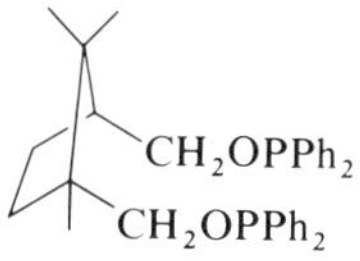

camphinite

the monodendate BMPP ligand, application of monohydrosilanes was more favorable. In contrast to the hydrogenation of ketones, no characteristic difference could be detected between ionic and covalent rhodium complexes as catalysts of hydrosilylation.

Benes prepared some new chiral phosphines suitable as ligands for this reaction.[60] Capka anchored (C$_2$H$_5$O)$_3$Si(CH$_2$)$_3$P(Ph)Men to silica (6). Complexed with rhodium, a catalyst for the enantioselective hydrosilylation of ketones like acetophenone and propiophenone was obtained.[64]

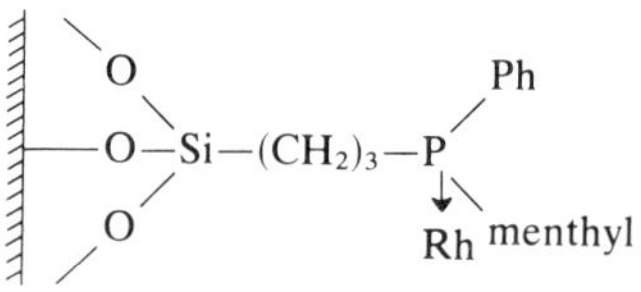

$$\text{O—Si—(CH}_2)_3\text{—P}\overset{\text{Ph}}{\underset{\text{Rh menthyl}}{\diagdown}} \quad (6)$$

Chiral phosphinites were also employed as ligands in hydrosilylation.[55]

Based on kinetic and spectroscopic studies, it was suggested[65,66] that in hydrosilylation of $t$-butylphenylketone the interaction of the silyl hydrido-rhodium phosphine species with the ketone is rate determining.

If prochiral ketones are hydrosilylated with dihydrosilanes carrying different alkyl groups ($R^1 \neq R^2$) and the siloxane is treated with a Grignard reagent, optically active silanes are produced (7):

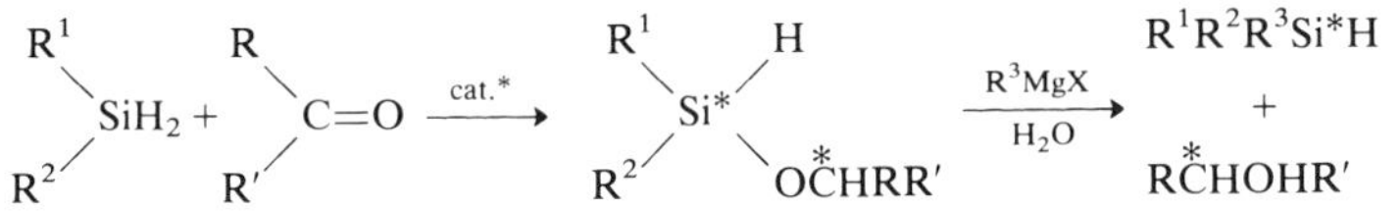

$$\begin{array}{c} R^1 \\ R^2 \end{array}\!\!SiH_2 + \begin{array}{c} R \\ R' \end{array}\!\!C\!=\!O \xrightarrow{cat.^*} \begin{array}{c} R^1 \\ R^2 \end{array}\!\!Si^*\!\!\begin{array}{c} H \\ O\overset{*}{C}HRR' \end{array} \xrightarrow[H_2O]{R^3MgX} R^1R^2R^3Si^*H + R\overset{*}{C}HOHR' \quad (7)$$

In this way, $\alpha$-NpPhMeSiH of 82% e.e. was formed by hydrosilylation of ($-$)-menthone with a RhCl($+$)-DIOP catalyst,[67] a reaction in which two chiral centers were acting in a diastereoselective procedure.

## 4. HYDROGENATION OF KETONES CONTAINING OTHER FUNCTIONAL GROUPS

### 4.1. Selective Hydrogenation of Unsaturated Ketones

By hydrogenating ketones containing C=C and C=O double bonds with rhodium phosphine catalysts, the olefinic bond is saturated first. RhCl(PPh$_3$)$_3$ is inactive in hydrogenation of the keto group (except activated groups in ketoesters, as discussed in Section 4.2.). This fact made the selective reduction of mesityl oxide to methyl-isobutylketone possible.[68] By adding H$_2$O$_2$ to the catalyst it becomes heterogeneous, but its activity is enhanced enormously. Using it for more than 500 hours, turnover numbers above 100,000 were measured.[69] The Wilkinson compound hydrogenates the olefinic bond in $\Delta^4$-$\alpha,\beta$-unsaturated 3-keto steroids stereoselectively, and the $5\alpha$-form is produced. A very strong influence of reaction parameters was detected.[70]

Solodar studied the reduction of piperitenone (8), a compound containing two different olefinic bonds and a prochiral carbon–oxygen double bond.[71] He found that mainly the sterically less hindered C=C bond was favored in the hydrogenation reaction, and thus pulegone was the major product, but also other compounds were formed in minor amounts.

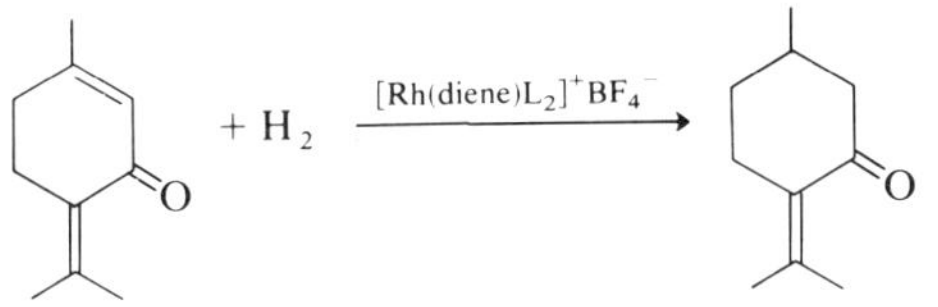

$$\text{piperitenone} + H_2 \xrightarrow{[Rh(diene)L_2]^+ BF_4^-} \text{pulegone} \quad (8)$$

$L_2$ = DIOP, MDPP, PCH$_3$(2—R$^1$O—C$_6$H$_4$)R$^2$

(R$^1$ = Me, $i$Pr, Bz; R$^2$ = Ph, cHex, $i$Pr)

Reducing α,β-unsaturated ketones with catalysts prepared from [Rh(diene)Cl]$_2$ and $PR_3$ *in situ* again the C=C bond is saturated preferentially, but performing the reaction in primary alcohols hydrogen transfer as a side reaction may take place, and the aldehyde thus formed is decarbonylated and the catalyst deactivated.[72]

No difference in selectivity was detected reducing unsaturated ketones in a hydrogen-transfer reaction. Beaupere and co-workers reported experiments in α-phenylethanol with HRh(PPh$_3$)$_4$ as catalyst[73] and found that electron-attracting groups (X) on substrates like $XC_6H_5CH=CHCOR$ increase the reaction rate. In a similar reaction, ruthenium complexes were the catalysts, and hydrogen was transferred from isopropylidene- or cyclohexylidene-1,2-α-D-glucofuranose to benzalacetophenone, forming the saturated ketone with 100% yield.[74]

In contrast to the results discussed above, hydrosilylation of unsaturated ketones with rhodiumphosphine catalysts makes the selective saturation of either the C=C or the C=O bond possible. Using monohydrosilanes, 1,4-addition takes place and saturated ketones are formed, while dihydrosilanes prefer the 1,2-addition and unsaturated alcohols are the products (9):

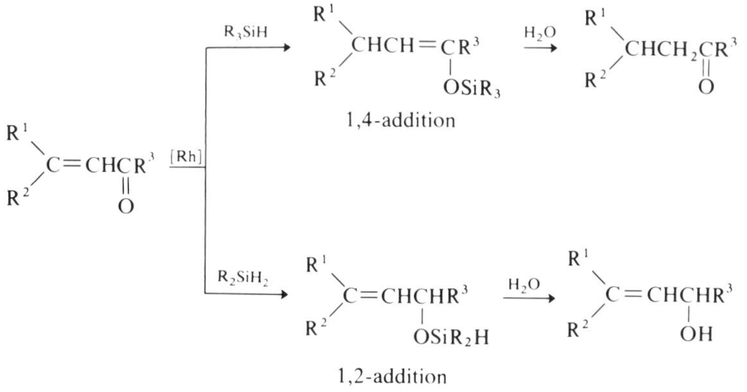

$$\begin{array}{c} R^1 \\ R^2 \end{array} C=CHCR^3 \quad \xrightarrow{[Rh]} \quad \begin{cases} \xrightarrow{R_3SiH} \quad \begin{array}{c} R^1 \\ R^2 \end{array} CHCH=CR^3 \; (OSiR_3) \xrightarrow{H_2O} \begin{array}{c} R^1 \\ R^2 \end{array} CHCH_2CR^3 \text{ (1,4-addition)} \\ \xrightarrow{R_2SiH_2} \quad \begin{array}{c} R^1 \\ R^2 \end{array} C=CHCHR^3 \; (OSiR_2H) \xrightarrow{H_2O} \begin{array}{c} R^1 \\ R^2 \end{array} C=CHCHR^3 \text{ (OH) (1,2-addition)} \end{cases} \quad (9)$$

Accordingly α-ionone(**a**) and citral(**d**) (α,β-unsaturated carbonyl compounds containing an additional isolated double bond) could be transformed with 96–97% selectivity[74] into dihydroionone(**c**) and citronellal(**f**) in the presence of RhCl(PPh$_3$)$_3$ and Et$_3$SiH (10, 11):

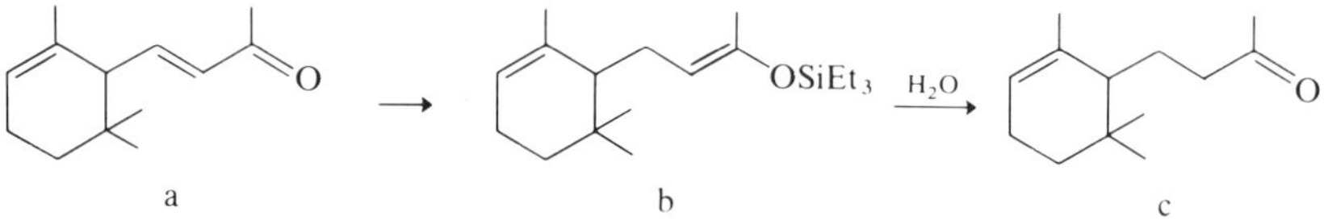

a → b (OSiEt$_3$) $\xrightarrow{H_2O}$ c

(10)

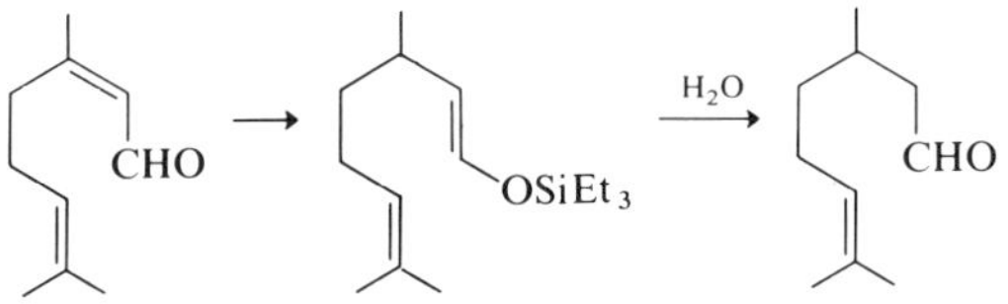

$$\text{d} \xrightarrow{} \text{e} \xrightarrow{H_2O} \text{f} \qquad (11)$$

Structures: d = unsaturated CHO; e = OSiEt$_3$ silyl enol ether; f = CHO product.

Similar observations were made reducing β-ionone and pulegone: using PhMe$_2$SiH mainly, saturation of the α,β-olefinic bond was achieved, while in the presence of Et$_2$SiH$_2$ or Ph$_2$SiH$_2$, the unsaturated alcohol was formed exclusively from both substrates.

Later, Ojima reported on the hydrosilylation of α,β-unsaturated ketones with chiral rhodium catalysts (Rh-BMPP and Rh-DIOP) and α-NpPhSiH$_2$. As models, β-ionone and 2-methylcyclohexenone were reduced to the corresponding alcohols with 33.5 and 43% optical selectivity.[76]

## 4.2. Hydrogenation of Keto Acids and Keto Esters

In contrast to the hydrogenation of simple ketones, RhCl(PPh$_3$)$_3$ was found to be an active catalyst[77] for the reduction of keto esters. In addition, several ionic or covalent rhodium complexes containing other types of phosphines have been used successfully.

Some characteristic results in asymmetric reduction of keto acids and keto esters are summarized in Table 5. It should be mentioned here that there are some remarkable differences between the hydrogenation of simple ketones and keto esters. For example, the optical selectivities are generally much higher with keto esters. This is probably due to the bidentate ligand character of these compounds.

Recent results suggest that small amounts of water decrease the reaction rate of hydrogenation of propyl-pyruvate.[77] It should be mentioned, however, that Schrock[13] and Solodar[47] made their earlier opposite observations while working with monophosphine ligands. Ojima obtained the highest optical selectivities in benzene and THF with neutral rhodium complexes, while methanol decreased both optical selectivity and the reaction rate.[77] The fact that neutral and ionic rhodium-DIOP complexes produce enantiomers of opposite configuration, and further, the finding that enantioselectivity of keto ester reduction depends on the covalent or ionic character of the catalytic system, is in accordance with earlier observations in asymmetric hydrogenation of acetophenone.[20]

While these and other reactions discussed earlier were carried out in different organic solutions, several water-soluble transition-metalphosphine complexes have also been synthesized and used as catalysts in the

Table 5. Asymmetric Reduction of Keto Acids and Keto Esters

| Substrate | Reducing agent | Phosphine | Solvent | Optical yield, % | Reference |
|---|---|---|---|---|---|
| $CH_3COCH_2COOMe$ | $H_2$ | ICMP | EtOH | 71 | 48 |
| $CH_3COCOOH$ | $H_2$ | BPPFOH | MeOH | 83 | 39 |
| $CH_3COCOOEt$ | $H_2$ | CPPM | $c\text{-}C_6H_{12}$ | 67.3 | 78 |
| $CH_3COCOOEt$ | $H_2$ | BPPM | $C_6H_6$ | 65.3 | 78 |
| $CH_3COCOOPr^n$ | $H_2$ | BPPM | $C_6H_6$ | 75.8 | 77 |
| $CH_3COCOOBu^n$ | $H_2$ | CPPM | $c\text{-}C_6H_{12}$ | 62.8 | 78 |
| $CH_3COCOOPr^n$ | $\alpha$-NpPhSiH$_2$ + H$_2$O | (−)DIOP | $C_6H_6$ | 85.4 | 79 |
| PhCOCOOMen | $\alpha$-NpPhSiH$_2$ + H$_2$O | (+)DIOP | $C_6H_6$ | 77 | 79 |
| PhCOCOOMen | $\alpha$-NpPhSiH$_2$ + H$_2$O | PPh$_3$ | $C_6H_6$ | 17 | 79 |
| $CH_3COCOOMen$ | $\alpha$-NpPhSiH$_2$ + H$_2$O | (+)DIOP | $C_6H_6$ | 85.6 | 79 |
| $CH_3COCH_2CH_2COOBu^i$ [a] | $\alpha$-NpPhSiH$_2$ + H$_2$O | (+)DIOP | $C_6H_6$ | 84.4 | 79 |
| $CH_3COCOOPr^n$ | $H_2$ | (−)DIOP | THF | 41.9 | 77 |

[a] The product: 4-methyl-$\gamma$-butyrolactone.

last years. Joó prepared ruthenium complexes containing m-sulphophenyl-diphenylphosphine as ligand and found that among them $HRuCl(mSPPh_2)_3$ and $HRu(OAc)(mSPPh_2)_3$ are active in hydrogenation of keto-acids.[80]

## 4.3. Hydrogenation of Aminoketones and Other Biologically Active Derivatives

Increasing interest has been shown in recent years for the synthesis of biologically active secondary alcohols by homogeneous catalytic stereoselective and enantioselective hydrogenation reactions.

The first result in this field was the reduction of 2-(6-carbomethoxy-hexyl)cyclopentane-1,3,4-trione with $[Rh(COD)(ACMP)_2]^+BF_4^-$ as catalyst to 2-(6-carbomethoxy-hexyl)-4-(R)-hydroxycyclopentane-1,3-dione,[81] an intermediate of the $E_1$ prostaglandine synthesis (12):

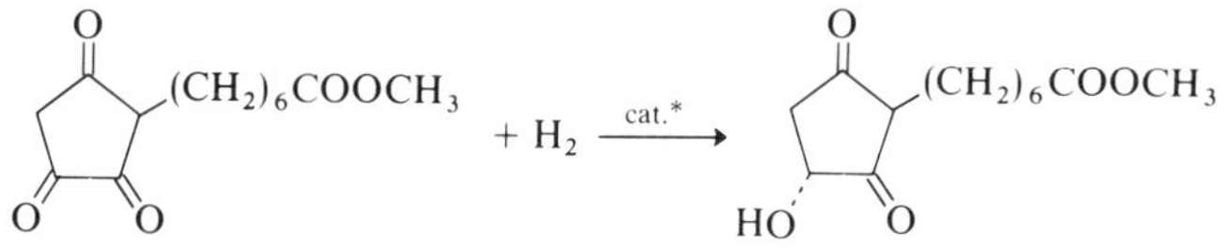

$$\text{structure} + H_2 \xrightarrow{cat.^*} \text{structure} \quad (12)$$

Japanese authors reported[82] on the synthesis of (R)(−)-pantolactone (a precursor of pantothenic acid, which is an important constituent of coenzyme A) by hydrogenation of α-keto-β,β-dimethyl-γ-butyrolactone with rhodium catalysts containing different chiral pyrrolidinediphosphines (13):

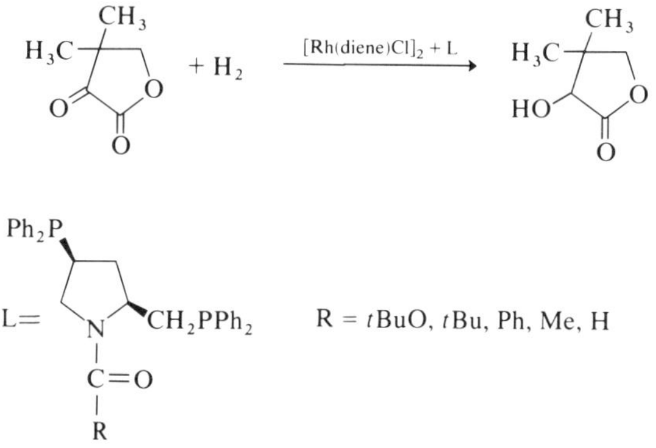

$$\text{structure} + H_2 \xrightarrow{[Rh(diene)Cl]_2 + L} \text{structure} \quad (13)$$

$L = $ (pyrrolidine structure with $Ph_2P$, $CH_2PPh_2$, N–C(=O)–R)

R = tBuO, tBu, Ph, Me, H

They found that the enantioselectivity of the reaction is highly influenced by the steric bulkiness of the N-substituents in the phosphine, while electronic factors play only a minor role. Working with the Rh–BPPM catalytic system ($R = t$BuO), 86.7% optical yield was achieved under optimum reaction conditions,[83] enabling the preparation of pure D(−)-pantoyl lactone in a single recrystallization step.

Table 6. Enantioselective Reduction of Aminoketones

| Substrate | Reducing agent | Catalyst | Reaction time, h | Conversion, % | Optical yield, % | Reference |
|---|---|---|---|---|---|---|
| $PhCOCH_2NHCH_2Ph$ | $H_2$ | $[Rh(COD)(ACMP)_2]^+BF_4^-$ | 6 | 80 | $-1.7^{a}$ | 37 |
| $CH_3COCH(COOH)NHCOCH_3^{b}$ | $H_2$ | $[Rh(NBD)(MDPP)_2]^+BF_4^-$ | 6 | 80 | $95^c$ | 84 |
| $3,4\text{-}(MeO)_2C_6H_3COCH_2NH_2 \cdot HCl$ | $H_2$ | $[Rh(HD)Cl]_2 + BPPFOH$ | 48–96 | 100 | 89 | 85 |
| $3,4\text{-}(HO)_2C_6H_3COCH_2NHMe \cdot HCl$ | $H_2$ | $[Rh(NBD)(BPPFOH)]^+ClO_4^-$ | 168 | 100 | 95 | 85 |
| $PhCOCH_2NEt_2$ | $H_2$ | $[Rh(NBD)Cl]_2 + (+)DIOP$ | 20 | 70 | 93 | 86 |
| $2\text{-}NpCOCH_2NEt_2$ | $H_2$ | $[Rh(NBD)Cl]_2 + (+)DIOP$ | 20 | 93 | 95 | 86 |
| $PhCOCH_2NBu_2^n$ | $H_2$ | $[Rh(NBD)Cl]_2 + (+)DIOP$ | 20 | 52 | 90 | 86 |
| $4\text{-}Et\text{-}C_6H_4COCH_2NEt_2$ | $H_2$ | $[Rh(NBD)Cl]_2 + (+)DIOP$ | 20 | 39 | 90 | 86 |
| $MeCOCONHCH(CH_2Ph)COOMe$ | $H_2$ | $[Rh(COD)Cl]_2 + (+)DIOP$ | 20 | 100 | $26^c$ | 87 |
| $MeCOCONHCH(CH_2Ph)COOMe$ | $H_2$ | $[Rh(COD)Cl]_2 + BPPM$ | 64 | 100 | $28^c$ | 87 |
| $MeCOCONHCH(CH_2Ph)COOMe$ | $H_2$ | $RhCl(PPh_3)_3$ | 64 | 100 | $20^c$ | 87 |
| $PhCOCONHCH(CH_2Ph)COOMe$ | $\alpha\text{-}NpPhSiH_2$ | $[Rh(COD)Cl]_2 + (+)DIOP$ | 24 | 79 | $82^c$ | 87 |
| $PhCOCONHCH(CH_2Ph)COOMe$ | $\alpha\text{-}NpPhSiH_2$ | $RhCl(PPh_3)_3$ | 24 | 62 | $56^c$ | 87 |

[a] $[\alpha]_D^{25}$ value.
[b] 90% d, 10% l.
[c] diastereomeric excess.

Biological and therapeutic effects of 2-amino-1-arylketones and their derivatives are well known. Some results in asymmetric hydrogenation of these compounds are summarized in Table 6. High optical yields were observed also in these cases, apparently again pointing to the simultaneous coordination of the carbonyl group and the nitrogen atom of the substrate to the central rhodium atom in the active intermediate.

It has been shown by Ojima[87] in the synthesis of depsipeptides (see the last five lines in Table 6) that optical selectivities are much higher in the hydrosilylation reactions, where asymmetric induction by the chiral catalyst predominates over that of the chiral center already present in the $N$-($\alpha$-ketoacyl)-$\alpha$-aminoester substrates.

Special biological effects are often displayed only by given stereoisomers of secondary alcohols. This is the case for example with tropines, constituents of some well-known biologically active derivatives like cocaine and scopolamine. In the preparation of tropine stereoisomers, rhodiumphosphine catalysts prepared *in situ* could be used with success.[21] The structure of alcohols is mainly determined by the structure of phosphines in the catalyst: both $\alpha$- and $\beta$-hydroxy derivatives could be synthesized as major products. With PAr$_3$ type phosphines, the reaction rate is higher at P/Rh = 3 than at P/Rh = 2.2, and $\alpha$-hydroxy tropines are formed with up to 98% yield. Using basic alkyl phosphines as ligands, the catalytic system shows highest activity at P/Rh = 2.2, and the thermodynamically more stable $\beta$-isomers can be prepared in yields as high as 89%. Selectivity of the reduction is not depending on the quality of the $R$ group in nortropinones. Substantial similarities can be observed between the main characteristics of the hydrogenation of these derivatives and those of 4-$t$Bu-cyclohexanone (see Table 1, Section 3.1.).

## 5. HOMOGENEOUS HYDROGENATION OF CARBON–NITROGEN DOUBLE BONDS

Although both [Rh(diene)P$_2$]$^+$A$^-$ type complexes,[36] RhCl(PPh$_3$)$_3$,[88] and catalysts prepared *in situ*[88] from [Rh(diene)Cl]$_2$ + PR$_3$ are active in hydrogenation of the C=N bond, only a few papers have been published in this field. In contrast to ketone reduction, catalytic activity increases in the order PBu$_3^n$ < PEt$_2$Ph < PEtPh$_2$ < PPh$_3$ and has its optimum at a P/Rh = 2.2 mole ratio. While increasing amounts of ligand inhibit the catalyst,[88] the system remains homogeneous and active even at lower mole ratios. Phosphites are also suitable as ligands in this system.[89]

Reduction of Schiff-bases has been performed also with hydrogen transfer reactions[90] catalyzed by RhCl(PPh$_3$)$_3$, RhCl(CO)(PPh$_3$)$_2$, and

$RuCl_2(PPh_3)_2$ complexes using $i$-propyl alcohol as hydrogen donor. It was found that Me or MeO groups on the ring of $RC_6H_4R^1C=NC_6H_4R^2$ type substrates increase the reaction rate. KOH and minor amounts of water promote the hydrogenation of $PhCH=NPh$ with $RhCl(PPh_3)_2$ as catalyst.[91]

Aldimines have also been hydrogenated by hydrogen transfer from $i$PrOH catalyzed by $RhCl(PPh_3)_3$, $RuCl_2(PPh_3)_4$, $HOsCl(CO)(PPh_3)_3$, and $RhCl(CO)(PPh_3)_2$ in the presence of sodium carbonate.[92] The formation of hydrido-metal complexes as active species was suggested. To prove this assumption, $HRh(PPh_3)_4$ was prepared and used for reduction of $PhCH=NPh$ with a high reaction rate. Catalysts prepared *in situ* from $RhCl_3$, $PPh_3$, and sodium carbonate were less active. $RuCl_3$, $IrCl_3$, or $OsCl_3$ gave systems which were essentially inactive.

Ojima reported on the reduction of $R^1R^2C=NR^3$ type Schiff bases by hydrosilylation[93] with several catalysts. The reaction was also affected by the quality of hydrosilane. Using dihydrosilanes, which were found to be more active than monohydrosilanes and trihydrosilanes, the activity of the catalysts used decreased in the order $RhCl(PPh_3)_3 \gg Rh(CO)Cl(PPh_3)_2 > Py_2RhCl(dmf)BH_4 > [Rh(HD)Cl]_2 > [Rh(COD)Cl]_2 > PdCl_2 > PdCl_2(PPh_3)_2$.

The reduction of $C=N$ bonds and immonium ions was suggested to be involved in the $N$-alkylation of amines by $CO + H_2O$ with different rhodium catalysts like $RhCl_3$, $[Rh(COD)Cl]_2$, $RhCl(PPh_3)_3$, and $RhCl(CO)(PPh_3)_2$[94] (14, 15):

$$\geq NH + OHCR \rightarrow \geq N^+=CHR \rightarrow \geq N-CH_2R \quad (14)$$

$$-NH_2 + OHCR \rightarrow -N=CHR \rightarrow -NHCH_2R \quad (15)$$

Asymmetric reduction of prochiral imines, ketoximes, and several dihydroisoquinoline derivatives was successfully performed with chiral rhodium and ruthenium catalysts (Table 7).

Italian scientists reported[36] on asymmetric hydrogenation of $N$-($\alpha$-methylbenzylidene)benzylamine to the corresponding $N$-benzyl-1-phenylethylamine with $[Rh(NBD)(-)DIOP]^+ClO_4^-$ as catalyst and achieved 22% e.e. The same substrate was hydrogenated also with a catalyst prepared *in situ* from $[Rh(NBD)Cl]_2$ and DIOP, but the optical induction was much lower (<3.3% e.e.).[88] By preparing the catalyst with $(R)(+)$-$Ph_2PCH(iPr)CH_2PPh_2$ as the chiral ligand,[96] however, optical selectivities up to 65% have been achieved.[97]

Moderate optical inductions were found in the hydrogenation of ketoximes with $H_4Ru(CO)_8[(-)-DIOP]_2$ as catalyst. By reducing $t$-butylphenylketoxime, the optical yield was 14.5%.[38]

Kagan reported on asymmetric hydrosilylation of prochiral compounds with carbon–nitrogen double bonds.[95] $N$-($\alpha$-methylbenzylidene)benzyl-

## Table 7. Enantioselective Reduction of Compounds Containing C=N Double Bond

| Substrate | Catalyst | Reducing agent | Optical yield, % | Reference |
|---|---|---|---|---|
| PhMeC=NCH$_2$Ph | [Rh(NBD)(−)DIOP]$^+$Cl$_4^-$ | H$_2$ | 22 | 36 |
| PhMeC=NCH$_2$Ph | [Rh(NBD)Cl]$_2$ + (+)DIOP | H$_2$ | 3.3 | 88 |
| PhMeC=NCH$_2$Ph | [Rh(NBD)Cl]$_2$ + (s)BMPP | H$_2$ | 1.5 | 88 |
| PhMeC=NCH$_2$Ph | [Rh(NBD)Cl]$_2$ + valphos | H$_2$ | 65 | 97 |
| MeEtC=NOH | H$_4$Ru$_4$(CO)$_8$[(−)DIOP]$_2$ | H$_2$ | 2.6 | 38 |
| MePhC=NOH | H$_4$Ru$_4$(CO)$_8$[(−)DIOP]$_2$ | H$_2$ | 4 | 38 |
| Ph(tBu)C=NOH | H$_4$Ru$_4$(CO)$_8$[(−)DIOP]$_2$ | H$_2$ | 14.5 | 38 |
| PhMeC=NCH$_2$Ph | RhCl(+)DIOP | Ph$_2$SiH$_2$ | 65 | 95 |
| DHIQ A | RhCl(+)DIOP | Ph$_2$SiH$_2$ | 22.5 | 95 |
| DHIQ B | RhCl(+)DIOP | Ph$_2$SiH$_2$ | 5.7 | 95 |
| DHIQ C | RhCl(+)DIOP | Ph$_2$SiH$_2$ | 38.7 | 95 |
| MePhC=NPh | RhCl(+)DIOP | Ph$_2$SiH$_2$ | 47 | 95 |
| Me(CH$_2$Ph)C=NCH$_2$Ph | RhCl(+)DIOP | Ph$_2$SiH$_2$ | 11.5 | 95 |

DHIQ

A: R$^1$ = H, R$^2$ = CH$_2$Ph

B: R$^1$ = OCH$_3$, R$^2$ = CH$_3$

C: R$^1$ = OCH$_3$, R$^2$ = CH$_2$–C$_6$H$_3$(OCH$_3$)$_2$

amine was reduced with Ph$_2$SiH$_2$ in the presence of RhCl(DIOP) as catalyst with 65% optical selectivity. In a similar reaction, 1,2,3,4-tetrahydropapaverine was synthesized with 38% e.e. from the corresponding dihydroisoquinoline.

## ABBREVIATIONS FOR LIGANDS

| Abbreviation | Name of Ligand |
|---|---|
| ACMP | o-Anisylcyclohexylmethylphosphine |
| ICMP | (2-Isopropyloxyphenyl)cyclohexylmethylphosphine |
| BMPP | (R)-(+)-Benzylmethylphenylphosphine |
| EMPP | (R)-(−)-Ethylmethylphenylphosphine |
| IMPP | (+)-Isopropylmethylphenylphosphine |

| | |
|---|---|
| BPPM | (2s,4s)-N-(t-Butoxycarbonyl)-4-diphenylphosphino-2-diphenylphosphinomethylpyrrolidine |
| CPPM | (2s,4s)-N-(Cholesteryloxycarbonyl)-4-diphenylphosphino-2-diphenylphosphinomethylpyrrolidine |
| PPPM | (2s,4s)-N-(t-Butylcarbonyl)-4-diphenylphosphino-2-diphenylphosphinomethylpyrrolidine |
| BPPFOH | (R)-α-[(S)-1,2-Bis(diphenylphosphino)ferrocenyl]ethylalcohol |
| MPFA | (R)-α-[(S)-2-Dimethylphosphinoferrocenyl]ethyldimethylamine |
| DPM | Bis(diphenylphosphino)methane |
| DPE | Bis(1,2-diphenylphosphino)ethane |
| DPP | Bis(1,3-diphenylphosphino)propane |
| DPB | Bis(1,4-diphenylphosphino)butane |
| DPET | cis-Bis(1,2-diphenylphosphino)ethylene |
| DIOP | 2,3-O-Isopropylidene-2,3-dihydroxy-1,4-bis(diphenylphosphino)butane |
| prophos | (R)-(+)-Bis(1,2-diphenylphosphino)propane |
| valphos | (R)-(+)-1,2-Bis(diphenylphosphino)-3-methylbutane |
| cycphos | (R)-1,2-Bis(diphenylphosphino)cyclohexylethane |
| cycDIOP | (2R,3R)-2,3-O-cyclohexylidene-2,3-dihydroxy-1,4-bis(cyclohexylphenylphosphino)butane |
| naphos | (S)-(−)-2,2'-Bis(diphenylphosphinomethyl)-1,1'-binaphtyl |
| BDPCP | d-trans-1,2-Bis(diphenylphosphinoxy)cyclopentane |
| glup | Methyl-4,6-O-benzylidene-2,3-bis-O-(diphenylphosphino)-α-D-glucopyranoside |
| DPPEA | Diphenylphosphino-N-(α-phenylethyl)-acetamide |
| PNNP | N,N'-Bis(diphenylphosphino)-N,N'-bis((S)-(−)-α-methyl-benzyl)-ethylenediamine |
| MDPP | (−)-Menthyldiphenylphosphine |
| NMDPP | (+)-Neomenthyldiphenylphosphine |
| mSPPh$_2$ | m-Sulphophenyl-diphenylphosphine |
| HD | 1,5-Hexadiene |
| COD | 1,5-Cyclooctadiene |
| NBD | Bicyclo[2.2.1]hepta-2,5-diene |
| Phen | 1,10-Phenantroline |
| PPEI | 2-Pyridinal-α-phenylethylimine |

## REFERENCES

1. K. Ohkubo, T. Ohgushi, and K. Yoshinaga, *Chem. Lett.* **1976**, 775–778.
2. C. Fragale, M. Gargano, and M. Rossi, *J. Mol. Catal.* **5**, 65–73 (1979).
3. D. H. Doughty and L. H. Pignolet, *J. Am. Chem. Soc.* **100**, 7083–7085 (1978).
4. H. Fujitsu, S. Shirahama, E. Matsumara, K. Takeshita, and I. Mochida, *J. Org. Chem.* **46**, 2287–2290 (1981).
5. R. S. Coffey, *Chem. Commun.* **1967**, 923.
6. W. Strohmeier and H. Steigerwald, *J. Organomet. Chem.* **129**, C43–C46 (1977).
7. J. Tsuji and H. Suzuki, *Chem. Lett.* **1977**, 1085–1086.
8. R. A. Sanchez-Delgado and O. L. de Ochoa, *J. Mol. Catal.* **6**, 303–305 (1979).

9. R. A. Sanchez-Delgado, A. Andriollo, O. L. de Ochoa, T. Suarez, and N. Valencia, *J. Organomet. Chem.* **209**, 77–83 (1981).
10. W. Strohmeier and L. Weigelt, *J. Organomet. Chem.* **145**, 189–194 (1978).
11. H. Imai, T. Nishiguchi, and K. Fukuzumi, *J. Org. Chem.* **41**, 665–671 (1976).
12. S. Meguro, T. Mizoroki, and A. Ozaki, *Chem. Lett.* **1975**, 943–946.
13. R. R. Schrock and J. A. Osborn, *J. Chem. Soc. D.* **1970**, 567–568.
14. R. H. Crabtree, *J. Chem. Soc., Chem. Commun.* **1975**, 647–648.
15. R. H. Crabtree, A. Gautier, G. Giordano, and T. Khan, *J. Organomet. Chem.* **141**, 113–121 (1977).
16. B. Heil, S. Tőrös, S. Vastag, and L. Markó, *J. Organomet. Chem.* **94**, C47–C48 (1975).
17. S. Vastag, B. Heil, and L. Markó, *J. Mol. Catal.* **5**, 189–195 (1979).
18. M. Gargano, P. Giannoccaro, and M. Rossi, *J. Organomet. Chem.* **129**, 239–242 (1977).
19. B. Heil, S. Tőrös, J. Bakos and L. Markó, *J. Organomet. Chem.* **175**, 229–232 (1979).
20. S. Tőrös, B. Heil, and L. Markó, *J. Organomet. Chem.* **159**, 401–407 (1978).
21. B. Heil, L. Kollár, L. Markó, and S. Tőrös, *J. Organomet. Chem.*, in press.
22. G. Strukul, M. Bonivento, M. Graziani, E. Cernia, N. Palladino, *Inorg. Chim. Acta* **12**, 15–21 (1975).
23. B. Heil, S. Tőrös, H. J. Kreuzfeld, and H. Pracejus, *React. Kinet. Catal. Lett.*, in press.
24. P. Frediani, U. Matteoli, M. Bianchi, F. Piacenti, and G. Menchi, *J. Organomet. Chem.* **150**, 273–278 (1978).
25. W. Strohmeier and L. Weigelt, *J. Organomet. Chem.* **171**, 121–129 (1979).
26. B. Graser and H. Steigerwald, *J. Organomet. Chem.* **193**, C67–C70 (1980).
27. R. Spogliarich, G. Zassinovich, G. Mestroni, and M. Graziani, *J. Organomet. Chem.* **179**, C45–C47 (1979).
28. ———, *J. Organomet. Chem.* **198**, 81–86 (1980).
29. R. Uson, L. A. Oro, R. Sariego, and M. A. Esteruelas, *J. Organomet. Chem.* **214**, 399–404 (1981).
30. G. Zassinovich, G. Mestroni and A. Camus, *J. Organomet. Chem.* **168**, C37–C38 (1979).
31. A. Camus, G. Mestroni, and G. Zassinovich, *J. Mol. Catal.* **6**, 231–233 (1979).
32. P. Svoboda and J. Hetflejs, *Collect. Czech. Chem. Commun.* **42**, 2177–2181 (1977).
33. J. Ishiyama, Y. Senda, I. Shinoda, and S. Imaizumi, *Bull. Chem. Soc. Jpn.* **52**, 2353–2355 (1979).
34. P. Bonvicini, A. Levi, G. Modena, and G. Scorrano, *J. Chem. Soc., Chem. Commun.* **1972**, 1188–1189.
35. M. Tanaka, Y. Watanabe, T. Mitsudo, H. Iwane and Y. Takegami, *Chem. Lett.* **1973**, 239–240.
36. A. Levi, G. Modena, and G. Scorrano, *J. Chem. Soc., Chem. Commun.* **1975**, 6–7.
37. J. Solodar, *Ger. Offen.* 2 306 222 (1973); *Chem. Abstr.* **79**, 146179t (1973).
38. C. Botteghi, M. Bianchi, E. Benedetti, and U. Matteoli, *Chimia* **29**, 256–258 (1975).
39. T. Hayashi, T. Mise, and M. Kumada, *Tetrahedron Lett.* **1976**, 4351–4354.
40. T. Hayashi, M. Tanaka, and I. Ogata, *Tetrahedron Lett.* **1977**, 295–296.
41. M. Fiorini, F. Marcati, and G. M. Giongo, *J. Mol. Catal.* **3**, 385–387 (1977).
42. K. Ohkubo, M. Setoguchi, and K. Yoshinaga, *Inorg. Nucl. Chem. Lett.* **15**, 235–238 (1979).
43. S. Tőrös, B. Heil, L. Kollár, and L. Markó, *J. Organomet. Chem.* **197**, 85–86 (1980).
44. D. P. Riley and R. E. Shumate, *J. Org. Chem.* **45**, 5187–5193 (1980).
45. K. Ohkubo, M. Haga, K. Yoshinaga, and Y. Motozato, *Inorg. Nucl. Chem. Lett.* **16**, 155–158 (1980).
46. F. Joó and E. Trócsányi, *J. Organomet. Chem.* **231**, 63–70 (1982).
47. J. Solodar, CHEMTECH **1975**, 421–423.
48. J. Halpern, D. P. Riley, A. S. C. Chan, and J. J. Pluth, *J. Am. Chem. Soc.* **99**, 8055–8057 (1977).

49. U. Matteoli, P. Frediani, M. Bianchi, C. Botteghi, and S. Gladiali, *J. Mol. Catal.* **12**, 265–319 (1981).
50. M. Bianchi, U. Matteoli, G. Menchi, P. Frediani, S. Pratesi, F. Piacenti, and C. Botteghi, *J. Organomet. Chem.* **198**, 73–80 (1980).
51. G. Zassinovich, A. Camus, and G. Mestroni, *J. Mol. Catal.* **9**, 345–347 (1980).
52. I. Ojima, K. Yamamoto, and M. Kumada, *Aspects of Homogeneous Catalysis*, edited by R. Ugo, (D. Reidel Publishing Co., Dordrecht, 1977), Vol. 3, pp. 185–228.
53. W. Dumont, J. C. Poulin, T. P. Dang, and H. B. Kagan, *J. Am. Chem. Soc.* **95**, 8295–8299 (1973).
54. K. Tamao, H. Yamamoto, H. Matsumoto, N. Miyake, T. Hayashi, and M. Kumada, *Tetrahedron Lett.* **1977**, 1389–1392.
55. M. Capka, J. Hetflejs, and R. Selke, *React. Kinet. Catal. Lett.* **10**, 225–228 (1979).
56. K. Achiwa, *Fundamental Research in Homogeneous Catalysis* **3**, 549–564 (1979).
57. T. H. Johnson, K. C. Klein, and S. Thomen, *J. Mol. Catal.* **12**, 37–40 (1981).
58. H. B. Kagan, J. F. Peyronel, and T. Yamagishi, *Adv. Chem. Ser.* **173**, 50–66 (1979).
59. R. J. P. Corriu and J. J. E. Moreau, *J. Organomet. Chem.* **85**, 19–33 (1975).
60. J. Benes and J. Hetflejs, *Collect. Czech. Chem. Commun.* **41**, 2264–2272 (1976).
61. I. Ojima and Y. Nagai, *Chem. Lett.* **1974**, 223–228.
62. T. Hayashi, K. Yamamoto, K. Kasuga, H. Omizu, and M. Kumada, *J. Organomet. Chem.* **113**, 127–137 (1976).
63. I. Ojima, T. Kogure, M. Kumagai, S. Horiuchi, and T. Sato, *J. Organomet. Chem.* **122**, 83–97 (1976).
64. M. Capka, *Collect. Czech. Chem. Commun.* **42**, 3410–3416 (1977).
65. I. Kolb and J. Hetflejs, *Collect. Czech. Chem. Commun.* **45**, 2224–2239 (1980).
66. ———, *Collect. Czech. Chem. Commun.* **45**, 2808–2816 (1980).
67. R. J. P. Corriu and J. J. E. Moreau, *J. Organomet. Chem.* **91**, C27–C30 (1975).
68. W. Strohmeier and E. Hitzel, *J. Organomet. Chem.* **91**, 373–377 (1975).
69. ———, *J. Organomet. Chem.* **102**, C37–C41 (1975).
70. W. Voelter and C. Djerassi, *Chem. Ber.* **101**, 58–68 (1968).
71. J. Solodar, *J. Org. Chem.* **43**, 1787–1789 (1978).
72. L. Kollár, S. Tőrös, B. Heil, and L. Markó, *J. Organomet. Chem.* **192**, 253–256 (1980).
73. D. Beaupere, P. Bauer, and R. Zan, *Can. J. Chem.* **57**, 218–221 (1979).
74. G. Descotes and D. Sinou, *Tetrahedron Lett.* **1976**, 4083–4086.
75. I. Ojima, T. Kogure, and Y. Nagai, *Tetrahedron Lett.* **1972**, 5035–5038.
76. ———, *Chem. Lett.* **1975**, 985–988.
77. I. Ojima, T. Kogure, and K. Achiwa, *J. Chem. Soc., Chem. Commun.* **1977**, 428–430.
78. K. Achiwa, *Tetrahedron Lett.* **1977**, 3735–3738.
79. I. Ojima, T. Kogure, and M. Kumagai, *J. Org. Chem.* **42**, 1671–1679 (1977).
80. Z. Tóth, F. Joó, and M. Beck, *Magy. Kem. Foly.* **86**, 173–177 (1980).
81. C. J. Sih, J. Heather, G. P. Peruzzotti, P. Price, R. Sood, and L. H. Lee, *J. Am. Chem. Soc.* **95**, 1676–1677 (1973).
82. K. Achiwa, T. Kogure, and I. Ojima, *Chem. Lett.* **1978**, 297–298.
83. I. Ojima, T. Kogure, T. Terasaki, and K. Achiwa, *J. Org. Chem.* **43**, 3444–3446 (1978).
84. J. Solodar, *Ger. Offen.* 2 312 924 (1973); *Chem. Abstr.* **80**, 3672h (1974).
85. T. Hayashi, A. Katsumura, M. Konishi, and M. Kumada, *Tetrahedron Lett.* **1979**, 425–428.
86. S. Tőrös, L. Kollár, B. Heil, and L. Markó, *J. Organomet. Chem.* **232**, C17–C18 (1982).
87. I. Ojima, T. Tanaka, and T. Kogure, *Chem. Lett.* **1981**, 823–826.
88. S. Vastag, B. Heil, S. Tőrös, and L. Markó, *Transition Met. Chem.* **2**, 58–59 (1977).
89. Z. Nagy-Magos, S. Vastag, B. Heil, and L. Markó, *Transition Met. Chem.* **3**, 123–124 (1978).
90. J. S. Shekoyan, G. V. Varnakova, U. N. Krutii, K. I. Karpeiskaya, and V. Z. Sharf, *Izv. Akad. Nauk SSSR, Ser. Khim.* **1975**, 2811–2813.

91. V. Z. Sharf, L. H. Freidlin, J. S. Portjakova, U. N. Krutii, *Izv. Akad. Nauk SSSR, Ser. Khim.* **1979**, 1414–1415.
92. R. Grigg, T. R. Mitchell and N. Tongpenyai, *Synthesis*, **1981**, 442–444.
93. I.-Ojima, T. Kogure, and Y. Nagai, *Tetrahedron Lett.* **1973**, 2475–2478.
94. Y. Watanabe, M. Yamamoto, T. Mitsudo, and Y. Takegami, *Tetrahedron Lett.* **1978**, 1289–1290.
95. H. B. Kagan, N. Langlois, and T. P. Dang, *J. Organomet. Chem.* **90**, 353–365 (1975).
96. W. Bergstein, A. Kleemann, and J. Martens, *Synthesis* **1981**, 76–78.
97. S. Vastag, J. Bakos, S. Tőrös, N. E. Takach, R. B. King, B. Heil, and L. Markó, *J. Mol. Catal.*, in press.

# 11

# Decarbonylation Reactions Using Transition Metal Complexes

*Daniel H. Doughty and Louis H. Pignolet*

## 1. INTRODUCTION

The decarbonylation of aldehydes, acyl halides, aroyl halides, alcohols, or ketones is a useful and important reaction in organic synthesis.[1,2,3] Although several methods not utilizing transition metals are known[4,5] (including various deformylation reactions and thermal and photochemical decarbonylations), they are not general and not usually applicable under mild conditions where undesirable side reactions are minimized.[1]

Several transition metal complexes have been reported to function as stoichiometric homogeneous decarbonylation reagents under mild conditions. The following discussion will show that the organic products observed upon decarbonylation of a given substrate depend strongly on the catalyst and type of substrate to be decarbonylated. Completely different reaction products are observed, for example, when different types of acid chlorides are used with the same catalyst. This fact must be considered when trying to formulate a general reaction mechanism for these decarbonylation reactions. The earliest reported metal-promoted decarbonylation reactions are

---

*Daniel H. Doughty.* • Exploratory Chemistry, Div. 8315, Sandia National Laboratories, Livermore, California 94550.
*Dr. Louis H. Pignolet* • Department of Chemistry, University of Minnesota, Minneapolis, Minnesota 55455.

the formation of ruthenium carbonyl compounds upon refluxing ruthenium trichloride hydrate in alcoholic solvents containing tertiary phosphines and base. Using this method, complexes such as $RhHCl(CO)(PPh_3)_3$ were prepared by Chatt and Shaw[6,7] and Vaska[8,9] in 1960 and 1961. In 1964, Chatt, Shaw, and Field proposed a mechanism[10] that involved a metal-promoted oxidation of the alcohol to the corresponding aldehyde and subsequent decarbonylation of the aldehyde. In support of this mechanism, they reported that methane and acetaldehyde were observed when ethanol was decarbonylated.

Another early report of this class of reactions was by Vaska,[11] showing that iridium halides, in the presence of triphenylphosphine, abstract CO from alcohols to give $IrCl(CO)(PPh_3)_2$. Dimethylformamide is also decarbonylated by this reaction scheme.[12] Similarly, $OsBr_2(PPh_3)_3$ reacts with alcohols to yield $OsHBr(CO)(PPh_3)_3$ and the corresponding alkane.[13]

In 1966, Prince and Raspin reported the decarbonylation of saturated aldehydes by $Ru_2Cl_3(PEt_2Ph)_6{}^+Cl^-$.[14,15] The metal product formed in the absence of oxygen was $RuCl_2(CO)(PEt_2Ph)_3$.[6] These decarbonylation reactions were slow, generally requiring approximately 80 hours to undergo several transformations upon heating in solution[16,18] and so the actual reactive species is not known. The organic products depended upon the aldehyde that was used. With acetaldehyde, the product was methane. With propanal and butanal, however, a majority of the hydrocarbon product was the alkene. For example, with butanal at 70–80°C, the ratio of propene to propane was 9:1. The overall stoichiometry of the reaction was such that one mole of ruthenium produced one mole of hydrocarbon and one mole of $RuCl_2(CO)(PEt_2Ph)_3$.

The hydrogen that is liberated when the olefins are formed is consumed by the reduction of butyraldehyde to butyl alcohol, but this cannot account for all the hydrogen formed. Labeling studies using deuterated aldehyde $C_2H_5CDO$ produced a mixture of products: $C_2H_6$ (40–45%), $C_2H_5D$ (5–10%), and $C_2H_4$ (50%). When another deuterated aldehyde, $CD_3CH_2CHO$, was used, many different products were observed. Intermolecular hydrogen transfer was postulated to account for the lack of product selectivity.

Vaska's complex, $IrCl(CO)(PPh_3)_2$, decarbonylates acyl halides catalytically at 78°C to give mixtures of olefins.[19] Aroyl halides are not decarbonylated, even at high temperatures, although oxidative addition is observed. The difference in behavior between the two kinds of acid halides is explained in terms of the greater stability of the metal–carbon bond in the aryl complex and the absence of a $\beta$-hydrogen in the aryl complex, making elimination of olefins impossible. A mechanism for the catalytic decarbonylation of acyl halides has been proposed[19] and is given in Scheme 1.

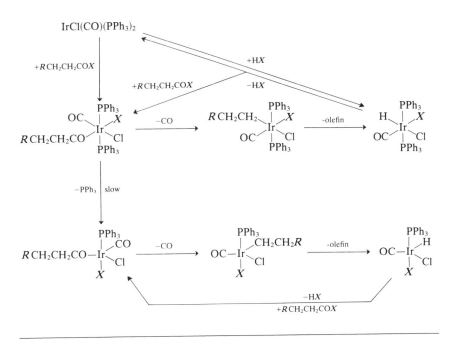

Scheme 1

The main reaction path is believed to be the second route in which PPh$_3$ has been eliminated. This accounts for the relatively long induction times. Acyl- and alkyl-iridium intermediates have been isolated, including some containing only one phosphine ligand.[19]

The best studied decarbonylation reagent is chlorotris(triphenylphosphine)rhodium(I), RhCl(PPh$_3$)$_3$, which decarbonylates aldehydes and acyl and aroyl halides under mild conditions (i.e., thermally in solution at temperatures below 100°C). The reactions are *stoichiometric* and are summarized by Equations 1–4.[20-24]

$$R\overset{O}{\overset{\|}{C}}H + RhCl(PPh_3)_3 \rightarrow RH + RhCl(CO)(PPh_3)_2 + PPh_3 \qquad (1)$$

$$R_2CHCR_2\overset{O}{\overset{\|}{C}}H + RhCl(PPh_3)_3 \rightarrow R_2C=CR_2 + H_2 + RhCl(CO)(PPh_3)_2 + PPh_3 \qquad (2)$$

$$R\overset{O}{\overset{\|}{C}}Cl + RhCl(PPh_3)_3 \rightarrow RCl + RhCl(CO)(PPh_3)_2 + PPh_3 \qquad (3)$$

$$R_2CHCR_2\overset{O}{\underset{\|}{C}}Cl + RhCl(PPh_3)_3 \rightarrow R_2C=CR_2 + HCl + RhCl(CO)(PPh_3)_2 + PPh_3 \quad (4)$$

The olefin producing reactions, Equations 2 and 4, occur when a $\beta$-H is present, but they are not equally important. While reaction 2 is of minor importance (e.g., decarbonylation of heptanal using $RhCl(PPh_3)_3$ yields 86% hexane and 14% 1-hexene),[22] acid chlorides will be completely converted to olefin if a $\beta$-hydrogen is present (Equation 4). Hence, $RhCl(PPh_3)_3$ is useful for the stoichiometric conversion of aldehydes into alkanes and olefins (if $\beta$-H is present) and of acid halides into alkyl or aryl halides and olefins (if $\beta$-H is present). Stoichiometric decarbonylation of general ketones using transition metal complexes has not been reported, although a brief mention of this possibility has appeared,[25] and several activated ketones have presumably been decarbonylated stoichiometrically.[26]

Decarbonylation of transition-metal carbonyl complexes is also affected by $RhCl(PPh_3)_3$. For example, CO is abstracted from $Mo(CO)_6$,[27,28] $W(CO)_6$,[27] and $Fe(CO)_5$[28] according to Equation 5:

$$RhCl(PPh_3)_3 + M(CO)_x \rightarrow M(CO)_{x-1}PPh_3 + RhCl(CO)(PPh_3)_2 \quad (5)$$

Alexander and Wojcicki[29,30] have used $^{13}CO$ to show that when an acyl transition metal carbonyl complex is decarbonylated, abstraction of the carbonyl ligand occurs according to Equation 6:

$$RhCl(PPh_3)_3 + 2(\eta^5\text{-}C_5H_5)Fe(CO)_2{}^{13}COR \rightarrow RhCl(CO)(PPh_3)_2$$
$$+ (\eta^5\text{-}C_5H_5)Fe(CO)(^{13}CO)R + (\eta^5\text{-}C_5H_5)Fe(PPh_3)(CO)^{13}COR \quad (6)$$

The product distribution depends on the ability of the $R$ group to migrate. A recent review[31] discusses the many other catalytic and stoichiometric reactions of $RhCl(PPh_3)_3$.

Catalytic decarbonylation of aldehydes using rhodium complexes of chelating diphosphine ligands has been studied (*vide infra*). These complexes give good catalytic activities (over 300 turnovers/hr for heptanal at 150°C) and are highly selective. The catalysts are stable for days, and turnovers in excess of $10^5$ have been achieved. Experiments on a variety of aldehydes have established the general usefulness of this reaction in organic synthesis (see Chapter 11, Section 3).

Even the observation of catalytic decarbonylation under ambient conditions has been reported.[32] The catalyst precursor, $Ru(PPh_3)_2tpp$ (tpp = dianion of tetraphenylporphyrin), is reported to decarbonylate phenylacetaldehyde at very rapid rates, giving in excess of $5 \times 10^4$ turnovers/hr at 50°C. Further research may yield even more active catalysts.

## 2. DISCUSSION OF DECARBONYLATION MECHANISM WITH $RhCl(PPh_3)_3$

A considerable amount of work has been performed over the last 15 years to determine the mechanism of acid chloride decarbonylation with $RhCl(PPh_3)_3$.[19,22–24,31,33,34] Although the discovery of aldehyde decarbonylation preceded that of acid chlorides,[35,36] much more time has been spent on the acid chloride system because it is more easily studied. Many intermediates have been isolated and characterized (see Table 1). Even though the mechanism of the catalytic reaction is not well understood, the mechanism for the stoichiometric decarbonylation of acid chlorides has been proposed. However, the generally accepted mechanism has recently been challenged.[37,38] In this section, we will first review the stoichiometric decarbonylation mechanism for acid chlorides, followed by the stoichiometric decarbonylation of aldehydes. Finally, the mechanism of catalytic decarbonlyation of acid chlorides and aldehydes will be discussed.

### 2.1. Stoichiometric Decarbonylation of Acid Chlorides

The general mechanism proposed for acid chloride stoichiometric decarbonylation[33,34,39,40] under mild conditions is given in Equations 7a–7e.

$$RhCl(PPh_3)_3 \underset{}{\overset{K \ll 1}{\rightleftharpoons}} RhCl(PPh_3)_2 + PPh_3 \quad (7a)$$

$$RhCl(PPh_3)_2 + RCOX \rightarrow cis\text{-}RCO(X)RhCl(PPh_3)_2 \quad (7b)$$

$$cis\text{-}RCO(X)RhCl(PPh_3)_2 \rightarrow trans\text{-}RCO(X)RhCl(PPh_3)_2 \quad (7c)$$

$$trans\text{-}RCO(X)RhCl(PPh_3)_2 \rightarrow R(X)RhCl(CO)(PPh_3)_2 \quad (7d)$$

$$R(X)RhCl(CO)(PPh_3)_2 \rightarrow RX + RhCl(CO)(PPh_3)_2 \quad (7e)$$

When $\beta$-hydrogens are present, i.e., $R = R'CH_2CH_2$, olefins and HCl are predominantly formed, as shown in Equation 7f:

$$(R'CH_2CH_2)(X)RhCl(CO)(PPh_3)_2 \rightarrow R'CH=CH_2 + HX + RhCl(CO)(PPh_3)_2 \quad (7f)$$

The only known exceptions (*vide infra*) to the above mechanism, Equations 7a–7f, occur when $R$ is an aryl group[37,38] or when $R$ is an $\alpha, \beta$-unsaturated species.[41] In the former case, decarbonylation is not observed under mild conditions, and in the latter case, phosphonium salts are produced. Recent work[38,39] has, however, substantiated the proposed mechanism for aliphatic and substituted aliphatic acid chlorides.

Table 1. Isolated or Observed Intermediates during the Stoichiometric Decarbonylation of Acid Chlorides Using $RhCl(PPh_3)_3$

| Acid chloride | Aroyl (or acyl) complex | Aryl (or alkyl) complex | Reference |
|---|---|---|---|
| MeCOCl | $RhCl_2(COMe)(PPh_3)_2$ | — | 20b |
| $CH_2ClCOCl$ | $RhCl_2(COCH_2Cl)(PPh_3)_2$ | — | 45 |
| EtCOCl | $RhCl_2(COEt)(PPh_3)_2$ | — | 20b |
| PrCOCl | $RhCl_2(COPr)(PPh_3)_2$ | — | 45 |
| $Me(CH_2)_4COCl$ | $RhCl_2[CO(CH_2)_4Me](PPh_3)_2$ | — | 1 |
| PhCOCl | $RhCl_2(COPh)(PPh_3)_2$ | $RhCl_2(CO)(C_6H_5)(PPh_3)_2$ | 20b, 24a |
| $p$-$ClC_6H_4COCl$ | $RhCl_2(p\text{-}COC_6H_4Cl)(PPh_3)_2$ | $RhCl_2(CO)(C_6H_4Cl)(PPh_3)_2$ | 24a |
| $p$-$NO_2C_6H_4COCl$ | $RhCl_2(p\text{-}COC_6H_4NO_2)(PPh_3)_2$ | $RhCl_2(CO)(C_6H_4NO_2)(PPh_3)_2$ | 24a |
| $p$-$MeOC_6H_4COCl$ | $RhCl_2(p\text{-}COC_6H_4OMe)(PPh_3)_2$ | $RhCl_2(CO)(C_6H_4OCH_3)(PPh_3)_2$ | 24a |
| $PhCH_2COCl$ | $RhCl_2(COCH_2Ph)(PPh_3)_2$ | $RhCl_2(CO)(CH_2C_6H_5)(PPh_3)_2$ | 24a |
| $PhCH_2CH_2COCl$ | $RhCl_2(COCH_2CH_2Ph)(PPh_3)_2$ | $RhCl_2(CO)(CH_2CH_2Ph)(PPh_3)_2$ | 46 |
| $threo$-$PhCHDCHDCOCl$ | $RhCl_2(threo\text{-}COCHDCHDPh)(PPh_3)_2$ | — | 46 |
| $C_6D_5CD_2CH_2COCl$ | $RhCl_2(COCH_2CD_2C_6D_5)(PPh_3)_2$ | — | 24c |
| $p$-$ClC_6H_4CH_2COCl$ | $RhCl_2(p\text{-}COCH_2C_6H_4Cl)(PPh_3)_2$ | $RhCl_2(CO)(CH_2C_6H_4Cl)(PPh_3)$ | 24a |
| $p$-$NO_2C_6H_4CH_2COCl$ | $RhCl_2(p\text{-}COCH_2C_6H_4NO_2)(PPh_3)_2$ | $RhCl_2(CO)(CH_2C_6H_4NO_2)(PPh_3)_2$ | 24a |
| $p$-$MeOC_6H_4CH_2COCl$ | $RhCl_2(p\text{-}COCH_2C_6H_4OMe)(PPh_3)_2$ | $RhCl_2(CO)(CH_2C_6H_4OCH_3)(PPh_3)_2$ | 24a |
| $C_8H_{17}COCl$ | $RhCl_2(COC_8H_{17})(PPh_3)_2$ | — | 22 |
| $Me(CH_2)_nCOCl$[a] | $RhCl_2[CO(CH_2)_nMe](PPh_3)_2$ | — | 1, 22 |
| $erythro$-MeCHPhCHPhCOCl | $RhCl_2(erythro\text{-}COCHPhCHPhMe)(PPh_3)_2$ | — | 24c |
| $threo$-MeCHPhCHPhCOCl | $RhCl_2(threo\text{-}COCHPhCHPhMe)(PPh_3)_2$ | — | 24c |
| (s)-(−)-α-$CF_3$CHPhCOCl | (s)-$RhCl_2(CO\text{-}\alpha\text{-}CF_3CHPh)(PPh_3)_2$ | (s)-$RhCl_2(CO)(\alpha\text{-}CF_3CHPh)(PEt_2Ph)_2$ | 24b |

[a] $n = 4, 5, 14, 16$

The stoichiometric decarbonylation reaction begins with the oxidative addition of acid chloride to $RhCl(PPh_3)_2$ (Equation 7b), which is presumably a solvent-stabilized, very reactive intermediate.[31] Tolman[42] and Halpern[43] have presented kinetic evidence for the importance of $RhCl(PPh_3)_2$ in the catalytic hydrogenation of olefins by $RhCl(PPh_3)_3$. In addition, the solvated species, $RhCl(S)(PPh_3)_2$ (where S = DMF, acetonitrile), was observed in the stoichiometric decarbonylation of aldehydes[44] (*vide infra*).

The 5-coordinate acyl complex has been the subject of many studies since it is isolable from the reaction mixture (see Table 1). Proton and phosphorus nmr studies show that the acyl complex isomerizes from a *cis*- to a *trans*-phosphine stereochemistry which have different chemical shifts and coupling constants[38,45] as shown in Table 2. In addition, the infrared spectra of the isomers are different in the Rh–Cl and carbonyl regions (e.g., when $R = CH_2Ph$, $\nu_{co} = 1720$ cm$^{-1}$ for the *cis* isomer and $\nu_{co} = 1770$ cm$^{-1}$ for the *trans* isomer).[38]

The conversion of the *cis* isomer to the *trans* isomer has been clearly demonstrated by monitoring the $^1$H and $^{31}$P nmr as a function of time.[34,38] However, without more detailed kinetic information, it is not certain that the migration of the carbonyl group to form the alkyl complex, as shown in Equation 7d, occurs exclusively from the *trans* isomer.

The detailed structure of these isomers remains somewhat clouded, however. Preliminary results from single crystal X-ray structural determinations of the intermediates are not in agreement. $RhCl_2(COCH_2CH_2Ph)(PPh_3)_2$ is reported to have square pyramidal geometry,[34] whereas $RhCl_2(COCH_2Ph)(PPh_3)_2$ is reported to have a trigonal bipyramidal structure.[33] The similarity in $^{31}$P nmr spectra of these two compounds makes it unlikely that the differences in reported solid-state structure are maintained in solution. In addition, both reports of structural determination were made in the preliminary stages of refinement and as such are subject

Table 2. $^1$H and $^{31}$P nmr data on Acyl-Rhodium Complexes, $RhCl_2(COR)(PPh_3)_2$

| | | $^1$H nmr | $^{31}$P nmr | | |
|---|---|---|---|---|---|
| R | Isomer | δ, ppm | δ, ppm | J(Rh–P, Hz) | Reference |
| $CH_3$ | cis | 3.37 (s) | 29.8 (d) | 145 | 34, 45 |
| $CH_3$ | trans | 2.49 (s) | 23.6 (d) | 108 | 34, 45 |
| $CH_2Ph$ | cis | 5.29 (s) | 27.3 (d) | 145 | 38, 45 |
| $CH_2Ph$ | trans | 4.26 (s) | 23.4 (d) | 108 | 38, 45 |

to significant uncertainty. Since no further reports of these structures have appeared from either group, this area merits further investigation.

Solution molecular weight determinations indicate that the acyl complex is monomeric,[24] not dimeric with chloro bridges. These 5-coordinate complexes must have an open coordination site for the next step in the reaction, alkyl migration, to occur, as given in Equation 7d.

Generally, migration occurs with retention of configuration at the α-carbon.[1,24,46,47] For example, decarbonylation of an optically active acid chloride, (S)-α-$^2$H-phenylacetyl chloride, yields S-benzyl-α-$^2$H-chloride with net retention of configuration.[47]

The elimination reaction, given in Equation 7e, is accompanied by a formal reduction in the oxidation state of the rhodium from (III) to (I). The stereochemistry of this reaction has also attracted substantial interest. If β-hydrogens are absent, the alkyl halide is eliminated with retention of configuration at the carbon.[47] If β-hydrogens are present, β-hydride elimination occurs as shown in Equation 7f, giving an alkene and hydrogen halide.[24,33,46,47] Investigations into the β-hydride elimination have shown that Saytzeff (rather than Hoffman) elimination is preferred.[24]

In order to further study the stereochemistry of this elimination, acid chlorides having asymmetric carbons at the α- and β-position were reacted with RhCl(PPh$_3$)$_3$. The alkyl complexes formed from erythro- and threo-2,3-diphenylbutanoyl chloride are given in Equation 8.[24c]

$$\begin{array}{cc} \text{erythro} & \text{threo} \end{array} \tag{8}$$

Product analysis shows that the E isomer is produced from erythro-2,3-diphenylbutanoyl chloride, and the Z isomer is produced from the threo-acid chloride. Both these results are consistent with *cis* β-hydride elimination.[24,33,46] Further studies on the elimination reaction have shown a large kinetic isotope effect ($k_H/k_D \approx 7$) for decarbonylation of PhCH$_2$CH$_2$COCl and PhCD$_2$CH$_2$COCl.[24] This result suggests that β-hydride elimination is the rate-determining step and that C–H bond scission is involved.[47] Failure to observe the alkyl complex led the authors to postulate a concerted elimination from the acyl complex.[24] However, the complex RhCl$_2$(CO)(CH$_2$CH$_2$C$_6$H$_5$)(PPh$_3$)$_2$ has been observed by another group[46] when a toluene solution of RhCl$_2$(COCH$_2$CH$_2$C$_6$H$_5$)(PPh$_3$)$_2$ was analyzed at 80°C by infrared spectroscopy ($\nu_{co} = 1996 \text{ cm}^{-1}$). Thus, it seems that *cis* β-hydride elimination as given in Equation 7e is most consistent with the available data.

The proposed mechanism (Equation 7) has been challenged recently[37,38] on two points. First, benzoyl chloride did not produce chlorobenzene under reaction conditions that should lead to stoichiometric decarbonylation, in contradiction of earlier work.[24a,48] Under mild conditions (30–80°C) benzoyl chloride did react with $RhCl(PPh_3)_3$, but the major product was an aryl-rhodium phosphine complex, $Rh(C_6H_5)Cl_2(PPh_3)_2$ (see Equation 9), not $RhCl(CO)(PPh_3)_2$ and $C_6H_5Cl$, as expected in Equation 7e:

$$Rh(C_6H_5)Cl_2(CO)(PPh_3)_2 \rightleftarrows Rh(C_6H_5)Cl_2(PPh_3)_2 + CO \qquad (9)$$

The previous workers have confirmed this result.[49]

The earlier kinetic investigation[24a] had used infrared spectroscopy and UV-VIS electronic absorption spectroscopy to monitor the concentration of metal carbonyl compounds. They had often observed the production of alkyl chlorides[24] or olefins plus HCl[24c,40] from alkyl-rhodium carbonyl complexes at low temperatures and had relied[49] on an earlier incorrect report of chlorobenzene production from benzoyl chloride.[48] Thus, they misinterpreted the products formed on disappearance of the aryl-rhodium carbonyl complex.[49]

This seemingly anomalous result can be understood by the increased stability of the metal–carbon bond in aryl complexes.[50,51] This added stability may give the aryl-rhodium carbonyl complex, $Rh(C_6H_5)Cl_2(CO)(PPh_3)_2$, another reaction path that is unfavorable for alkyl-rhodium carbonyl complexes. In fact, the stability of the metal–carbon bond was invoked to explain the difference in reactivity between alkyl- and aryl-acid chlorides with $IrCl(CO)(PPh_3)_2$.[19] Only the former were decarbonylated, even at elevated temperature. In addition, aryl-rhodium phosphine complexes of the type observed in the above report have been postulated[21] as intermediates in the catalytic decarbonylation of *para*-substituted benzoyl chlorides. For example, when *p*-chlorobenzoyl chloride was heated briefly with $RhCl(PPh_3)_3$, yellow rhodium complexes were obtained (but not isolated in analytical purity) that did not show any carbonyl absorptions in the infrared[21] and were postulated to be the aryl-rhodium complex $Rh(C_6H_4Cl)Cl_2(PPh_3)_2$.

The other exception to the general mechanism occurs when an $\alpha,\beta$-unsaturated acid chloride is used as the substrate. The presence of a $\beta$-hydrogen would allow the possibility of $\beta$-hydride elimination to occur, but the product would have a terminal carbon–carbon triple bond, which has not been observed previously. Apparently, reductive elimination of $R$–Cl is also too slow, and as a result phosphonium salts are the major product (Equation 10):

$$R\text{-CH}=CH_2\text{-}RhCl_2(CO)(PPh_3)_2 + PPh_3$$
$$\to RhCl(CO)(PPh_3)_2 + R\text{-CH}=CH_2\text{-}PPh_3Cl^- \qquad (10)$$

Both of these results are important in that they point out that the reductive elimination step is poorly characterized for these complexes. It is clear that the reaction will be sensitive to the relative bond strengths of reactants and products. The strength of the metal–carbon and metal–$X$ bond must be compared to the strength of bonds formed during the elimination reaction. Very little thermodynamic data exist for organometallic complexes, and so these comparisons are difficult to make. The general trend is that reductive elimination becomes less favored as one progresses in the series of products: olefin + $HX$ > $R$-Cl > phosphonium salts > ArCl. The study of reductive elimination reactions from these types of complexes merits further investigation.

## 2.2. Stoichiometric Decarbonylation of Aldehydes

The mechanism of aldehyde decarbonylation is thought to follow the established mechanism for acyl halide decarbonylation discussed in the previous section (Equation 7, where $X = H$). Several observations support this idea, even though intermediates are much more labile than those of the acid chloride system.

The only Rh(III) intermediate that has been isolated from the reaction sequence is given below.[52] This compound was prepared by reacting an aldehyde, 8-quinoline carboxaldehyde, with RhCl(PPh$_3$)$_3$. The ability of the aldehyde to form a chelate after oxidative addition has occurred (termed "chelate trapping" by the author) imparted sufficient stability to the compound to allow isolation and characterization. Prolonged heating in refluxing xylene yields the expected decarbonylation products. Other examples of oxidative addition of aldehydes to Rh(I) complexes are presented in Chapter 7, Section 4.

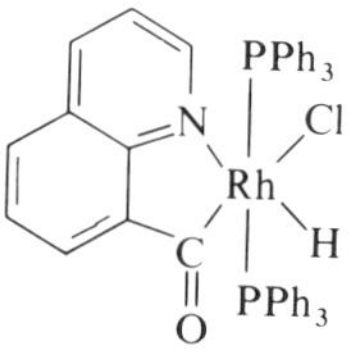

The stereochemistry of aldehyde decarbonylation has received much attention. Walborski and Allen[53] have shown that the decarbonylation of optically active aldehydes proceeds with 93% retention of configuration, as shown in Equation 11:

$$C_2H_5 - \underset{\underset{C_6H_5}{|}}{\overset{\overset{CH_3}{|}}{C}} - CHO + RhCl(PPh_3)_3 \longrightarrow C_2H_5 - \underset{\underset{C_6H_5}{|}}{\overset{\overset{CH_3}{|}}{C}} - H$$

$$+ RhCl(CO)(PPh_3)_2 + PPh_3 \tag{11}$$

Retention of configuation at the α-carbon was also observed with acid chlorides (*vide supra*), but aldehydes give a product of higher optical purity.[47,53]

Decarbonylation of deuteroaldehydes has been used for specific deuteration of compounds. The incorporation of deuterium into the products as given in reaction 12 occurs with high yield.[53,54]

$$\text{Ph} \underset{\text{Ph}}{\overset{}{\triangle}} \text{CDO} \atop \text{Me} + \text{RhCl(PPh}_3)_3 \rightarrow \text{Ph} \underset{\text{Ph}}{\overset{}{\triangle}} \text{D} \atop \text{Me} + \text{RhCl(CO)(PPh}_3)_2 + \text{PPh}_3 \quad (12)$$

The decarbonylation of other deuterated aldehydes provides deuterated products according to Equations 13 and 14.[15]

$$\text{RhCl(PPh}_3)_3 + \text{EtCDO} \rightarrow \text{EtD} + \text{RhCl(CO)(PPh}_3)_2 + \text{PPh}_3 \quad (13)$$

$$\text{RhCl(PPh}_3)_3 + \text{CD}_3\text{CH}_2\text{CHO} \rightarrow \text{CD}_3\text{CH}_3 + \text{RhCl(CO)(PPh}_3)_2 + \text{PPh}_3 \quad (14)$$

RhCl(PPh$_3$)$_3$ is known to isomerize double bonds during decarbonylation.[22] For example, α-methyl cinnamaldehyde is converted to *cis*- and *trans*-β-methylstyrene, as shown in Equation 15:

$$\underset{H}{\overset{Ph}{>}}C=C\underset{CHO}{\overset{CH_3}{<}} + \text{RhCl(PPh}_3)_3 \xrightarrow{140°C} \underset{H}{\overset{Ph}{>}}C=C\underset{H}{\overset{CH_3}{<}} + \underset{H}{\overset{Ph}{>}}C=C\underset{CH_3}{\overset{H}{<}}$$
$$\text{(91\%) cis} \qquad \text{(9\%) trans}$$
$$+ \text{RhCl(CO)(PPh}_3)_2 + \text{PPh}_3 \quad (15)$$

Isomerization does not occur.[54]

When β-hydrogens are present in the substrate aldehyde, olefins and H$_2$ are produced, but in substantially smaller quantities than observed with analogous acid chlorides. For example, decarbonylation of heptanal yields 14% hexene,[22] whereas exclusive formation of alkene is observed from similar acyl chlorides. This difference is due to the ease with which hydride transfer (as opposed to chloride transfer) occurs in the respective alkyl intermediates.[30]

The results discussed above lead to the conclusion that the mechanism of decarbonylation of aldehydes is very similar to that postulated for the decarbonylation of acid chlorides. However, kinetic studies of the reaction show that a different rate-limiting step is operative with aldehydes. With acid chlorides, the rate-limiting step is thought to be migration or reductive elimination, depending on the *R*-group.[24] A detailed kinetic study on the stoichiometric decarbonylation of aldehydes with RhCl(PPh$_3$)$_3$[44] has shown a different kinetic behavior. This result is consistent with the observation that Rh(III) intermediates are insoluble in the acid chloride system but not with aldehydes.

The kinetic study[44] was performed on $RhCl(PPh_3)_3$ and the chlorine-bridged dimer, $Rh_2Cl_2(PPh_3)_4$, with a variety of aldehydes. The progress of the reaction was monitored by UV-VIS electronic absorption spectroscopy at 25°C, and decarbonylation products $RhCl(CO)(PPh_3)_2$ and alkane were analyzed.

The study shows that the decarbonylation reaction is second-order, and a rapid pre-equilibrium step is observed when $RCHO$ forms an association complex with $RhCl(PPh_3)_3$, as given in Equation 16:

$$RCHO + RhCl(PPh_3)_3 \xrightleftharpoons{K_1} RhCl(RCHO)(PPh_3)_2 + PPh_3 \qquad (16)$$

$$RhCl(RCHO)(PPh_3)_2 \xrightarrow{k_2} RhCl(CO)(PPh_3)_2 + RH$$

For the reaction of propanal with $RhCl(PPh_3)_3$ at 25°C in benzene, $K_1 = 3.7 \pm 0.6 \times 10^{-3}$ and $k_2 = 8.7 \pm 2 \times 10^{-4}$ sec$^{-1}$. Phosphine inhibition was measured and the observed kinetics agree with the rate law given in Equation 17:

$$\frac{d[RhCl(CO)(PPh_3)_2]}{dt} = \left[\frac{k_2 K_1 [RCHO]}{K_1[RCHO] + [PPh_3]}\right][Rh]_{total} \qquad (17)$$

Two isobestic points were observed during the course of the reaction.

In order to study the reaction without excess phosphine, the rhodium dimer, $Rh_2Cl_2(PPh_3)_4$, was used as the reagent. Coordinating solvents (*DMF* or $CH_2CN = S$) were added to stabilize the formation of the solvated species, $RhCl(S)(PPh_3)_2$. Decarbonylation of propanal by this solution yielded the expected products, ethane and $RhCl(CO)(PPh_3)_2$. Again, the solution was monitored by UV–VIS spectroscopy, and isobestic points were observed that corresponded to mixtures of $RhCl(S)(PPh_3)_2$ and $RhCl(CO)(PPh_3)_2$ with no indication of intermediates.[44] The kinetic behavior was consistent with either a rapid pre-equilibrium step or the use of a steady-state approximation on $RhCl(RCHO)(S)(PPh_3)_2$.

The postulated mechanism is shown in Equations 18 and 19.

$$RhCl(S)(PPh_3)_2 + RCHO \underset{k_{-1}}{\overset{k_1}{\rightleftharpoons}} RhCl(RCHO)(S)(PPh_3)_2 \qquad (18)$$

$$RhCl(RCHO)(S)(PPh_3)_2 \xrightarrow{k_2} RhCl(CO)(PPh_3)_2 + S + RH \qquad (19)$$

Since the rate laws for the two mechanisms (rapid pre-equilibrium or steady-state approximation) are indistinguishable, the mechanism could not clearly be established.

The work[44] also provided activation parameters: for propanal, $\Delta H^{\pm} = 13.2 \pm 0.3$ kcal/mol, $\Delta S^{\pm} = -18 \pm 1$ e.u.; for $n$-butanal, $\Delta H^{\pm} = 12.9 \pm 1$ kcal/mol, $\Delta S^{\pm} = -19 \pm 3$ e.u. In addition, the kinetic isotope effect was determined for propanal: $k_H/k_D = 1.55$. These activation parameters are very similar to values measured for the oxidative addition of $H_2$ and $O_2$ to $IrCl(CO)(PPh_3)_2$.[55-57]

The conclusions, therefore, are that oxidative addition is the rate-limiting step and that C–H(D) bond breaking (or making) is involved to a small extent. This conclusion is further supported by the fact that electronic effects measured by analysis of a Hammett plot for *para*-substituted benzaldehydes gave $\rho = 1.5$, nearly identical to the value ($\rho = 1.4$) obtained for oxidative addition of *para*-substituted benzoylchlorides to $IrCl(CO)(PPh_3)_2$.[56]

Recent work by Kampmeier[39] has excluded the possibility of free-radical intermediates taking part in the decarbonylation of aldehydes. Decarbonylation of exo- and endo-5-norbornene-2-carboxaldehyde gives norbornene and nortricyclane, respectively. Reaction of citronella with $RhCl(PPh_3)_3$ gives only 2,6-dimethyl-2-heptene. Thus, the individual steps in the decarbonylation mechanism do not exhibit free-radical character and must be intramolecular. This work also reported a kinetic isotope effect, $k_H/k_D = 1.86$, for the decarbonylation of $C_6H_5CH_2CH(D)O$ by $RhCl(PPh_3)_3$. These results imply that the oxidative addition step is a concerted insertion of the metal into the C–H bond. These results are also consistent with the previous work[44] which suggests that oxidative addition is the rate-limiting step.

## 2.3. Catalytic Decarbonylation of Acid Chlorides and Aldehydes

At elevated temperature, $RhCl(PPh_3)_3$ becomes a catalyst (or catalyst precursor) for the decarbonylation reaction. Acid chlorides are reported to be decarbonylated catalytically at 170°C.[21,58,59] A recent report has lowered the temperature where catalytic activity is observed to 111°C,[38] but this appears to be more a result of lower detection limits than an observation of different reactivity. Aldehydes are reported to be catalytically decarbonylated above 200°C[21,59] by $RhCl(PPh_3)_3$ or $RhCl(CO)(PPh_3)_2$.

The essential question to be addressed is whether the rhodium complex re-enters the decarbonylation sequence previously discussed (Equation 7) or if a different pathway is available at elevated temperature. Loss of CO from $RhCl(CO)(PPh_3)_2$ is not observed even at high temperature (300°C) or under UV irradiation.[60] Therefore, mechanisms that have been proposed involve either oxidative addition to this complex[24,31,59] or loss of CO from the alkyl-rhodium carbonyl complex.[21] These two mechanisms

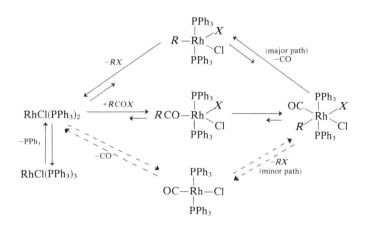

Scheme 2[24,31,59]

are illustrated in Schemes 2 and 3, respectively. Scheme 2 seems more consistent with experimental observations than does Scheme 3. Evidence supporting Scheme 2 includes:

a. $Rh(RCO)Cl_2CO(PPh_3)_2$ is observed upon reaction of $Rh(RCO)Cl_2(PPh_3)_2$ with CO at room temperature for 90 minutes.[38] Replacing the CO atmosphere with nitrogen causes dissociation of CO from the complex followed by reductive elimination of the product, $RCl$ ($R = C_6H_5CH_2$), to produce $RhCl(CO)(PPh_3)_2$.

Scheme 3[21]

b. $RhCl(CO)(PPh_3)_2$ is a catalyst for the carbonylation of certain alkyl halides,[59] as shown in Equation 20.

$$PhCH_2Cl \xrightarrow[\substack{100 \text{ atm, } 150°C \\ RhCl(CO)(PPh_3)_2}]{CO} PhCH_2COCl \qquad (20)$$

c. Oxidative addition of acid chlorides (e.g., acetyl chloride[61]) occurs, albeit slowly, at room temperature to $RhCl(CO)(PPh_3)_2$.

Scheme 3 is supported by the isolation from the reaction mixture of a yellow rhodium complex that does not exhibit carbonyl stretching frequencies in the infrared.[21] This compound, originally postulated to be $RhCl_2(Ar)(PPh_3)_2$, has been observed in recent work by Kampmeier, where $Ar = C_6H_5$ and $p\text{-}CH_3C_6H_4$.[38,39] It is possible that both mechanisms are correct for certain specific aldehydes or acid chlorides. Additional work on phosphine inhibition of the catalytic reaction, however, leads to the conclusion that another mechanism may be important at elevated temperature.[38]

Phosphine inhibition is observed in two types of experiments. Catalytic activities were measured for several catalyst precursors having different $PPh_3/Rh$ ratios.[38] Compounds studied were: $RhCl(PPh_3)_3$, $[PPh_3/Rh] = 3$; $RhCl(CO)(PPh_3)_2$ and $Rh_2Cl_2(PPh_3)_4$, $[PPh_3/Rh] = 2$; and $Rh_2Cl_2(CO)_2(PPh_3)_2$, $[PPh_3/Rh] = 1$. The second method was to simply add excess $PPh_3$. Results[38] for decarbonylation of phenylacetyl chloride at 111°C are listed in Table 3. The results show a consistent trend of increasing catalytic activity as the $PPh_3/Rh$ ratio decreases to 1.0.

This inhibition by $PPh_3$ can be accommodated better by Scheme 3 than by Scheme 2. Phosphine inhibition only occurs in the original

Table 3. The Catalytic Activity of Decarbonylation of Phenylacetyl Chloride at 111°C[a]

| Catalyst | $PPh_3^b$ (eq.) | $PPh_3/Rh$ | Time (h) | $PhCH_2Cl^c$ | Catalytic activity (turnovers/hr) |
|---|---|---|---|---|---|
| $RhCl(PPh_3)_3$ | — | 3 | 5.5 | 25 | 4.5 |
| $RhCl(CO)(PPh_3)_2$ | 1 | 3 | 6 | 24 | 4.0 |
| $RhCl(CO)(PPh_3)_2$ | — | 2 | 2 | 41 | 20.5 |
| $Rh_2Cl_2(PPh_3)_4$ | — | 2 | 2 | 37 | 18.5 |
| $Rh_2Cl_2(CO)_2(PPh_3)_2$ | — | 1 | 1.25 | 44 | 35.2 |

[a] Taken from Reference 38, $[\text{Catalyst}]_0 = 0.06\,M$; mol $[PhCH_2COCl]$/mol $[\text{Catalyst}]_0 = 45$; nitrogen sweep.
[b] Equivalents of $PPh_3$ added per mol $[\text{Catalyst}]_0$.
[c] Mol $[PhCH_2Cl]$/mol $[\text{Catalyst}]_0$.

dissociation of $PPh_3$ from $RhCl(PPh_3)_3$, not in the ensuing catalytic sequence given in Scheme 2. Scheme 3, on the other hand, postulated the catalytically active intermediate $RhCl(PPh_3)_2$, and the concentration of this species is related to the concentrations of $RhCl(PPh_3)_3$ and $PPh_3$ through the dissociation equilibrium constant. Increasing the concentration of $PPh_3$ would diminish the concentration of $RhCl(PPh_3)_2$, thus lowing the reaction rate. However, since $RhCl(PPh_3)_3$ is not observed upon quenching the reaction, this explanation seems tenuous.

These results are best explained if dissociation of $PPh_3$ from one of the intermediates in the above mentioned schemes led to another more reactive intermediate. This type of mechanism was proposed for decarbonylation of $RCH_2CH_2COX$ with $IrCl(CO)(PPh_3)_2$[19] and is presented in the lower pathway of Scheme 1. In order to explain phosphine inhibition observed with the rhodium system, a similar mechanism has been tentatively proposed.[39]

It seems reasonable that some alternate pathway may exist for the following reasons:

a. Migration reactions and reductive elimination are not inhibited by excess phosphine.[24,38]
b. The observed phosphine inhibiition is too large to be attributed to reversible association of $PPh_3$ to $RhCl(CO)(PPh_3)_2$,[38] which would have the result of diminishing the reaction rate by lowering the concentration of the metal complex required for oxidative addition.
c. While it is possible that excess $PPh_3$ may retard CO dissociation from $RhCl_2(RCO)(PPh_3)_2$, there is no precedent to show that this reaction should occur.

A plausible, but speculative, reaction scheme for the catalytic decarbonylation of acid chlorides (and aldehydes) that involves phosphine dissociation from one of the intermediates has been postulated.[24c,38] It is clear, though, that there are presently not enough facts to substantiate this hypothesis. Future work on the mechanism of catalytic decarbonylation using $RhCl(CO)(PPh_3)_2$ and other catalysts which investigates phosphine inhibition could be very informative.

## 3. CATALYTIC DECARBONYLATION OF ALDEHYDES WITH CATIONIC DIPHOSPHINE COMPLEXES OF Rh(I)

As discussed in the previous sections, the stoichiometric decarbonylation of aldehydes is effectively carried out by using $RhCl(PPh_3)_3$. It would be extremely advantageous if this reaction made catalytic at mild temperatures (i.e., ≤100°C), especially when the high cost of rhodium is considered. Catalytic decarbonylation of aldehydes using trans-RhCl(CO)(PPh$_3$)$_2$ requires very high temperatures, and at 178°C a catalytic activity of only 10 turnovers per hour is observed.[62] This temperature is too high to be of practical synthetic use. Early attempts at making this reaction catalytic at mild temperatures involved methods that were designed to labilize CO from trans-RhCl(CO)(PPh$_3$)$_2$, thereby regenerating the reactive catalyst "RhCl(PPh$_3$)$_2$" (see Equation 7). For example, photolysis of trans-RhCl(CO)(PPh$_3$)$_2$ was examined, but unfortunately the CO ligand was not photolabilized.[60] Other attempts that met with some success involved the use of solvent-stabilized cationic complexes of the types [Rh(PPh$_3$)$_2$]$^+$ and [Rh(PPh$_3$)$_2$(CO)]$^+$.[65] The latter complex reversibly binds CO. The increased lability of CO compared with that of the neutral complex trans-RhCl(CO)(PPh$_3$)$_2$ is presumably due in part to the lowered effective basicity of rhodium. [Rh(PPh$_3$)$_2$]$^+$ was shown to decarbonylate benzaldehyde at 100°C catalytically and with a rate considerably faster than that obtained using trans-RhCl(CO)(PPh$_3$)$_2$.[66] Unfortunately, this catalyst system decomposed readily at 100°C and therefore the reaction was not synthetically useful.[66]

Cationic complexes of Rh(I) with chelating diphosphine ligands have been known for some time but only recently have their catalytic properties been examined (see Chapter 4).[62,64,66–70] Complexes of the types [Rh(P–P)$_2$]$X$ and [Rh(P–P)]$X$, where $X$ = Cl or BF$_4$ and P–P = Ph$_2$P(CH$_2$)$_n$PPh$_2$ (named dppm, dppe, dppp, and dppb for $n$ = 1, 2, 3, and 4, respectively), are known to bind CO weakly and reversibly, and in the case of [Rh(dppe)$_2$]$^+$ the reaction with CO does not even give a detectable adduct. The fact that these complexes require *cis* phosphine stereochemistry is important and results in major reactivity differences compared to their triphenylphosphine analogues.[71] In the case of aldehyde decarbonylation, this difference is quite remarkable. For example, the catalytic decarbonylation of benzaldehyde at 150°C using [Rh(dppp)$_2$]BF$_4$ gives a catalytic activity for benzene production of $1.1 \times 10^2$ turnovers per hour, whereas an activity of only 0.60 turnovers per hour was obtained using trans-RhCl(CO)(PPh$_3$)$_2$.[64,66] Data for the catalytic decarbonylation of benzaldehyde using diphosphine and triphenylphosphine complexes of Rh(I) is presented in Table 4.

The major points concerning the catalytic decarbonylation of benzaldehyde using diphosphine complexes are: (i) the activities are significantly larger than with RhCl(PPh$_3$)$_3$; (ii) the activities show a marked dependence on chelate ring size with the order in activity dppp > dppe > dppb > dppm observed at all temperatures; (iii) the activities using [Rh(P–P)$_2$]$^+$ are approximately two times larger than with [Rh(P–P)]$^+$; (iv) the activiites are independent of the counter ion type (Cl$^-$, BF$_4^-$, or PF$_6^-$); (v) the activities

Table 4. Catalytic Decarbonylation of Benzaldehyde Using Rhodium Catalysts[a]

| Catalyst | Catalytic activity (moles benzene/moles catalyst/hour)[b] | Temperature °C |
|---|---|---|
| RhCl(PPh$_3$)$_3$ | Stoichiometric | 115 |
| [Rh(dppm)$_2$]$^+$ | 0.40 | 115 |
| [Rh(dppe)$_2$]$^+$ | 3.6 | 115 |
| [Rh(dppp)$_2$]$^+$ | 11 | 115 |
| [Rh(dppb)$_2$]$^+$ | 1.2 | 115 |
| RhCl(PPh$_3$)$_3$ | 0.60 | 150 |
| [Rh(dppp)$_2$]$^+$ | $1.1 \times 10^2$ | 150 |
| Rh(dppp)Cl | 30 | 145 |
| [Rh(dppp)]BF$_4$ | 40 | 145 |
| [Rh(dppp)$_2$]$^+$ | $1.1 \times 10^3$ | 178 |
| [Rh(dppe)$_2$]$^+$ | $2.1 \times 10^2$ | 178 |
| [Rh(dppm)$_2$]$^+$ | $7.4 \times 10^1$ | 178 |
| RhCl(PPh$_3$)$_3$[c] | $1.0 \times 10^1$ | 178 |

[a] Catalyst concentrations between $1 \times 10^{-4}$ and $1 \times 10^{-3}$ M and neat benzaldehyde as the solvent. (See References 62 and 66 for details.)
[b] Values for benzene production are averaged over a 30–40 hr period and first-order dependence on [catalyst] is assumed.
[c] Same results using trans-RhCl(CO)(PPh$_3$)$_2$.

for the [Rh(P–P)$_2$]$^+$ complexes remain approximately constant for at least several days, whereas the [Rh(P–P)]$^+$ catalyst systems are less stable; and (vi) the [Rh(dppp)$_2$]$^+$ system shows catalytic behavior at temperatures as low as 100°C (activity = 3 turnovers hour$^{-1}$). Since the [Rh(dppp)$_2$]$^+$ complex is the best diphosphine catalyst, the remaining discussion will concentrate on this system. A wide variety of aldehydes have been decarbonylated using this catalyst. Results are presented in Table 5.

In order to achieve the activities tabulated, it is important to run the reactions under a continuous purge of dinitrogen. If the concentration of CO is allowed to build up, the reactions become significantly inhibited. This observation has important mechanistic implications (*vide infra*) and is of great practical importance. Also, oxygen irreversibly destroys the catalyst, so the gas purge must utilize nitrogen that has been purified (passage through an activated BASF catalyst column works well). The data

Table 5. Catalytic Decarbonylation of Aldehyde Using $[Rh(dppp)_2]BF_4$ [66]

| Aldehyde | Observed volatile product | Temperature, °C | Catalytic activity (turnovers/hr) |
|---|---|---|---|
| Heptanal | Hexane | 150 | $3.3 \times 10^2$ |
| 2-Ethylbutanal | Pentane | 115 | 5.0 |
| 2-Phenyl-2-methylbutanal | 2-Phenylbutane | 180 | 1.3 |
| 2-Phenylpropanal | Ethylbenzene | 150 | $6.7 \times 10^1$ |
| Cinnamaldehyde | Styrene | 150 | $1.4 \times 10^2$ |
| α-Methylcinnamaldehyde | cis-β-Methylstyrene | 150 | $1.6 \times 10^1$ |
| Benzaldehyde | Benzene | 150 | $9.0 \times 10^1$ |
| para-Tolylbenzaldehyde | Toluene | 150 | $3.4 \times 10^1$ |
| para-Methoxybenzaldehyde | Anisole | 150 | $1.0 \times 10^1$ |
| para-Chlorobenzaldehyde | Chlorobenzene | 150 | 1.0 |

in Table 5 illustrate the general usefulness of this catalyst. Important points are that the reaction shows a high degree of selectivity and that the activity is much lower for sterically hindered tertiary aldehydes. The selectivity is best illustrated by noting that heptanal and 2-phenylpropanal yield only saturated products. Decarbonylation of these aldehydes using $RhCl(PPh_3)_3$ at similar temperatures produces significant amounts of hexene and styrene, respectively. Importantly, decarbonylation of heptanal using the monodiphosphine complex $[Rh(dppp)]BF_4$ is not selective. In this case hexene and hexane are produced in a 1:2 mole ratio at 150°C. The catalytic decarbonylation of the optically active aldehyde (−)-(R)-2-phenyl-2-methylbutanal using $[Rh(dppp)]BF_4$ at 165°C produced (+)-(S)-2-phenylbutane with 100% retention of absolute configuration.[72]

For most aldehydes studied, the decarbonylation reaction occurs with high product yields (yields based on aldehyde). Typical data are shown in Table 6 using $[Rh(dppp)_2]^+$ as the catalyst. However, the saturated aldehydes, heptanal and 2-ethylbutanal, are decarbonylated to hexane and pentane, respectively, in rather low yield (but with good catalytic activities, see Table 5). Further experiments revealed that the low yields observed for these aldehydes are due to thermal decomposition, which also occurs in the absence of catalyst. Even lower yields (based on aldehyde) are obtained for these aldehydes using $RhCl(PPh_3)_3$ at the same temperature, since with this complex decarbonylation activities are much smaller.

Table 6. Yield and Conversion for Various Aldehydes[a]

| Aldehyde (mmol) | Time | Temp., °C | Product | Yield, %[b] | Conversion, %[b] |
|---|---|---|---|---|---|
| Benzaldehyde (2.45) | 2 days | 140 | Benzene | 79 | 100 |
| Cinnamaldehyde (4.00) | 2 days | 118 | Styrene | 95 | 100 |
| 2-Phenylpropanal (64.0) | 1.5 days | 175 | Ethylbenzene | 83 | 93 |
| Heptanal (1.85) | 2 days | 118 | Hexane | 39 | 46 |
| 2-Ethylbutanal (60.8) | 1.5 days | 110 | Pentane | 15 | 43 |

[a] 0.02 to 0.05 mmol catalyst, $Rh(dppp)_2^+$ in solvent $m$-xylene or $\alpha$-methyl-naphthalene.
[b] % yield = (moles decarbonylation product/moles aldehyde added) × 100; % conversion = (moles decarbonylation product/moles aldehyde consumed) × 100.

### 3.1. Discussion of Mechanism with $[Rh(P-P)_2]^+$ Complexes

In order to gain insight into the mechanism of this catalytic reaction a variety of experiments have been carried out. Monitoring of the reaction during actual catalytic conditions by $^{31}P$ nmr and UV-VIS absorption spectroscopy reveals that only the complex $[Rh(P-P)_2]^+$ is present in any significant concentration (only cases where P-P = dppp and dppe and aldehyde = benzaldehyde have been examined in detail). Upon quenching the reaction by the rapid precipitation of rhodium species (via addition of cold pentane), the complexes $[Rh(dppp)_2]^+$ and $[Rh(dppp)_2CO]^+$ are observed, and in the case of the dppe system only $[Rh(dppe)_2]^+$ is observed. The original catalytic activity can be restored by reusing the complexes isolated from the reaction in a new experiment. During at least a two-day period of running the catalytic reaction (benzaldehyde → benzene), there are no signs of Rh(III) intermediates and catalyst decomposition is minimal. Identical activities are obtained using $[Rh(dppp)_2CO]^+$ or $[Rh(dppp)_2]^+$ as the catalyst.

Deuterium-labeling experiments have been carried out. Deuterobenzaldehyde, $C_6H_5CDO$, and $p$-tolualdehyde, $CH_3C_6H_4CHO$, were mixed in equimolar amounts and decarbonylated at 140°C using $[Rh(dppp)_2]^+$. Toluene and benzene were produced simultaneously, separated by GLC and analyzed by mass spectroscopy (MS). The isotopic purities of $C_6H_5CH_3$ and $C_6H_5D$ were determined to be 100 ± 1% and 100 ± 4%, respectively, and therefore all possible processes that are intermolecular in aldehyde can be excluded. This experiment, along with the one that used an optically active aldehyde (vide supra), argues against a free-radical chain mechanism. The deuterium isotope effect ($k_H/k_D$) for $C_6H_5C(H, D)O$ was 1.6 ± 0.1 using $[Rh(dppp)_2]^+$ at 140°C. A deuterium isotope effect of 1.55 has been observed for the stoichiometric decarbonylation of propionaldehyde at 25°C with RhCl(PPh$_3$)$_2$(*DMF*), and this result was used to support the theory that oxidative addition of the aldehyde is the rate-determining step.[44]

The above results are consistent with a mechanism that involves oxidative addition of the aldehyde, *R*-group migration (deinsertion), and reductive elimination of RH. The active catalyst is then regenerated by rapid CO loss from [Rh(dppp)$_2$CO]$^+$. Loss of CO from this complex is known to be a facile process.[66] Also, since Rh(III) intermediates are not observed, the migration and reductive elimination steps cannot be rate-limiting. There are several problems with this simple mechanistic picture. First, the *para*-substituted benzaldehyde substituent effect on activity does not follow a simple trend as would be expected if oxidative addition was rate-limiting. With [Rh(dppp)$_2$]$^+$ the trend in $k_{obs}$, where rate = $k_{obs}$[complex], is *para*-H > –CH$_3$ > –OCH$_3$ > –Cl. In cases where oxidative addition is known to be rate-limiting, the observed trend in rate constant is *para*-Cl > –H > –CH$_3$ > –OCH$_3$.[44,73] Also, the decarbonylation of benzaldehyde is inhibited by the presence of added reagents such as CO, benzonitrile, PEtPh$_2$, and dppp. This is not unexpected for [Rh(dppp)$_2$]$^+$, where the 5-coordinate adducts [Rh(dppp)$_2$L]$^+$ are formed (*vide infra*),[64] thereby diminishing the active catalyst concentration. However, inhibition also occurs using [Rh(dppe)$_2$]$^+$ as catalyst, although this complex does not form 5-coordinate adducts with these ligands. Finally, the absence of a large kinetic isotope effect suggests that C–H bond breaking or making may not be important in the transition state.

The rate of benzene production from neat benzaldehyde solution was measured as a function of [Rh(dppp)$_2$]$^+$ concentration. The reaction is first-order such that

$$\text{rate} = d[\text{benzene}]/dt = k_{obs}[\text{Rh(dppp)}_2{}^+].$$

The rate of benzene production was also measured as a function of benzaldehyde concentration while keeping the concentration of the catalyst constant. A plot of the rate constant, $k_{obs}$, vs. benzaldehyde concentration is shown in Figure 11.1. At low aldehyde concentrations the rate is first-order in (benzaldehyde), but as the aldehyde concentration is increased a saturation effect is observed. This behavior can be interpreted in terms of a rapid pre-equilibrium followed by a slow rate-limiting step as shown by Equation 21.[64]

$$\text{Rh(dppp)}_2{}^+ + \text{PhCHO} \underset{\text{fast}}{\overset{K}{\rightleftharpoons}} (\text{PhCHO})\text{Rh(dppp)}_2{}^+ \tag{21}$$

$$(\text{PhCHO})\text{Rh(dppp)}_2{}^+ \xrightarrow[\text{slow}]{k} (\text{PhCO})(\text{H})\text{Rh(dppp)}_2{}^+$$

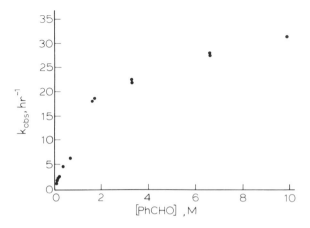

Figure 11.1 Plot of $k_{obs}$ for benzene production vs. benzaldehyde concentration at 150°C using [Rh(dppp)$_2$]BF$_4$ as catalyst.

The rate expression for this mechanism is

$$k_{obs} = \frac{kK[\text{PhCHO}]}{K[\text{PhCHO}] + 1}$$

which requires

$$\frac{1}{k_{obs}} = \frac{1}{k} + \frac{1}{kK}\left(\frac{1}{[\text{PhCHO}]}\right)$$

A plot of $1/k_{obs}$ vs. $1/[\text{PhCHO}]$ is linear and is shown in Figure 11.2. An analysis of the data in terms of this rate law gives $k = 1.1 \times 10^2$ hr$^{-1}$ and $K = 0.12\ M^{-1}$ at 150°C. Additional support for the mechanism shown by Equation 21 is the direct observation of the aldehyde adduct (PhCHO)Rh(dppp)$_2^+$ by low-temperature $^{31}$P nmr.[64,74] At temperatures above 25°C, this adduct is not observable by $^{31}$P nmr due to its nonrigid nature; however, its presence may be inferred.

Although the direct observation of [(PhCHO)Rh(dppp)$_2$]$^+$ lends support to Equation 21, the possibility exists that this species is unimportant kinetically. For example, it is possible that the kinetically important intermediate contains a monodentate dppp ligand such as [(PhCHO)Rh(dppp)-(dppp*)]$^+$, where dppp* is monodentate. The rate law for this case would be identical, and since the establishment of the pre-equilibrium is fast such an intermediate is likely to be unobservable by $^{31}$P nmr. The actual isolation or observation of "intermediates" in catalytic reactions is known to often lead to erroneous mechanistic predictions (see Chapter 4).[75]

The mechanism of Equation 21 is plausible and may indeed be operative for [Rh(dppp)$_2$]$^+$; however, several pieces of data are a cause for

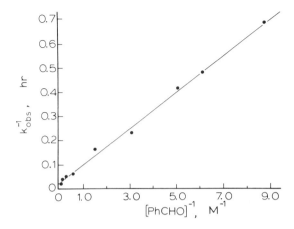

Figure 11.2. Plot of $k_{obs}^{-1}$ vs. $[PhCHO]^{-1}$ for benzaldehyde decarbonylation using $[Rh(dppp)_2]BF_4$ as catalyst at 150°C.

concern. First, the lack of a large kinetic isotope effect may be inconsistent with C–H bond breaking being involved in the rate-limiting step as required by Equation 21. Second, if the reaction mechanism is the same for $[Rh(dppp)_2]^+$ and $[Rh(dppe)_2]^+$, the rate inhibition by CO (see Table 7) is inconsistent with Equation 21 for the dppe system. $[Rh(dppp)_2]^+$ forms a

Table 7. Inhibition in the Rate of Catalytic Decarbonylation of Benzaldehyde[a]

| Catalyst | Temperature, °C | Inhibitor | Catalytic activity (turnovers/hour) |
| --- | --- | --- | --- |
| $[Rh(dppp)_2]^+$ | 145 | None | 60 |
| $[Rh(dppp)_2]^+$ | 145 | Benzonitrile (5.0 $M$) | 3.3 |
| $[Rh(dppp)_2]^+$ | 145 | CO (saturated) | 0.1 |
| $[Rh(dppp)_2]^+$ | 145 | $PEtPh_2$ (0.2 $M$) | 0.2 |
| $[Rh(dppp)_2]^+$ | 115 | None | 11 |
| $[Rh(dppp)_2]^+$ | 115 | $PPh_3$ (0.05 $M$) | 7 |
| $[Rh(dppp)_2]^+$ | 115 | dppp (0.03 $M$) | 0.1 |
| $[Rh(dppe)_2]^+$ | 145 | None | 20 |
| $[Rh(dppe)_2]^+$ | 145 | CO (saturated) | 1.5 |

[a] For conversion of benzaldehyde into benzene, using [catalyst] = 1 mM and benzaldehyde as solvent, see Reference 66 for details.

carbonyl adduct, and the inhibition observed in the rate of benzaldehyde decarbonylation can be explained by the competitive equilibria shown in Equation 22:

$$Rh(dppp)_2^+ + PhCHO \underset{}{\overset{K_1}{\rightleftharpoons}} (PhCHO)Rh(dppp)_2^+ \quad (22)$$

$$Rh(dppp)_2^+ + CO \underset{}{\overset{K_2}{\rightleftharpoons}} (CO)Rh(dppp)_2^+$$

$K_2$ is larger than $K_1$, since the carbonyl stretch in the IR is very intense when recorded using benzaldehyde saturated with CO as solvent. The problem is that in the case of $[Rh(dppe)_2]^+$, a CO adduct is not observed by IR. Also, the low-temperature $^{31}P$ nmr of $[Rh(dppe)_2]^+$ recorded using benzaldehyde and acetone as solvents shows no evidence for adduct formation. Therefore, the mechanism shown by Equation 21 is inconsistent with the facts for $[Rh(dppe)_2]^+$. An alternate mechanism, which is consistent with all of the data for both catalysts, is shown in Equation 23.

$$Rh(p-p)_2^+ \underset{k_{-1}}{\overset{k_1}{\rightleftharpoons}} Rh(p-p)(p-p^*)^+ \quad (23)$$

$$Rh(p-p)(p-p^*)^+ + PhCHO \underset{k_{-2}}{\overset{k_2}{\rightleftharpoons}} (PhCHO)Rh(p-p)(p-p^*)^+$$

$$(PhCHO)Rh(p-p)(p-p^*)^+ \underset{fast}{\overset{k_3}{\longrightarrow}} Rh(PhCO)(H)(p-p)(p-p^*)^+$$

where p-p* = monodentate diphosphine (dppp or dppe)

The rate expression for this mechanism, assuming a steady-state concentration of the undetectable intermediate $[Rh(dppp)(dppp^*)]^+$ and assuming that $k_{-2}$ is negligibly small, is

$$k_{obs} = k_1 k_2 [PhCHO]/k_{-1} + k_2 [PhCHO]$$

which requires

$$\frac{1}{k_{obs}} = \frac{1}{k_1} + \frac{k_{-1}}{k_1 k_2} \frac{1}{[PhCHO]}$$

This rate expression is experimentally indistinguishable from the one derived from Equation 21, since a plot of $1/k_{obs}$ vs. $1/[PhCHO]$ is linear for both mechanisms. From the data shown in Figure 11.2 and assuming the mechanism in Equation 23, the values of $k_1$ and $k_{-1}/k_2$ are $1.1 \times 10^2$ hr$^{-1}$ and 8.5 $M$, respectively.

The major difference between the mechanisms of Equations 21 and 23 is that in the former the rate-determining step is oxidative addition and the aldehyde adduct concentration builds up, while in the latter Rh–P bond rupture becomes the slow step as [aldehyde] increases. The absence of a large kinetic effect for the $[Rh(dppp)_2]^+$ system favors the latter mechanism; however, the magnitude of such effects is not well understood in organometallic chemistry. Also, the inhibition by CO in the $[Rh(dppe)_2]^+$ system is explained by the latter mechanism as a competition between CO and aldehyde for the undetectable intermediate $[Rh(dppe)(dppe^*)]^+$. Therefore, the failure to observe a CO adduct in this system is explained, since the concentration of $[(CO)Rh(dppe)(dppe^*)]^+$ needed for effective inhibition would be very low and unobservable. It should be realized that Equation 23 incorporates the features of Equation 21. For example, if $k_3 \ll k_1, k_{-1}, k_2, k_{-2}$, we have the essential features of Equation 21, except that a monodentate dppp ligand is implicated in the aldehyde adduct. The relative magnitudes of these rate constants may depend significantly on the ligand and aldehyde type, and it is possible that $k_3 \ll k_2, k_{-2}$, so that the plot of Figure 11.2 will give values of $k_1$ and $k_{-1}/k_3$. In fact, this latter case is probably the most reasonable by analogy to results on the stoichiometric decarbonylation of aldehydes using $RhCl(PPh_3)_3$.[44] In this case the initial $PPh_3$ dissociation and aldehyde adduct formation occur much faster than oxidative addition (rapid pre-equilibrium and slow rate-limiting step).

In principle, the relative magnitudes of the specific rate constants can be determined by complete kinetic analyses for the dppe and dppp systems using several *para*-substituted benzaldehydes. Providing that all of these systems exhibit the same mechanism, this series of experiments should permit determinations of the specific rate and/or equilibrium constants of Equation 23. Although the observed *para*-substituent effect on $k_{obs}$ for the decarbonylation using $[Rh(dppp)_2]^+$ (Table 5) can be rationalized by either mechanism, a detailed comparison of the specific rate constants and equilibrium constants should permit a mechanistic distinction. These experiments have not yet been carried out.

In light of the above arguments, an overall mechanistic scheme may be postulated. Such a scheme is shown in Figure 11.3. Note that in this scheme the steps of both Equations 21 and 23 are incorporated. For Equation 21, $k_3$ is rate-limiting whereas for Equation 23, $k_1$ and $k_2$ are the slow steps. The 5-coordinate CO and aldehyde adducts are only observed in the dppp system, while all species with monodentate diphosphine ligands are unobservable. However, there are many examples in the literature where metal complexes having monodentate diphosphine ligands have been isolated and characterized.[76-79] This scheme provides an hypothesis for future experiments.

There is good precedence for the oxidative addition of aldehydes to metal complexes.[52,76,80] The migration and reductive elimination steps are fast and are consistent with the high selectivity observed with the dppp system (*vide supra*) and by analogy to known organometallic reactions.[81]

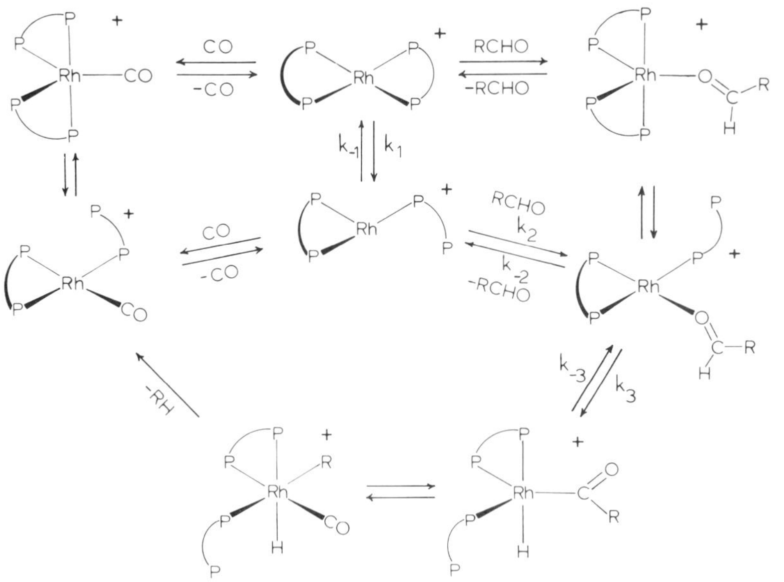

Figure 11.3. Possible mechanism for the catalytic decarbonylation of aldehydes using [Rh(P–P)$_2$]$^+$ as catalyst where (P–P) = dppp or dppe.

For example, the decarbonylation of heptanal using [Rh(dppp)$_2$]$^+$ gives hexane as the only volatile product, whereas both hexane and 1-hexene are observed using [Rh(dppp)]$^+$ or RhCl(PPh$_3$)$_3$. In order for hexene to be produced, a $\beta$-hydride abstraction must occur from an intermediate that contains Rh–R. As shown in Figure 11.3, the species that contains the Rh–R grouping [Rh(R)(H)(CO)(dppp)(dppp*)]$^+$ is coordinatively saturated, and a $\beta$-hydride abstraction is highly unlikely. Even if CO dissociation occurs or another Rh–P bond ruptures in this intermediate, the dangling P atom would quickly resaturate the coordination. For monodiphosphine complexes or for PPh$_3$ complexes, the analogous reaction intermediates are coordinatively unsaturated and $\beta$-H abstraction may easily occur, leading to olefin and H$_2$ production as shown in Equation 24.

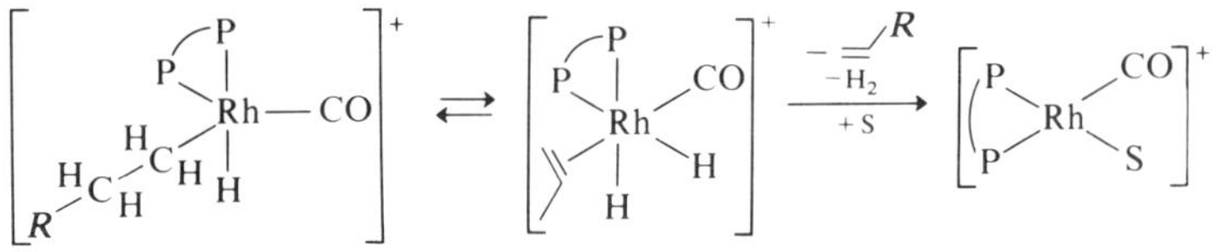

$$\left[ \begin{array}{c} P \\ P-Rh-CO \\ H_2C-CH \\ R-CH \end{array} \right]^+ \rightleftharpoons \left[ \begin{array}{c} P \\ P-Rh \\ H \end{array} \begin{array}{c} CO \\ H \end{array} \right]^+ \xrightarrow[+S]{-\overset{R}{=}} \left[ \begin{array}{c} P \\ P-Rh \\ \end{array} \begin{array}{c} CO \\ S \end{array} \right]^+ \quad (24)$$

S = solvent

The marked rate dependence on $n$, the number of methylene groups in the chelate backbone (Table 5), warrants some discussion. The rate increases as a function of $n$ up to $n = 3$ but falls off significantly for $n = 4$. A similar trend has been noted in other catalytic reactions using diphosphine ligands,[70,82,83] and an understanding of this effect will obviously involve both steric and electronic factors. For the catalytic decarbonylation of aldehydes the rate of Rh–P bond rupture may be important (Figure 11.3). The available evidence suggests that 5-membered chelate rings are more inert to bond rupture than 6-membered chelate rings.[84] This could explain the increase in rate ongoing from dppe to dppp. Additional data which supports this is that $[Rh(dppe)_2]^+$ does not react with $H_2$ (1 atm pressure), while $[Rh(dppp)_2]^+$ reacts with $H_2$ readily, giving a dihydride.[70] It is expected that the 4- and 7-membered chelate ring cases will be even more labile. In the case of dppb, bimetallic complexes with a dppb ligand bridging both metal centers are easily formed.[85] Indeed, the true form of $[Rh(dppb)_2]^+$ in aldehyde or acetone solution is complex, and high concentrations of bimetallic species have been implicated.[74] The single crystal X-ray structure of $[Rh(dppb)_2]BF_4$ shows a highly distorted coordination geometry (twisted $ca.$ 44% towards tetrahedral) with one unusually long Rh–P bond.[74] Also, the reaction of this complex with CO gives the bimetallic dppb-bridged complex $[Rh_2(CO)_4(dppb)_3]^{2+}$.[85] Clearly, the slow rate of decarbonylation using $[Rh(dppb)_2]^+$ could be due to the presence of bimetallic species. Although it is expected that $[Rh(dppm)_2]^+$ should be less stable than its dppe and dppp analogues, it is possible that a monodentate dppm ligand lacks the flexibility to swing away from the metal, thus enabling the transformations shown in Figure 11.3. It is likely that a delicate balance is needed between Rh–P bond lability, chelate ring flexibility, and thermodynamic stability of less reactive species.

## 4. CATALYTIC DECARBONYLATION OF ALDEHYDES USING CATIONIC DIPHOSPHINE COMPLEXES OF Ir(I)

Catalytic decarbonylation of benzaldehyde using several iridium complexes has also been examined.[72] Results of these experiments are shown in Table 8. The main points to be made here are: (i) $[Ir(P-P)_2]^+$ catalysts have activities that are $ca.$ twenty times lower than their Rh analogs; (ii) the iridium mono-diphosphine catalysts are better than the *bis*-diphosphine Ir catalysts (opposite trend noted using rhodium, see Table 5); and (iii) $IrCl(CO)(PPh_3)_2$ is a much better catalyst than $RhCl(CO)(PPh_3)_2$ and is also better than most of the iridium diphosphine catalysts. The results for the $[M(P-P)_2]^+$ catalysts may be explained in terms of the proposed mechanistic scheme in Figure 11.3. Since Ir–P bonds should be stronger than Rh–P bonds, the value of $k_1$ will be smaller for the Ir catalysts, thus

Table 8. Catalytic Decarbonylation of Benzaldehyde Using Iridium Catalysts[a]

| Catalyst | Catalytic activity (moles benzene/moles catalyst/hr) | Temperature (°C) |
|---|---|---|
| $[Ir(dppe)_2]^+$ | 2.5 | 178 |
| $[Ir(dppp)_2]^+$ | 4.4 | 178 |
| $[Ir(dppb)_2]^+$ | 4.2 | 178 |
| $[Ir(dppp)_2CO]^+$ | 4.6 | 178 |
| $[Ir(dppp)(cod)]BF_4$[b] | 55 | 178 |
| $[Ir(dppp)(cod)]BF_4$[b] | 27 | 150 |
| $[Ir(dppp)]BF_4$[c] | $1.3 \times 10^2$ | 178 |
| $[Ir(dppp)]BF_4$[c] | 39 | 150 |
| $IrCl(CO)(PPh_3)_2$ | 66 | 178 |
| $RhCl(CO)(PPh_3)_2$ | 10 | 178 |

[a] Catalyst concentration is 10–15 mg in 25 ml of neat benzaldehyde as solvent (see Reference 72 for details).
[b] Initial values observed during first several hours of reaction.
[c] Made *in situ* by first purging the system with $H_2$ to hydrogenate the cod (1,5-cyclooctadiene) to cyclooctane.

accounting for their lower activities. Also, since $k_3$ should be larger for the Ir system,[86] $k_1$ is likely to be the rate-limiting step. Evidence which supports faster oxidative addition reactions for Ir over Rh is that benzaldehyde is decarbonylated much faster by $IrCl(CO)(PPh_3)_2$ than by its Rh analogue (Table 8). In these complexes, oxidative addition is expected to be rate-limiting.[87] An alternate explanation for the slow rates with $[Ir(P-P)_2]^+$ is that the rate of CO loss from $[Ir(P-P)_2CO]^+$ is rate-limiting. Further experiments are needed in order to determine the slow step.

The rates of catalytic decarbonylation of benzaldehyde using mono-diphosphine complexes of Rh and Ir provide an interesting comparison. First of all, the mono-diphosphine complexes of Rh and Ir are not robust under the conditions of the catalytic reaction and therefore are of little practical use. However, they do provide useful data for mechanistic arguments. With Rh, the bis-diphosphine catalysts $[Rh(P-P)_2]^+$ are always more active than their mono-diphosphine analog $[Rh(P-P)]^+$ when neat aldehyde is used as solvent.[66] Although Rh–P bond rupture is not necessary with the coordinatively unsaturated mono-diphosphine complexes, the rhodium may not be "electron-rich" enough to promote facile oxidative addition. In support of this argument, the presence of the diolefin cod in the coordination core, $[Rh(cod)(dppp)]^+$, increased the activity of decarbonylation by a factor of 6 compared with $[Rh(dppp)]^+$.[66] With Ir there is no problem with oxidative addition to $[Ir(dppp)]^+$, since Ir complexes are well known to undergo oxidative addition much more readily than their Rh analogs. Therefore, we see that the decarbonylation activity of $[Ir(dppp)]^+$ is *ca.* 30 times greater than that of $[Ir(dppp)_2]^+$, and that the presence of cod slows down the activity by a factor of 2. These comparisons show the delicate balancing of effects that is required in order to tune a catalyst system. They also provide supporting evidence for the importance of *M*–P bond rupture as shown in Figure 11.3.

## 5. ADDITIONAL STUDIES WITH BIS-CHELATE COMPLEXES OF Rh(I)

A good example of the difficulties that can be encountered in attempts to improve the performance of a catalyst by ligand changes is the use of chelating P–N ligands. Since the rate of Rh–P bond breaking appears to be an important factor in determining the activity of catalytic decarbonylation (Figure 11.3), a complex that contains a potentially more labile ligand was synthesized—$[Rh(P-N)_2]BF_4$, where

$$PN = (C_6H_5)_2PCH_2CH_2-\underset{N}{\underset{|}{\bigcirc}}$$

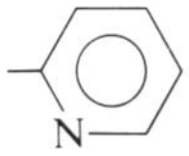

has been characterized by single crystal X-ray diffraction and has a distorted square-planar *cis* geometry.[88] This ligand is expected to undergo facile Rh–N bond rupture by analogy to other P–N complexes.[84] However, this complex is not effective as a decarbonylation catalyst and exhibits an activity of only 10 turnovers/hour at 178°C for benzene production from benzaldehyde.[88] The reason for the inactivity of this complex is unknown; however, since Rh(III) species are not observed during the reaction and $[Rh(P-N)_2CO]^+$ readily loses CO, the inactivity must result from slow oxidative addition or initial bond dissociation. Examination of crystal structure of complexes with diphosphine ligands reveals that interligand steric interactions play a major role in determining geometric distortions.[74,89] For example, in $[Rh(dppb)_2]BF_4$ the distortion toward tetrahedral geometry and the unusual lengthening of one Rh–P bond is due to phenyl . . . phenyl interaction.[74] This complex readily forms dppb-bridged bimetallic species in solution.[85] It is likely that the desired bond lability in *bis*-chelate complexes of Rh is affected more by steric forces than by electronic factors. More experiments on various P–N complexes are needed in order to determine the relative importance of these factors.

## 6. DECARBONYLATION OF BENZOYLCHLORIDE WITH $[Rh(dppp)_2]^+$

Benzoylchloride cannot be decarbonylated catalytically or stoichiometrically by $[Rh(dppp)_2]^+$ or $[Rh(dppp)]^+$ under the same conditions used for aldehyde decarbonylation. In fact, the addition of benzoylchloride to active aldehyde decarbonylation systems completely and irreversibly stops the aldehyde decarbonylation.[64] It has been found that PhCOCl quickly reacts with [Rh(dppp)]Cl at 25°C or with $[Rh(dppp)_2]Cl$ at 150°C to produce the oxidative addition product $Rh(Cl)_2(PhCO)$-(dppp).[90] This Rh(III) complex has a square pyramidal geometry, with the benzoyl ligand occupying the axial position. Surprisingly, the benzoyl phenyl group in this complex will not migrate even at 180°C and, therefore, chlorobenzene cannot be produced. A similar observation has been made by Baird.[45] It is interesting that the bis-triphenylphosphine analog $Rh(Cl)_2(PhCO)(PPh_3)_2$, which has *trans* phosphine ligands, exhibits facile Ph group migration but not reductive elimination.[37] Obviously, there are subtle effects at work here. In light of recent results by Kampmeier (*vide supra*),[37] it would be interesting to examine the decarbonylation of alkyl acid chlorides using diphosphine catalysts. An important observation with respect to the decarbonylation of benzoylchloride using $[Rh(dppp)_2]^+$ is that both chlorobenzene and benzene are catalytically produced using $H_2$ as the purging gas.[91] Catalytic activities at 195°C for chlorobenzene and benzene production are 38 and 27 turnovers per hour, respectively, using neat benzoylchloride as solvent and a catalyst concentration of *ca.* 1 mM. The mechanism of this reaction has not been elucidated; however, radical intermediates are suspected.[91]

ACKNOWLEDGEMENTS

The research on catalytic decarbonylation using diphosphine complexes was supported by the National Science Foundation. The Johnson Matthey Company is acknowledged for generous loans of rhodium and iridium trichloride.

## REFERENCES

1. J. Tsuji and K. Ohno, *Synthesis* **1**, 157 (1969).
2. C. W. Bird, *Transition Metal Intermediates in Organic Synthesis* (Logos Press, London, 1967), pp. 112, 239.

3. A. Kozikowski and H. Wetter, *Synthesis* **1976**, 561.
4. W. M. Schubert and R. R. Kinther, *The Chemistry of the Carbonyl Group* (Wiley-Interscience of John Wiley and Sons, New York, 1966), p. 695.
5. J. N. Pitts and J. K. S. Wan, *The Chemistry of the Carbonyl Group* (Wiley-Interscience of John Wiley and Sons, New York, 1966), p. 823.
6. J. Chatt and B. L. Shaw, *Chem. and Ind.* **1960**, 931.
7. ———, *Chem. and Ind.* **1961**, 290.
8. L. Vaska and J. W. DiLuzio, *J. Am. Chem. Soc.* **83**, 1262 (1961).
9. L. Vaska, *Chem. and Ind.* **1961**, 1402.
10. J. Chatt, B. L. Shaw, and A. E. Field, *J. Chem. Soc.* **1964**, 3466.
11. L. Vaska and J. W. DiLuzio, *J. Am. Chem. Soc.* **83**, 2784 (1961).
12. J. P. Collman, C. I. Sears, and M. Kubota, *Inorg. Synthesis* **11**, 101 (1968).
13. L. Vaska, *J. Am. Chem. Soc.* **86**, 1943 (1964).
14. R. H. Prince and K. A. Raspin, *J. C. S., Chem. Comm.* **1966**, 156.
15. ———, *J. Chem. Soc. (A)* **1969**, 612.
16. ———, *J. Inorg. Nucl. Chem.* **31**, 695 (1969).
17. K. A. Raspin, *J. Chem. Soc. (A)* **1969**, 461.
18. N. W. Alcock and K. A. Raspin, *J. Chem. Soc. (A)* **1968**, 2108.
19. J. Blum, S. Kraus, and Y. Pickholtz, *J. Organomet. Chem.* **33**, 18 (1971).
20. (a) M. C. Baird, C. J. Nyman, and G. W. Wilkinson, *J. Chem. Soc. (A)* **1969**, 346; (b) M. C. Baird, J. T. Mague, J. A. Osborn, and G. Wilkinson, *J. Chem. Soc. (A)* **1967**, 1347.
21. J. Blum, E. Oppenheimer, and E. D. Bergman, *J. Am. Chem. Soc.* **89**, 2338 (1967).
22. K. Ohno and J. Tsuji, *J. Am. Chem. Soc.* **90**, 99 (1968).
23. J. K. Stille, M. T. Regan, R. W. Fries, F. Huang, and T. McCarley, *Adv. Chem. Ser.* **132**, 181 (1974).
24. (a) J. K. Stille and M. T. Regan, *J. Am. Chem. Soc.* **96**, 1508 (1974); (b) J. K. Stille and R. W. Fries, *J. Am. Chem. Soc.* **96**, 1514 (1974); (c) J. K. Stille, F. Huang, and M. T. Regan, *J. Am. Chem. Soc.* **96**, 1518 (1974).
25. A. Rusina and A. A. Vlcek, *Nature* **206**, 295 (1965).
26. E. Muller, A. Segnitz, and E. Langer, *Tetrahedron Lett.* **1969**, 1169.
27. Y. S. Varshavskii, E. P. Shestakova, N. A. Buzina, T. G. Cherkasova, N. V. Kiseleva, and V. A. Kormer, *Koord. Khim.* **2**, 1410 (1976); through *Chem. Abstr.* **86**, 37002 (1977).
28. Y. S. Varshavskii, E. P. Shestakova, N. V. Kiseleva, T. G. Cherkasova, N. A. Buzina, L. S. Bresler, and V. A. Kormer, *J. Orgnomet. Chem.* **170**, 81 (1979).
29. J. J. Alexander and A. Wojcicki, *Inorg. Chem.* **12**, 74 (1973).
30. ———, *J. Organomet. Chem.* **15**, C23 (1968).
31. F. H. Jardine, *Progress in Inorganic Chemistry*, edited by S. J. Lippard, Vol. 28, pp. 63–202 (Wiley-Interscience of John Wiley and Sons, New York 1981).
32. G. Gomazetis, B. Tarpey, D. Dolphin, B. R. James, *J. C. S., Chem. Commun.* **1980**, 939.
33. K. S. Y. Lau, Y. Becker, F. Huang, N. Baenziger, and J. K. Stille, *J. Am. Chem. Soc.* **99**, 5664 (1977).
34. D. L. Egglestone, M. C. Baird, C. J. L. Lock, and G. Turner, *J. C. S., Dalton* **1977**, 1576.
35. J. A. Osborn, F. H. Jardine, and G. Wilkinson, *J. Chem. Soc. (A)* **1966**, 1711.
36. M. C. Baird, C. J. Nyman, and G. Wilkinson, *J. Chem. Soc. (A)* **1968**, 348.
37. J. A. Kampmeier, R. M. Rodehorst, and J. B. Philip, Jr., *J. Am. Chem. Soc.* **103**, 1847 (1981).
38. J. B. Philip, Jr., Ph.D. Thesis, University of Rochester, 1980.
39. J. A. Kampmeier, S. H. Harris, and D. K. Wedegaertner, *J. Org. Chem.* **45**, 315 (1980).
40. J. K. Stille, M. T. Regan, R. W. Fries, F. Huang, and T. McCarley, *Adv. Chem. Series* **132**, 181 (1974).
41. J. A. Kampmeier, S. H. Harris, and R. M. Rodehorst, *J. Am. Chem. Soc.* **103**, 1478 (1981).

42. C. A. Tolman, P. Z. Meakin, D. L. Lindner, and J. P. Jesson, *J. Am. Chem. Soc.* **96,** 2762 (1974).
43. J. Halpern and C. S. Wong, *J. C. S., Chem. Commun.* **1973,** 629.
44. C. S. Wong, Ph.D. Thesis, University of Chicago, 1973.
45. D. A. Slack, D. L. Egglestone, and M. C. Baird, *J. Organomet. Chem.* **146,** 71 (1978).
46. N. A. Dunham and M. C. Baird, *J. C. S., Dalton* **1975,** 774.
47. J. K. Stille, *Ann. New York Academy of Science* **295,** 52 (1977).
48. J. Blum, *Tetrahedron Lett.* **1966,** 1605.
49. J. K. Stille, private communication.
50. P. J. Davidson, M. F. Lappert, and R. Pearce, *Chem. Rev.* **76,** 219 (1976).
51. M. C. Baird, *J. Organomet. Chem.* **64,** 289 (1974).
52. J. W. Suggs, *J. Am. Chem. Soc.* **100,** 640 (1978).
53. H. M. Walborsky and L. E. Allen, *J. Am. Chem. Soc.* **93,** 5465 (1971).
54. ———, *Tetrahedron Lett.* **1970,** 823.
55. P. B. Chock and J. Halpern, *J. Am. Chem. Soc.* **88,** 3511 (1966).
56. J. Y. Chen, Ph.D. Thesis, The University of Chicago, 1972.
57. R. Ugo, A. Pasini, A. Fusi, and S. Cenini, *J. Am. Chem. Soc.* **94,** 7364 (1972).
58. W. Strohmeier and P. Prohler, *J. Organomet. Chem.* **108,** 393 (1976).
59. J. Tsuji and K. Ohno, *Tetrahedron Lett.* **1966,** 4713.
60. G. L. Geoffroy, D. A. Denton, M. E. Keeney, and R. R. Bucks, *Inorg. Chem.* **15,** 2382 (1976).
61. J. Chatt and B. L. Shaw, *J. Chem. Soc. (A)* **1966,** 1437.
62. D. H. Doughty and L. H. Pignolet, *J. Am. Chem. Soc.* **100,** 7083 (1978).
63. D. H. Doughty, M. F. McGuiggan, H. Wang, and L. H. Pignolet, *Fundamental Research in Homogeneous Catalysis* **3,** 909 (1979).
64. D. H. Doughty, M. P. Anderson, A. L. Casalnuovo, M. F. McGuiggan, C. C. Tso, H. H. Wang, and L. H. Pignolet, *Adv. Chem. Ser.* **196,** 65 (1982).
65. R. R. Schrock and J. A. Osborn, *J. Am. Chem. Soc.* **93,** 2397 (1971).
66. D. H. Doughty, Ph.D. Thesis, University of Minnesota, 1979.
67. A. S. C. Chan, J. J. Pluth, and J. Halpern, *J. Am. Chem. Soc.* **102,** 5952 (1980).
68. M. D. Fryzuk and B. Bosnich, *J. Am. Chem. Soc.* **99,** 6262 (1977).
69. W. S. Knowles, M. J. Sabacky, and B. D. Vineyard, *Adv. Chem. Ser.* **132,** 274 (1974).
70. B. R. James and D. Mahajan, *Can. J. Chem.* **57,** 180 (1979).
71. J. Halpern, D. P. Riley, A. C. S. Chan, and J. J. Pluth, *J. Am. Chem. Soc.* **99,** 8055 (1977).
72. H. H. Wang, Ph.D. Thesis, University of Minnesota, 1981.
73. J. Y. Chen, Ph.D. Thesis, University of Chicago, 1972.
74. M. P. Anderson and L. H. Pignolet, *Inorg. Chem.* **20,** 4101 (1981).
75. P. S. Chua, N. K. Roberts, B. Bosnich, S. J. Okrasinski, and J. Halpern, *J. Chem. Soc., Chem. Comm.* **1981,** 1278.
76. S. D. Ittle, C. A. Tolman, A. D. English, and J. P. Jesson, *Adv. Chem. Ser.* **173,** 67 (1979).
77. C. A. Tolman, S. D. Ittle, A. D. English, and J. P. Jesson, *J. Am. Chem. Soc.* **100,** 4080 (1978).
78. A. R. Sanger, *J. C. S., Dalton* **1977,** 120.
79. C. L. U. Su, *Adv. Organomet. Chem.* **17,** 269 (1979).
80. T. B. Rauchfuss, *J. Am. Chem. Soc.* **101,** 1045 (1979).
81. J. P. Collman and L. S. Hegedus, *Principles and Applications of Organotransition Metal Chemistry* (University Science Books, Mill Valley, CA, 1980).
82. Y. Kawabata, T. Hayashi, and I. Ogata, *J. C. S., Chem. Commun.* **1975,** 462.
83. J.-C. Poulin, T.-P. Dang, and H. B. Kagan, *J. Organomet. Chem.* **84,** 87 (1975).
84. W. J. Knebel and R. J. Angelici, *Inorg. Chem.* **13,** 627, 632 (1974).
85. L. H. Pignolet, D. H. Doughty, S. C. Nowicki, M. P. Anderson, and A. L. Casalnuovo, *J. Organomet. Chem.* **202,** 211 (1980).

86. J. P. Collman and L. S. Hegedus, *Principles and Applications of Organotransition Metal Chemistry* (University Science Books, Mill Valley, CA, 1980), p. 211.
87. ———, *Principles and Applications of Organotransition Metal Chemistry* (University Science Books, Mill Valley, CA, 1980), p. 205.
88. M. P. Anderson, B. J. Johnson, and L. H. Pignolet, preliminary results.
89. L. H. Pignolet, D. H. Doughty, S. C. Nowicki, and A. L. Casalnuovo, *Inorg. Chem.* **19,** 2172 (1980).
90. M. F. McGuiggan, D. H. Doughty, and L. H. Pignolet, *J. Organomet. Chem.* **185,** 241 (1980).
91. M. F. McGuiggan, Ph.D. Thesis, University of Minenesota, 1982.

# 12

# Homogeneous Catalysis of Oxidation Reactions Using Phosphine Complexes

## D. Max Roundhill

### 1. SIGNIFICANCE OF METAL CATALYZED OXIDATION REACTIONS

Oxidation is one of the most important reactions in chemistry and biochemistry. Combustion of hydrocarbons drives much of our economy and transportation, and biological oxidation processes are fundamental to life and ecology. From an industrial viewpoint, oxidation reactions occupy a pivotal role in the conversion of hydrocarbons into required products, and indeed it has been estimated that over 50% of such processes involve hydrocarbon oxidations.[1]

In choosing a particular oxidant one must be particularly cognizant of the required selectivity for the desired transformation. Among the available reagents for oxidations are metal compounds in high-oxidation states, molecular oxygen, and peroxides or peracids.[2-15] Both oxygen and the peroxidic reagents can be used in conjunction with metal complexes, whereby the oxygen containing reagent is "activated" to reactivity, or the selectivity of the reagent is modified. Several review articles have discussed the structural and chemical aspects of dioxygen binding to transition metal compounds.[16-21] Furthermore, alkyl peroxides in conjunction with *high-valent* metal compounds have been used to advantage for selective alkene epoxidation.[22]

---

*Dr. D. Max Roundhill* • Department of Chemistry, Tulane University, New Orleans, Louisiana 70118.

## 1.1. Mechanistic Features of Metal Catalyzed Oxidations

The primary focus of research using transition metal phosphine complexes for oxidations is in the complexation and activation of molecular oxygen. These oxygen complexes have been variously regarded as complexes of coordinated peroxide, superoxide, or singlet oxygen, and their reactivity with reduced substrate has been interpreted on such a basis. In this chapter, we will focus on the chemical reactivity of these compounds for oxygen atom transfer oxidation reactions, with a particular emphasis on the mechanistic features of these processes.

Atom-transfer-type oxidation reactions can occur by electrophilic, nucleophilic, or radical pathways. Electrophilic mechanisms are prevalent with metal compounds in high-oxidation states. Common among these reagents are oxo complexes of Mn(VII), Cr(VI), V(V), as well as organic reagents such as peracids.[8–10,15] Recently, a proposal has been forwarded with respect to this chemistry that metallocycle formation between M=O and an alkene can occur, provided that a second spectator oxo group is available that can simultaneously convert from a metal oxygen double to a triple bond.[23] Reagents such as hydrogen peroxide and organic peroxides have been previously considered to follow nonselective, free-radical pathways. Recent work by Sharpless, however, using *tert*-butyl hydroperoxide as oxidant in conjunction with high-valent metal compounds has shown that stereospecific alkene epoxidation can be effected.[24]

Catalyzed oxidations using molecular oxygen and transition metal phosphine complexes appear to follow two pathways. Coordination of dioxygen to the metal center results in a formal transfer of one or two electrons from the metal to give a complexed superoxo or peroxo ligand, which then undergoes reaction with the reduced substrate. A second pathway that is frequently followed involves initial oxygen attack at the hydrocarbon substrate to form an organic peroxide, which is then catalytically converted into oxygenated product by reaction with the phosphine metal complex. Both of these courses are prevalent in catalytic oxidations with transition metal phosphine complexes, and diagnostic experiments are necessary to deduce the function of any metal phosphine complex used as an oxidation catalyst.

## 2. TRANSITION-METAL PHOSPHINE OXYGEN COMPLEXES

### 2.1. Synthesis and Structure

Synthetic complexes with a coordinated dioxygen have been known for many years, mainly because of their relevance to reversible oxygen carriers.[16,17] Much of the earlier work centered around formally divalent cobalt complexes having nitrogen or oxygen donor ligands. In 1963, Vaska reported that the $d^8$ iridium(I) compound $IrCl(CO)(PPh_3)_2$ would reversibly add molecular oxygen,[25] and this result instigated considerable interest in the chemistry of low-valent, electron rich, metal ions with oxygen. Since that time many similar compounds have been obtained by addition of oxygen to complexes that can undergo a formal two-electron oxidative addition reaction. Examples of these compounds are ones found with Ir(I), Rh(I), Ni(O), Pd(O), Pt(O), Ru(O). Since substituted phosphines are commonly used to stabilize these precursor complexes in their low-valent oxidation state, the complexes formed by oxygen addition are predominantly ones with such phosphine ligands. Examples of these compounds that are formed by oxygen addition are $IrCl(CO)(PPh_3)_2O_2$,[25] $Pt(PPh_3)_2O_2$,[26,27] $Ni(CNPh)_2O_2$,[28,29] $[Ir(dppe)_2O_2]X$,[30] and $RuX(NO)(PPh_3)_2O_2$.[31]

$$IrCl(CO)(PPh_3)_2 + O_2 \rightarrow IrCl(CO)(PPh_3)_2O_2 \qquad (1)$$

$$Pt(PPh_3)_4 + O_2 \rightarrow Pt(PPh_3)_2O_2 + 2PPh_3 \qquad (2)$$

$$Ni(CNPh)_4 + O_2 \rightarrow Ni(CNPh)_2O_2 + 2CNPh \qquad (3)$$

$$[Ir(dppe)_2]X + O_2 \rightarrow [Ir(dppe)_2O_2]X \qquad (4)$$

$$RuX(NO)(PPh_3)_2 + O_2 \rightarrow RuX(NO)(PPh_3)_2O_2 \qquad (5)$$

The reversibility of the oxygenation reaction is dependent on the nature of the ligands. Thus, whereas the compound $IrCl(CO)(PPh_3)_2$ will reversibly add oxygen, the analogs with more basic phosphines, such as $PMePh_2$ or $PMe_2Ph$, bind oxygen more strongly, and the $O_2$ is not readily dissociated. Structurally, it has been found that each of these compounds has a $\eta^2$-bonded oxygen ligand(**I**), and, from electronic considerations, the coordinated dioxygen resembles a complexed peroxide ligand.[16] This binding mode differs significantly from that of type **II** found for metals such as Fe(**II**) and Co(**II**), which undergo one-electron oxidation.

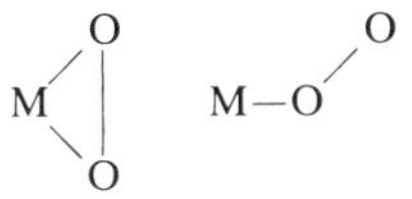

$$\underset{\mathbf{I}}{M\overset{O}{\underset{O}{\big\langle\big|}}} \qquad \underset{\mathbf{II}}{M-O\overset{O}{\diagup}}$$

Structural and infrared ($\nu(O-O)$) studies have been published showing that the complexed O–O bond lengths are in the range 1.41–1.52 Å, which compares closely with the sum of 1.46 Å for the O–O single-bond covalent radii. The infrared stretch $\nu$(O–O) for these compounds I is in the 800–900 cm$^{-1}$ range.

New palladium dioxygen complexes have been prepared from reaction of superoxide ion and [PdCl(methoxydicyclopentadienyl-$\eta^1$)]$_2$.[32] This dioxygen-bridged complex [(MeO-DiCp)PdO)]$_2$ reacts with methanol to yield the methoxy-bridged compound [(MeO-DiCp)PdOMe]$_2$. Subsequently, the oxygen-bridged compound Rh$_2$O$_2$(1,5-COD)$_2$ has also been prepared. On treatment with cyclohexanone this compound gives dehydrogenation products cyclohexene-3-one and phenol.[33] On pyrolysis of Rh$_2$O$_2$(1,5-COD)$_2$, cyclooctanone is formed quantitatively in the presence of cyclohexene.[34]

Photolysis of PtO$_2$(PPh$_3$)$_2$ leads to the formation of Pt(PPh$_3$)$_2$ and O$_2$. Irradiation in the presence of the singlet oxygen traps 2,2,6,6-tetraethylpiperidine, and 1,3-diphenylisobenzofurane confirms that the oxygen is dissociated in a singlet state. Carrying out similar experiments with [Ir(dppe)$_2$O$_2$]$^+$ shows that oxygen dissociates in the triplet state. The difference is related to the observation that the lowest energy transition in PtO$_2$(PPh$_3$)$_2$ is an O$_2 \rightarrow$ Pt charge transfer band, whereas in [Ir(dppe)$_2$O$_2$]$^+$ the lowest energy band is assigned as Ir $\rightarrow$ dppe charge transfer.[35]

## 2.2. Reactions with Electrophiles

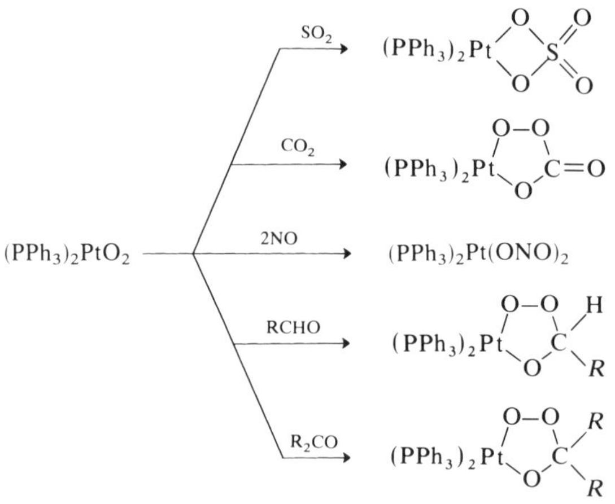

The reactivity of these $\eta^2$-oxygen compounds can be rationalized on the basis of the coordinated dioxygen having nucleophilic character. Thus, the compound Pt(PPh$_3$)$_2$O$_2$ reacts with electrophiles such as SO$_2$,[36] CO$_2$,[37] NO,[36] R$_2$CO,[37,38] and RCHO[37] to give cyclic or acyclic adducts,[6–10] where the electrophilic center of the adduct molecules are bonded to the oxygen atom from the coordinated dioxygen. The reaction with $SO_2$ has been studied using oxygen-18 labeling techniques, where it is concluded that the reaction pathway involves an intermediate peroxo compound **III**, which subsequently transforms to the sulfate complex[39,40]:

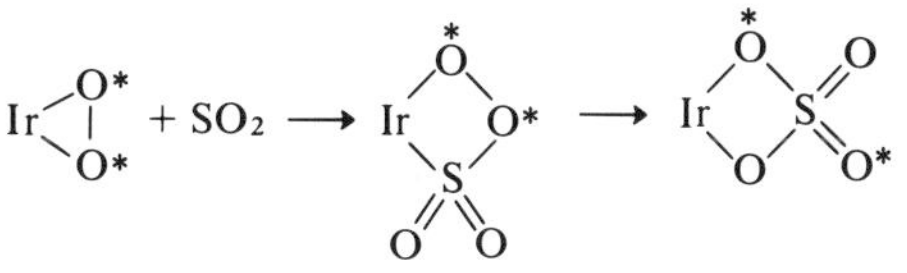

$$Ir\begin{matrix}O^*\\|\\O^*\end{matrix} + SO_2 \longrightarrow Ir\begin{matrix}\overset{*}{O}\\ \diagdown\\ O^*\\ \diagup S \diagdown\\ O \quad O\end{matrix} \longrightarrow Ir\begin{matrix}\overset{*}{O}\diagdown\diagup O\\ S\\ \diagup \diagdown\\ O \quad O^*\end{matrix} \quad (11)$$

**III**

Kinetic studies have been made on the addition of ketones to both $PtL_2O_2$ and $IrX(CO)L_2O_2$ ($L$ = substituted phosphine, $X$ = halide).[41,42] The reaction with $PtL_2O_2$ and ketones ($R_2CO$) follows a two-term rate law (12).

$$\text{Rate} = (k_A + k_B[R_2CO])[PtL_2O_2] \quad (12)$$

These data have been interpreted in support of a two pathway mechanism, one being independent, and the other first-order, in ketone concentration. The proposed intermediates are shown in Figure 12.1. The faster pathway is binuclear with the ketone binding to platinum to give a pentacoordinate intermediate. The slower pathway, which is ketone independent, is proposed to have a slow step to give an "activated" form of $PtL_2O_2$, which may structurally resemble the $\eta^1$-form II rather than the $\eta^2$-form I. In agreement with the concept that the coordinated dioxygen has nucleophilic character, the reaction rate increases for ketones having electron withdrawing groups.

The reactivity of $IrX(CO)L_2O_2$ to ketones is less than that of $PtL_2O_2$. A similar metallocycle is formed (Equation 13) but only with the electrophilic ketone hexafluoroacetone. Now the oxygen complex

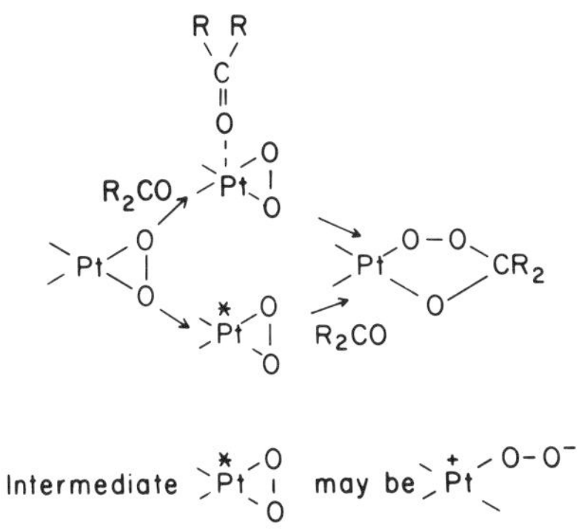

Figure 12.1. Proposed pathways for the reaction of $PtL_2O_2$ with ketones.

$IrX(CO)L_2O_2$ is coordinately saturated, and intermediate adduct formation with ketone is unlikely. Since the reaction does not involve prior ligand dissociation, the results provide a kinetic measurement of the direct reactivity of a coordinated oxygen molecule to an external electrophile.[42]

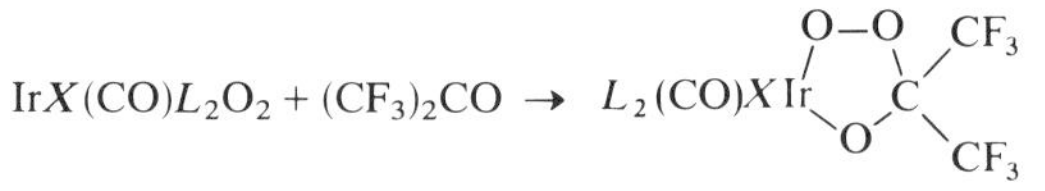

$$IrX(CO)L_2O_2 + (CF_3)_2CO \rightarrow L_2(CO)XIr\begin{matrix}O-O\\|\phantom{xx}|\\O\phantom{xx}\end{matrix}\begin{matrix}CF_3\\C\\CF_3\end{matrix} \quad (13)$$

The reaction between $Pt(PCy_3)_2O_2$ and hexafluorobutyne-2 or dimethyl acetylenedicarboxylate forms an adduct (14) with formal cleavage of the O–O bond:

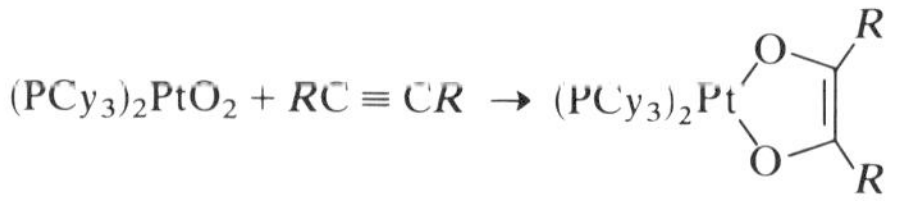

$$(PCy_3)_2PtO_2 + RC \equiv CR \rightarrow (PCy_3)_2Pt\begin{matrix}O\\|\\O\end{matrix}\begin{matrix}R\\C\\\|\\C\\R\end{matrix} \quad (14)$$

The mechanism proceeds by two steps. The first step is first-order in both platinum oxygen complex and alkyne, and the rate of the second step is independent of alkyne concentration ($R = CF_3$, $CO_2Me$). The intermediate may be an adduct with the alkyne coordinated to platinum, or the reaction may proceed in the unsymmetrical metallocycle as is suggested for the $SO_2$ insertion.[43]

The most definitive study for establishing the nucleophilic character of the complexed dioxygen in form I comes from the reaction between $PtL_2O_2$ ($L = PPh_3$, $PCy_3$) and alkyl or aroyl halides.[44] The products from reacting triphenylmethyl bromide and benzoyl bromide are the corresponding alkyl-peroxo platinum(II) complexes $L_2PtBr(OOR)$ ($L = PPh_3$; $R = CPh_3$, $COPh$)[15].

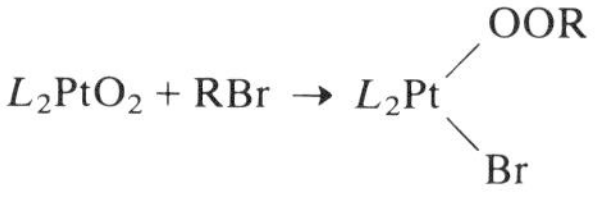

$$L_2PtO_2 + RBr \rightarrow L_2Pt\begin{matrix}OOR\\\\Br\end{matrix} \quad (15)$$

Both an $S_N2$ type (IV) and a cyclic (V) transition state are considered plausible. Discrimination between the mechanistic types was achieved with

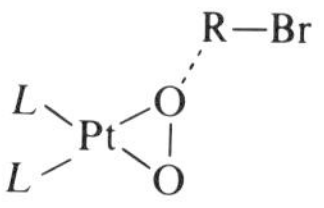

$$\begin{matrix}&R-Br\\L\diagdown&\diagup O\\&Pt&|\\L\diagup&\diagdown O\end{matrix} \qquad \begin{matrix}&Br\diagdown R\\L\diagdown&\diagup O\\&Pt&|\\L\diagup&\diagdown O\end{matrix}$$

IV 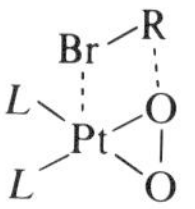 $\qquad\qquad$ V

(S)-(−)-$\alpha$-phenethyl bromide. Treatment of this alkyl halide with $(PPh_3)_2PtO_2$, followed by O–O cleavage of the peroxo complex with $LiAlH_4$, gives S-phenethyl alcohol having the $R$ configuration. This result is in agreement with an $S_N2$-type mechanism involving the coordinated dioxygen as nucleophile. Again the concept of conversion of form **I** to **II** being involved in the reaction pathway is considered. These authors suggest that this conversion represents the extreme of an asymmetric vibration which can lead to a transient species **II**, which will undergo preferential electrophilic attack at the distal oxygen atom. Such a species has not been isolated for a $d^8$ or $d^{10}$ metal center, but is suggested on the basis of a variety of kinetic measurements.[41,42,44,45] The concept of a nucleophilic peroxidic dioxygen ligand when bonded to a formal $d^8$ or $d^{10}$ metal center appears to be well founded. Furthermore, these reactions are supported by calculations on $Pt(PH_3)_2O_2$, which are in agreement with the concept of a $Pt^{2+}(PH_3)_2O_2^{2-}$ model.[46]

The formation of metallocyclic complexes by addition of ketones to dioxygen complexes resembles a three-step Criegee mechanism for ozonolysis.[47,48] In each case one envisages the formation of a dipolar O–O bond that leads to electrocyclic ring closure with the ketone. No proven oxidation of aldehydes or ketones has yet been achieved with these metallocycles. Nevertheless, in the catalytic conversion of cyclic ketones to lactones using hydrogen peroxide and a molybdenum complex as catalyst, it is believed that such an intermediate metallocycle is first formed and that it subsequently converts to lactone and molybdenyl products (16).[49]

$$Mo\begin{matrix}O\\|\\O\end{matrix} + \text{cyclohexanone} \longrightarrow Mo\begin{matrix}O-O\\O\end{matrix}\text{(cyclic)} \longrightarrow Mo=O + \text{lactone} \quad (16)$$

## 3. OXIDATION OF ALKENES

From the initial discovery of these $\eta^2$-dioxygen complexes, attempts have been made to use them for alkene oxidation. The introduction of oxygen functionality into an alkene hydrocarbon is a particularly significant goal, especially if the source is molecular oxygen and the reaction can be achieved catalytically with high selectivity. Since dioxygen will readily coordinate to $d^8$ and $d^{10}$ metal complexes, it has been speculated that one can modify its chemical reactivity to undergo selective reaction with alkenes. In the presence of catalytic amounts of $RhCl(CO)(PPh_3)_2$, $Pt(PPh_3)_2O_2$, $IrI(CO)(PPh_3)_2$, $IrCl(N_2)(PPh_3)_2$, or $RhCl(PPh_3)_3$, cyclohexene and cyclopentene undergo autoxidation:

$$\text{cyclohexene} + O_2 \longrightarrow \text{cyclohexenone} + \text{cyclohexene oxide} \quad (17)$$

The major product from the former is cyclohexen-3-one, along with minor amounts of cyclohexene oxide (17). Epoxide formation has also been identified as a minor product from cyclopentene autoxidation. The intermediacy of 3-cyclohexene hydroperoxide was proposed in this report but not verified.[50] Subsequent work on the autoxidation of cyclohexene using $RhCl(PPh_3)_3$ verified this premise,[51] and a number of review articles have emphasized this conclusion.[52–55] The involvement of preformed hydroperoxide has been verified by comparing the rate of cyclohexene oxidation both with hydroperoxide present, and also when the cyclohexene is purified free from peroxide. In the former case the reaction is rapid and there is no induction period. Under conditions where the cyclohexene is peroxide free the reaction proceeds more slowly, and there is an induction period of close to three hours (using $IrCl(CO)(PPh_3)_2$ as catalyst) as the hydroperoxide intermediate is being performed (18):

$$\bigcirc + O_2 \longrightarrow \bigcirc\!\!-\!\!OOH \qquad (18)$$

The mechanistic details of the subsequent chemistry are unclear, but it is apparent that the catalytic decomposition of hydroperoxide follows a chain process that resembles a Haber–Weiss pathway.[56] It has also been found that $d^8$ complexes show a greater activity for cyclohexene autoxidation that do $d^{10}$ complexes, but no rationale has been provided to explain this observation.[51]

This pathway involving catalytic decomposition of preformed peroxides from alkenes and oxygen, whereby the transition metal phosphine is the catalyst, is a frequently encountered situation. Any oxidation study using metal complexes as catalyst should involve a careful check for induction periods, low selectivity, inhibition by added free-radical chain traps, partial order kinetics, and observed zeroth order in the rate law for added oxygen. Examples of such processes are found in the transition metal phosphine complex catalyzed autoxidation of cyclohexene,[51,57–65] styrene,[66–68] tetramethylethylene,[69] and butadiene.[70] Among the various complexes that have been used are $Pt(PPh_3)_4$, $RhCl(PPh_3)_3$, $RuCl_2(PPh_3)_3$, $IrCl(CO)(PPh_3)_2$, $RhCl(CO)(PPh_3)_2$, $Ru(CO)_3(PPh_3)_2$, and $OsBr_2(PPh_3)$, or their oxygen adducts. The radical chain mechanism has been generally accepted for these reactions, except for a case where the reactions are carried out at the lower temperature of 30°. In this reaction the final product is 2-cyclohexene-1-yl hydroperoxide (18).[71] The catalyst used is $RhCl(PPh_3)_3$, and it is considered that a concerted coordinated sphere mechanism can be applied to fit the data equally well.

As a final generalization, it should be mentioned that these radical chain processes are common when transition metal compounds other than phosphine complexes are used. Catalytic oxidations have been carried out with a wide range of metal complexes, and a comparison has been observed between reactions catalyzed by phosphine complexes and those with acetylacetonate or other ligands.[55,57]

## 4. OXIDATION OF OTHER SUBSTRATES

### 4.1. Isocyanides

The dioxygen complexes $MO_2(CNBu-t)_2$ ($M$ = Ni, Pd) have been isolated in the pure state.[28,29,72] Treating the nickel complexes with excess *tert*-butyl isocyanide at 30°C results in the formation of $Ni(CNBu-t)_4$ and $t$-BuNCO(19):

$$NiO_2(CNBu-t)_2 + 4t-BuNC \rightarrow Ni(CNBu-t)_4 + 2t-BuNCO \qquad (19)$$

When, however, triphenylphosphine is added to the nickel complex, this compound is preferentially oxidized to triphenylphosphine oxide (20):

$$NiO_2(CNBu-t)_2 + 4PPh_3 \rightarrow Ni(CNBu-t)_2(PPh_3)_2 + 2OPPh_2 \qquad (20)$$

The reaction of methyl or cyclohexyl isocyanide with $NiO_2(CNBu-t)_2$ produces a mixture of *tert*-butyl and either methyl or cyclohexyl isocyanate, and it is considered important that the formed isocyanate complex be kinetically labile for achieving successful isocyanide oxidation.[29]

### 4.2. Carbon Monoxide

Although the primary method of choice for the metal-catalyzed oxidation of carbon monoxide to carbon dioxide involves nucleophilic attack of water or hydroxide on a coordinated carbonyl,[73-76] there are an increasing number of cases where a transition metal phosphine complex will catalyze this conversion.

Following the initial reports that the compound $Rh_6(CO)_{16}$ will catalyze the oxidation, with molecular oxygen, of carbon monoxide to carbon dioxide,[77-79] it has been found that this process is more favorable when phosphines are added.[80-81] The addition of triphenylphosphine and CO to a mixture of $Rh_6(CO)_{16}$ and benzene in the presence of oxygen leads to the formation of both $CO_2$ and $OPPh_3$. The stoichiometric yield of $CO_2$

is greater than that of $OPPh_3$, and the addition of water increases the quantity of $CO_2$ formed. Labeling studies have been carried out to probe the mechanism, and further details are discussed in Section 5.3.

The addition of triphenylphosphine causes the $Rh_6$ cluster to convert to a lower homologue. Among the solution species identified are the compounds $Rh_4(CO)_{10}(PPh_3)_2$ and $Rh_2(CO)_4(PPh_3)_4$. These species presumably react with oxygen to give adducts in a manner similar to that found with $Rh_2(CO)_4(PPh_3)_4$ [21] when prepared from $Rh_4(CO)_{12}$ and $PPh_3$,[82] and also with the compound $Rh_4Cl_4(CO)_4(PPh_3)_2(O_2)_2$ (**22**)[83]:

$$Rh_2(CO)_4(PPh_3)_4 + O_2 \rightarrow Rh_2(CO)_2(CO_2)(PPh_3)_3 + 2CO + OPPh_3 \quad (21)$$

$$Rh_4Cl_4(CO)_4(PPh_3)_2(O_2)_2 + 6PPh_3 + 3H_2 \rightarrow$$
$$[RhHCl(PPh_3)_2]_2 + 2RhCl(CO)(PPh_3)_2 + 2H_2O + 2CO_2 \quad (22)$$

A zerovalent platinum bimetallic complex $Pt_2(dppm)_3$ will also catalyze the oxidation of CO to $CO_2$ using either $O_2$ or NO as oxidant.[84] Again, no comment can be made regarding the reaction mechanism. Oxidation of CO on platinum metal surfaces involves the reaction of chemisorbed atomic oxygen,[85] but it is unlikely that such species are formed under the mild conditions of temperature used in these examples. In the conversion of the coordinated carbonyl in IrCl(CO)PN (PN = $o$-(diphenylphosphino)-$N,N$-dimethylaniline) to $CO_2$ though it has been tentatively suggested that a metallocyclic intermediate is formed (**23**)[86]:

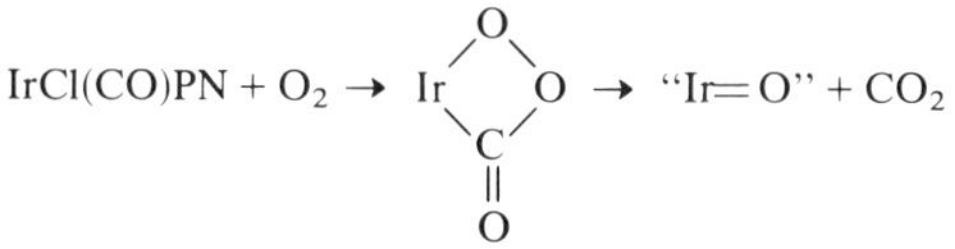

$$IrCl(CO)PN + O_2 \rightarrow Ir\overset{O}{\underset{\underset{\parallel}{C}}{\diagup\diagdown}}O \rightarrow \text{"Ir=O"} + CO_2 \quad (23)$$

Such an intermediate can be formed because of the nucleophilic reactivity of a dioxygen molecule when complexed to iridium(I) and because of the known propensity of a coordinated carbonyl ligand to undergo such nucleophilic attack.

### 4.3. Aldehydes and Ketones

In benzene solvent a series of $d^8$ and $d^{10}$ phosphine complexes catalyze the oxidation of benzaldehyde to benzoic acid and perbenzoic acid under an oxygen atmosphere. The catalytic activity follows the order $RhCl(CO)(PPh_3)_2 > PdO_2(PPh_3)_2 \simeq Pd(PPh_3)_4 > RhCl(PPh_3)_3 > [RhCl(PPh_3)_2]_2 > IrCl(CO)(PPh_3)_2O_2 \simeq PtO_2(PPh_3)_2$.[87] These authors propose that a nonradical pathway is followed, but subsequent work showed the oxidation to occur by a free-radical mechanism.[88] This correction was based on a number of new experiments. The kinetics of the oxidation using $PdO_2(PPh_3)_2$ as catalyst resembles that found using cobalt(III) catalysts, and also the reaction is retarded by adding the free-radical inhibitor 2,6-di-*tert*-butyl-*p*-cresol. Furthermore, the presence of free radicals is verified using the spin traps nitroso-*tert*-butane and phenyl-*tert*-butyl nitrone.

Subsequent to the finding that the compounds $Rh_6(CO)_{16}$ and $Re_2(CO)_{10}$ will catalyze the autoxidation of ketones,[77,89,90] it was also found that the phosphine complexes $IrCl(CO)(PPh_3)_2$ and $Pt(PPh_3)_3$ were effective catalysts for this oxidation.[91,92] The organic oxidation products are carboxylic acids (Equations 24–27), but the phosphine compounds are not recovered unchanged. Inhibition by free-radical scavengers is observed, and a reaction pathway is followed whereby the phosphine complex acts to accelerate the conversion of preformed hydroperoxides and peracids to carboxylic acid.[90]

$$CH_3COCH_2CH_3 \xrightarrow{[O]} CH_3CO_2H \qquad (24)$$

$$(CH_3CH_2)_2CO \xrightarrow{[O]} CH_3CH_2CO_2H + CH_3CO_2H \qquad (25)$$

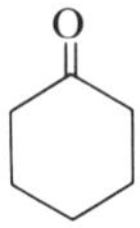

$$\text{cyclohexanone} \xrightarrow{[O]} HO_2C(CH_2)_4CO_2H \qquad (26)$$

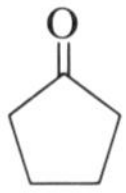

$$\text{cyclopentanone} \xrightarrow{[O]} HO_2C(CH_2)_3CO_2H \qquad (27)$$

### 4.4. Cumene

The compound $Pd(PPh_3)_4$ can be used as a liquid-phase autoxidation catalyst for cumene oxidation.[93,94] It is again concluded that the role of the transition metal compound is in its reaction with preformed cumene hydroperoxides.[94]

### 4.5. Tertiary Phosphines

Tertiary phosphines are among the easiest molecules to catalytically oxidize with molecular oxygen. The reactions can usually be effected under ambient temperature and pressure conditions. Metal complexes of both the platinum metal group and from the first row transition series have been used as catalysts, although in only a few cases have detailed mechanistic

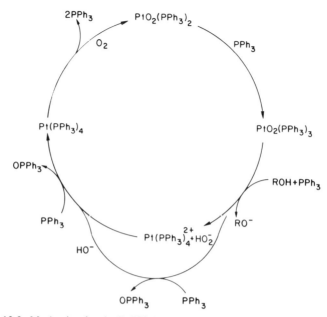

Figure 12.2. Mechanism for the Pt(PPh$_3$)$_3$ catalyzed oxidation of triphenylphosphine.

studies been carried out. Among the most studied is the compound Pt(PPh$_3$)$_3$.[95-98] Under protolytic conditions, Halpern suggests that the compound PtO$_2$(PPh$_3$)$_2$ forms free hydroperoxide ion, which then converts the tertiary phosphine to oxide (Figure 12.2). The resulting compound Pt(PPh$_3$)$_4^{2+}$, in conjunction with OH$^-$, is then proposed to further oxidize PPh$_3$ to yield a second mole of OPPh$_3$, with concomitant reduction of platinum(II) to the zerovalent oxidation state. The selectivity to oxidation by PtO$_2$(PPh$_3$)$_2$ is related to the nucleophilicity of the ligand to platinum, and hence its ability to undergo substitution at platinum and generate free hydroperoxide ions. It is suggested, but not verified, that the oxidation of triphenylphosphine with Ru(NCS)(NO)(PPh$_3$)$_2$[31,99] as catalyst may follow a similar pathway. Other platinum metal complexes that have been used in a similar manner for triphenylphosphine oxidation are RhCl(PPh$_3$)$_3$[100-102] and IrX(CO)(PPh$_3$)$_2$.[66]

Complexes of cobalt have also been used as catalysts for the oxidation of tertiary phosphines with molecular oxygen. In methanol solvent the compound Co$_2$(CN)$_4$(PMe$_2$Ph)$_5$O$_2$ reacts with PMe$_2$Ph, converting it into the oxide.[103] Since the oxygen compound is readily formed again from the product complex Co(CN)$_2$(PMe$_2$Ph)$_3$, a catalytic cycle can be obtained (28, 29) for the phosphine oxidation.

$$2\text{Co(CN)}_2(\text{PMe}_2\text{Ph})_3 + \text{O}_2 \rightarrow \text{Co}_2(\text{CN})_4(\text{PMe}_2\text{Ph})_5\text{O}_2 + \text{PMe}_2\text{Ph} \qquad (28)$$

$$Co_2(CN)_4(PMe_2Ph)_5O_2 + 3PMe_2Ph \rightarrow 2Co(CN)_2(PMe_2Ph)_3 + 2Me_2PhPO \qquad (29)$$

Studies on the reactions of $CoCl_2(PEt_3)_2$,[104] and also on a mixture of $Co(acac)_2$ and $PBu_3$,[105] show the reactions to form *only* $OPEt_3$ and $OPBu_3$, respectively. The reactions show no inhibition upon addition of free radical scavengers. For $CoCl_2(PEt_3)_2$, the rate is first-order in both $CoCl_2(PEt_3)_2$ and $O_2$, but the kinetics of the latter reaction show first-order dependence on $Co(acac)_2(PBu_3)$ and only half-order on oxygen concentration. This difference between the two studies is shown in the latter intermediate being a bimetallic dioxygen complex, whereas the former is proposed to be a monomer $CoCl_2(PEt_3)_2O_2$. Extension of the work to phosphinite and phosphite esters[106] of type $Et_nP(OEt)_{3-n}$ shows conversion to oxide with *no* change in the value of $n$. Autoxidation of uncomplexed $Et_nP(OEt)_{3-n}$ by a free radical pathway does cause a change in the value of $n$.[107] Thus the oxidation of the *coordinated* phosphorus ligand must be considerably faster and cannot involve prior dissociation, which would lead to oxidation of free phosphine and a change in the value of $n$. For triphenylphosphine oxidation, the reaction is catalytic using the compound $CoCl_2(PPh_3)_2$ with oxygen.[108]

Although the oxidation of tertiary phosphines by these catalytic processes has minimal useful application, it needs to be considered as a problematic side reaction in homogeneous catalysis. Much effort is being currently expended to immobilize platinum metal phosphine complexes on heterogenized tertiary phosphine supports, and irreversible oxidation at phosphorus on these supports effectively destroys the supported catalyst. Recent observations that the compound $Rh_6(CO)_{16}$ catalyzes the oxidation of tertiary phosphines[80] correlate with the report that phosphine oxidation occurs with molecular oxygen on $Rh_6(CO)_{16}$ bound to diphenylphosphino-functionalized poly(styrenedivinylbenzene).[109] Thus, in order to use these phosphinated polymer-supported rhodium catalysts, one needs either to rigorously exclude oxygen, or to find a way to inhibit the simultaneous catalyzed phosphine oxidation.

## 5. CO-OXIDATIONS

A challenge for chemists designing homogeneously catalyzed oxidation reactions is the frequent necessity to incorporate multiple substrates in the reaction mixture for selective co-oxidation. If the oxygen atom transfer reagent is molecular oxygen, it is unlikely that *both* oxygen atoms will be transferred to reduced substrate in a single step. Following transfer, therefore, of a single oxygen atom from dioxygen to reduced substrate in the coordination sphere of the metal, the second oxygen will remain coordinated to the metal. To complete the catalytic cycle, it is therefore necessary to incorporate a second reaction step whereby this bonded oxo ligand is transferred to another substrate or a second mole of the same reduced substrate.

## 5.1. Alkenes and Tertiary Phosphines

Read and co-workers found that the complex $RhCl(PPh_3)_3$ will selectively catalyze the co-oxidation of both terminal alkenes to methyl ketones, and triphenylphosphine to triphenylphosphine oxide[30]:

$$RCH=CH_2 + PPh_3 + O_2 \xrightarrow{RhCl(PPh_3)_3} R-\underset{\underset{O}{\|}}{C}-CH_3 + OPPh_3 \qquad (30)$$

The oxidation of alkene is stoichiometric unless excess triphenylphosphine is present.[110–112] In the reaction mechanism, one oxygen atom from the dioxygen is transferred to olefin, while the triphenylphosphine acts as a co-reducing agent for reaction with the second oxygen atom to produce $OPPh_3$. The proposed reaction pathway involves metallocycle formation (31) followed by reductive elimination to form the oxetane. The oxetane can undergo subsequent reaction with triphenylphosphine to yield $OPPh_3$ and methyl ketone (32):

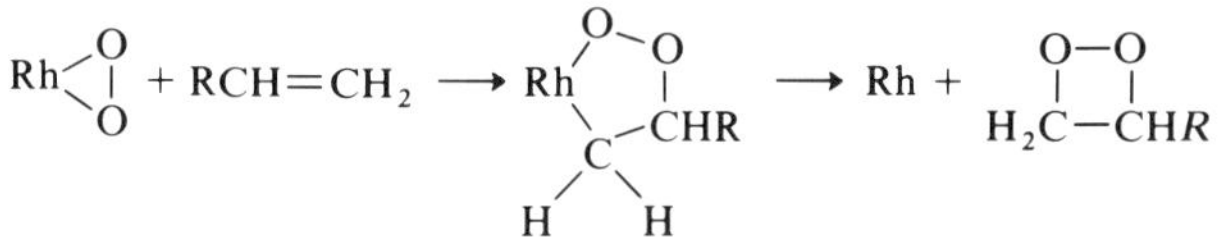

$$Rh\underset{O}{\overset{O}{<}}| + RCH=CH_2 \longrightarrow Rh\underset{H\ \ H}{\overset{O-O}{\underset{C}{|}}}\!\!\underset{}{\overset{}{-}}CHR \longrightarrow Rh + \underset{H_2C-CHR}{\overset{O-O}{|\ \ |}} \qquad (31)$$

$$\underset{O-O}{\overset{RHC-CH_2}{|\ \ |}} + PPh_3 \longrightarrow R-\underset{\underset{O}{\|}}{C}-CH_3 + OPPh_3 \qquad (32)$$

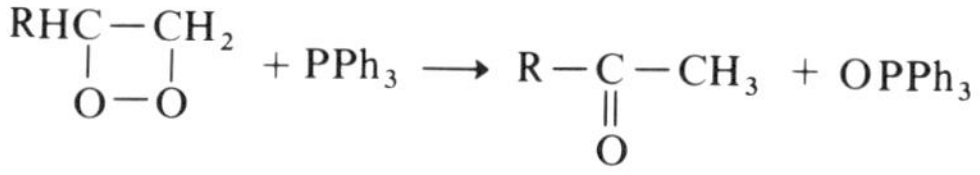

The reaction is inhibited by added water. The formation of such a metallocyclic intermediate resembles the isolation of such compounds from the reaction of $PtO_2(PPh_3)_2$ with electron-deficient alkenes such as tetracyanoethylene [33][113]:

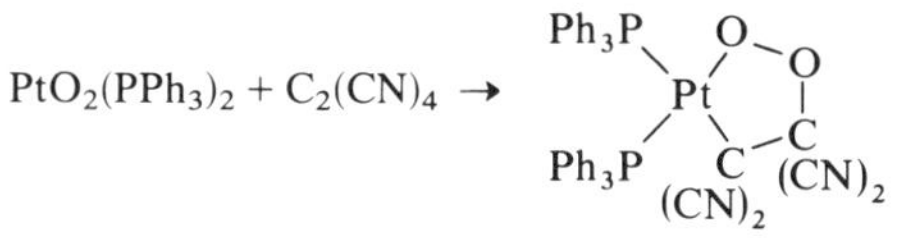

$$PtO_2(PPh_3)_2 + C_2(CN)_4 \longrightarrow \underset{Ph_3P}{\overset{Ph_3P}{>}}Pt\underset{\underset{(CN)_2}{C}}{\overset{O-O}{<}}\!\!\underset{}{\overset{}{-}}\underset{(CN)_2}{C} \qquad (33)$$

Mares has also added support to this mechanism by showing that octene-1 is oxidized to unlabeled 2-octanone using $RhCl(AsPh_3)_3$ as catalyst in the presence of $^{16}O_2$ and $^{18}OH_2$,[114] thereby making it unlikely that a Wacker-type mechanism is operable.

In an elegant extension of this work, Mimoun has used the compound $[RhO_2L_4]X$ ($L = AsPh_3$, $AsPhMe_2$; $X = ClO_4$, $PF_6$) to convert octene-1 into 2-octanone, and he has shown that oxygen from $^{18}O_2$ is exclusively incorporated into the methyl ketone.[115] Mimoun has also shown that $RhCl_3$ and $Cu(ClO_4)_2$, or even $Rh(ClO_4)_3$ alone, will convert terminal alkenes into methyl ketones, and he proposed that this mechanism involves a peroxymetalation pathway coupled with a Wacker cycle.[116] The hydroxyl group for the Wacker step in the cycle is produced in the protonation of the oxorhodium species remaining at the termination of the peroxymetalation cycle (Figure 12.3). A similar chemistry has been explored using Rh(I) and Cu(II) immobilized on site-separated organosulfide complexes. As the gel's surface sulfide concentration is decreased the catalyst activity increases, a result interpreted to support the concept that the active catalyst precursor is a monomer.[117]

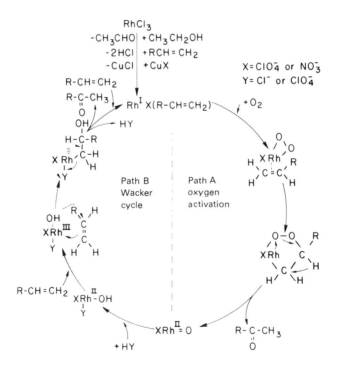

Figure 12.3. Rhodium-catalyzed conversion of alkenes into methyl ketones with molecular oxygen involving both a peroxymetalation and a Wacker cycle.

## 5.2. Alkenes and Hydrogen

This co-oxidation has been used by James[52] to convert a mixture of hydrogen, oxygen, and cyclooctene into water and cyclooctanone. The catalytic intermediate is proposed to be $[IrHCl_2(C_8H_{12})]_2$. The reaction is accompanied by the catalytic hydrogenation of cyclooctene to cyclooctane. No definitive mechanistic details have been given, and no data yet produced to confirm whether the two oxidations occur by a coupled pathway.

## 5.3. Isocyanides and Carbon Monoxide

A brief report on the reaction of CO with $NiO_2(tert\text{-BuNC})_2$ at 20° in chlorobenzene solvent shows that both $CO_2$ and $t$-BuNCO are formed (34).[72] The yield of isocyanate is substantial, but again no mechanistic details are given as to whether the oxygen transfer reactions occur by separate or integrated pathways.

$$NiO_2(t\text{-BuNC})_2 + 4CO \rightarrow Ni(CO)_2(t\text{-BuNC})_2 + 2CO_2 + \text{``}t\text{-BuNCO''} \quad (34)$$

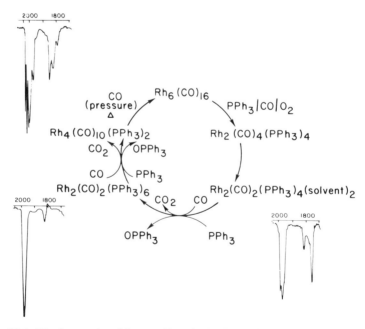

Figure 12.4. Rhodium carbonyl intermediates in the $Rh_6(CO)_{16}$ catalyzed co-oxidation of carbon monoxide and triphenylphosphine.

## 5.4. Triphenylphosphine and Carbon Monoxide

The compound $Rh_6(CO)_{16}$, in the presence of triphenylphosphine, carbon monoxide, and oxygen, will catalyze the oxidation of reduced substrates to form triphenylphosphine oxide and carbon dioxide.[80-81] The solution changes color variously through red and yellow during the catalyzed oxidation, and using infrared spectroscopy in the carbonyl region, the solution species $Rh_2(CO)_4(PPh_3)_4$, $Rh_2(CO)_2(PPh_3)_4(solvent)_2$, $Rh_2(CO)_2(PPh_3)_6$, and $Rh_4(CO)_{10}(PPh_3)_2$ have been identified (Figure 12.4). The oxidation of CO only occurs when $PPh_3$ is present, and after complete conversion to $OPPh_3$ the formation of $CO_2$ ceases. Decreasing the pressure of oxygen while maintaining constant carbon monoxide and other concentrations results in a drop in the quantity of $CO_2$ produced. There is no 1:1 correspondence in the quantity of carbon dioxide and triphenylphosphine oxide formed—the amounts vary with time—however, the ratio of $CO_2:OPPh_3$ always remains above unity. Addition of water to the reaction mixture produces an increase in the quantity of $CO_2$ formed. Carrying out the reaction in the presence of $^{18}O_2$ and $^{16}O_2/^{18}OH_2$ confirms that dioxygen is the only oxygen atom source for $OPPh_3$. Analysis of the formed carbon dioxide concludes that no oxygen-18 label is incorporated into $CO_2$ from $^{18}OH_2$, provided the contact time between the $CO_2$ formed and the $^{18}OH_2$ present in the reaction mixture is short. This latter condition is necessary because the rhodium complexes in solution will catalyze the oxygen exchange between $CO_2$ and water, even under the neutral pH conditions of the reaction (35, 36):

$$CO + PPh_3 + O_2 \xrightarrow{Rh(O)} CO_2 + OPPh_3 \quad (35)$$

$$CO_2 + {}^{18}OH_2 \underset{}{\overset{Rh(O)}{\rightleftharpoons}} CO_2 + {}^{18}OCO + {}^{18}OC{}^{18}O \quad (36)$$

The suggested mechanistic pathways for these observations are shown in Figure 12.5.

## 6. OXYGEN ATOM TRANSFER FROM METAL PHOSPHINE HYDROPEROXIDES AND SUPEROXIDES

The involvement of hydroperoxy metal complexes in oxidation reactions has been variously postulated and discussed.[4,6,11,118,119] Metal hydroperoxides and alkylperoxides can be potentially prepared from hydrogen peroxide or alkylperoxides. Alternately, it is apparent that since a dioxygen molecule coordinated to a low-valent phosphine metal complex is an

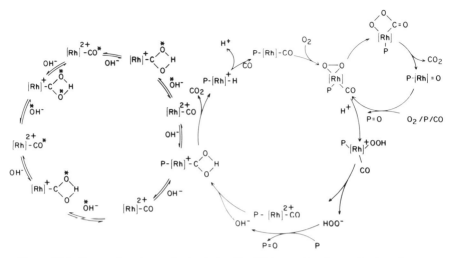

Figure 12.5. Proposed pathways for both the $Rh_6(CO)_{16}$ catalyzed co-oxidation of carbon monoxide and triphenylphosphine, and the catalyzed oxygen atom exchange between carbon dioxide and water.

electron rich center, it can also be protonated or alkylated by electrophile to produce metal hydroperoxides or alkylperoxides. Such compounds can be anticipated to show behavior resembling peroxides, and thus allow one to do oxygen atom transfer chemistry from peroxidic compounds where the initial oxygen source is molecular oxygen (37):

$$M\underset{O}{\overset{O}{\diagup\!\!\!\!|\!\!\!\!\diagdown}} \quad \begin{array}{c} \xrightarrow{H^+} \overset{+}{M}OOH \\ \xrightarrow{R^+} \overset{+}{M}OOR \end{array} \qquad (37)$$

### 6.1. Metal Hydroperoxides and Alkylperoxides

When the compounds $Pt(OH)(CF_3)L_2$ and $PtH(CF_3)L_2$ ($L_2$ = dppe, $2PPh_2Me$) are treated with hydrogen peroxide, the hydroperoxy platinum(II) compound $Pt(OOH)(CF_3)L_2$ is formed (38).[120,121]

$$PtH(CF_3)L_2 + H_2O_2 \rightarrow Pt(OOH)(CF_3)L_2 + H_2 \qquad (38)$$

The hydroperoxide compounds having a range of ligands $L$ are isolable, and they react with $PPh_3$, CO, and NO to form the oxidized products $OPPh_3$, $CO_2$, and $HNO_2$. No reaction is observed with added cyclohexene.

The compound $PtO_2(PPh_3)_2$ under acidic conditions undergoes protonation at oxygen with the formation of hydrogen peroxide.[122,123] The reaction with $HX$ ($X$ = $ClO_4$, $BF_6$, $NO_3$) proceeds in a stepwise manner, and the complexes $[Pt_2(O_2)(OH)(PPh_3)_4]X$ and $[Pt_2(OH)_2(PPh_3)_4]X_2$ have been isolated (39, 40).

$$2PtO_2(PPh_3)_2 + HX + H_2O \rightarrow [Pt_2(O_2)(OH)(PPh_3)_4]Y + H_2O_2 \qquad (39)$$

$$[Pt_2(O_2)(OH)(PPh_3)_4]X + HX + H_2O \rightarrow [Pt_2(OH)_2(PPh_3)_4]X + H_2O_2 \qquad (40)$$

The structure of the formed peroxo-bridged compound shows an O–O separation of 1.547(21) Å, and treatment of the compound with $SO_2$ yields a product having a bidentate sulfate ligand. Similarly, when $MO_2(PPh_3)_2$ ($M$ = Pd, Pt) is treated with PhC(O)NHOH at room temperature, hydrogen peroxide is formed along with an aroylhydroxylamido complex of platinum (41).[124]

$$MO_2(PPh_3)_2 + PhC(O)HOH \rightarrow M(ONC(O)Ph)(PPh_3)_2 + H_2O_2 \qquad (41)$$

$$(M = Pd, Pt)$$

A similar reaction occurs with the hydrazine derivative $PhC(O)NHNH_2$.

Addition of a stoichiometric amount of $MeSO_3H$ to an anhydrous solution of $PdO_2(PPh_3)_2$ in $CH_2Cl_2$ and octene-1 gives 2-octanone and $OPPh_3$.[125] The quantity of 2-octanone formed is proportional to the quantity of added $MeSO_3H$, until a unit ratio of $MeSO_3H : PdO_2(PPh_3)_2$ is obtained, after which the yield of 2-octanone remains constant. Other terminal alkenes such as hexene-1 behave similarly, but internal alkenes such as norbornene are unreactive. A mechanism is proposed involving a cationic palladium hydroperoxide intermediate (Figure 12.6). Further work

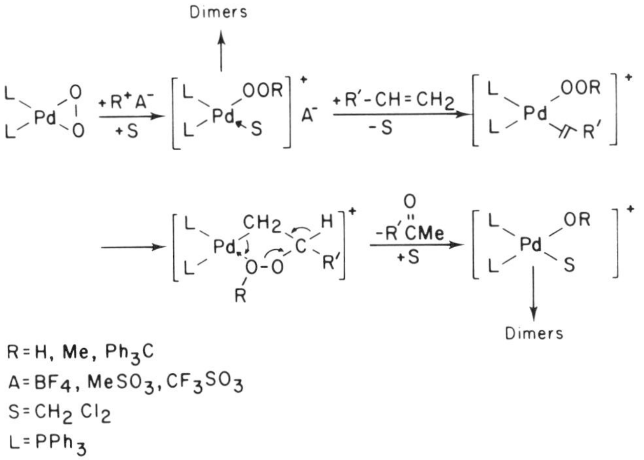

R = H, Me, $Ph_3C$
A = $BF_4$, $MeSO_3$, $CF_3SO_3$
S = $CH_2Cl_2$
L = $PPh_3$

Figure 12.6. Proposed mechanism for the $PdO_2(PPh_3)_2$ catalyzed conversion of terminal alkenes to methyl ketones with oxygen in acid medium.

on this project by Mimoun has shown that *tert*-butylperoxypalladium(II) complexes can also be used for the selective oxidation of terminal alkenes to methyl ketones,[126,127] and he presents a case whereby pseudocyclic peroxymetalation can lead to either epoxidation or ketonization (42).[128,129] Strong electrophilic metals such as Mo(VI) and V(V) direct the electron transfer toward the alkyl carbon atom, producing epoxide, while less electrophilic Group VIII metals direct electron transfer toward the $\beta$-carbon atom, from which $\beta$-hydride migration produces methyl ketone.

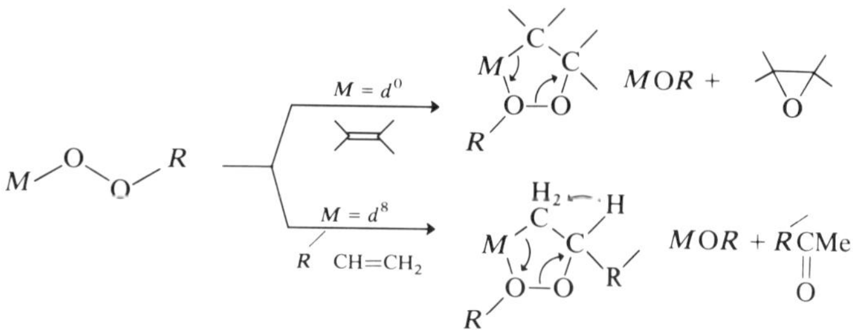

$$M\diagdown O\diagdown O\diagdown R \quad \begin{cases} \xrightarrow{M=d^0} & \text{epoxide pathway} \to MOR + \text{epoxide} \\ \xrightarrow[R\ CH=CH_2]{M=d^8} & \text{ketone pathway} \to MOR + R\overset{O}{C}Me \end{cases} \quad (42)$$

## 6.2. Metal Peracyls

When the compound $PtO_2(PPh_3)_2$ is reacted with PhCOCl at $-78°$, a peroxy-benzoate complex $PtCl(OOCOPh)(PPh_3)_2$ is formed (43).[130]

$$PtO_2(PPh_3)_2 + PhCOCl \xrightarrow{-78°} PtCl(OOCOPh)(PPh_3)_2 \qquad (43)$$

This thermally unstable compound reacts with triphenylphosphine to form the oxide, and with alkenes to form epoxides. Yields of up to 50% epoxide have been obtained using the alkenes norbornene and cyclohexene. Otsuka *et al.* have recently isolated the bromo analog compound $PtBr(OOCOPh)(PPh_3)_2$ in a pure state and have shown that it will oxidize norbornene to *exo*-norbornene epoxide in 40% yield.[44]

## 7. OXYGEN ATOM TRANSFER FROM COORDINATED NITRITE LIGANDS

### 7.1. Transfer from Metal Nitrites to Carbon Monoxide and Triphenylphosphine

The reaction of nitrite complexes with carbon monoxide to yield the metal nitrosyl complex and carbon dioxide has been known for a considerable time.[131-134] The initial work with $Ni(NO_2)_2(PEt_3)_2$ and carbon monoxide yielded the nitrosyl complex $Ni(NO)(NO_2)(PEt_3)_2$ (44).[134]

$$Ni(NO_2)_2(PEt_3)_2 + CO \rightarrow Ni(NO)(NO_2)(PEt_3)_2 + CO_2 \qquad (44)$$

The reaction is rapid under ambient conditions, and further work has shown that the $PEt_3$ ligand can be replaced by dppe, $PMe_2Ph$, $Ph_2PCH = CHPPh_2$.[135] The rate law for the reaction is $-d[Ni(NO_2)_2L_2]/dt = k_2[Ni(NO_2)_2L_2][CO]$, and an associative mechanism (45) is proposed.

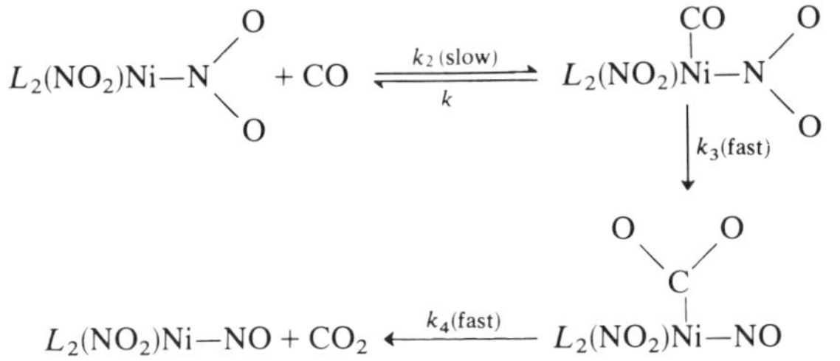

$$L_2(NO_2)Ni-N\begin{matrix}O\\\\O\end{matrix} + CO \underset{k}{\overset{k_2 (slow)}{\rightleftharpoons}} L_2(NO_2)Ni-N\begin{matrix}CO\\\\O\end{matrix}\begin{matrix}O\\\\O\end{matrix} \\ \downarrow k_3 \text{(fast)} \\ L_2(NO_2)Ni-NO + CO_2 \xleftarrow{k_4 \text{(fast)}} L_2(NO_2)Ni-N\begin{matrix}O\\\\C\\\\O\end{matrix} \qquad (45)$$

A similar reaction with $Pd(NO_2)_2(PMePh_2)_2$ and CO produces $CO_2$ and the cluster compound $Pd_4(CO)_5(PMePh_2)_4$.[136] Further work on this reaction using oxygen-18-labeled nitrite has confirmed this ligand to be the source of oxygen, and a cyclic intermediate is proposed for the oxygen transfer step.[137]

Similarly, complexes $NiL_4$ ($L$ = tertiary phosphine) react with aliphatic and aromatic nitro compounds $RNO_2$ to yield the complex $Ni(RNO)L_2$ and phosphine oxide (46).[138] The ligand RNO is kinetically labile and readily displaced by phosphine $L$ (47), thereby making it possible to design a catalytic oxidation cycle. For aliphatic nitroalkanes, oxygen transfer is retarded both

$$NiL_4 + RNO_2 \rightarrow Ni(RNO)L_2 + L + L=O \qquad (46)$$

$$Ni(RNO)L_2 + 2L \rightarrow NiL_4 + RNO \qquad (47)$$

by branching at the $\alpha$ carbon and by electron release from the alkyl group. It is believed that the mechanism involves an electron-transfer process to form $Ni^I L_3 \cdot RNO_2^-$ prior to oxygen-atom transfer, and that neither N–O bond-breaking nor P–O bond-making steps are rate-limiting. A cyclic intermediate (48) is considered to be possible, but this suggestion is not proven.

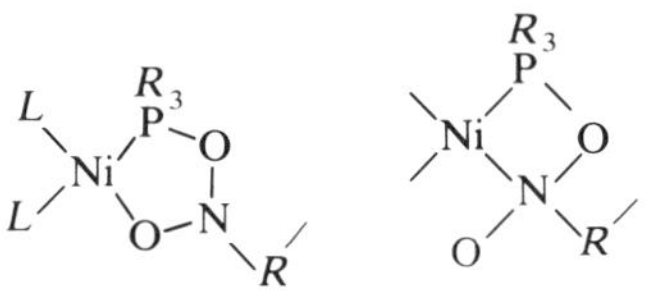

$$\begin{matrix} L & R_3 \\ & P-O \\ Ni & \\ L & O-N \\ & R \end{matrix} \qquad \begin{matrix} & R_3 \\ & P \\ Ni & O \\ & N \\ O & R \end{matrix} \qquad (48)$$

Johnson, in a series of articles, has investigated oxygen-atom transfer from nitrite complexes of Ni, Pd, and Pt.[139–142] In addition to the catalytic oxidation of CO to $CO_2$ with molecular oxygen, using $Ni(NO_2)_2$dppe as catalyst, these workers also observe some isocyanate formation on treating $Pt(NO_2)_2(PEt_3)_2$ with CO (49).

$$Pt(NO_2)_2(PEt_3)_2 + 2CO \rightarrow Pt(NCO)(NO_2)(PEt_3)_2 + CO_2 \qquad (49)$$

## 7.2. Transfer from Metal Nitrites to Alkenes

Other metal complexes that will undergo oxygen-atom transfer from a coordinated nitrite are $M(NO_2)_2(CO)_2(PPh_3)_2$ ($M$ = Ru, Os), which oxidizes CO;[143] Co(N,N-Bisalicylidene-o-phenylenediamino)$py \cdot NO_2$, which oxidizes $PPh_3$;[144] $Ni(NO_2)_2(PBu_3)_2$, which catalyzes the oxidation of $t$-BuNC to $t$-BuNCO;[145] Co(tetraphenylpophyrin)$py \cdot NO_2$, which catalyzes alkene oxidation;[146,147] and $PdCl(MeCN)_2NO_2$, which also catalyzes alkene oxidation.[148] The complex Co(tetraphenylporphyrin)$py \cdot NO_2$, which reacts by transfer of an oxygen atom to the alkene, can be considered as having a weak, oxygen-centered nucleophile on the $NO_2^-$ ligand. The alkene is complexed to palladium(II), such that it is susceptible to nucleophilic attack. Using this concept, catalytic amounts of acetaldehyde and acetone have been obtained from ethylene and propylene, respectively. No palladium precipitates from the reaction mixture, since the formed Pd(O) species are being reoxidized by the cobalt nitrite complexes. Thus, the process involves oxygen transfer from the nitro ligand of cobalt-nitro complexes to palladium(II)-bound alkene, followed by reoxidation of the reduced nitrosyl ligand by molecular oxygen. The palladium serves exclusively as a cocatalyst.[146] In the conversion of decene-1 into 2-decanone using $PdCl(MeCN)_2NO_2$, an oxygen-18-labeling study verifies that the ketonic oxygen atom originates from the nitro group. Spectroscopic evidence is presented for two intermediates, an alkene complex and a species derived from attack by the nitro group on coordinated alkene.[148] These alkene oxidations using oxygen-atom transfer from a coordinated nitrite ligand offer advantages of selectivity over catalytic pathways involving complexes having dioxygen directly bonded to the metal center.

## ACKNOWLEDGMENTS

We thank M. A. Andrews, R. D. Feltham, W. A. Goddard III, B. R. James, H. Mimoun, and J. Valentine for making articles available prior to publication. Our own research in this field is supported by the Division of

Chemical Sciences of the U.S. Department of Energy, to whom we owe thanks.

## REFERENCES

1. S. Benson and P. S. Nangia, *Accts. Chem. Res.* **12**, 223–228 (1979).
2. O. Hayaishi, *Molecular Mechanisms of Oxygen Activation* (Academic Press, New York, 1974).
3. T. G. Spiro, ed., *Metal Ion Activation of Dioxygen* (John Wiley and Sons, New York, 1980). Vol. 2 in Metal Ions in Biology.
4. J. Lyons, *Aspects of Homogeneous Catalysis*, edited by R. Ugo (Reidel, Boston, 1977), Vol 3, pp. 1–136.
5. ———, *Fundamental Research in Homogeneous Catalysis*, edited by M. Tsutsui and R. Ugo (Plenum Publishing, New York, 1977), pp. 1–52.
6. G. Modena, K. B. Sharpless, G. Costa, J. Halpern, Y. Ishii, B. R. James, J. E. Lyons, H. Minoun, P. Rossi, R. A. Sheldon, and P. Teyessie, *Fundamental Research in Homogeneous Catalysis*, edited by M. Tsutsui and R. Ugo (Plenum, New York, 1977), pp. 193–205.
7. R. Stewart, *Oxidation in Organic Chemistry*, Part A, edited by K. B. Wiberg (Academic Press, New York, 1965), pp. 1–68.
8. K. B. Wiberg, *Oxidation in Organic Chemistry*, Part A, edited by K. B. Wiberg (Academic Press, New York, 1965), pp. 69–184.
9. W. A. Waters and J. S. Littler, *Oxidation in Organic Chemistry*, Part A, edited by K. B. Wiberg (Academic Press, New York, 1965), pp. 185–241.
10. D. G. Lee and M. van der Engh, *Oxidation in Organic Chemistry*, Part B, edited by W. S. Trahanovsky (Academic Press, New York, 1973), pp. 177–227.
11. R. A. Sheldon and J. K. Kochi, *Catalyzed Oxidations of Organic Compounds* (Academic Press, New York, 1981).
12. L. H. Chinn, *Selection of Oxidants in Synthesis* (Marcel Dekker, New York, 1971).
13. P. M. Henry, *Palladium Catalyzed Oxidation of Hydrocarbons* (Riedel, Boston, 1980).
14. L. Reich and S. S. Stivala, *Autoxidation of Hydrocarbons and Polyolefins* (Marcel Dekker, New York, 1969).
15. B. Plesnicar, *Oxidation in Organic Chemistry*, Part C, edited by W. S. Trahanovsky (Academic Press, New York, 1978), pp. 211–294.
16. L. Vaska, *Accts. Chem. Res.* **9**, 175–183 (1976).
17. L. H. Vogt, Jr., H. M. Faigenbaum, and S. E. Wiberley, *Chem. Rev.* **63**, 269–277 (1963).
18. J. Valentine, *Chem. Rev.* **73**, 235–245 (1973).
19. V. J. Choy and C. J. O'Connor, *Coord. Chem. Rev.* **9**, 145–170 (1972).
20. R. S. Drago and B. B. Corden, *Acc. Chem. Res.* **13**, 353–360 (1980).
21. F. Basolo, B. M. Hoffman, and J. A. Ibers, *Acc. Chem. Res.* **8**, 384–392 (1975).
22. K. B. Sharpless and T. R. Verhoeven, *Aldrichimica Acta.* **12**, 63–74 (1979).
23. A. K. Rappe and W. A. Goddard III, *J. Am. Chem. Soc.* **104**, 3287–3294 (1982).
24. T. Katsuki and K. B. Sharpless, *J. Am. Chem. Soc.* **102**, 5974–5976 (1980).
25. L. Vaska, *Science* **140**, 809–810 (1963).
26. C. D. Cook and G. S. Jauhal, *Inorg. Nucl. Chem. Lett.* **3**, 31–33 (1967).
27. R. Ugo, G. LaMonica, F. Cariati, S. Cenini, and F. Conti, *Inorg. Chim. Acta* **4**, 390–394 (1970).
28. S. Otsuka, A. Nakamura, and Y. Tatsuno, *J. Chem. Soc., Chem. Comm.* 836 (1967).

29. ———, *J. Am. Chem. Soc.* **91**, 6994–6999 (1969).
30. M. J. Nolte, E. Singleton, and M. Laing, *J. Am. Chem. Soc.* **97**, 6396–6400 (1975).
31. B. W. Graham, K. R. Laing, C. J. O'Connor, and W. R. Roper, *J. Chem. Soc., Chem. Comm.* 1272 (1970).
32. H. Suzuki, K. Mizutani, Y. Moro-oka, and T. Ikawa, *J. Am. Chem. Soc.* **101**, 748–749 (1979).
33. F. Sakurai, H. Suzuki, Y. Moro-oka, and T. Ikawa, *J. Am. Chem. Soc.* **102**, 1749–1751 (1980).
34. Y. Moro-oka, H. Suzuki, R. Sugimoto, F. Sakurai, and T. Ikawa, *Abstr. Xth. International Conf. Organometallic Chemistry*, Toronto, Abstr. 5DO4.
35. A. Vogler and H. Kunkely, *J. Am. Chem. Soc.* **103**, 6222–6223 (1981).
36. J. J. Levison and S. D. Robinson, *J. Chem. Soc.* (A) 762–767 (1971).
37. P. J. Hayward, D. M. Blake, G. Wilkinson, and C. J. Nyman, *J. Am. Chem. Soc.* **92**, 5873–5878 (1970).
38. R. Ugo, F. Conti, S. Cenini, R. Mason, and G. B. Robertson, *J. Chem. Soc., Chem. Commun.* 1498–1499 (1968).
39. R. W. Horn, E. Weissberger, and J. P. Collman, *Inorg. Chem.* **91**, 2367–2371 (1970).
40. J. S. Valentine, D. Valentine, and J. P. Collman, *Inorg. Chem.* **10**, 219–225 (1971).
41. R. Ugo, G. M. Zanderighi, A. Fusi, and D. Carreri, *J. Am. Chem. Soc.* **102**, 3745–3751 (1980).
42. W. B. Beaulieu, G. D. Mercer, and D. M. Roundhill, *J. Am. Chem. Soc.* **100**, 1147–1152 (1978).
43. H. C. Clark, A. B. Goel, and C. S. Wong, *J. Am. Chem. Soc.* **100**, 6241–6243 (1978).
44. Y. Tatsuno and S. Otsuka, *J. Am. Chem. Soc.* **103**, 5832–5839 (1981).
45. W. J. Louw, T. I. A. Gerber, and D. J. A. de Waal, *J. Chem. Soc., Chem. Commun.* 760–761 (1980).
46. J. G. Norman, Jr. *Inorg. Chem.* **16**, 1328–1335 (1977).
47. R. Criegee, *Angew. Chemie. Int. Ed. Engl.* **14**, 745–752 (1975).
48. D. D. Fong and R. L. Kuczkowski, *J. Am. Chem. Soc.* **102**, 4763–4768 (1980).
49. S. E. Jacobson, R. Tang, and F. Mares, *J. Chem. Soc., Chem. Commun.* 888–889 (1978).
50. J. P. Collman, M. Kubota, and J. W. Hosking, *J. Am. Chem. Soc.* **89**, 4809–4811 (1967).
51. A. Fusi, R. Ugo, A. Pasini, and S. Cenini, *J. Organomet. Chem.* **26**, 417–430 (1971).
52. B. R. James, *Adv. Chem. Ser.* **191**, 253–276 (1980).
53. R. VaAtta, J. Burstyn, and J. S. Valentine, *Reactions of Coordinated Ligands*, edited by P. S. Braterman (Plenum, New York, 1983), in press.
54. H. Mimoun, *Pure Appl. Chem.* **53**, 2389–2399 (1981).
55. R. A. Sheldon and J. K. Kochi, *Adv. Catal.* **25**, 272–413 (1976).
56. F. Haber and J. Weiss, *Proc. Roy. Soc.* **147**, 332–351 (1934).
57. A. Fusi, R. Ugo, and G. M. Zanderighi, *J. Catal.* **34**, 175–190 (1974).
58. S. Cenini, A. Fusi, and G. Capparella, *J. Inorg. Nucl. Chem.* **33**, 3576–3579 (1971).
59. ———, *Inorg. Nucl. Chem. Lett.* **8**, 127–131 (1972).
60. V. P. Kurkov, J. Z. Pasky, and J. B. Lavigne, *J. Am. Chem. Soc.* **90**, 4743–4744 (1968).
61. H. Arzoumanian, A. Blanc, U. Hartig, and J. Metzger, *Tetrahedron Lett.* 1011–1014 (1974).
62. H. Arzoumanian, A. A. Blanc, J. Metzger, and J. E. Vincent, *J. Organomet. Chem.* **82**, 261–270 (1974).
63. A. A. Blanc, H. Arzoumanian, E. J. Vincent, and J. Metzger, *Bull. Chim. Soc. Fr.* 2175–2179 (1974).
64. W. Strohmeier and E. Eder, *J. Organomet. Chem.* **94**, C14–C19 (1975).
65. S. Cenini, A. Fusi, and F. Porta, *Gazz. Chim. Ital.* **108**, 109–114 (1978).
66. K. Takao, Y. Fujiwara, T. Imanaka, and S. Teranishi, *Bull. Chem. Soc. Jpn.* **43**, 3898–3900 (1970).

67. K. Takao, M. Wayaku, Y. Fujiwara, T. Imanaka, and S. Teranishi, *Bull. Chem. Soc. Jpn.* **43**, 3898-3900 (1970).
68. J. E. Lyons and J. O. Turner, *Tetrahedron Lett.* 2903-2906 (1972).
69. ———, *J. Org. Chem.* **37**, 2881-2884 (1972).
70. F. Mares and R. Tang, *J. Org. Chem.* **43**, 4631-4632 (1978).
71. B. H. van Vugt and W. Drenth, *Recl. Trav. Chim. Pays-Bas.* **96**, 225-229 (1977).
72. S. Otsuka, A. Nakamura, Y. Tatsuno, and M. Miki, *J. Am. Chem. Soc.* **94**, 3761-3767 (1972).
73. J. Halpern and A. C. Harkness, *J. Am. Chem. Soc.* **83**, 1258-1259 (1961).
74. J. Halpern and S. Nakamura, *J. Am. Chem. Soc.* **83**, 4102-4103 (1961).
75. B. R. James and G. L. Rempel, *J. Chem. Soc. A.*, 78-84 (1969).
76. J. E. Bercaw, L.-Y. Goh, and J. Halpern, *J. Am. Chem. Soc.* **94**, 6534-6536. (1972).
77. G. D. Mercer, J. S. Shu, T. B. Rauchfuss, and D. M. Roundhill, *J. Am. Chem. Soc.* **97**, 1967-1968 (1975).
78. C. S. Chin, M. S. Sennett, and L. Vaska, *J. Mol. Catal.* **4**, 375-378 (1978).
79. K. L. Watters, R. F. Howe, T. P. Chojnacki, C.-M. Fu, R. L. Schneider, and N. B. Wong, *J. Catal.* **66**, 424-440 (1980).
80. D. M. Roundhill, M. K. Dickson, N. S. Dixit, and B. P. Sudha-Dixit, *Adv. Chem. Ser.* **196**, 291-301 (1982).
81. M. K. Dickson, Ph.D. Thesis, Washington State University (1982).
82. Y. Iwashita and A. Hayata, *J. Am. Chem. Soc.* **91**, 2525-2528 (1969).
83. W. R. Cullen, B. R. James, and G. Strukul, *Canad. J. Chem.* **56**, 1965-1969 (1978).
84. C.-S. Chin, M. S. Sennett, P. J. Wier, and L. Vaska, *Inorg. Chim. Acta.* **31**, L443-L444 (1978).
85. T. Engel and G. Ertle, *Adv. Catal.* **28**, 1-78 (1979).
86. D. M. Roundhill, G. H. Allen, R. A. Bechtold, and W. B. Beaulieu, *Inorg. Chim. Acta.* **54**, L99-L100 (1981).
87. J.-I. Hojo, S. Yuasa, N. Yamazoe, I. Mochida, and T. Seiyama, *J. Catal.* **36**, 93-98 (1975).
88. H. Sakamoto, T. Funabiki, and K. Tarama, *J. Catal.* **48**, 427-429 (1977).
89. M. K. Dickson, B. P. Sudha, and D. M. Roundhill, *J. Organomet. Chem.* **190**, C43-C46 (1980).
90. D. M. Roundhill, M. K. Dickson, N. S. Dixit, and B. P. Sudha-Dixit, *J. Am. Chem. Soc.* **102**, 5538-5542 (1980).
91. D. M. Roundhill, *Proc. of the 2nd International Workshop on Fundamental Research in Homogeneous Catalysis*, edited by Y. Ishii and M. Tsutsui (Plenum, New York, 1978), pp. 11-23.
92. G. D. Mercer, W. B. Beaulieu, and D. M. Roundhill, *J. Am. Chem. Soc.* **99**, 6551-6554 (1977).
93. E. W. Stern, *J. Chem. Soc., Chem. Commun.* 736 (1970).
94. R. A. Sheldon, *J. Chem. Soc., Chem. Commun.* 788-789 (1971).
95. G. Wilke, H. Schott, and P. Heimbach, *Angew. Chemie Int. Ed. Engl.* **6**, 92-93 (1967).
96. J. P. Birk, J. Halpern, and A. L. Pickard, *J. Am. Chem. Soc.* **90**, 4491-4492 (1968).
97. J. Halpern and A. L. Pickard, *Inorg. Chem.* **9**, 2798-2800 (1970).
98. A. Sen and J. Halpern, *J. Am. Chem. Soc.* **99**, 8337-8339 (1977).
99. B. W. Graham, K. R. Laing, C. J. O'Connor, and W. R. Roper, *J. Chem. Soc. Dalton Trans.* 1237-1243 (1972).
100. R. K. Poddar and U. Agarwala, *Inorg. Nucl. Chem. Lett.* **9**, 785-789 (1973).
101. R. L. Augustine and J. Van Peppen, *J. Chem. Soc., Chem. Commun.* 497-498 (1970).
102. B. H. Van Vugt, N. J. Kooke, W. Drenth, and F. P. J. Kiujpers, *Recl. Trav. Chim. Pays-Bas.* **92**, 1321-1325 (1973).

103. J. Halpern, B. L. Goodall, G. P. Khare, H. S. Lim, and J. J. Pluth, *J. Am. Chem. Soc.* **97**, 2301–2303 (1975).
104. D. D. Schmidt and J. T. Yoke, *J. Am. Chem. Soc.* **93**, 637–640 (1971).
105. R. P. Hanzlik and D. Williamson, *J. Am. Chem. Soc.* **98**, 6570–6573 (1976).
106. W.-S. Hwang, I. B. Joedicke, and J. T. Yoke, *Inorg. Chem.* **19**, 3225–3229 (1980).
107. W.-S. Hwang and J. T. Yoke, *J. Org. Chem.* **45**, 2088–2091 (1980).
108. J. Drapier and A. J. Hubert, *J. Organomet. Chem.* **64**, 385–391 (1974).
109. M. S. Jarrell, B. C. Gates, and E. D. Nicholson, *J. Am. Chem. Soc.* **100**, 5727–5732 (1978).
110. C. Dudley and G. Read, *Tetrahedron Lett.* **52**, 5273–5276 (1972).
111. C. W. Dudley, G. Read, and P. J. C. Walker, *J. Chem. Soc., Dalton Trans.* 1926–1931 (1974).
112. G. Read and P. J. C. Walker, *J. Chem. Soc., Dalton Trans.* 883–888 (1977).
113. R. A. Sheldon and J. A. Van Doorn, *J. Organomet. Chem.* **94**, 115–129 (1975).
114. R. Tang, F. Mares, N. Neary, and D. E. Smith, *J. Chem. Soc., Chem. Commun.* 274–275 (1979).
115. F. Igersheim and H. Mimoun, *Nouv. J. de Chimie.* **4**, 161–166 (1980).
116. H. Mimoun, M. M. Perez Machirant, and I. Seree de Roch, *J. Am. Chem. Soc.* **100**, 5437–5444 (1978).
117. E. D. Nyberg and R. S. Drago, *J. Am. Chem. Soc.* **103**, 4966–4968 (1981).
118. A. G. Davies, *Organic Peroxides* (Butterworths, London, 1961).
119. R. A. Sheldon, *Aspects of Homogeneous Catalysis*, edited by R. Ugo (Reidel, Boston, 1981), Vol. 4, pp. 3–70.
120. R. A. Michelin, R. Ros, and G. Strukul, *Inorg. Chim. Acta.* **37**, L491–L492 (1979).
121. G. Strukul, R. Ros, and R. A. Michelin, *Abstr. Xth International Conference Organometallic Chemistry*, Toronto Abstract 5DO6 (1981).
122. S. Bhaduri, L. Casella, R. Ugo, P. R. Raithby, C. Zuccaro, and M. B. Hursthouse, *J. Chem. Soc., Dalton Trans.* 1624–1629 (1979).
123. R. Ugo, A. Fusi, G. M. Zanderighi, and L. Casella, *J. Mol. Catal.* **7**, 51–57 (1980).
124. P. L. Bellon, S. Cenini, F. Demartin, M. Manassero, M. Pizzotti, and F. Porta *J. Chem. Soc., Dalton Trans.* 2060–2067 (1980).
125. F. Igersheim and H. Mimoun, *Nouv. J. de Chimie.* **4**, 711–713 (1980).
126. M. Roussel and H. Mimoun, *J. Org. Chem.* **45**, 5387–5390 (1980).
127. J.-M. Bregeault and H. Mimoun, *Nouv. J. de Chimie.* **5**, 287–289 (1981).
128. H. Mimoun, R. Charpentier, A. Mitschler, J. Fischer, and R. Weiss, *J. Am. Chem. Soc.* **102**, 1047–1054 (1980).
129. H. Mimoun, *J. Mol. Catal.* **7**, 1–29 (1980).
130. M. J. Y. Chen and J. K. Kochi, *J. Chem. Soc., Chem. Commun.* 204–205 (1977).
131. W. Manchot and A. Waldmuller, *Chem. Ber.* **59**, 2363–2366 (1926).
132. W. Hieber and J. S. Anderson, *Z. Anorg. Allgem. Chem.* **208**, 238–248 (1932).
133. W. Hieber and H. Beutner, *Z. Naturforsch. B*, **15**, 323–324 (1960).
134. G. Booth and J. Chatt, *J. Chem. Soc.* 2099–2106 (1962).
135. R. D. Feltham and J. C. Kriege, *J. Am. Chem. Soc.* **101**, 5064–5065 (1979).
136. J. Dubrawski, J. C. Kreige-Simondsen, and R. D. Feltham, *J. Am. Chem. Soc.* **102**, 2089–2091 (1980).
137. D. T. Doughty, G. Gordon, and R. P. Stewart Jr., *J. Am. Chem. Soc.* **101**, 2645–2648 (1979).
138. R. S. Berman and J. K. Kochi, *Inorg. Chem.* **19**, 248–254 (1980).
139. S. Bhaduri, B. F. G. Johnson, A. Khair, A. Pickard, Y. Ben-Taarit, and R. Ugo, *J. Chem. Soc., Chem. Commun.* 694–695 (1976).
140. B. F. G. Johnson, C. J. Savory, J. A. Segal, and R. H. Walter, *J. Chem. Soc., Chem. Commun.* 809–810 (1974).

141. S. Bhaduri, B. F. G. Johnson, A. Khair, I. Ghatak, and D. M. P. Mingos, *J. Chem. Soc., Dalton Trans.* 1582–1576 (1980).
142. S. A. Bhaduri, I. Bratt, B. F. G. Johnson, A. Khair, J. A. Segal, R. Walters, and C. Zuccaro, *J. Chem. Soc., Dalton Trans.* 234–239 (1981).
143. K. R. Grundy, K. R. Laing, and W. R. Roper, *J. Chem. Soc., Chem. Commun.* 1500–1501 (1970).
144. B. S. Tovrog, S. E. Diamond, and F. Mares, *J. Am. Chem. Soc.* **101,** 270–272 (1979).
145. M. A. Andrews, personal communication.
146. B. S. Tovrog, F. Mares, and S. E. Diamond, *J. Am. Chem. Soc.* **102,** 6616–6618 (1980).
147. F. Mares, B. S. Tovrog, and S. E. Diamond, *Abstr. Xth International Conference Organometallic Chemistry*, Toronto, Abstract 5DO5 (1981).
148. M. A. Andrews and K. P. Kelly, *J. Am. Chem. Soc.* **103,** 2894–2896 (1981).

# 13

# Catalysis of Nitrogen-Fixing Model Studies

### T. Adrian George

## NOTATION

| | |
|---|---|
| dppe: | 1,2-Bisdiphenylphosphinoethane, $Ph_2PCH_2CH_2PPh_2$ |
| THF: | Tetrahydrofuran |
| dmpe: | 1,2-Bisdimethylphosphinoethane, $Me_2PCH_2CH_2PMe_2$ |
| Cp: | Penta*hapto*cyclopentadienyl, $\eta^5\text{-}C_5H_5$ |
| dppm: | Bisdiphenylphosphinomethane, $Ph_2PCH_2PPh_2$ |
| dptpe: | 1,2-Bisdi-*p*-tolylphosphinoethane, $(p\text{-tolyl})_2PCH_2CH_2P(totyl\text{-}p)_2$ |
| depe: | 1,2-Bisdiethylphosphinoethane, $Et_2PCH_2CH_2PEt_2$ |
| triphos: | Bis(2-diphenylphosphinoethyl)phenylphosphine, $PhP(CH_2CH_2PPh_2)_2$ |
| Cy: | Cyclohexyl |
| DME: | 1,2-Dimethoxyethane |
| dppp: | 1,3-Bisdiphenylphosphinopropane, $Ph_2PCH_2CH_2CH_2PPh_2$ |

## 1. INTRODUCTION

The catalytic fixation of dinitrogen is accomplished biologically and abiologically. In nature, both symbiotic and nonsymbiotic nitrogen-fixing microorganisms reduce dinitrogen from the atmosphere or soil to ammonia.[1-4] A common denominator among these microorganisms is the

---

*T. Adrian George* • Department of Chemistry, University of Nebraska-Lincoln, Lincoln, Nebraska 68588.

enzyme nitrogenase (comprising a molybdenum-iron [MoFe] protein and an iron protein), which has been isolated from a number of diverse microorganisms: anaerobic species (e.g., *Clostridium pasteurianum*), strict aerobes (e.g., *Azobacter vinelandii*), and facultative aerobes (e.g., *Klebsiella pneumoniae*).[5,6] The requirements for *in vitro* reduction of dinitrogen to ammonia by nitrogenase are (i) a reducing agent (such as $Na_2S_2O_4$); (ii) a divalent cation (usually $Mg^{+2}$); (iii) ATP; (iv) an anaerobic environment; and (v) a controlled pH.[7] Of special interest to the chemist is the nature of the active site, the site at which dinitrogen is reduced to ammonia in the enzyme, and the mechanism of the reaction. A number of significant steps toward the elucidation of the nitrogenase reaction have been made recently. Among these are (i) the isolation of an iron-molybdenum cofactor (FeMoco) with a molecular mass of less than 5000 Daltons that will reconstitute the cofactorless MoFe protein from a number of mutant organisms;[8] (ii) application of X-ray Absorption Spectroscopy, particularly the analysis of the Extended X-ray Absorption Fine Structure, to determine the coordination environment of molybdenum in FeMoco and the MoFe proteins isolated from nitrogenase of a number of organisms;[9] (these data have established both sulfur and iron atoms as nearest neighbors for the molybdenum atom in the MoFe proteins and FeMoco) and (iii) detection of an intermediate dinitrogen hydride, which is formed on the enzyme when it is actively fixing dinitrogen.[10] The biological fixation of dinitrogen is best represented by Equation (1), in which dihydrogen production is a specific part of the nitrogenase reaction[11]:

$$N_2 + 8e^- + 8H^+ \rightarrow 2NH_3 + H_2 \qquad (1)$$

Abiologically, dinitrogen is catalytically converted to ammonia in the Haber Process.[12] Dinitrogen and dihydrogen react to form ammonia on a promoted ($K_2O$, $MgO$, $CaO$, $BaO$, $MoO_3$, etc.) iron-alumina catalyst at around 350–1000 atmospheres pressure and 300–400°C. While the catalyst in this process is cheap, an incentive to discover a better catalyst is provided by the dramatic increases in the cost of energy, which is required for the high temperatures and pressures, in particular of natural gas, from which most dihydrogen is obtained for this process by the reforming reaction (methane and steam). Economic advantages would be gained by either finding a catalyst that will enable the Haber Process to run under milder conditions or by discovering a cheaper source of dihydrogen, or both.

## 1.1. Scope and Limitations

Nitrogen fixation can be defined as any chemical reaction in which dinitrogen is a reagent, although it is generally thought of in terms of reactions in which ammonia, hydrazine, or organonitrogen compounds are formed. Nitrogen fixation may involve oxidation or reduction, although only examples of the latter will be encountered in this chapter since no examples of oxidation have been reported within the limits imposed in this chapter (see below). A logical starting point for any discussion of nitrogen fixation is the preparation of complexes containing ligating dinitrogen that have been formed in reactions in which dinitrogen is a reagent (e.g., Equation 2)[13]:

$$[Ru(NH_3)_5(H_2O)]^{+2} + N_{2(aq)} \rightarrow [Ru(NH_3)_5(N_2)]^{+2} + H_2O \qquad (2)$$

This follows naturally from (i) the assumption that nitrogenase binds dinitrogen at a metal center prior to its reduction to ammonia, and (ii) the fact that the rate-determining step in the Haber Process is the binding of dinitrogen at the surface of the catalyst.[14] In many cases, direct interaction of dinitrogen with metals or highly reduced metal complexes results in a nitriding reaction. Subsequent hydrolysis yields ammonia and/or hydrazine. However, reactions in which dinitrogen is generated during the reaction and then becomes ligated do not fit within this definition of nitrogen fixation (e.g., Equation 3[15]):

$$trans\text{-}[IrCl(CO)(PPh_3)_2] + RN_3 \xrightarrow[\text{EtOH}]{\text{CHCl}_3} trans\text{-}[IrCl(N_2)(PPh_3)_2] + R\text{NHCOOEt} \qquad (3)$$

Discussion of nitrogen fixation within this chapter is limited by the key words *homogeneous catalysis* and *metal phosphine complexes*. Currently, there are no examples of homogeneous, catalytic nitrogen fixation involving metal-phosphine complexes. However, there are examples of both heterogeneous and homogeneous catalysis involving other metal complexes that will be reviewed in Section 3. Justification for devoting a chapter to the subject of nitrogen fixation is not difficult to find. First, a number of vitally important, naturally occurring microorganisms exist that effect homogeneous catalytic nitrogen fixation. This provides a stimulus to try and emulate the natural system and, with an understanding of the mechanism of nitrogenase, contributes to the general knowledge of catalysis. Secondly, the large strides that have been made in stoichiometric homogeneous nitrogen fixation contribute toward a unifying view of reductive nitrogen fixation. From these studies it is hoped that realistic, catalytic nitrogen-fixing systems will evolve. Thirdly, there are economic, political, and social pressures to supply sufficient food to feed the ever increasing world population; these pressures provide an incentive to develop alternative nitrogen-fixing systems, regardless of whether they are oxidative, reductive, heterogeneous, or homogeneous.

In limiting this chapter to metal-phosphine complexes, many significant compounds and nitrogen-fixing systems will not be discussed. However, interested readers are referred to a number of relatively recent reviews of the chemistry of nitrogen fixation.[16-18] Similarly, adequate descriptions of the bonding and physical properties of the metal dinitrogen complexes have appeared elsewhere and will only be mentioned in this chapter where relevant.

This chapter is organized in the following manner. It is assumed that coordination of dinitrogen to a transition metal is the initial step in all nitrogen-fixing reactions in which a transition metal is present. Therefore, only those systems in which it has been clearly demonstrated that chemistry beyond that point has occurred will be discussed. Well-characterized complexes containing coordinated dinitrogen and phosphine ligands will be listed in Table 1, but only if dinitrogen is the source of the ligating dinitrogen. The preparation of these complexes will be referenced in Table 1 but not discussed. Reactions of coordinated dinitrogen will be presented and discussed according to the metal to which it is bonded. Reactions that have been observed and will be reviewed include the formation of nitrogen–hydrogen and nitrogen–carbon bonds, as well as coordination of ligated dinitrogen to a second metal. In a few cases nitrogen fixation has been shown to occur without isolation of an intermediate complex containing coordinated dinitrogen and phosphine ligands. These reactions will be included because in a successful homogeneous nitrogen-fixing system it is unlikely that any intermediate dinitrogen complex will be isolated (or identified). The major part of this chapter will be devoted to the chemistry of dinitrogen complexes of molybdenum, rhenium, and tungsten, together with recent developments of tantalum. To date, most of the chemistry of dinitrogen coordinated to other metals is limited to displacement of dinitrogen or reaction at the metal. Reactions of ligated nitrogen-hydrides and similar species (e.g., Equation 4[19]) will be omitted from consideration unless they bear directly upon the mechanism of nitrogen-fixing reactions.

$$[MoF(NNH_2)_2(dppe)_2]^+ + RCHO \xrightarrow{H^+} [MoF(NNCHR)(dppe)_2]^+ + H_2O \quad (4)$$

## 2. NITROGEN-FIXING REACTIONS

The majority of nitrogen-fixing reactions begin with a preformed dinitrogen complex. These complexes show a wide range of stabilities. Among the most stable are the molybdenum and tungsten complexes containing the dppe ligand; $trans\text{-}M(N_2)_2(dppe)_2$, ($M$ = Mo, W). These two complexes have been the most thoroughly studied and together with similar complexes containing monodentate ligands have displayed the most diverse chemistry of coordinated dinitrogen.

Briefly, the dinitrogen complexes listed in Table 1 are prepared by a number of different methods. The most common method involves the reduction in an ether solvent under a dinitrogen atmosphere of either a phosphine complex of a metal halide or a mixture of metal salt and appropriate phosphine. The only other frequently used method is ligand displacement by dinitrogen. To illustrate these methods, a number of high-yield preparations of $trans$-$Mo(N_2)_2(dppe)_2$ are given in Equations 5-8[20-22]:

$$MoCl_3(THF)_3 + 2\ dppe \xrightarrow[THF,N_2]{Mg} trans\text{-}[Mo(N_2)_2(dppe)_2] \quad (5)$$

$$MoCl_5 + 2\ dppe \xrightarrow[THF,N_2]{Na/Hg} trans\text{-}[Mo(N_2)_2(dppe)_2] \quad (6)$$

$$MoCl_2(dppe)_2 \xrightarrow[THF,N_2]{Mg/Hg} trans\text{-}[Mo(N_2)_2(dppe)_2] \quad (7)$$

$$MoH_4(dppe)_2 + 2N_2 \xrightarrow{THF} trans\text{-}[Mo(N_2)_2(dppe)_2] + 2H_2 \quad (8)$$

In the following sections, nitrogen-fixing reactions will be reviewed element by element rather than according to the type of reaction. All reactions are reductive. There are no examples to date of oxidative nitrogen fixation within the confines of this chapter.

## 2.1. Titanium, Zirconium, and Hafnium

There are no dinitrogen complexes of any of these metals that contain a phosphine ligand. The considerable amount of nitrogen-fixing chemistry reported for titanium, and to a lesser extent zirconium, involves alkoxide, cyclopentadienyl, and halide complexes.[23-25] The catalytic nitriding reactions of titanium will be reviewed in Section 3.

Titanium(IV) chloride forms a number of different adducts with $trans$-$[ReCl(N_2)(PMe_2Ph)_4]$ (see Table 1) that are believed to involve dinitrogen bridging between rhenium and titanium. Hydrolysis or treatment of these adducts with ethanol results in the virtual quantitative recovery of the original rhenium-dinitrogen complex.[26]

## 2.2. Vanadium, Niobium, and Tantalum

A large amount of work has been carried out, primarily by Shilov, on the chemical fixation of dinitrogen using various vanadium-containing systems.[24] However, no stable dinitrogen complexes of vanadium have been reported or $in\ situ$ nitrogen fixation observed with phosphines present.

Table 1. Dinitrogen Complexes Prepared from Dinitrogen

| Complex | Text No. | Reference |
|---|---|---|
| $[TiCl_4\{(N_2)ReCl(PMe_2Ph)_4\}_2]$ | 1[a] | 26 |
| $[(THF)TiCl_4\{(N_2)ReCl(PMe_2Ph)_4\}]$ | 2[a] | 26 |
| $[Ti_2Cl_6O(OEt_2)\{(N_2)ReCl(PMe_2Ph)_4\}]$ | 3[a] | 26 |
| $[\{NbCl(dmpe)_2\}_2(N_2)]$ | 4[b] | 27 |
| $[\{Ta(CHCMe_3)(PMe_3)_2Cl\}_2(N_2)]$ | 5 | 30 |
| $[\{Ta(CHCMe_3)(PMe_3)_2R\}_2(N_2)]$ <br> ($R$ = Me, $CH_2CMe_3$[c]) | 6[a,c] | 30, 31[c] |
| $[\{TaCl(PMe_3)_3(C_2H_4)\}_2(N_2)]$ | 7 | 30 |
| $cis$-$[Cr(N_2)_2(PMe_3)_4]$ | 8 | 33 |
| $[\{Cr(dppe)_2\}_2(N_2)]$ | 9 | 34 |
| $[(THF)_2Cl_3Cr\{(N_2)ReCl(PMe_2Ph)_4\}]$ | 10[a] | 35 |
| $[MoAr(N_2)(PR_3)_2]$ <br> (Ar = $C_6H_5Me$, $C_6H_3Me_3$; $PR_3$ = $PPh_3$, $PPh_2Me$) | 11 | 36, 37 |
| $[\{Mo(\eta^6\text{-}C_6H_6)(PPh_3)_2\}_2(N_2)]$ | 12 | 37 |
| $[\{Mo(\eta^6\text{-}C_6H_3Me_3)(dmpe)\}_2(N_2)]$ | 13[c] | 38, 39[c] |
| $[Ar(PPh_3)_2Mo(N_2)FeCp(dmpe)]BF_4$ <br> (Ar = $C_6H_5Me$) | 14[a] | 37 |
| $trans$-$[Mo(N_2)_2(dppe)_2]$ | 15[c] | 20, 21, 40, 41[c], 42–46 |
| $trans$-$[Mo(N_2)_2(dppm)_2]$ | 16[b] | 40 |
| $trans$-$[Mo(N_2)_2(dppp)_2]$ | 17 | 40 |
| $trans$-$[Mo(N_2)_2(dptpe)_2]$ | 18 | 22 |
| $trans$-$[Mo(N_2)_2(Ar_2PCH_2CH_2PAr_2)_2]$ <br> (Ar = p-$C_6H_4CF_3$, p-$C_6H_4Cl$, p-$C_6H_4Me$, p-$C_6H_4OMe$) | 19 | 47 |
| $trans$-$[Mo(N_2)_2(dpe)_2]$ | 20 | 20 |
| $trans$-$[Mo(N_2)_2(Ph_2PCH=CHPPh_2)_2]$ | 21 | 48 |
| $trans$-$[Mo(N_2)_2(dppe)(PMe_2Ph)_2]$ | 22 | 49, 50 |
| $trans$-$[Mo(N_2)(CO)(dppe)_2]$ | 23[c] | 51 |
| $trans$-$[Mo(N_2)_2(PPh_2Me)_4]$ | 24 | 45 |
| $cis$-$[Mo(N_2)_2(PPhMe_2)_4]$ | 25 | 45, 52 |
| $trans$-$[Mo(N_2)_2(PEt_2Ph)_4]$ | 26 | 52 |
| $trans$-$[Mo(N_2)_2(triphos)(PR_3)]$ <br> [$PR_3$ = $PPh_3$, P(p-tolyl)$_3$, P(p-$C_6H_4OMe)_3$] | 27 | 50, 53 |

Table 1. (Continued)

| Complex | Text No. | Reference |
|---|---|---|
| $[M(N_2)(CO)_3(PCy_3)_2]$ ($M$ = Mo, W) | 28 | 54 |
| $[Cl_4(MeO)Mo\{(N_2)ReCl(PMe_2Ph)_4\}]$ | 29[a,c] | 55, 56[c] |
| $[(THF)_2Cl_3Mo\{(N_2)ReCl(PMe_2Ph)_4\}]$ | 30[a] | 57 |
| $[(PMe_2Ph)Cl_4W\{(N_2)ReCl(PMe_2Ph)_4\}]$ | 31[a] | 57 |
| $[Cl_4M\{(N_2)ReCl(PMe_2Ph)_4\}_2]$ ($M$ = Mo[c], W) | 32[a,c] | 57, 58[c] |
| trans-$[W(N_2)_2(dppe)_2]$ | 33 | 20 |
| trans-$[W(N_2)_2(PPh_2Me)_4]$ | 34 | 49 |
| cis-$[W(N_2)_2(PMe_2Ph)_4]$ | 35 | 20 |
| trans-$[W(N_2)_2(depe)_2]$ | 36 | 47 |
| $[ReH(N_2)(dppe)_2]$ | 37 | 111 |
| $[Re(NHPh)(N_2)(PMe_3)_4]$ | 38 | 112 |
| $[\{Fe(\eta^5\text{-}C_5H_5)(dppe)\}_2(N_2)](BF_4)_2$ | 39 | 115 |
| $[\{Fe(\eta^5\text{-}C_5H_5)(dmpe)\}_2(N_2)](BF_4)_2$ | 40 | 116 |
| $[FeH_2(N_2)(PR_3)_3]$ ($PR_3$ = $PPh_3$, $PBu^nPh_2$, $PEtPh_2$, $PMePh_2$) | 41 | 117–119 |
| $[FeH(N_2)L]BPh_4$ [$L$ = $N(CH_2CH_2PPh_2)_3$] | 42 | 120 |
| $[FeH(N_2)L]Y$, [$L$ = $P(CH_2CH_2PPh_2)_3$, Y = Br, I, $BPh_4$] | 43[c] | 120, 121[c] |
| $[FeH(N_2)(depe)_2]BPh_4$ | 44 | 122 |
| $[FeH(Pr^i)(PPh_3)_2(N_2)Fe(Pr^i)(PPh_3)_2]$ | 45 | 123 |
| $[\{(depe)_2ClFe\}_2(N_2)][BPh_4]_2$ | 46 | 124 |
| $[Fe(N_2)(dppe)_2]$ | 47 | 125 |
| $[RuH_2(N_2)(PR_3)_3]$ ($R$ = Ph, p-$C_6H_4CH_3$) | 48 | 126, 127 |
| $[OsHCl(N_2)(PR_3)_3]$ ($PR_3$ = $PEt_2Ph$, $PMe_2Ph$, $PPh_2Et$) | 49 | 128 |
| $[CoH(N_2)(PPh_3)_3]$ | 50[c] | 129 |
| $Na[Co(N_2)(PR_3)_3]$ ($PR_3$ = $PEt_2Ph$, $PPh_3$) | 51 | 130 |
| $K[Co(N_2)(PMe_3)_3]$ | 52[c] | 132 |
| $[\{(PR_3)_3Co\}_2(N_2)]$ ($PR_3$ = $PPh_3$, $PEt_2Ph$) | 53 | 130 |

Table 1. (Continued)

| Complex | Text No. | Reference |
|---|---|---|
| [{(PMe$_3$)$_3$Co(N$_2$)}$_2$Mg(THF)$_4$] | 54$^c$ | 132 |
| [Co(N$_2$)(P$R_3$)$_3$Mg(THF)$_2$]<br>(P$R_3$ = PPh$_3$, PPh$_2$Et) | 55$^a$ | 133 |
| [RhCl(N$_2$)(PCy$_3$)$_2$] | 56 | 134 |
| [RhH(N$_2$)(PPhBu$_2^t$)$_2$] | 57$^c$ | 135 |
| [RhCl(N$_2$)(PPr$_3^i$)$_2$] | 58$^c$ | 136 |
| [RhH(N$_2$)(P$R_3$)$_2$]<br>($R$ = Cy, Bu$^t$) | 59 | 137–139 |
| [{(PCy$_3$)$_2$HRh}$_2$(N$_2$)] | 60 | 137–139 |
| [{(PPr$_3^i$)$_2$HRh}$_2$(N$_2$)] | 61$^c$ | 138 |
| [{(PCy$_3$)$_2$Ni}$_2$(N$_2$)] | 62$^c$ | 144 |
| [Ni(N$_2$)(P$R_3$)$_3$]<br>(P$R_3$ = PEt$_3$, PBu$_3^n$, PE$t_2$Ph) | 63$^b$ | 142 |

$^a$ Prepared using a preformed metal-dinitrogen complex.
$^b$ Quoted as being impure.
$^c$ Crystal structure determined.

An impure dinitrogen complex of niobium has been reported by Leigh et al.$^{(27)}$ Complex 4 (see Table 1) is formulated as a dinuclear complex with a bridging dinitrogen. Treatment with acid yields no ammonia, but ca. 20% of the nitrogen is converted to hydrazine, the remainder being evolved as N$_2$.

In 1980, Schrock et al. reported the first stable dinitrogen complexes of a group 5 metal.$^{(28-30)}$ Of the more than ten complexes synthesized, two are prepared directly using dinitrogen (Table 1, structures 5, 7), while the remainder are prepared by a metathesis-like reaction of an alkylidene complex with a diimine (Equation 9):

$$2[Ta(CHCMe_3)(THF)_2Cl_3] + PhCH=NN=CHPh \rightarrow [\{TaCl_3(THF)_2\}_2(N_2)] + 2PhCH=CHCMe_3 \quad (9)$$

and further reactions thereof. Each of the complexes contains one dinitrogen in a bridging position and a wide variety of other ligands. X ray structure determinations of 6 and [{TaCl$_3$(THF)(PBz$_3$)}$_2$(N$_2$)]-(Bz=CH$_2$Ph)$^{(31,32)}$ not only confirm a linear Ta-N-N-Ta unit but also reveal very long N-N bonds (1.298(12) and 1.282(6) Å, respectively) and extremely short Ta–N bonds. The N–N bonds are the longest observed in a simple bridging dinitrogen complex. The structural data and chemical behavior (*vide infra*) of the bridging dinitrogen lead the authors to propose a $\mu$-$N_2^{4-}$ formalism (*sp* hybridized N atoms) with the linkage best described as [Ta=N–N=Ta]. It appears that in these complexes prepared using dinitrogen tantalum is able to bind and reduce dinitrogen. Addition of two more electrons into this unit would produce $\mu$-$N_2^{6-}$ which of course would not exist but would give two nitride ligands: [Ta=NN=Ta] + $2e^- \rightarrow 2[Ta\equiv N]^-$. These electron-deficient complexes appear to contain the most activated dinitrogen to date in a simple bridging system and should renew peoples' interest in the preparation and chemistry of bridging dinitrogen complexes.

## 2.3. Chromium, Molybdenum, and Tungsten

### 2.3.1. Chromium

Few dinitrogen complexes of chromium are known. Complex **8** decomposes at 20°C with loss of both $N_2$ and $PMe_3$.[33] Treatment of complex **9** with acid produces small quantities of ammonia (7%) and hydrazine (1%).[34] Reaction of the polynuclear complex **10** with water regenerates the starting rhenium complex *trans*-[ReCl($N_2$)($PMe_2Ph$)$_4$], and with dioxygen produces the cation [ReCl($N_2$)($PMe_2Ph$)$_4$]$^+$.[35]

### 2.3.2. Molybdenum and Tungsten

Following the discovery of *trans*-Mo($N_2$)$_2$(dppe)$_2$, **15**, by Hidai in 1969,[40] a series of four landmark reactions in the area of the chemistry of coordinated dinitrogen were reported by Chatt and co-workers over a period of three years. These are (i) the reaction of **33** with acyl halides to form nitrogen–carbon bonds (Equation 10);[59] (ii) protonation of **15** and **33** to form nitrogen–hydrogen bonds (Equation 11);[60] (iii) alkylation of **15** and **33** to form nitrogen–carbon bonds using alkyl halides (Equation 12);[61] and (iv) protonolysis to produce ammonia (and hydrazine) (Equation 13)[62]:

$$[W(N_2)_2(dppe)_2] + RCOCl \xrightarrow{HCl} [WCl(NN\overset{H}{|}COR)(dppe)_2]Cl + N_2 \quad (10)$$

$$[Mo(N_2)_2(dppe)_2] \xrightarrow{HBr} [MoBr(NNH_2)(dppe)_2]Br + N_2 \quad (11)$$

$$[Mo(N_2)_2(dppe)_2] + RBr \rightarrow [MoBr(NNR)(dppe)_2] + N_2 \quad (12)$$

$$\textit{cis}\text{-}[Mo(N_2)_2(PMe_2Ph)_4] \xrightarrow[MeOH]{H_2SO_4} N_2 + [NH_4]^+ + \cdots \quad (13)$$

This latter reaction only occurs when there is at least one monodentate phosphine coordinated to the metal. For the purpose of discussion, the remainder of this section will be divided into reactions that (i) form nitrogen–carbon bonds; (ii) form nitrogen–hydrogen bonds; and (iii) reactions of polynuclear complexes.

*2.3.2a. Formation of Nitrogen–Carbon Bonds.* A wide variety of organic halides react directly with the bisdinitrogen complexes of molybdenum and tungsten. However, nitrogen–carbon bonds are only formed when the phosphine ligands are bidentate. In all the other complexes that contain one or more monodentate phosphines both dinitrogen ligands are lost and phosphine–metal–halide complexes are formed.

Successful reactions of complexes **15** and **33**, some occurring in very high yield, with various organic halides (Equations 14–19) will be presented in approximately chronological order. Acyl and aroyl chlorides react to form hydrazido and diazenido complexes, respectively. The latter complexes were treated with one mole of hydrogen chloride to form the hydrazido complexes.[63] The hydrazido complexes can be deprotonated reversibly to give the corresponding diazenido complexes (Equation 14):

$$[M(N_2)_2(dppe)_2] + RCOCl + HCl \rightarrow [MCl(NNHCOR)(dppe)_2]Cl + N_2$$

$$HCl \updownarrow base$$

$$[MCl(NNCOR)(dppe)_2] \qquad (14)$$

The X ray structure of $[MoCl(NNCOPh)(dppe)_2]$ confirms the presence of a nitrogen–carbon bond.[64] Important bond angles (°) and lengths (Å) are shown here:

Simple alkyl bromides and iodides react at room temperature to give alkyldiazenido complexes (Equation 15):[63,65–67]

$$[M(N_2)_2(dppe)_2] + RX \rightarrow [MX(NNR)(dppe)_2] + N_2 \qquad (15)$$

$$(M = Mo, W; R = Alkyl; X = Br, I)$$

although they are often isolated as the corresponding alkylhydrazido complex and subsequently converted to the alkyldiazenido complex by deprotonation. A number of X ray structure determinations have established the bonding mode for both diazenido[65,68,69] and hydrazido[70,71] complexes of molybdenum and tungsten. The metal–nitrogen bond becomes progressively shorter and the nitrogen–nitrogen bond longer as more atoms are attached to the end nitrogen. $[Mo(N_2)_2(dppe)_2]$,[41] $[MoI(NNOct^n)(dppe)_2]$,[69] and $[MoI(NNHOct^n)(dppe)_2]I$[70] are compared below:

|  | $(N_2)Mo-N-N$ | $IMo-N-N{\overset{R}{\diagup}}$ | $IMo-N-N{\overset{R}{\underset{H}{\diagup}}}$ |
|---|---|---|---|
| Mo-N(Å) | 2.014(5) | 1.850(12) | 1.801(5) |
| N-N(Å) | 1.118(8) | 1.146(13) | 1.259(8) |
| NN$R$(°) | — | 128(1) | 120(1) |

Normally alkyl chlorides and aryl halides do not give nitrogen-containing products, though "activated" alkyl chlorides such as $ClCH_2COOEt$[72] and $ClCOOEt$[73] (Equation 16) do:

$$[Mo](N_2)_2(dppe)_2] + RCl \rightarrow [MoCl(NNR)(dppe)_2] + N_2 \quad (16)$$

$$(R = CH_2COOEt, COOEt)$$

$n$-Butyl chloride has been reported to react with **33** in benzene solution under irradiation at 45°C to give the corresponding butyldiazenido complex.[74] Reaction of racemic 2-bromooctane with the pure diasteriomer *trans*-$[Mo(N_2)_2\{(s,s)\text{-chiraphos}\}_2]$, where s,s-chiraphos is $(-)$-(2s, 3s)-*bis*(diphenylphosphino)butane, produces the 2-octyldiazenido complex, which is a mixture of two diasteriomers, in 78% yield.[75] The proton nmr spectra recorded at 100, 200, and 396 MHz show an excess (*ca.* 10%) of one diasteriomer over the other.

*gem*-Dibromoalkanes react with **33** to give a novel series of cationic diazoalkane complexes (Equation 17)[76]:

$$[W(N_2)_2(dppe)_2] + R^1R^2CBr_2 \rightarrow [WBr(NN=CR^1R^2)(dppe)_2]Br + N_2 \quad (17)$$

that react with hydride ion and carbanions to form the corresponding neutral alkyldiazenido complexes. On the other hand, the reactions of **15** with *gem*-dibromides are more complex, and diazoalkane complexes are, at best, minor products.[76] For example, $CH_2Br_2$ reacts with **15** to give a complex formulated as *trans-,trans*-$[(dppe)_2BrMo(NNHCH_2NHN)$-$MoBr(dppe)_2]Br_2$. When the mother liquor from this reaction is treated with HBr, a further complex formulated as *trans*-$[MoBr(NNHCH_2CH_2Br)(dppe)_2]Br$ is isolated. The X ray crystal structures of three diazoalkane complexes have been determined: $[WBr(NN=CHCH_2CH_2CH_2OH)(dppe)_2]PF_6$,[77] $[WBr(NN=CMe_2)$-

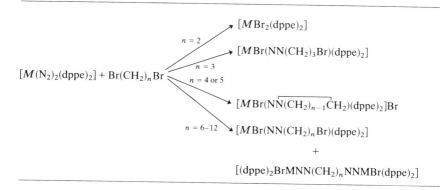

Scheme 1

(dppe)$_2$]Br,[77] and [WF(NN=CMeCH$_2$COMe)(dppe)$_2$]BF$_4$.[78] The W–N–N units are linear, with N–N–C angles of 116–125° and N–C bond lengths that are consistent with a bond order of two.

α,ω-Dibromoalkanes, Br(CH$_2$)$_n$Br, undergo a variety of reactions with **15** and **33** depending upon the value of $n$.[79] The case where $n = 1$ has been discussed above and for $n = 2$, ethene is eliminated and [MBr$_2$(dppe)$_2$] formed. For $n = 3$, complexes containing the group NN(CH$_2$)$_3$Br are formed, and for $n = 4$ or 5, complexes with rings $\overline{\text{NN(CH}_2)_n}$ are formed. For $n = 6$–12, two series of complexes containing either NN(CH$_2$)$_n$NN in which the alkyl group bridges two molecules of substrate, or NN(CH$_2$)$_n$Br are isolated. These reactions are summarized in Scheme 1. The dinuclear species are isolated as either hydrazido or diazenido complexes. Difficulties are experienced in working up the products when $M$ = Mo and $n$ = 6–12.

Interestingly, 1-bromo-4-chlorobutane also cyclizes with **15** to produce [MoBr($\overline{\text{NNCH}_2\text{CH}_2\text{CH}_2\text{CH}_2}$)(dppe)$_2$]Cl in which bromine is attached to molybdenum, and chloride is the anion.[80] Recently, the secondary alkylation of alkyldiazenido complexes to form dialkylhydrazido complexes has been reported.[47,81] However, when **20** is treated with MeBr or EtBr, no alkyldiazenido complex is isolated, but instead a dialkylhydrazido complex is produced (Equation 18):

$$[\text{Mo(N}_2)_2(\text{depe})_2] + 2\text{MeI} \rightarrow [\text{MoI(NNMe}_2)(\text{depe})_2]\text{I} + \text{N}_2 \qquad (18)$$

The alkylation of *trans*-[$M$(N$_2$)$_2$(dppe)$_2$], ($M$ = Mo or W) by primary alkyl bromides and iodides has been studied in considerable detail and a mechanism proposed for the reactions observed.[74,82] The mechanism is summarized in Scheme 2. The initial rate-determining dissociation of one

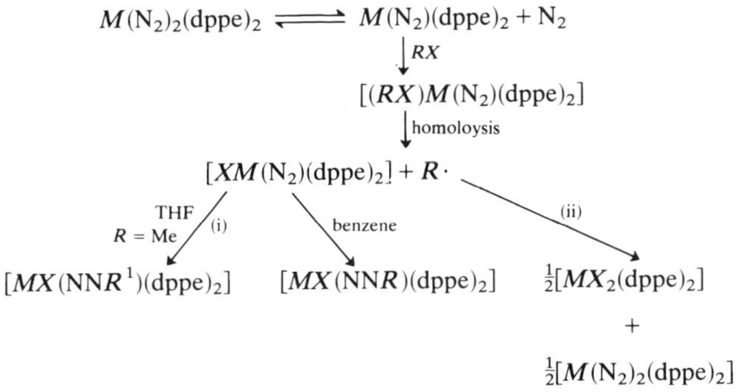

$$M(N_2)_2(dppe)_2 \rightleftharpoons M(N_2)(dppe)_2 + N_2$$
$$\downarrow RX$$
$$[(RX)M(N_2)(dppe)_2]$$
$$\downarrow \text{homoloysis}$$
$$[XM(N_2)(dppe)_2] + R\cdot$$

```
         THF  /(i)         \ benzene      \ (ii)
  R = Me /                  \              \
[MX(NNR¹)(dppe)₂]   [MX(NNR)(dppe)₂]   ½[MX₂(dppe)₂]
                                           +
                                       ½[M(N₂)₂(dppe)₂]
```

(i) $R^1 = \overline{CHOCH_2CH_2CH_2}$;  (ii) $R\cdot$ couples or disproportionates

*Scheme 2*

dinitrogen ligand is followed by formation of an adduct between the coordinatively unsaturated metal species and the alkyl halide, $[M(N_2)(RX)(dppe)_2]$. Following carbon–halogen bond homolysis (inner-sphere electron transfer), the halogen atom remains bonded to the metal and the organic radical either (i) attacks the remaining coordinated dinitrogen to form an alkyldiazenido complex, (ii) attacks the solvent (e.g., THF when $R = Me$) to produce a solvent-derived radical which then attacks the dinitrogen ligand to form a different diazenido complex, or (iii) dimerizes (e.g., $R = PhCH_2$) or disproportionates. The initial dissociation of dinitrogen occurs in the dark for **15** at room temperature, but for **33** a source of visible light is necessary (without which no reaction occurs at room temperature). The presence of light accelerates the alkylation reactions of **15**. A number of flash photolysis studies of **15** and **33** have been carried out[83,84] and lend support to the above mechanism.

It is assumed that all of the other nitrogen–carbon bond-forming reactions mentioned so far proceed according to Scheme 2 with loss of one dinitrogen being the initial step.[74,82] It can be argued that in the 5-coordinate intermediate, $[M(N_2)(dppe)_2]$, the remaining dinitrogen will be more tightly held due to increased $\pi$-back bonding from the metal and therefore less likely to be lost from the coordination sphere.

Dinitrogen can also be alkylated in complexes containing only one dinitrogen and a labile ligand. A series of complexes of the type *trans*-$[M(N_2)(NCR)(dppe)_2]$, ($M = Mo$ or W), have been prepared[85] by the reaction[86] of an organic nitrile with **15** or **33**. Subsequent reactions with alkyl bromides and iodides proceed with slow loss of RCN (slower than loss of $N_2$ from **15** or **33**) and formation of an alkyldiazenido complex (Equation 19)[82]:

$$[M(N_2)(NCR)(dppe)_2] + RX \rightarrow [MX(NNR)(dppe)_2] + RCN \quad (19)$$

In these reactions loss of $RCN$ creates the 5-coordinate intermediate that then follows the inner-sphere redox pathway depicted in Scheme 2 to form product.

The presence of a labile ligand is not an absolute requirement for alkylation of coordinated dinitrogen. Thus reactions of **15** with a series of tetrabutylammonium halides and pseudohalides produce a number of anionic monodinitrogen complexes (Equation 20):

$$[Mo(N_2)_2(dppe)_2] + [Bu_4^nN]X \rightarrow [Bu_4^nN][MoX(N_2)(dppe)_2] + N_2 \quad (20)$$
$$(X = SCN, CN, N_3)$$

some too unstable to isolate.[82] Interestingly, reaction of $[Mo(SCN)(N_2)(dppe)_2]^-$ with $Bu^nI$ shows a rate which is first-order in complex concentration and first-order in $Bu^nI$ concentration. In addition, the product retained the thiocyanate ligand. These data are inconsistent with rate-controlling ligand loss. Instead, it is proposed that the reaction proceeds by an outer-sphere electron-transfer reaction (Equation 21)[82]:

$$[Mo(SCN)(N_2)(dppe)_2]^- + RX \rightarrow [Mo(SCN)(N_2)(dppe)_2] + R\cdot + X^- \quad (21)$$

The radical $R\cdot$ couples with the molybdenum(I) radical to form product, $[Mo(SCN)(NNR)(dppe)_2]$.

Those reactions that give dialkylated-dinitrogen products, e.g., Equation (18) and Scheme 1 ($n = 4$ or 5) are believed to proceed in two steps. The first is monoalkylation according to Scheme 2, and the second is $S_N2$ displacement at a halogen-bound carbon atom by the carbon-bound nitrogen atom of the diazenido ligand. Typical $S_N2$ reactions have been demonstrated for the reactions of MeI and EtI with preformed methyldiazenido complexes $[MBr(NNMe)(dppe)_2]$.[47] Complex **20**, which contains the basic depe ligand, reacts with MeI to form the dimethylhydrazido complex (Equation 18). It is suggested that in this case the basic depe has the effect of (i) slowing down the rate of loss of dinitrogen, and (ii) increasing the nucleophilicity of the NNMe group with the net result that second alkylation occurs faster than first alkylation.

So far in this section all reactions have resulted in organonitrogen ligands that are firmly attached to the metal. Do subsequent reactions of these species form amines, for example? The answer is yes, but only in reactions in which the integrity of the metal complex is destroyed. There is no evidence for protonation of alkyldiazenido complexes beyond the alkylhydrazido form. Reactions of tungsten diazoalkane complexes with HBr cause protonolysis of the nitrogen–carbon bond to form [WBr(NNH$_2$)(dppe)$_2$]Br with no evidence for organonitrogen compounds.[87] The reaction of a tenfold excess of sodium borohydride with [MoBr(NNBu$^n$)(dppe)$_2$] in benzene–methanol solution at 100°C for 10 h in an autoclave produces NH$_3$ (51%), $n$-butylamine (54%), and butylmethylamine (9%).[66,88] Under the reaction conditions, sodium borohydride is completely converted to Na[B(OMe)$_4$] before amine formation occurs. The molybdenum complex is completely degraded during the reaction and ca. 90% dppe is recovered as free ligand. It is proposed that butylmethylamine arises from the reaction of butylamine with formaldehyde that is generated from methoxide ion (or methanol) during the reaction. When ethanol, rather than methanol, is used as co-solvent, only small amounts of volatile base are obtained. However, MoH$_4$(dppe)$_2$ is isolated from the reaction in 25% yield.

Amines have also been generated from dialkylhydrazido complexes by a variety of destructive methods involving both strong acid and strong base.[89] For example, treatment of [WBr(N$_2$Me$_2$)(dppe)$_2$]$^+$ with LiAlH$_4$ in ether followed by base distillation yielded half the available nitrogen as dimethylamine (0.95 mol), but the other one equivalent is unaccounted for. In a few cases tungsten-containing products are isolated; e.g., [WH$_4$(dppe)$_2$] and [WBr$_2$H$_2$(dppe)$_2$].

It is not immediately obvious how any of the nitrogen–carbon bond-making reactions discussed in this section could be adapted to provide a catalytic reaction for the formation of organonitrogen compounds. A prerequisite for any catalytic nitrogen-fixing process is the continual regeneration of a dinitrogen-binding site. In fact, this has been accomplished recently in a report of the *electrochemical* reduction of *trans*-[MoBr{NNCH$_2$(CH$_2$)$_3$CH$_2$}(dppe)$_2$]$^+$ at a platinum electrode in THF-0.2 $M$ [Bu$_4$N][BF$_4$] under dinitrogen that yields free N-aminopiperidine, H$_2$NNCH$_2$(CH$_2$)$_3$CH$_2$, and *trans*-Mo(N$_2$)$_2$(dppe)$_2$.[90] Under carbon monoxide, the reaction proceeds similarly with *cis*- and *trans*-[Mo(CO)$_2$(dppe)$_2$] being formed. The starting dialkylhydrazido complex is prepared in a one-step reaction from **15** (see Scheme 1, $n$ = 4 or 5), and the authors propose a nitrogen-fixing cycle based upon these initial results, which is summarized in Scheme 3. Solvent is presumed to be the proton source. This work provides the most encouraging results toward the development of a catalytic process for the synthesis of organonitrogen compounds reported to date.

*2.3.2b. Formation of Nitrogen–Hydrogen Bonds.* The bisdinitrogen complexes of molybdenum and tungsten that contain two bidentate ligands such as **15, 20,** and **33** react with strong acid to produce hydrazido complexes

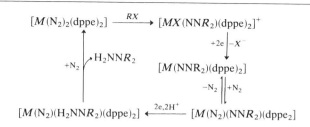

Scheme 3

with the loss of one mol of dinitrogen (e.g., Equation 11).[20,60,91] Ammonia and hydrazine are not products in these reactions. By contrast, those complexes that have at least one monodentate phosphine react with strong acid to yield an array of products that may include ammonia and/or hydrazine depending upon (i) the acid, (ii) the solvent, and (iii) the metal (Mo or W).[62,49,92-97] Discussion will focus primarily upon those reactions of bisdinitrogen complexes that produce ammonia and/or hydrazine.

A clear distinction between the stoichiometry of the ammonia-forming reactions of molybdenum and tungsten was noted from the outset by Chatt and co-workers.[62] The yield of ammonia is essentially 2 mol per tungsten atom but only ca. 0.7 mol per molybdenum atom. Typically, complexes 24, 25, 34, and 35 are treated with an excess of $H_2SO_4$ in methanol at 20°C. One mol of dinitrogen is rapidly evolved. Beyond this point, the reaction pathway differs depending upon whether the metal is tungsten or molybdenum. The reaction is shown for complex 35 in Equation 22.

$$cis\text{-}[W(N_2)_2(PMe_2Ph)_4] \xrightarrow[\text{MeOH}]{H_2SO_4} N_2 + 2NH_3 + W(IV) \text{ products} + 4[HPMe_2Ph]HSO_4$$

(22)

By varying the acid and particularly the solvent (e.g., THF, benzene, N-methyl-pyrrolidone, or dichloromethane), increasing yields of hydrazine are obtained at the expense of ammonia: 0.6 mol of $N_2H_4$ per tungsten atom for 35/HCl/DME.[94]

Richards and co-workers have followed the reaction of excess $H_2SO_4$ with 35 in THF solution by $^{15}N$ nmr spectroscopy.[96] Following the disappearance of 35, resonances due to $^{15}N_2$ and a series of six hydrazido ($NNH_2$) complexes are observed in solution. Finally, after 120 min, a signal due to $^{15}NH_4^+$ was the only resonance remaining. An analogous experiment with 25 shows the appearance of resonances due to $^{15}N_2$ and two hydrazido complexes but no $^{15}NH_4^+$. These and other data clearly indicate that the hydrazido unit, $NNH_2$, is an intermediate formed during the reduction of coordinated dinitrogen to ammonia. In the case of the tungsten complex 35, cited above, the hydrazido group remains intact through a series of changes in the coordination environment around tungsten before ammonia formation. No other nitrogen hydride intermediates are isolated or detected spectroscopically.

Treatment of **34** with HCl in dichloromethane[93] affords a complex formulated[100] as a hydrazido–tungsten hydride, [WHCl$_3$(NNH$_2$)(PMePh$_2$)$_2$]. Similar complexes are prepared by treating [WX$_2$(NNH$_2$)(PMe$_2$Ph)$_3$] (X = Cl or Br) with HX in DME[94] or THF.[98] Although originally formulated as complexes containing an NHNH$_2$ group,[93,94] subsequent $^{15}$N[98] and $^1$H[99] NMR studies strongly suggest that the hydrazido–tungsten hydride formulation is correct. Upon further treatment with acid, these complexes produce ammonia and hydrazine; the latter in yields significantly greater than those obtained when the original complexes, **34** and **35**, are treated with acid.[93,94] Hidai proposes that the {WH(NNH$_2$)} stage of reduction is on the route to hydrazine formation.[99]

Returning now to the importance of the hydrazido ligand as an intermediate in dinitrogen reduction, it is noted that hydrazido complexes can be isolated from many reactions that ultimately give ammonia if allowed to continue, e.g., Equations 23–25[96,93,20]:

$$[M(N_2)_2(PMe_2Ph)_4] \xrightarrow[\text{(ii) pentane}]{\text{(i) H}_2\text{SO}_4/\text{THF}/30\text{ min}} [M(HSO_4)_2(NNH_2)(PMe_2Ph)_3]$$
$$+ N_2 + [HPMe_2Ph]HSO_4 \quad (23)$$

$$[M(N_2)_2(PMe_2Ph)_4] \xrightarrow{\text{HX/MeOH}} [MX_2(NNH_2)(PMe_2Ph)_3] + N_2 + [HPMe_2Ph]X \quad (24)$$

$$[W(N_2)_2(PMe_2Ph)_4] \xrightarrow{\text{HX/CH}_2\text{Cl}_2} [WX(NNH_2)(PMe_2Ph)_4]X + N_2 \quad (25)$$

The X ray structure of a number of hydrazido complexes of molybdenum and tungsten have been determined: [WBr(NNH$_2$)(dppe)$_2$][BPh$_4$],[100] [MoF(NNH$_2$)(dppe)$_2$][BF$_4$],[91] [WHBrCl(NNH$_2$)(PMe$_2$Ph)$_3$][BPh$_4$],[94] [WCl$_3$(NNH$_2$)(PMe$_2$Ph)$_2$],[98] and [M(NNH$_2$)(quin)(PMe$_2$Ph)$_3$]X, where quin = quinolin-8-olate; M = Mo, X = Br, I; M = W, X = I.[101] In all these complexes the metal–nitrogen bonds are very short, indicating considerable multiple bonding. The bonding data for the NNH$_2$ unit are similar in all the above complexes despite the variation in coligands. The N–N bond lengths indicate an order greater than one with varying degrees of noncoplanarity of the atoms in the hydrazido group, NNH$_2$. These data support the following bonding representation [$M\doteq N\doteq NH_2$] in which there is considerable delocalization within the unit.

Hydrazido complexes with two bidentate phosphine ligands can be deprotonated reversibly to give the corresponding neutral diazenido complex (Equation 26)[102]:

$$[MX(NNH_2)(dppe)_2]X \underset{\text{HX}}{\overset{\text{NEt}_3}{\rightleftharpoons}} [MX(NNH)(dppe)_2] \quad (26)$$

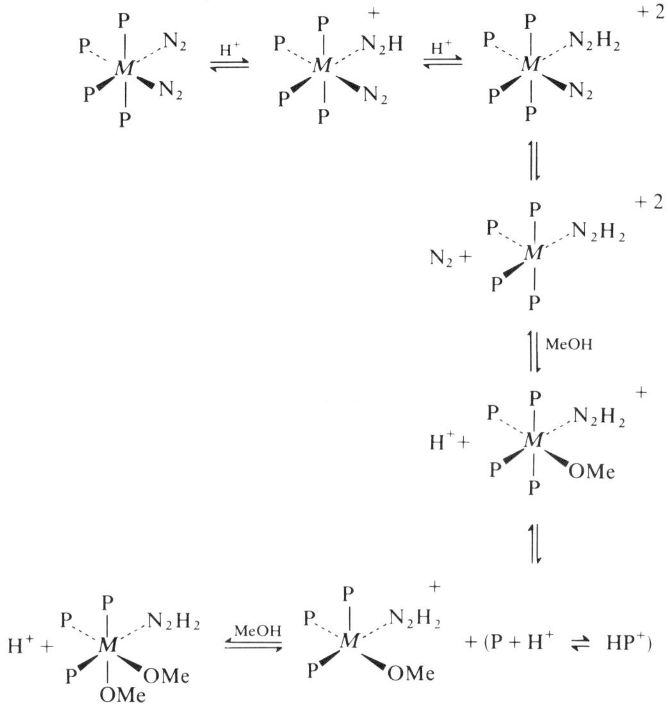

*Scheme 4*[a]

[a] P = PMe₂Ph

Recently, the first mechanistic study of the conversion of a coordinated dinitrogen to a hydrazido complex was reported.[103] Treatment of cis-$[M(N_2)_2(PMe_2Ph)_4]$ with excess HCl, HBr, or $H_2SO_4$ in methanol yields $[M(NNH_2)(OMe)_2(PMe_2Ph)_3]$ in a reaction that would ultimately give ammonia. The stoichiometry of the reaction is shown in Equation (27):

$$[M(N_2)_2(PMe_2Ph)_4] + H^+ + 2CH_3OH \rightarrow$$
$$[M(NNH_2)(OCH_3)_2(PMe_2Ph)_3] + N_2 + [HPMe_2Ph]^+ \quad (27)$$

The kinetics of the reaction exhibit a first-order dependence in complex concentration, a second-order dependence in total acid concentration, and no dependence upon the type of anion. A mechanism is proposed to fit the data, and it is shown in Scheme 4 (P=PMe₂Ph). Loss of phosphine is suggested as the rate-determining step. It is methoxide ion rather than the anion derived from the "strong acid" that coordinates to the metal. Note also that methanol effectively provides both protons in the reaction and the added acid scavenges the liberated phosphine. It has been shown that in methanol alone either when heated to reflux or irradiated with visible light the tungsten complexes **34** and **35** yield $NH_3$ in high yield.[92] This does not occur for the molybdenum complexes **24** and **25**. The greater basicity of dinitrogen attached to tungsten is shown by the greater reactivity of **35** compared with **25**; $k_W/k_{Mo} = 9.2 \times 10^2$. An isotope effect observed in the reaction of **25** ($k_H/k_D = 0.3$) is consistent with a mechanism involving protolytic-equilibria prior to the rate-limiting step.[103]

It seems clear that in reactions of the tungsten complexes of dinitrogen that yield ammonia, reduction is occurring at one metal center that is also the source of the required six electrons. Hence, tungsten is oxidized from W(O) to W(VI), although no tungsten (VI)-containing product has yet been isolated. This overall process is shown in Scheme 5.[92] An important part of this scheme that is omitted is the incorporation of the necessary electron-releasing ligands such as $MeO^-$ and $HSO_4^-$ ($SO_4^{-2}$) into the coordination sphere to assist tungsten in reducing coordinated dinitrogen.[92] Unless there are labile ligands already attached to tungsten (such as $N_2$ and $PR_3$), the reaction stops at some intermediate step, e.g., $[WBr(NNH_2)(dppe)_2]Br$. This latter complex can be converted to ammonia (0.40 mol) and hydrazine (0.44 mol) under conditions ($HBr/CH_2Cl_2/80°C/15$ h) vigorous enough to displace at least one dppe ligand.[87]

In those reactions in which hydrazine is formed, tungsten is probably behaving as a four-electron reducing agent. Hydrazine may result if the hydrazido intermediate is converted into a side-on bonded $NH-NH_2$ species in the presence of acid. Complexes of this type (e.g., $[Mo(\eta^2-NHNMePh)(NNMePh)(S_2CNMe_2)_2]^{+[104]}$) have been shown to yield organohydrazines quantitatively upon treatment with acid. An alternate source of hydrazine is from the decomposition of $N_2H_2$. However, mechanistic studies are required before anything definitive can be said about the pathway to hydrazine formation.

The identification and isolation of a molybdenum(III) complex, $MoBr_3(triphos)$, in almost quantitative yield (>90%) by the author and

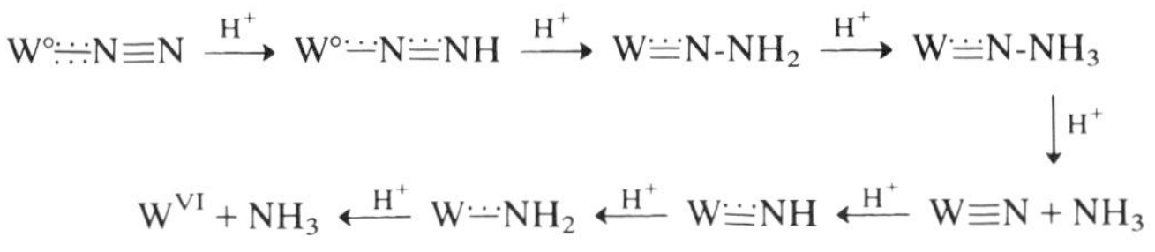

$$W^\circ\!\!=\!\!N\!\equiv\!N \xrightarrow{H^+} W^\circ\!\!-\!\!N\!\equiv\!NH \xrightarrow{H^+} W\!\equiv\!N\text{-}NH_2 \xrightarrow{H^+} W\!\equiv\!N\text{-}NH_3$$
$$\downarrow H^+$$
$$W^{VI} + NH_3 \xleftarrow{H^+} W\!-\!NH_2 \xleftarrow{H^+} W\!\equiv\!NH \xleftarrow{H^+} W\!\equiv\!N + NH_3$$

Scheme 5

co-worker from the reaction of *trans*-[Mo(N$_2$)$_2$(triphos)(PR$_3$)], **27**, with a large excess of anhydrous HBr in THF (Equation 28)[95]

$$2[\text{Mo}(\text{N}_2)_2(\text{triphos})(\text{P}R_3)] \xrightarrow[\text{THF}]{\text{HBr}} 2\text{MoBr}_3(\text{triphos}) + 2[R_3\text{PH}]\text{Br} + 3\text{N}_2 + 2[\text{NH}_4]\text{Br} \quad (28)$$

underscores the difference in behavior between molybdenum and tungsten in these types of reactions. Whereas the tungsten complexes liberate one mol of N$_2$ and two mol of NH$_3$ per metal atom, complexes of molybdenum yielded 1.5 mol of N$_2$ and 1 mol of NH$_3$ per metal atom. The latter stoichiometry has been observed for all molybdenum complexes containing at least one monodentate phosphine; **22**,[47,97] **24**,[49] **25**,[49] and **27**.[95] From three of these complexes, a molybdenum(III) product has been isolated.[95,97,105] The reaction stoichiometry indisputably establishes the formation of one NH$_3$ per molybdenum atom, a net transfer of 3 electrons per molybdenum atom [Mo(O) → Mo(III)], and loss of 1.5 mol of N$_2$ per molybdenum atom.

In general, the reactions of molybdenum complexes result in a poor nitrogen balance. Although 1.5 mol of N$_2$ are liberated routinely, the ammonia yields have tended to be around 70%. There is a report of 88% NH$_3$ from an experiment involving **25** and aqueous HBr in propylene carbonate for 48 hr followed by solvent removal at 140°C *in vacuo* and base distillation.[49] The low yields of ammonia appear to be artifacts of the experimental conditions. The molybdenum reactions are considerably slower than their tungsten analogs. During the course of these reactions, acid reacts with the solvent as well as the complex so that before the ammonia-forming reaction is complete the acid concentration is effectively zero. Recently, it has been shown that either addition of increments of HBr during the reaction, or addition of acetic acid and LiBr after the loss of the first mol of N$_2$, provides conditions under which the nitrogen balance is quantitative.[106]

Bisdinitrogen complexes of molybdenum react rapidly with strong acids to liberate 1.0 mol of N$_2$ per molybdenum atom, totally analogous to the similar tungsten complexes. The product at this stage is a hydrazido complex. The first step appears to require a certain minimum effective concentration of acid, otherwise protonation of the metal occurs, which results in loss of all dinitrogen from the molybdenum complex.[20,105,106] The presence of a hydrazido complex after the loss of 1.0 mol of dinitrogen is clearly demonstrated by isolation and characterization of the complex and by *in situ* $^{15}$N NMR spectral studies.[92,96] A recent study using $^{31}$P nmr spectroscopy to follow the reaction of HBr with **27** in THF reveals the presence of two isomeric hydrazido complexes, shown below. The differences arise in the relative positions of Br and PPh$_3$. Isomer **A** isomerizes to **B** which reacts further. A low concentration of a non-PPh$_3$-

containing complex is also observed. This is believed to be [MoBr$_2$(NNH$_2$)(triphos)]. Eventually, an intense resonance due to PPh$_3$ is observed as the spectral lines begin to broaden after 1 hr.[106]

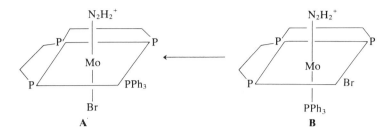

How are the hydrazido complexes of molybdenum converted to ammonia? First, the stoichiometry of this latter reaction must be reexamined (Equation 29), since it implies N–N bond cleavage to

$$[MoBr(NNH_2)(triphos)(PPh_3)]^+ \xrightarrow{HBr} [MoBr_3(triphos)] + \tfrac{1}{2}N_2 + [HPPh_3]Br + [NH_4]^+ \quad (29)$$

generate one [NH$_4$]$^+$ and $\tfrac{1}{2}$N$_2$. That this does not occur as written is proved by an isotopic-labeling experiment that establishes that no nitrogen atom scrambling occurs.[95] Specifically, the reaction of a mixture of **27**-($^{14}$N$_2$)$_2$ and **27**-($^{15}$N$_2$)$_2$ and HBr in THF produces dinitrogen-28 and dinitrogen-30 but no dinitrogen-29. Therefore, one-half of the {Mo(NNH$_X$)} units are being oxidized to dinitrogen (giving 0.5 mol of N$_2$ per Mo atom) and the other half of the {Mo(NNH$_X$)} units are being reduced to ammonia (to give 1 mol of NH$_4^+$ per Mo atom) and therefore Equation 29 should be rewritten as shown in Equation 30.

$$2[MoBr(NNH_2)(triphos)(PPh_3)]^+ \xrightarrow{HBr}$$
$$2MoBr_3(triphos) + N_2 + 2[NH_4]^+ + 2[HPPh_3]Br \quad (30)$$

Secondly, since there is no nitrogen atom scrambling, there is a net requirement of 6 electrons per two molybdenum atoms to accomplish the reduction of a dinitrogen to 2[NH$_4$]$^+$. This is summarized in Equation 31.

$$\{Mo^0(N_2)_2\} + \{Mo^0(N_2)_2\} \xrightarrow{H^+} 2\{Mo^{III}\} + 3N_2 + 2[NH_4]^+ \quad (31)$$

Assuming that a mononuclear hydrazido complex is reduced to 2[NH$_4$]$^+$, then intermolecular electron transfer from another molybdenum complex is necessary to account for all molybdenum appearing in the product as molybdenum(III). Intermolecular electron transfer can be ruled out, however, for two reasons: (i) Electrochemical studies of a variety of hydrazido complexes that will produce ammonia in the presence of acid show no reduction waves above that of the solvent ($CH_2Cl_2$) discharge;[92] (ii) Protonolysis of a mixture of **15** and **27** produces no increase in the yield of $[NH_4]^+$; i.e., $[MoBr(NNH_2)_2(dppe)_2]^+$, which is formed in this reaction and does not afford $[NH_4]^+$, under these conditions could still behave as reducing agents towards a $Mo(NNH_X)$(triphos) intermediate if intermolecular electron transfer is a requirement for ammonia formation.[106]

Thirdly, the necessary requirement that one of the ligands of the hydrazido complex be labile in order that the reaction proceeds to generate a further 0.5 mol of $N_2$ and $2[NH_4]^+$ can be satisfied by $Br^-$ or $PPh_3$ in the triphos system. The dissociation of $PPh_3$ from isomer **B** being the more important step is supported by the following data: (i) The fluoride complex, $[MoF(NNH_2)(triphos)(PPh_3)]^+$, in which the fluorine atom is *trans* to $NNH_2$, upon treatment with anhydrous HBr in THF for 6 days produces no ammonia. The strong Mo–F bond precludes facile $F^-$ dissociation. (ii) The addition of $PMe_2Ph$ to a THF solution that contained **A** and **B** showed

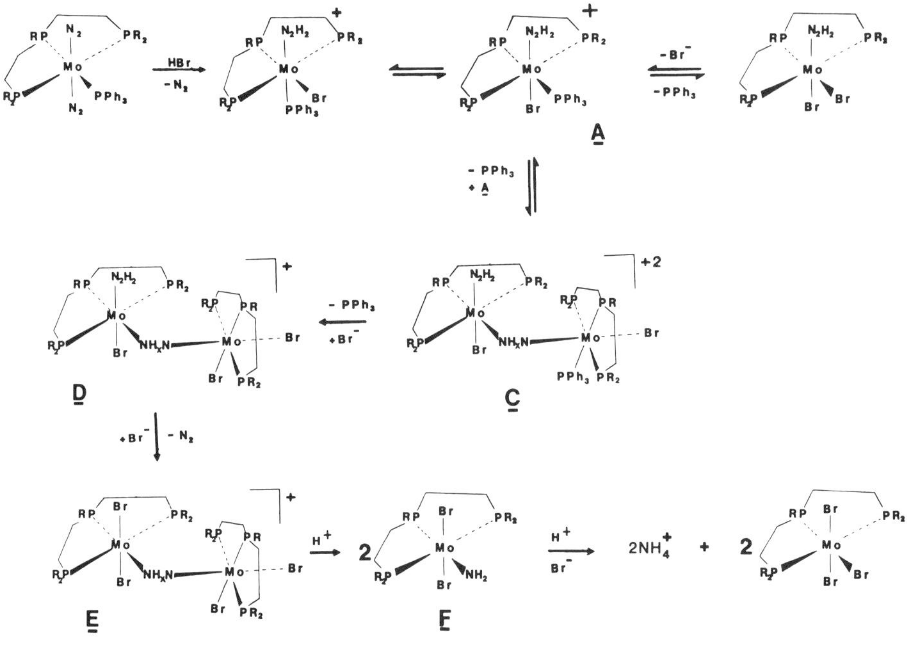

Scheme 6

incorporation of PMe$_2$Ph into **B** (but not **A**) and no appreciable amount of ammonia was produced.[106]

A proposed reaction pathway by which ammonia is formed is shown in Scheme 6. After the dissociation of phosphine, attack by another hydrazido complex (e.g., **A**) to form a bridged species, **C**, is proposed. Hydrazido complexes such as **A** are known to be good nucleophiles. For example, they react in acid catalyzed reactions with aldehydes and ketones to form diazoalkane complexes (e.g., Equations 4 and 32[107]).

$$[\text{MoF(NNH}_2)(\text{triphos})(P\text{Ph}_3)]^+ + [(\eta^5\text{-C}_5\text{H}_5)\text{Fe}(\eta^5\text{-C}_5\text{H}_4\text{CHO})] \xrightarrow{H^+}$$
$$[(\eta^5\text{-C}_5\text{H}_5)\text{Fe}\{\eta^5\text{-C}_5\text{H}_4\text{CHNNMoF(triphos)}(P\text{Ph}_3)\}]^+ + \text{H}_2\text{O} \quad (32)$$

The driving force for this step must be the formation of a strong Mo–N bond. The bridging ligand in the proposed intermediates **C**, **D**, and **E** may be of the diimine (NHNH) type, similar to those isolated and characterized by Sellmann.[108] Formation of an intermediate with a bridging [N$_2$H$_x$] unit and a terminal [N$_2$H$_x$] unit is consistent with the isotopic-labeling experiment. No firm evidence is available to say whether it is the terminal or bridging [N$_2$H$_x$] unit that is oxidized to give dinitrogen; 0.5 mol per molybdenum atom.

The formation of bridging nitrogen-hydride ligands as intermediates on route to the reduction of dinitrogen to ammonia and hydrazine has been suggested by many authors.[7,57,109] The unique feature of this molybdenum system is the pathway by which the bridge is constructed. Loss of dinitrogen from an intermediate such as **D** followed by N–N bond homolysis leads eventually to a molybdenum(III) amide **F** that should be readily converted to ammonia in the presence of acid. In no step is intermolecular electron transfer a requirement.

In principle, the molybdenum system could be developed into a catalytic cycle for ammonia formation (see Scheme 7). The molybdenum(III) product is the normal starting point for synthesis of **27** in a one-step, high-yield ($>60\%$) reaction. An attractive feature of the system is the stability introduced by the tridentate ligand that helps to retain the structural integrity of the coordination environment of the metal. The major drawback of the proposed cycle is the difficulty of regenerating the N$_2$-binding site without reducing protons to dihydrogen.

Treatment of **35** with alcohols in the presence of KOH at 50°C produces ammonia and hydrazine in moderate yields.[99] Ammonia is the major product with all alcohols studied except for 2-propanol in which hydrazine is predominant. The molybdenum analog **25** gives no ammonia or hydrazine under these conditions. Hidai *et al.* have also carried out similar reactions with metal hydrides. Hydridic hydrides such as [($\eta^5$-C$_5$H$_5$)$_2$ZrHCl] and

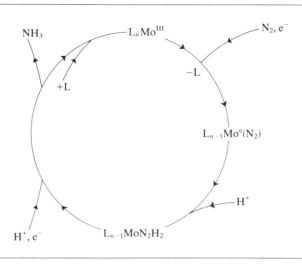

Scheme 7

[NaAlH$_2$(OCH$_2$CH$_2$OCH$_3$)$_2$] produce ammonia in moderate to high yield when reacted with **25** and **35** in benzene solution.[99]

*2.3.2c. Reactions of Polynuclear Complexes.* A number of polynuclear complexes of molybdenum and tungsten are known in which dinitrogen bridges two metals: e.g., [Mo(N$_2$)Mo], **12** and **13**; [Mo(N$_2$)Fe], **14**; [Mo(N$_2$)Re], **29**, **30**, and **32**; and [W(N$_2$)Re], **31** and **32**. Of those that have been structurally characterized by X ray diffraction studies (**13**, **29**, and **32** (M = Mo)), none shows significant lengthening of the nitrogen–nitrogen bond compared with that in gaseous dinitrogen and dinitrogen coordinated in mononuclear complexes. This is to be contrasted with the long nitrogen–nitrogen bonds observed in dinuclear tantalum complexes, e.g., **6**.

There is very little published data on the reactivity of this class of complex. The molybdenum complex **12** does not produce any nitrogen-containing products upon reaction with H$_2$O, LiAlH$_4$, or BuLi.[37] Reaction of **13** with HBF$_4$ causes protonation of the metal to form [{HMo($\eta^6$-C$_6$H$_3$Me$_3$)(dmpe)}$_2$(N$_2$)][BF$_4$]$_2$, which upon reaction with dilute HCl or with H$_2$SO$_4$/FeSO$_4$ produces no ammonia.[38]

Some of these complexes appear to be good candidates for chemical investigation. For example, **29** has a very low $\nu$(N$_2$) at 1660 cm$^{-1}$ and in the X ray photoelectric spectrum the N(1s) binding energy of 398.6 eV (the two nitrogen atom peaks are not resolved) is similar to those for **15** (399.6 and 398.6 eV), a complex that can be both alkylated and protonated, and the rhenium precursor [ReCl(N$_2$)(PMe$_2$Ph)$_4$] (400.1 and

398.4 eV).[119] A short nitrogen–nitrogen bond does not mean that there is no activation of coordinated dinitrogen; witness the chemistry of **15** (N–N = 1.118(8) Å, cf. **29**, N–N = 1.18(3) Å and $N_2$, N–N = 1.0976(1) Å).

### 2.4. Manganese, Technetium, and Rhenium

No reactions of coordinated dinitrogen have been reported for complexes **37** and **38**. However, mention should be made of trans-[ReCl($N_2$)(PMe$_2$Ph)$_4$] and its analogs even though they are not normally prepared directly from dinitrogen.[113] Dinitrogen coordinates particularly strongly to rhenium(I) as evidenced by the most extensive series of dinitrogen complexes known for any metal to date.[18] A few examples are given including the only stable complexes containing both coordinated dinitrogen and sulfur: [ReCl($N_2$)(P$R_3$)$_4$](P$R_3$ = PMe$_2$Ph, PPh$_2$Me, Me$_2$NPF$_2$, PPh$_2$H, P(CH$_2$OMe)$_3$), [ReCl($N_2$){Ph$_2$P(CH$_2$)$_n$PPh$_2$}] ($n$ = 1, 2, 3), [ReCl(CO)$_2$($N_2$)(PPh$_3$)$_2$], [ReCl($N_2$)$L$(PMe$_2$Ph)$_3$] ($L$ = py, 3-picoline,4-picoline,pyrazine), and [Re$L$($N_2$)(PMe$_2$Ph)$_3$] ($L$ = S$_2$CNMe$_2$, S$_2$CNEt$_2$, S$_2$COEt, S$_2$CPPh$_2$).

The electron-richness of the rhenium complexes [ReCl($N_2$)(P$R_3$)$_4$] is manifest in their relatively low $\nu(N_2)$ of 1920–1980 cm$^{-1}$, oxidation potentials of 0.00 to 0.30 volts (vs. saturated calomel electrode), and ability to behave as Lewis bases via the noncoordinated nitrogen atom toward Lewis acids such as AlMe$_3$, CrCl$_3$(THF)$_3$, and other metal and nonmetal halides.[18] In fact, the basicity of this coordinated dinitrogen is similar to that of THF towards AlMe$_3$, when P$R_3$ = PMe$_2$Ph, and greater than for the other dinitrogen complexes studied.[114] The polynuclear complexes containing dinitrogen bridged to titanium and molybdenum have already been discussed.

Acids, acyl, and aroyl halides all react with [ReCl($N_2$)(PMe$_2$Ph)$_4$]. However, protonation occurs at the metal to give the hydride [ReClH($N_2$)(PMe$_2$Ph)$_4$]$^+$, whereas slow acylation and aroylation occur at the end nitrogen atom. Interestingly, this latter reaction is the reverse of the preparation of the rhenium dinitrogen complexes.[18] Alkyl halides do not react with rhenium complexes of dinitrogen.

### 2.5. Iron, Ruthenium, and Osmium

Although a large number of dinitrogen complexes of these metals have been reported (many containing ligands other than phosphines), only a few reactions have been reported. Complexes **39** and **40** react with LiAlH$_4$, NaBH$_4$, and acids to produce the corresponding hydride, e.g., [($\eta^5$-C$_5$H$_5$)FeH(dmpe)] with loss of $N_2$.[115,116] Complex **45** is prepared by the reduction of [FeCl$_3$(PPh$_3$)$_2$] with Pr$^i$MgBr in ether at −50°C under dinitrogen. A band at 1761 cm$^{-1}$ is attributed to $\nu(N_2)$ and indicates a N–N bond of low-bond order. Treatment of the complex with HCl in ether at $-50°C$ gave 10% hydrazine (based upon complexed $N_2$).[123]

### 2.6. Cobalt, Rhodium, and Iridium

Complex **50** is one of the most thoroughly studied dinitrogen complexes. Reaction with hydrohalic acids leads to loss of dinitrogen and dihydrogen.[140] Similar results occur with rhodium and iridium dinitrogen complexes. Complex **55** is prepared by the reaction of **50** with MgEt$_2$ in THF at 0°C and has $\nu(N_2)$ at 1840 cm$^{-1}$ (cf., $\nu(N_2)$ at 2088 cm$^{-1}$ for **50**). Complex **55** reacts with water to regenerate **50** and acids to give hydrazine and a trace of ammonia. With concentrated sulfuric acid, 0.29 mol of hydrazine and 0.03 mol of ammonia per mol of complex and dinitrogen are liberated.[133] In this complex, it is not known whether dinitrogen is bridging between cobalt and magnesium as it does in complex **54** or nonbridging as it is in the salt **52**. Complex **51** (PR$_3$ = PPh$_3$) reacts with water to give **50**[130] but no reactions of **54** have been reported. No reaction of dinitrogen coordinated to rhodium or iridium have been reported.

### 2.7. Nickel, Palladium, and Platinum

No reactions of complexes **62** and **63** have been reported. Palladium and platinum form no well-defined dinitrogen complexes. However, until relatively recently there were few carbonyl complexes of palladium and platinum. This has changed rapidly with the preparation of a wide variety of compounds such as dinuclear complexes, e.g., $[M_2Cl_2(\mu\text{-CO})(\mu\text{-dppm})_2]$ where $M$ = Pd, Pt,[143] and neutral and anionic polynuclear complexes such as $[Os_2(CO)_6\{\mu\text{-Pt-(CO)(PPh}_3)\}_2]$[144] and $[Pt_9(CO)_{18}]^{-2}$.[145] The absence of simple, mononuclear palladium and platinum dinitrogen complexes should not be construed as evidence that this is a barren area for research.

### 2.8. Copper

There is one report of nitrogen fixation involving copper complexes.[146] When dinitrogen is bubbled through either a pyridine or THF solution of copper(I) borohydride or THF solution of $[(PPh_3)_2CuBH_4]$ for long periods of time, ammonia is formed. Ammonia is removed from the reaction mixture either by sweeping it out with dinitrogen or by heating to ca. 100°C and collecting in an acid trap. Reaction conditions are mild with almost quantitative yields, and hydrolysis is not required to form ammonia. Blank reactions all give negative results for ammonia production.

These results have been successfully reproduced in one other laboratory.[147] Unfortunately, the crucial experiment using $^{15}N_2$ and analyzing for $^{15}NH_3$ has not been carried out yet.

## 3. NITRIDING REACTIONS

Many metals have the ability to undergo nitriding reactions with dinitrogen. Titanium has been studied particularly thoroughly because of its facile nitriding ability. Typically, metal salts are reduced with strong reducing agents such as Grignard reagent, organolithium, Mg and $LiAlH_4$ in ether and hydrocarbon solvent under a dinitrogen atmosphere, usually high pressure, over an extended period of time. Subsequent hydrolysis produces ammonia and/or hydrazine often in stoichiometric yields.[23] Reduction of $[TiCl_4(PPh_3)_2]$ and $[FeCl_3(PPh_3)_2]$ with EtMgBr in ether solution for 10 h under dinitrogen (150 atm) produces very little ammonia upon hydrolysis.[148] Low yields of organonitrogen compounds such as aniline are obtained (upon hydrolysis) in reactions in which either a lithium aryl is the reducing agent or a titanium-aryl precursor is employed, e.g., $[(\eta^5-C_5H_5)_2TiPh_2]$.[23]

Catalytic nitrogen-fixing behavior is found in the nitriding reactions of titanium. When $TiCl_4$, $TiBr_4$, or $Ti(OBu)_4$ is heated with a mixture of Al and $AlBr_3$ under dinitrogen, the yields of ammonia formed by hydrolysis constantly increase with increase in the amount of Al and $AlBr_3$ and can be brought up to over 200 mol/g-atom of Ti.[23] Aluminum is the reducing agent; $AlBr_3$ effectively removes the nitride ligand from titanium, and a reduced titanium complex is the catalyst. Intriguing solvents have been used in these reactions, such as benzene, molten $AlBr_3$ (m.p. 97.5°C), $AlBr_3/AlCl_3$ eutectic (2.17:1; m.p. 67°C), and the ternary low-melting mixture $AlBr_3/AlCl_3$/benzene. It is difficult to see how these systems could be developed into commercially viable catalysts unless dihydrogen could be used as the sole reducing agent, i.e., hydrogenolyze the metal-nitride bond and reduce the metal to its active form.

## 4. DINITROGEN BINDING AND REACTIVITY

Dinitrogen has been shown to adopt an end-on bonding arrangement in those phosphine-containing complexes that have been isolated and characterized. In most of these complexes, dinitrogen bonds in a linear end-on arrangement either to one metal **G** (e.g., **15, 23, 43, 50, 52,** and

57) or bridges between two metals **H** (e.g., **6, 13, 29, 32, 61,** and **62**). However, in complex **54** the bridging geometry **I** is found with CoNN and NNMg angles of 180° and 158°, respectively. The ability of a metal to bind dinitrogen depends upon a subtle balance between the energies of the appropriate metal and dinitrogen orbitals. The currently accepted bonding scheme for end-on dinitrogen ligation involves donation of $\sigma$-electron density from dinitrogen to the metal and accepting $\pi$-electron density from the metal into its $\pi^*$ orbitals. The extreme examples of no backing-donation and significant back-donation are shown in resonance forms **J** and **K**:

$$M\!\doteq\!N\!\equiv\!N \qquad M\!\doteq\!N\!\equiv\!N\!\doteq\!M \qquad M\!\doteq\!N\!\equiv\!N\diagdown_{M}$$

$$\text{G} \qquad\qquad \text{H} \qquad\qquad \text{I}$$

$$\overset{-}{\ddot{M}}\!-\!\overset{+}{N}\!\equiv\!N\!: \qquad M\!=\!\overset{+}{N}\!=\!\overset{-}{\ddot{N}}\!:$$

$$\text{J} \qquad\qquad \text{K}$$

Organophosphine ligands are particularly successful in helping the metal provide a favorable coordination site that is sufficiently electron-rich to provide strong back-donation into the vacant dinitrogen $\pi^*$-orbitals and help stabilize the metal-dinitrogen bond.

The extent of the dinitrogen-to-metal interaction varies considerably with changes in metal, oxidation number, and coligands. For example, dinitrogen forms a relatively weak bond with $d^6$ Fe(II), but with $d^6$ Re(I), Mo(O), and W(O), the bond is stronger, and dinitrogen is activated towards a variety of reagents. Ti(II) ($d^2$) is able to bind dinitrogen and, depending upon the reducing agent and temperature, the bound dinitrogen is reduced to either the "diazene-level" ($N_2^{-2}$), "hydrazine-level" ($N_2^{-4}$), or "nitride-level" ($N^{-3}$). The classification of "reduction-level" is based upon the nitrogen hydride(s) formed upon hydrolysis, which is not necessarily unambiguous in the case of diazene and hydrazine. Depending upon the reaction condition, diazene will decompose to a mixture of $N_2$ and $H_2$, $N_2$ and $N_2H_4$, and/or $N_2$ and $NH_3$, and hydrazine is readily decomposed by transition metal compounds. Ta(III) ($d^2$) binds dinitrogen and in so doing reduces it to the "hydrazine-level," and Ta(I) ($d^4$) behaves similarly. In a series of papers, Chatt and co-workers have attempted to define the dinitrogen binding site in terms of its electron-richness (or -poorness) and polarizability.[149]

So far, no example of a simple side-on bonded dinitrogen complex has been reported. A common problem in working with dinitrogen complexes is the facile loss of dinitrogen that is very frequently observed, particularly in solution. It is for this reason that great difficulty is often encountered when trying to isolate and purify this class of complex. In other words, there are experimental as well as electronic obstacles to the preparation of dinitrogen complexes.

The reactivity of coordinated dinitrogen toward electrophiles, nucleophiles, and radicals is dependent upon many factors, particularly the charge distribution within the dinitrogen-metal unit that results from the synergistic-bonding pattern. Attempts to determine the charge distribution[110] have been made using $X$ ray photoelectron spectroscopy (XPS),[110,150] infrared absorption spectroscopy,[151] dipole moment measurement,[110] and electrochemical measurements,[149,152] but no one technique can uniquely measure this term.

In the only documented example of nucleophilic attack at coordinated dinitrogen, LiPh (and LiMe) reacts with $[(\eta^5\text{-}C_5H_5)Mn(CO)_2(N_2)]$, $\nu(N_2) = 2165\ cm^{-1}$, to form, after protonation, a neutral phenyldiazene complex $[(\eta^5\text{-}C_5H_5)Mn(CO)_2(PhNNH)]$.[153] The evidence suggests that carbanion attack occurs at the metal-bound dinitrogen atom. The particularly high N(1s) binding energies (403.0 and 401.8 eV) and high $\nu$ ($N_2$) imply a relatively positive dinitrogen ligand (or electron-poor complex), with the bonding better described by resonance form **J** than **K**. Efforts to extend this work to other electron-poor dinitrogen-containing complexes have not been successful so far.[149]

The Lewis base behavior of coordinated dinitrogen has been studied and related to $\nu(N_2)$.[114] In many cases, stable Lewis base-Lewis acid adducts have been isolated (e.g., **1–3, 10, 14, 29–32**). Those Lewis acids that are both $\sigma$- and $\pi$-acceptors have been shown to cause considerable lengthening of the N–N bond (e.g., in **29** and **32**) due to depopulation of the $N_2$ $\pi$ bonds.[35] Resonance structure **K** is more important than **J** in describing the behavior of coordinated dinitrogen as a strong Lewis base and in the stable adducts. In some cases, bridging dinitrogen complexes are obtained directly from the preparation and isolation of dinitrogen complexes (e.g., **4–7, 9, 12, 13, 39, 40, 45, 53, 54**, and **60–62**). The noticeably long N–N bond observed in the tantalum complex **6** and concomitant short Ta–N bonds are undoubtably the result of the ability of tantalum (and certain other metals such as molybdenum, titanium, tungsten, and vanadium) to behave as a $\sigma$-acceptor and $\pi$-donor, as well as a $\pi$-acceptor.

The susceptibility of coordinated dinitrogen to attack by electrophiles and radicals and their relationship to the electron-richness of dinitrogen complexes has been addressed in an important series of articles using $E_{1/2}^{ox}$ and $\nu(N_2)$ data.[82,149,154] The successful attack of an electrophile or radical upon coordinated dinitrogen depends upon there being a certain degree of electron-richness. However, in many cases where alkylation or protonation of dinitrogen would be predicted to occur [based upon $E_{1/2}^{ox}$ and $\nu(N_2)$], instead oxidation or protonation of the metal or no reaction occurs. This arises because of the high degree of conjugation present in the [$M$–N–N] unit and our inability to quantitatively determine the charge distribution in the ground state and excited state(s).

How can you determine whether or not coordinated dinitrogen is "activated"? It is far from being as simple as just adding acid and looking for ammonia and hydrazine! For example, the protonation of coordinated dinitrogen may occur to give a {$M$NNH} or {$M$NNH$_2$} unit, which will be unstable unless strong metal-nitrogen bonding can occur, i.e., the metal can be oxidized. This readily occurs with Mo(O) and W(O) to form $M$(II) and $M$(IV) complexes but is much less likely in the case of Fe(II), for example. Unstable {$M$NNH} will decompose to give $N_2$ and $H^+$, giving the erroneous impression that coordinated dinitrogen did not react. {$M$NNH$_2$} may decompose to give $N_2H_2$, which would be unstable and would form $N_2$ and $H_2$, $N_2$ and $N_2H_4$, and/or $N_2$ and $NH_3$, or simple $N_2 + 2H^+$. In the former case, no apparent reaction of coordinated dinitrogen is the erroneous conclusion again, unless $N_2H_2$ is detected chemically or {$M$NNH$_2$} is detected spectroscopically. Stoichiometric formation of $N_2H_2$, $N_2H_4$, and $2NH_3$ requires 2, 4, and 6 electrons, respectively. Many metals can provide two electrons per metal (e.g., Co(I) → Co(III) + $2e^-$) or two per dinuclear metal unit, but the problems associated with recognizing dinitrogen reactivity in this case have been pointed out above. It is unrealistic to expect many metals that are coordinated to dinitrogen to be able to provide four or six electrons to produce $N_2H_4$ or $2NH_3$ at a single metal; only tungsten has unambiguously been shown to do this so far. In dinuclear tantalum(I), -(III),[29] titanium(II),[155] and zirconium(II)[156] complexes containing coordinated dinitrogen, protonation produces nearly quantitative yields of hydrazine. High yields of ammonia are formed upon protonation of molybdenum complexes such as **27** by virtue of a mechanism involving a bridging intermediate (Scheme 6), whereby three electrons come from each molybdenum atom. It is probably unreasonable to expect that many dinitrogen complexes will form $NH_3$, $N_2H_4$, or stable nitrogen hydride intermediates upon protonation, unless the metal has a number of stable oxidation states available to it or unless a reducing agent (other than the metal) is present.

## 5. FUTURE PROSPECTS

Despite the large amount of novel chemistry that has arisen from the chemical studies of nitrogen fixation, there is nothing to suggest an alternative to the Haber Process for the large-scale (>100 ton per day) production of ammonia. However, there may be a real need for alternatives to the Haber Process. A 400 million dollar plant to produce ammonia can be built in an underdeveloped country, but how will ammonia be distributed, even if it is converted into urea? Small scale, local production of nitrogen fertilizer would be of direct benefit to the food producers of underdeveloped countries. The use of local resources such as solar, water, and wind energy coupled with a low-technology ammonia-producing plant that would be working intermittently year round would produce an aqueous solution of ammonia (or ammonia nitrate if some of the ammonia is oxidized) that could either be run into the irrigation ditches or manually distributed.

The conversion of dinitrogen to ammonia is shown in Equations 33 and 34 to emphasize two approaches.

$$N_2 + 3H_2 \rightarrow 2NH_3 \tag{33}$$

$$N_2 + 6e^- + 6H^+ \rightarrow 2NH_3 \tag{34}$$

On the one hand, dihydrogen is the reducing agent, and, on the other, electrons and protons are provided separately. They could be classified as the chemical and electrochemical methods. It is going to be very difficult to develop a homogeneous "Haber Process" using metal-phosphine complexes or any other metal complexes. Although there are a few stable complexes known that contain both nitrogen and hydrogen or dihydrogen, generally dihydrogen displaces dinitrogen from a metal-binding site rather easily. One area that seems to have received no attention is dinitrogen complexes of metal cluster. It may be possible to assemble mixed-metal clusters that can coordinate both dihydrogen and dinitrogen and promote dinitrogen reduction. Phosphine rather than carbonyl ligands would be preferred to increase the electron-richness of the cluster, thus aiding dinitrogen binding.

The electrochemical activation of coordinated dinitrogen has received some attention. The moiety $[M(N_2)(dppe)_2]$ ($M$ = Mo or W) has been anchored to a $SnO_2$ electrode but with no chemistry of the coordinated dinitrogen reported.[157] In principle, it should be possible to anchor to an electrode those complexes that do produce ammonia or hydrazine upon treatment with acid. For the reaction to become catalytic, the dinitrogen binding site needs to be regenerated. In practice, strong reducing agents are used to generate these binding sites originally, strong enough to reduce protons to dihydrogen. However, not enough is known about the minimum potentials necessary for creation of the binding site, and this is information uniquely available by electrochemical studies.

The direct production of organic nitrogen compounds using dinitrogen has received little attention outside the reactions of organic halides with coordinated dinitrogen. Alkenes and alkynes would be viable carbon-atom sources, although like dihydrogen they often displace coordinated dinitrogen. If coordinated dinitrogen, in complexes such as 5–7 where dinitrogen is "activated," reacts with alkynes ($RC\equiv CH$) in an "olefin-methatesis-type" reaction, the product would be a nitrile ($RC\equiv N$) or hydrogen cyanide ($HC\equiv N$).

## 6. CONCLUDING REMARKS

To many people, nitrogenase is a Rosetta Stone: elucidation of the mechanism will provide the key to nitrogen fixation under mild conditions.[156] In the meantime, work is proceeding to find new dinitrogen-binding sites with the hope of discovering new chemistry. Certainly, organophosphine ligands have helped provide an environment to study the chemistry of coordinated dinitrogen. It is now time to investigate ligands containing other donor atoms and explore their ability to provide support for the binding and activation of dinitrogen. One of the new areas under intensive investigation is the [FeMoS] and [FeWS] clusters.[157] Within these systems lies the hope of modeling the nitrogenase cofactor, FeMoCo. All solutions to the problems of nitrogen-fixation under mild conditions lie in the future, but sufficient work has been done to *catalyze* further vigorous investigations.

#### ACKNOWLEDGMENTS

Thanks are due to Professor R. R. Schrock for informing me of results prior to publication. Published and unpublished work by the author reported in this chapter was supported by funds from the National Science Foundation and the University of Nebraska Research Council. Acknowledgment is made to the donors of the Petroleum Research Fund, administered by the American Chemical Society, for partial support of this work.

## REFERENCES

1. E. N. Mishustin and V. K. Shilnikova, *Biological Fixation of Atmospheric Nitrogen* (Plenum Publishing, London, 1971).
2. M. J. Dilworth, *Ann. Rev. Plant Physiol.* **25**, 81–114 (1974).

3. R. C. Burns and R. W. F. Hardy, *Molecular Biology, Biochemistry, and Biophysics* (Springer, Berlin, 1975), Vol. 12.
4. W. D. P. Stewart and J. R. Gallon, eds., *Nitrogen Fixation* (Academic Press, London, 1981).
5. J. R. Postgate, *The Chemistry and Biochemistry of Nitrogen Fixation*, edited by J. R. Postgate (Plenum Publishing, New York, 1971), pp. 161–190.
6. R. H. Burris, *The Chemistry and Biochemistry of Nitrogen Fixation*, edited by J. R. Postgate (Plenum Publishing, New York, 1971), pp. 105–160.
7. R. W. F. Hardy, R. C. Burns, and G. W. Parshall, *Bioinorganic Chemistry*, edited by R. F. Gould, *Adv. Chem. Ser.* (ACS Publications, Washington, D.C.), Vol. 100, pp. 219–245 (1971).
8. V. K. Shah and W. J. Brill, *Proc. Natl. Acad. Sci. U.S.A.* **74**, 3249–3253 (1977).
9. S. P. Cramer, W. O. Gillium, K. O. Hodgson, L. E. Mortenson, E. I. Stiefel, J. R. Chisnel, W. J. Brill, and V. K. Shah, *J. Am. Chem. Soc.* **100**, 3814–3819 (1978).
10. R. N. F. Thornley, R. R. Eady, and D. J. Low, *Nature (London)* **272**, 557–558 (1978).
11. K. L. Hadfield and W. A. Bullen, *Biochemistry* **8**, 5103–5108 (1969).
12. A. V. Slack, *Ammonia*, edited by A. V. Slack and G. R. James (Marcel Dekker, New York, 1973), Vol. 1, pp. 1–142.
13. D. E. Harrison and H. Taube, *J. Am. Chem. Soc.* **89**, 5706–5707 (1967).
14. M. I. Temkin, *Adv. Catal.* **28**, 250–263 (1979), and references therein.
15. J. P. Collman, M. Kubota, F. D. Vastine, J.-Y. Sun, and J. W. Kang, *J. Am. Chem. Soc.* **90**, 5430–5437 (1968).
16. R. W. F. Hardy, F. Bottomley, and R. C. Burns, eds., *A Treatise on Dinitrogen Fixation* (John Wiley and Sons, New York, 1979).
17. J. Chatt, L. M. da Camara Pina, and R. L. Richards, eds., *New Trends in the Chemistry of Nitrogen Fixation* (Academic Press, London, 1980).
18. J. Chatt, J. R. Dilworth, and R. L. Richards, *Chem. Rev.* **78**, 589–625 (1978).
19. M. Hidai, Y. Mizobe, and Y. Uchida, *J. Am. Chem. Soc.* **98**, 7824–7825 (1976).
20. J. Chatt, G. A. Heath, and R. L. Richards, *J. Chem. Soc., Dalton Trans.* **1974**, 2074–2082.
21. T. A. George and M. E. Noble, *Inorg. Chem.* **17**, 1678–1679 (1978).
22. L. J. Archer and T. A. George, *Inorg. Chem.* **18**, 2079–2082 (1979).
23. V. E. Vol'pin and V. B. Shur, *New Trends in the Chemistry of Nitrogen Fixation*, edited by J. Chatt, L. M. da Camara Pina, and R. L. Richards (Academic Press, London, 1980), pp. 67–100.
24. A. E. Shilov, *New Trends in the Chemistry of Nitrogen Fixation*, edited by J. Chatt, L. M. da Camara Pina, and R. L. Richards, (Academic Press, London, 1980), pp. 121–150.
25. J. H. Teuben, *New Trends in the Chemistry of Nitrogen Fixation*, edited by J. Chatt, L. M. da Camara Pina, and R. L. Richards (Academic Press, London, 1980), pp. 233–247.
26. R. Robson, *Inorg. Chem.* **13**, 475–479 (1974).
27. R. J. Burt, G. J. Leigh, and D. L. Hughes, *J. Chem. Soc., Dalton Trans.* **1981**, 793–799.
28. H. W. Turner, J. D. Fellmann, S. M. Rocklade, R. R. Schrock, M. R. Churchill, and H. J. Wasserman, *J. Am. Chem. Soc.* **102**, 7809–7811 (1980).
29. S. M. Rocklage and R. R. Schrock, *J. Am. Chem. Soc.* **104**, 3077–3081 (1982).
30. S. M. Rocklage, H. W. Turner, D. Fellmann, and R. R. Schrock, *Organometallics* **1**, 703–707 (1982).
31. M. R. Churchill and H. J. Wasserman, *Inorg. Chem.* **20**, 2899–2904 (1981).
32. ———, *Inorg. Chem.* **21**, 218–222 (1982).
33. H. H. Karsch, *Angew. Chem., Int. Ed. Engl.* **16**, 56–57 (1977).

34. P. Sobota and B. Jezowska-Trzebiatowska, *J. Organomet. Chem.* **131**, 341–345 (1977).
35. J. Chatt, R. C. Fay, and R. L. Richards, *J. Chem. Soc., Dalton Trans.* **1971**, 702–704.
36. M. Hidai, K. Tominari, Y. Uchida, and A. Misono, *J. Chem. Soc., Chem. Commun.* **1969**, 814.
37. M. L. H. Green and W. E. Silverthorn, *J. Chem. Soc., Dalton Trans.* **1973**, 301–306.
38. ———, *J. Chem. Soc., Dalton Trans.* **1974**, 2164–2166.
39. R. A. Forder and K. Prout, *Acta Crystallogr., Sect. B* **30**, 2778–2780 (1974).
40. M. Hidai, K. Tominari, Y. Uchida, and A. Misono, *J. Chem. Soc., Chem. Commun.* **1969**, 1392. M. Hidai, K. Tominari, and Y. Uchida, *J. Am. Chem. Soc.* **94**, 110–114 (1972).
41. Y. Uchida, T. Uchida, M. Hidai, and T. Komada, *Acta Crystallogr., Sect. B* **31**, 1197–1199 (1975).
42. L. K. Atkinson, A. H. Mawby, and D. C. Smith, *J. Chem. Soc., Chem. Commun.* **1971**, 157–158.
43. A. Frigo, G. Pousi, and A. Turco, *Gazz. Chim. Ital.* **101**, 637–638 (1971).
44. C. Miniscloux, G. Martino, and L. Sajus, *Bull. Soc. Chim. Fr.* **567**, 2183–2188 (1973).
45. T. A. George and C. D. Seibold, *J. Organomet. Chem.* **30**, C13–C14 (1971). T. A. George and C. D. Seibold, *Inorg. Chem.* **12**, 2544–2547 (1973).
46. J. R. Dilworth and R. L. Richards, *Inorg. Syn.*, edited by D. H. Busch (John Wiley and Sons, New York, 1980), pp. 119–124.
47. J. Chatt, W. Hussain, G. J. Leigh, H. Neukomm, C. J. Pickett, and D. A. Rankin, *J. Chem. Soc., Chem. Commun.* **1980**, 1024–1025.
48. M. W. Anker, J. Chatt, G. J. Leigh, and A. G. Wedd, *J. Chem. Soc., Dalton Trans.* **1975**, 2639–2645.
49. J. Chatt, A. J. Pearman, and R. L. Richards, *J. Chem. Soc., Dalton Trans.* **1977**, 1852–1860.
50. T. A. George and R. A. Kovar, *Inorg. Chem.* **20**, 285–287 (1981).
51. M. Sato, T. Tatsumi, T. Kodama, M. Hidai, T. Uchida, and Y. Uchida, *J. Am. Chem. Soc.* **100**, 4447–4452 (1978).
52. M. Aresta and A. Sacco, *Gazz. Chim. Ital.* **102**, 755–780 (1972).
53. J. A. Baumann, M. C. Davies, and T. A. George, unpublished results.
54. G. J. Kubas, *J. Chem. Soc., Chem. Commun.* **1980**, 61–62.
55. M. Mercer, R. H. Crabtree, and R. L. Richards, *J. Chem. Soc., Chem. Commun.* **1973**, 808–809.
56. M. Mercer, *J. Chem. Soc., Dalton Trans.* **1974**, 1637–1640.
57. J. Chatt and R. L. Richards, *J. Less-Common Metals* **54**, 477–484 (1977).
58. P. D. Chadwick, *J. Chem. Soc., Dalton Trans.* **1976**, 1934–1936.
59. J. Chatt, G. A. Heath, and G. J. Leigh, *J. Chem. Soc., Chem. Commun.* **1972**, 444–445.
60. J. Chatt, G. A. Heath, and R. L. Richards, *J. Chem. Soc., Chem. Commun.* **1972**, 1010–1011.
61. A. A. Diamantis, J. Chatt, G. J. Leigh, and G. A. Heath, *J. Organomet. Chem.* **84**, C11–C12 (1975).
62. J. Chatt, A. J. Pearman, and R. L. Richards, *Nature* **253**, 39–40 (1975).
63. J. Chatt, A. A. Diamantis, G. A. Heath, N. E. Hooper, and G. J. Leigh, *J. Chem. Soc., Dalton Trans.* **1977**, 688–697.
64. M. Sato, T. Kodama, M. Hidai, and Y. Uchida, *J. Organomet. Chem.* **152**, 239–254 (1978).
65. V. W. Day, T. A. George, and S. D. A. Iske, Jr., *J. Am. Chem. Soc.* **97**, 4127–4128 (1975).
66. G. E. Bossard, D. C. Busby, M. Chang, T. A. George, and S. D. A. Iske, Jr., *J. Am. Chem. Soc.* **102**, 1001–1008 (1980).

67. D. C. Busby, T. A. George, S. D. A. Iske, Jr., and S. D. Wagner, *Inorg. Chem.* **20**, 22–27 (1981).
68. C. S. Day, V. W. Day, T. A. George, and I. Tavanaiepour, *Inorg. Chim. Acta* **45**, L54 (1980).
69. V. W. Day and T. A. George, unpublished results.
70. V. W. Day, T. A. George, S. D. A. Iske, Jr., and S. D. Wagner, *J. Organomet. Chem.* **112**, C55–C58 (1976).
71. F. C. Marsh, R. Mason, and K. M. Thomas, *J. Organomet. Chem.* **96**, C43–C45 (1975).
72. D. C. Busby and T. A. George, *Inorg. Chem.* **18**, 3164–3167 (1979).
73. G. Butler, J. Chatt, G. J. Leigh, and D. L. Hughes, *Inorg. Chim. Acta* **30**, L287–L288 (1978).
74. J. Chatt, R. A. Head, G. J. Leigh, and C. J. Pickett, *J. Chem. Soc., Dalton Trans.* **1978**, 1638–1647.
75. G. E. Bossard and T. A. George, *Inorg. Chim. Acta* **54**, L241–L242 (1981).
76. R. Ben-Shoshan, J. Chatt, G. J. Leigh, and W. Hussain, *J. Chem. Soc., Dalton Trans.* **1980**, 771–775.
77. R. A. Head and P. B. Hitchcock, *J. Chem. Soc., Dalton Trans.* **1980**, 1150–1155.
78. M. Hidai, Y. Mizobe, M. Sato, T. Kodama, and Y. Uchida, *J. Am. Chem. Soc.* **100**, 5740–5748 (1978).
79. J. Chatt, W. Hussain, G. J. Leigh, and F. P. Terreros, *J. Chem. Soc., Dalton Trans.* **1980**, 1408–1415.
80. T. A. George and R. L. Turcotte, unpublished results.
81. G. E. Bossard and T. A. George, *Inorg. Chim. Acta* **54**, L239–L240 (1981).
82. J. Chatt, G. J. Leigh, H. Neukomm, C. J. Pickett, and D. R. Stanley, *J. Chem. Soc., Dalton Trans.* **1980**, 121–127.
83. R. J. W. Thomas, G. S. Laurence, and A. A. Diamantis, *Inorg. Chim. Acta* **30**, L353–L355 (1978).
84. A. Caruana, H. Hermann, and H. Kisch, *J. Organomet. Chem.* **187**, 349–359 (1980).
85. T. Tatsumi, M. Hidai, and Y. Uchida, *Inorg. Chem.* **14**, 2530–2534 (1975).
86. B. J. Carter, J. E. Bercaw, and H. B. Gray, *J. Organomet. Chem.* **181**, 105–116 (1979).
87. J. Chatt, R. A. Head, and G. J. Leigh, *J. Chem. Soc., Dalton Trans.* **1980**, 1129–1134.
88. D. C. Busby, C. D. Fendrick, and T. A. George, *Chemistry and Uses of Molybdenum*, Proceedings of the Third International Conference (Ann Arbor, 1979), pp. 290–295.
89. P. C. Bevan, J. Chatt, G. J. Leigh, and E. G. Leelamani, *J. Organomet. Chem.* **139**, C59–C62 (1977).
90. C. J. Pickett and G. J. Leigh, *J. Chem. Soc., Chem. Commun.* **1981**, 1033–1035.
91. M. Hidai, T. Kodama, M. Sato, M. Harakawa, and Y. Uchida, *Inorg. Chem.* **15**, 2694–2697 (1976).
92. J. Chatt, A. J. Pearman, and R. L. Richards, *J. Chem. Soc., Dalton Trans.* **1978**, 1766–1776.
93. ———, *J. Chem. Soc., Dalton Trans.* **1977**, 2139–2142.
94. T. Takahashi, Y. Mizobe, M. Sato, Y. Uchida, and M. Hidai, *J. Am. Chem. Soc.* **101**, 3405–3407 (1979).
95. J. A. Baumann and T. A. George, *J. Am. Chem. Soc.* **102**, 6153–6154 (1980).
96. S. N. Anderson, M. E. Fakley, R. L. Richards, and J. Chatt, *J. Chem. Soc., Dalton Trans.* **1981**, 1973–1980.
97. G. E. Bossard, T. A. George, and R. K. Lester, *Inorg. Chim. Acta Letters*, L241–L242 (1983).
98. J. Chatt, M. E. Fakley, P. B. Hitchcock, R. L. Richards, N. T. Luong-Thi, and D. J. Hughes, *J. Organomet. Chem.* **172**, C55–C58 (1979).
99. M. Hidai, *Current Perspectives in Nitrogen Fixation*, edited by A. H. Gibson and W. E. Newton (Australian Academy of Sciences, Canberra, 1981), pp. 26–29.

100. G. A. Heath, R. Mason, and K. M. Thomas, *J. Am. Chem. Soc.* **96,** 259–260 (1974).
101. I. R. Hanson and D. L. Hughes, *J. Chem. Soc., Dalton Trans.* **1981,** 390–399.
102. J. Chatt, A. J. Pearman, and R. L. Richards, *J. Chem. Soc., Dalton Trans.* **1976,** 1520–1524.
103. R. A. Henderson, *J. Organomet. Chem.* **208,** C51–54 (1981).
104. J. Chatt, J. R. Dilworth, P. L. Dahlstrom, and J. Zubieta, *J. Chem. Soc., Chem. Commun.* **1980,** 786–787.
105. J. R. Dilworth, R. A. Henderson, and R. L. Richards, *Current Perspectives in Nitrogen Fixation*, edited by A. H. Gibson and W. E. Newton (Australian Academy of Sciences, Canberra, 1981), p. 349.
106. G. E. Bossard, Ph.D. Thesis, University of Nebraska (1983).
107. M. C. Davies, T. A. George, and R. K. Lester, unpublished results.
108. D. Sellmann, R. Gerlach, and K. Jodden, *J. Organomet. Chem.* **178,** 433–447 (1979), and references therein.
109. C. R. Brûlet and E. E. van Tamelen, *J. Am. Chem. Soc.* **97,** 912–913 (1975).
110. J. Chatt, C. M. Elson, N. E. Hooper, and G. J. Leigh, *J. Chem. Soc., Dalton Trans.* **1975,** 2392–2401.
111. M. E. Tully and A. P. Ginsberg, *J. Am. Chem. Soc.* **95,** 2042–2043 (1973).
112. K. W. Chiu, W. K. Wong, and G. Wilkinson, *J. Chem. Soc., Chem. Commun.* **1981,** 451–452.
113. J. Chatt, J. R. Dilworth, and G. J. Leigh, *J. Chem. Soc., Dalton Trans.* **1973,** 612–618.
114. J. Chatt, R. H. Crabtree, E. A. Jeffery, and R. L. Richards, *J. Chem. Soc., Dalton Trans.* **1973,** 1167–1172.
115. D. Sellmann and E. Kleinschmidt, *J. Organomet. Chem.* **140,** 211–219 (1977).
116. W. E. Silverthorn, *J. Chem. Soc., Chem. Commun.* **1971,** 1310–1311.
117. M. Aresta, P. Giannoccaro, M. Rossi, and A. Sacco, *Inorg. Chim. Acta* **5,** 203–206 (1971).
118. D. H. Gerlach, W. G. Peet, and E. L. Muetterties, *J. Am. Chem. Soc.* **94,** 4545–4549 (1972).
119. Yu. G. Borod'ko, M. O. Broitman, L. M. Kachapina, A. K. Shilova, and A. E. Shilov, *Zh. Struct. Khim.* **12,** 545–546 (1971).
120. P. Stoppioni, F. Mani, and L. Sacconi, *Inorg. Chim. Acta* **11,** 227–230 (1974).
121. C. A. Ghilardi, S. Midollini, L. Sacconi, and P. Stoppioni, *J. Organomet. Chem.* **205,** 193–202 (1981).
122. G. M. Bancroft, M. J. Mays, B. E. Prater, and F. P. Stefanini, *J. Chem. Soc., A* **1970,** 2146–2149.
123. Yu. G. Borod'ko, M. O. Broitman, L. M. Kachapina, A. E. Shilov, and L. Yu. Ukhin, *J. Chem. Soc., Chem. Commun.* **1971,** 1185–1186.
124. J. M. Bellerby, M. J. Mays, and P. L. Sears, *J. Chem. Soc., Dalton Trans.* **1976,** 1232–1236.
125. R. A. Cable, M. Green, R. E. Mackenzie, P. L. Timms, and T. W. Turney, *J. Chem. Soc., Chem. Commun.* **1976,** 270–271.
126. W. H. Knoth, *J. Am. Chem. Soc.* **94,** 104–109 (1972).
127. S. Komiya and A. Yamamoto, *Bull. Soc. Chem. Japan.* **49,** 784–787 (1976).
128. J. Chatt, D. P. Melville, and R. L. Richards, *J. Chem. Soc., A* **1971,** 895–899.
129. Y. Yamamoto, S. Kitazume, L. S. Pu, and S. Ikeda, *J. Am. Chem. Soc.* **93,** 371–380 (1971).
130. M. Aresta, G. F. Nobile, M. Rossi, and A. Sacco, *J. Chem. Soc., Chem. Commun.* **1971,** 781.
131. R. Hammer, H.-F. Klein, P. Friedrich, and G. Huttner, *Angew. Chem., Int. Ed. Engl.* **16,** 485–486 (1977).
132. R. Hammer, H.-F. Klein, U. Schubert, A. Frank, and G. Huttner, *Angew. Chem., Int. Ed. Engl.* **15,** 612–613 (1976).

133. Y. Miura and A. Yamamoto, *Chem. Lett.* **1978,** 937–940.
134. H. L. M. van Gaal, F. G. Moers, and J. J. Steggerda, *J. Organomet. Chem.* **65,** C43–C45 (1974). H. L. M. van Gaal and F. L. A. van Bekerom, *J. Organomet. Chem.* **134,** 237–248 (1977).
135. P. R. Hofmann, T. Yoshida, T. Okano, S. Otsuka, and J. A. Ibers, *Inorg. Chem.* **15,** 2462–2466 (1976).
136. D. L. Thorn, T. H. Tulip, and J. A. Ibers, *J. Chem. Soc., Dalton Trans.* **1979,** 2022–2025.
137. T. Yoshida, T. Okano, and S. Otsuka, *J. Chem. Soc., Chem. Commun.* **1978,** 855–856.
138. T. Yoshida, T. Okano, D. L. Thorn, T. H. Tulip, S. Otsuka, and J. A. Ibers, *J. Organomet. Chem.* **181,** 183–201 (1979).
139. T. Yoshida, T. Okano, Y. Ueda, and S. Otsuka, *J. Am. Chem. Soc.* **103,** 3411–3422 (1981).
140. A. Sacco and M. Rossi, *Inorg. Chim. Acta* **2,** 127–132 (1968).
141. P. W. Jolly, K. Jonas, C. Kruger, and U.-H. Tsay, *J. Organomet. Chem.*, **33,** 109–122 (1971).
142. C. A. Tolman, D. H. Gerlach, J. P. Jesson, and R. A. Shunn, *J. Organomet. Chem.* **65,** C23–C26 (1974).
143. M. P. Brown, R. J. Puddephatt, M. Rashidi, and K. R. Seddon, *J. Chem. Soc., Dalton Trans.* **1978,** 1540–1544.
144. L. J. Farrugia, J. A. K. Howard, P. Mitrprachachon, J. L. Spencer, F. G. A. Stone, and P. Woodward, *J. Chem. Soc., Chem. Commun.* **1978,** 260–262.
145. J. C. Calabrese, L. F. Dahl, P. Chini, G. Longoni, and S. Martinengo, *J. Am. Chem. Soc.* **96,** 2616–2618 (1974).
146. C. A. Koerntgen, *Reduction of Nitrogen to Ammonia and Carbon Monoxide to Formaldehyde by Copper(I) Borohydride*, Ph.D. Thesis, Boston College, 1973.
147. B. D. James, personal communication.
148. M. E. Vol'pin, N. K. Chapovskaya, and V. B. Shur, *Izvest. Akad. Nauk SSSR, Ser. Khim.* **1966,** 1083–1084.
149. G. J. Leigh, R. H. Morris, C. J. Pickett, D. R. Stanley, and J. Chatt, *J. Chem. Soc., Dalton Trans.* **1981,** 800–804, and references therein.
150. B. Folkesson, *Acta Chem. Scand.* **27,** 287–302 (1973); **27,** 1441–1443 (1973). V. I. Nefedov, V. S. Lenenko, V. B. Shur, W. E. Vol'pin, J. E. Salyn, and M. A. Porai-Koshits, *Inorg. Chim. Acta* **7,** 499–502 (1973). P. Finn and W. L. Jolly, *Inorg. Chem.* **11,** 1434–1435 (1972). H. Binder and D. Sellmann, *Angew. Chem., Int. Ed. Engl.* **12,** 1017–1019 (1973). P. Brant and R. D. Feltham, *J. Less-Common Metals* **54,** 81–87 (1977).
151. D. J. Darensbourg, *Inorg. Chem.* **11,** 1436–1437 (1972). B. Folkesson, *Acta Chem. Scand.* **26,** 4008–4018 (1972); **27,** 276–286 (1973). M. S. Quinby and R. D. Feltham, *Inorg. Chem.* **11,** 2468–2476 (1972).
152. C. M. Elson, *J. Chem. Soc., Dalton Trans.* **1975,** 2401–2404.
153. D. Sellman and W. Weiss, *Angew Chem., Int. Ed. Engl.* **16,** 880–881 (1977); **17,** 269–270 (1978).
154. J. Chatt, C. T. Kan, G. J. Leigh, C. J. Pickett, and D. R. Stanley, *J. Chem. Soc., Dalton Trans.* **1980,** 2032–2038.
155. R. D. Sanner, D. M. Duggan, T. C. McKenzie, R. E. Marsh, and J. E. Bercaw, *J. Am. Chem. Soc.* **98,** 8358–8365 (1976).
156. J. M. Manriquez, R. D. Sanner, R. E. Marsh, and J. E. Bercaw, *J. Am. Chem. Soc.* **98,** 3042–3044 (1976).
157. G. J. Leigh and C. J. Pickett, *J. Chem. Soc., Dalton Trans.* **1977,** 1797–1800.
158. W. E. Newton, *Nitrogen Fixation*, Kirk-Othmer: Encyclopedia of Chemical Technology, 3rd Edition (John Wiley and Sons, New York, 1981), Vol. 15, pp. 942–968.
159. For example, D. Coucouvanis, *Acc. Chem. Res.* **14,** 201–209 (1981), and references therein.

# 14

# Polymer-Bound Phosphine Catalysts

## Norman L. Holy

### 1. INTRODUCTION AND SCOPE

While the theme of this volume is catalysis in a single phase, the principles and dynamics of homogeneous catalysis have impacted other areas of catalysis as well. Today, more than ever, there is an overlapping, a meshing of concepts from homogeneous and heterogeneous catalysis. One of the areas that interfaces both classical divisions is catalysis via polymer-bound transition metal complexes. It is an interface area because catalysts are prepared typically from an organic polymer such as polystyrene and then, after attachment of a ligand, a soluble metal complex is bound. The choice of metal complex is almost always based upon examples from homogeneous catalysis. One of the principal motivations for attaching a soluble complex to a polymer is that catalyst recovery becomes much easier. There are, however, motivations beyond simple recovery considerations for examining the reactions of polymer-bound catalysts. Though the characteristics of the polymer-bound phosphines may be described to a first approximation as being predictable on the basis of the properties, performances, and mechanistic interpretations of the homogeneous analogs, the support often has a significant influence on catalytic activity. For example, a polymer-bound catalyst may give a product distribution quite different from that of the soluble catalyst. Moreover, it is often observed that polymer-supported catalysts are less oxygen sensitive than the homogeneous ones. There is

---

*Dr. Norman L. Holy* • Department of Chemistry, Western Kentucky University, Bowling Green, Kentucky 42101.

also the very practical matter of solubility. Many transition metal complexes have but limited solubilities in desirable solvents, and the researcher is forced to utilize either very dilute solutions or expensive solvents. For potential commercial applications, the requirement of large solvent volumes is certainly deleterious. With polymer-bound catalysts the difficulty of establishing a high concentration of metal centers is circumvented, because the metal complex may be attached to a polymer in far higher concentrations than are possible in solution.

Polymer-bound catalysts share some of the limitations intrinsic in both classical homogeneous and heterogeneous catalysis. Like soluble catalysts, the range of reactions they facilitate is rather limited. As with heterogeneous catalysts it is often difficult to learn the intimate details of ligand coordination, mechanism, and catalyst poisoning. Normally, such interpretations rely heavily upon the characteristics of soluble complexes, though it is clear the correspondence is imperfect.

Attaching homogeneous transition metal complexes became an important discipline in the early 1970s. In addition to the use of phosphine-derived polymers, there are published accounts using virtually every conceivable type of attached ligand.[1,2]

The principal objectives of this review are to illustrate the range of reactions known to be catalyzed by phosphine-supported catalysts, to discuss critically performances of anchored catalysts in comparison to their soluble analogs, to consider the effects of the support itself, and to indicate a few areas where research activity might be fruitful. References are intended to be illustrative.

## 2. PREPARATION OF POLYMER-BOUND PHOSPHINE CATALYSTS

There are two quite different approaches to catalyst preparation. The most common route begins with a commercially available polymer to which a ligand and metal complex are anchored. Advantages of this route, in addition to the simplicity of using a commercially available material, are that the characteristics (surface area, pore size, swelling properties) of the polymer are already known. The alternative synthesis begins with a monomer already containing a phosphine; this monomer is then polymerized. A major advantage here is the greater flexibility in tailoring the support to optimize performance.

### 2.1. Modification of Preformed Polymers

Polystyrene is the most commonly used support because of its ready availability in several forms, its chemical inertness, and the ease with which ligands and metals may be anchored to it. For a variety of catalysts, three types of polystyrene are utilized. If a polymer lacks cross-linking, catalysts derived from such materials are normally soluble; that is, reactions with such polymeric catalysts are homogeneous. Even with a soluble polymer one of the real advantages of polymer-bound catalysts, the facility of catalyst recovery, is not negated. The catalyst may be recovered quite simply by filtration through a membrane[3] or by addition of a solvent in which the polymeric catalyst is insoluble.[4] A second type of polystyrene, indeed the customary one, is a gel-type (microporous) having a low degree (<2%) of cross-linking. The typical cross-linking agent is *p*-divinylbenzene (DVB). By proper choice of solvent, these resins swell and open all their internal volume to solvent and reagent. Microporous resins must be swollen prior to the anchoring of catalyst and also during their use as catalysts. The third type of polystyrene has a high degree of cross-linking (typically 20, 40, or 60% DVB); these "macroreticular" or "macroporous" resins have high surface areas.[5] The high cross-link density restricts diffusion into the interior, and attached catalysts are normally found in a thin layer at the internal surface. Before these macroporous resins are derivatized, it is important that the emulsifiers used in their preparation be removed; this requires a fairly lengthy washing procedure.[6]

Phosphine attachment to preformed polystyrene is accomplished typically by one of two routes. Bromination of the phenyl residues is followed by either metallation of the aryl moiety and treatment with chlorodiphenylphosphine, or by direct phosphination with diphenylphosphide. The second route to ligand attachment utilizes chloromethylation followed by diphenylphosphide treatment. Either method of phosphine

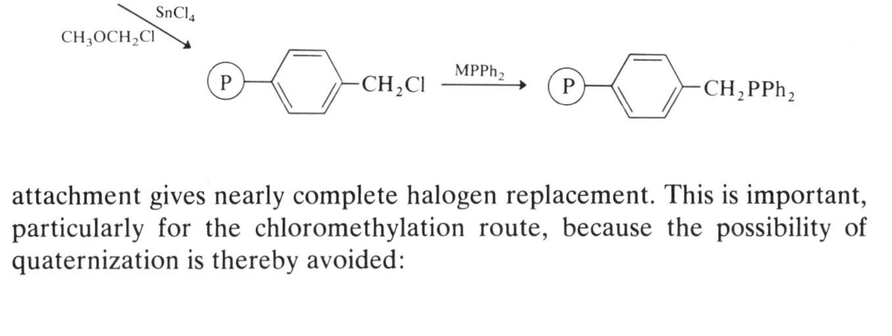

attachment gives nearly complete halogen replacement. This is important, particularly for the chloromethylation route, because the possibility of quaternization is thereby avoided:

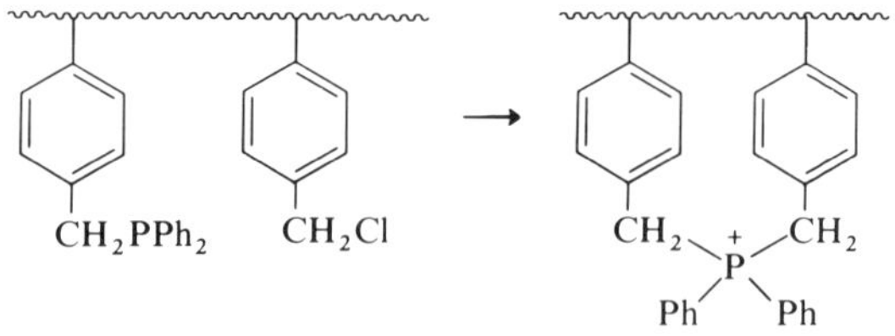

$$\text{Polymer-}C_6H_4\text{-}CH_2PPh_2 + \text{Polymer-}C_6H_4\text{-}CH_2Cl \longrightarrow \text{Polymer-}C_6H_4\text{-}CH_2\text{-}P^+(Ph)(Ph)\text{-}CH_2\text{-}C_6H_4\text{-Polymer}$$

The bidentate ligand *bis*(1,2-diphenylphosphine)ethane (DIPHOS) may be introduced in an analogous manner:

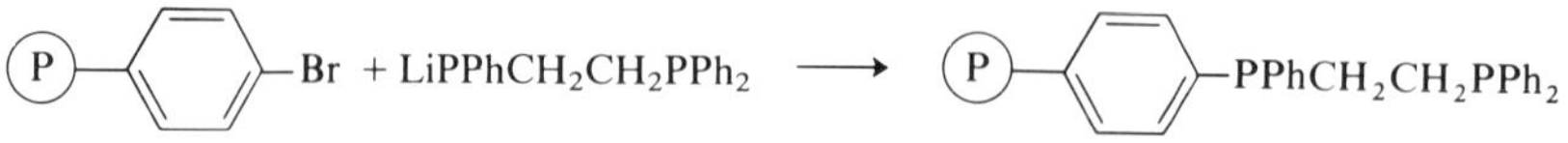

$$\text{(P)-C}_6H_4\text{-Br} + \text{LiPPhCH}_2\text{CH}_2\text{PPh}_2 \longrightarrow \text{(P)-C}_6H_4\text{-PPhCH}_2\text{CH}_2\text{PPh}_2$$

A similar approach was used to phosphenate noncross-linked polyvinylchloride.

$$-(CH_2-CH)_n- + n\text{LiPPh}_2 \rightarrow -(CH_2-CH)_n- + n\text{LiCl}$$
$$\phantom{-(CH_2-}|\phantom{H)_n-}\phantom{+ n\text{LiPPh}_2 \rightarrow -(CH_2-}|$$
$$\phantom{-(CH_2-}Cl\phantom{)_n- + n\text{LiPPh}_2 \rightarrow -(CH_2-}\text{PPh}_2$$

For complete chlorine replacement, it was necessary to use reflux times of up to 40 hr. Furthermore, a considerable breakdown of the polymer occurred.[7]

The distribution of phosphine ligands within the polymers depends upon the type of polymer. The distribution is quite uniform in gel-type (<2% cross-linking) polystyrenes but concentrated mainly at and near the surface in the macroreticular beads. A disadvantage of the highly cross-linked beads is that one is unsure of the precise phosphine distribution. Even though the catalyst distribution in macroreticular beads is limited, it is still possible to achieve high levels of catalyst loading because a high percentage of the phenyl residues are actually surface-exposed in macroreticular beads.

Attachment of the metal complex to the phosphinated polymer may be accomplished thermally or photochemically (Scheme 1).

Use of polystyrene as a support is limited by the mechanical and thermal stabilities of the support. Beads may fragment in stirred reactors. The upper thermal limit for most applications is 150–170°C.

While polystyrene is easily the most common organic support, alternatives are available. These include the use of polymethycrylate allyl chloride/DVB,[16] polyphenylene-isophthalamide,[17] soluble polysiloxanes,[18] and polyvinyl chloride.[7]

The use of oxide materials such as silica, alumina, silica-alumina, and molecular sieve zeolites offers an alternative support to organic polymers. These thermally stable supports operate without swelling, an advantage

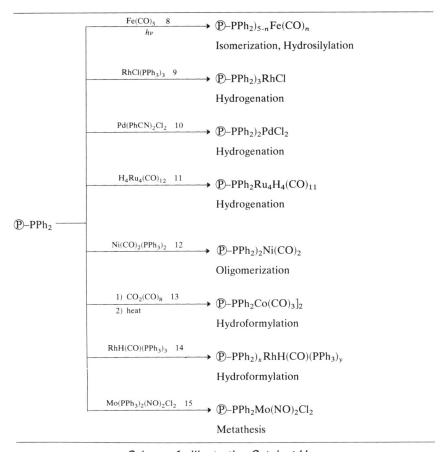

*Scheme 1. Illustrative Catalyst Uses*

from an engineering point of view. Many metal complexes are attached directly to these supports, but that is not the subject of this review. A phosphine may be grafted onto a support without great difficulty, and this is illustrated by the approach of the research group at British Petroleum[19-24]:

$$\begin{matrix} Si-OH \\ Si-OH \\ Si-OH \end{matrix} + (EtO)_3SiCH_2CH_2PPh_2 \rightarrow \begin{matrix} Si\diagdown O \\ Si-O-SiCH_2CH_2PPh_2 \\ Si\diagup O \end{matrix}$$

The metal is then attached. An alternative procedure involves coordination of the metal to the phosphine prior to the anchoring of the ligand. For some applications, it has proven advantageous to remove unreacted surface silanol groups (through addition of trimethylchlorosilane, for example) prior to the use of these materials as catalysts.[20,24]

Other research groups have adopted similar procedures for anchoring catalysts to silica[25-27] and to γ-alumina, glass, and molecular sieve zeolites.[27]

## 2.2. Polymerization of Phosphine Monomers

For support materials one is not limited to preformed products. The support can be synthesized from phosphinated monomers. A real advantage in this approach is that the uncertainties in understanding the support structure are reduced. Examples of phosphinated monomers include vinyldiphenylphosphine, allyldiphenylphosphine, and p-diphenylphosphinostyrene; these are copolymerized with styrene (and sometimes divinylbenzene).[28-34] By varying the amounts of monomers a wide range of ligand loadings and cross-linking densities are possible.

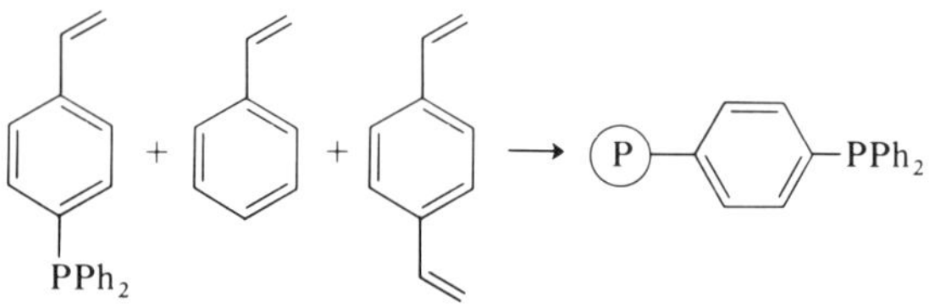

The flexibility of the polymer chain may be reduced by selecting a monomer that yields an unsaturated polymer. An example of a more rigid material is the example of the polymer formed by heating solid 1,6-*bis*-(toluenesulfonate)2,4-hexadiyne at 60°C and then treating it with lithium diphenylphosphide[35]:

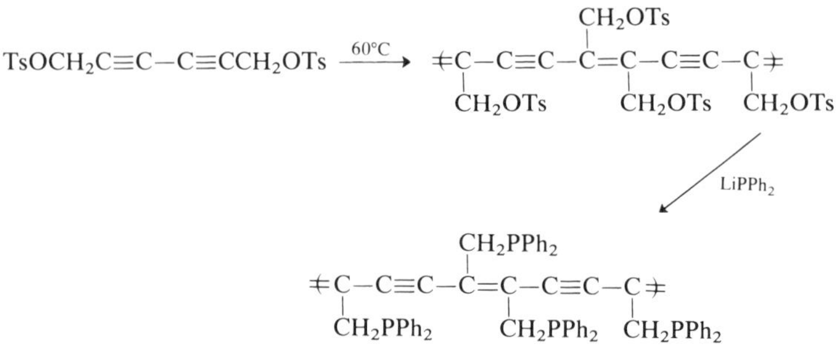

$$TsOCH_2C\equiv C-C\equiv CCH_2OTs \xrightarrow{60°C} \{C-C\equiv C-C=C-C\equiv C-C\}$$

with CH$_2$OTs substituents, then LiPPh$_2$ gives

$$\{C-C\equiv C-C=C-C\equiv C-C\}$$

with CH$_2$PPh$_2$ substituents.

Stille has prepared rather intricate hydrophilic polymers for use as chiral supports in reactions designed to induce asymmetry.[36,37] The first of these was based upon 2,3-O-isopropylidene-2,3-dihydroxy-1,4-*bis*(diphenylphosphino)-butane (DIOP):

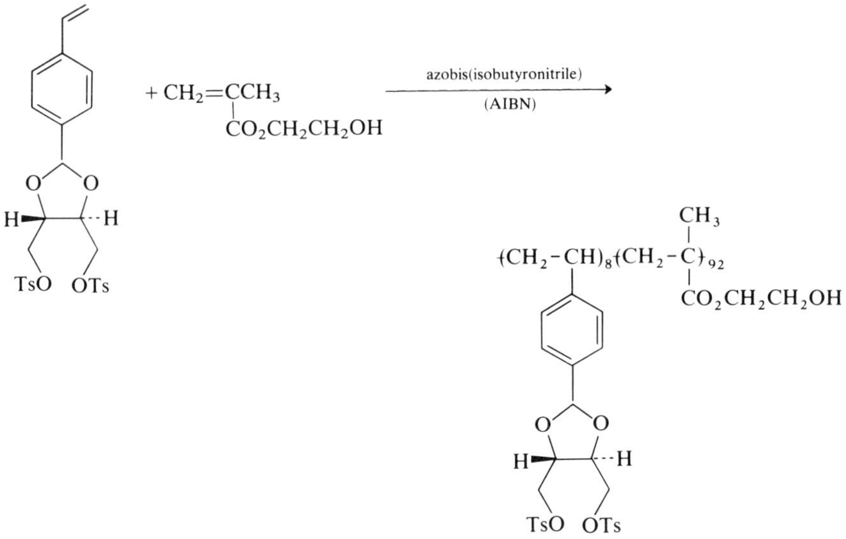

A more recent chiral support incorporated a pyrrolidine bisphosphine[38]:

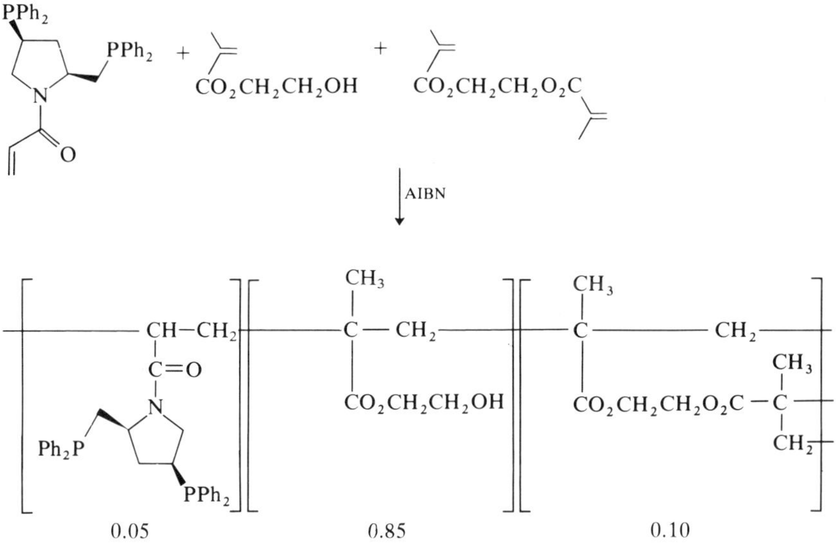

It is clear from extant hydrophobic and hydrophilic polymers that a wide range of phosphine-containing monomers could be prepared.

## 3. PHYSIOCHEMICAL CHARACTERIZATION OF CATALYSTS

One of the major difficulties with conventional heterogeneous catalysts is their characterization; certainly the problems posed by polymer-anchored catalysts are very similar. It would be fair to say that in some respects polymer-bound catalysts are rather poorly characterized as heterogeneous catalysts. Such basic determinations as total surface area, pore volume, and the pore distribution have been but rarely measured. It is often assumed that the mild conditions involved in catalyst preparation do not alter significantly the support structure. This assumption is not without some foundation. The textural properties of several styrene-DVV copolymers having from 10 to 60% cross-linking were examined by Hetfleis et al.[39] Surface areas for the organic polymers ranged from 25 (10% cross-linking) to $247 \, m^2 \, g^{-1}$ (60% cross-linking). Upon chloromethylation, treatment with dimethylamine, and attachment of $[Rh(CO)_2Cl]_2$, the surface areas became 58 and $287 \, m^2 \, g^{-1}$, respectively.

While the overall physical properties of the polymers have received rather cursory treatment, more attention has been devoted to the characterization of attached ligands and the coordination of the supported metal complexes. It is clear in this respect that polymer-anchored catalysts are more easily characterized than are conventional heterogeneous catalysts. Interpretations of catalyst performance rely upon both analysis of the polymer catalyst and a thorough knowledge of the homogeneous situation.

Information about the number of anchored ligands bonded to an attached metal complex is obtained normally from bulk properties, typically through elemental analysis. Supportive data for assignment of catalyst structure may also be garnered from the number of phosphines displaced during the attachment step,[40] and from molecular-weight determinations.[7,41,42] Additional insights concerning coordination and matrix effects are obtained by various spectroscopic tools.

*IR.* Most work is with supported carbonyl complexes.[13,42,43,44] Arguments for proposed structures are based upon close agreement between absorptions by polymer-bound catalysts and soluble models. The polymer matrix appears to have little influence on band energies or intensities, though it should be added that the subtle differences generally observed have not been interpreted. It is used rather infrequently for detection of metal halogen bands,[10] since these are normally too weak to be seen.

*$^1H$ nmr.* Doskocilova[45] et al. compared the spectra of a lightly crosslinked polystyrene with that of a linear polystyrene, using both conventional and magic angle rotation, and $^{13}C$nmr. The magic angle spectra of the cross-linked polymer were almost identical with those of the linear polymer in $CCl_4$ over a broad temperature range. Furthermore, the $^{13}C$ spin-lattice relaxation time ($T_1$) and nuclear Overhauser-effect parameters were equal

for both polymers, indicating little difference in the dynamics of internal motion.

*$^{31}$P nmr.* While $^{31}$P nmr would appear superficially to be the ideal probe into the chemistry of polymer-bound catalysts, in actuality there would appear to be rather severe limitations on the situations in which it can be applied usefully. There are just two major contributions in this area.[34,46] Both examined polymers prepared through the polymerization of phosphine monomers; this avoided the complicating effect of an uneven distribution of phosphine sites. These studies examined both cross-linked and noncross-linked polymers.

When noncross-linked poly(4-diphenylphosphinostyrene) was treated with an equimolar amount of [RhCl(COD)]$_2$ or [RhCl(C$_2$H$_4$)$_2$]$_2$ two spectra were taken: one was accumulated from 0.5-1 hr after adding the metal complex, the other from 8-24 hr.[34] Even with a phosphine function in each repeating unit, up to 200,000 pulses were required. The initial spectrum was different from the final one, apparently reflecting a 0.5-1 hr requirement for ligand exchange to reach equilibrium. Furthermore, the appearance of the spectrum was highly dependent on the Rh/P ratio (Figure 14.1). Spectra from Rh/P of 0.4-0.6 are most simply interpreted as the consequence of exchange in which apparently all phosphine ligands of this noncross-linked polymer participate. The spectral parameters of the polymer-bound complexes (Rh/P = 1) agreed very closely (Table 1) with those of monomeric phosphines, providing strong support for structural correspondence.

Stille *et al.* found that the greatest decreases in $^{31}$P spin-lattice relaxation times (T$_1$) occurred upon polymerization of a phosphinated monomer,

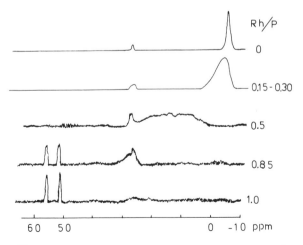

Figure 14.1. $^{31}$P nmr spectra during stepwise addition of [RhCl(C$_2$H$_4$)$_2$]$_2$ to Ⓟ-PPh$_2$.

Table 1. $^{31}P$ nmr Results of Monomeric and Polymeric Phosphines and Their Complexes

| Compound | $\delta$ | $^1J$(Rh–P), H$_2$ |
|---|---|---|
| PPh$_3$ | −5.6 | |
|  (P)–C$_6$H$_4$–PPh$_2$ | −5.3 | |
| Ph$_3$P–Rh(C$_2$H$_4$)$_2$–Cl | +53.3$d$ | 184.4 |
| 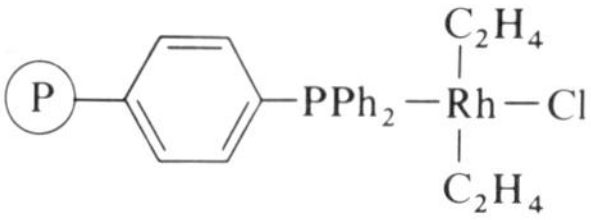 (P)–C$_6$H$_4$–PPh$_2$–Rh(C$_2$H$_4$)$_2$–Cl | +53.5$d$ | 184 |
| Ph$_3$P-Rh(COD)Cl | +31.5$d$ | 152 |
| 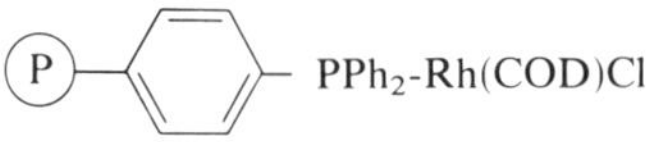 (P)–C$_6$H$_4$–PPh$_2$-Rh(COD)Cl | +30.9$d$ | 147 |

and that there were smaller decreases as cross-linking and metal complexation were introduced.[46] The volume swept by the bound phosphine was, therefore, smaller than for the soluble material, an observation which has important implications for asymmetric synthesis and regioselectivity.

*Electron Spectroscopy for Chemical Analysis (ESCA).* ESCA is a technique permitting the determination of the oxidation state of the metal.[47–50] For example, from the palladium 3$d$ binding energies listed in Table 2, it is clear that the oxidation state of palladium in the polymer catalysts is 2+. The method is also useful in determining the distribution of ligand and metal. Peak intensities are proportional to concentration. After making a surface determination, the polymer can be etched (argon) and the values determined again. An average for the whole polymer can be made after grinding the catalyst.

There are two limitations to the method. The most important of these is that the peaks are rather broad (often 2 eV at the $\frac{1}{2}$ height). This limits the usefulness of the technique to a determination of the bulk metal valence. If the "active" catalyst has a different oxidation state from that of the bulk metal and is present in minor amounts, it may go undetected. The second difficulty is that the etching process itself may effect a reduction of the metal.

*Extended X-ray Absorption Fine Structure (EXAFS).* Using a tunable X-ray synchrotron, the P/Cl and P/Br ratios were determined for

Table 2. Palladium 3d Binding Energies[50] (in eV)

| Complex | $3d_{3/2}$ | $3d_{5/2}$ |
|---|---|---|
| 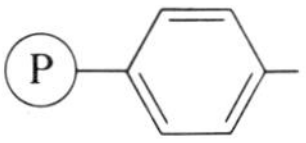 Ⓟ—⟨C₆H₄⟩—PPh₂)₂PdCl₂ | 344.0 | 338.6 |
| 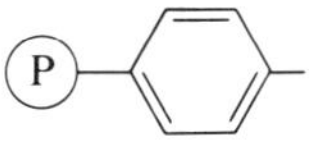 Ⓟ—⟨C₆H₄⟩—PPh₂)₂PdCl₂ | 343.5 | 338.2 |
| $(Ph_3P)_2PdCl_2$ | 343.6 | 338.3 |
| $PdCl_2$ | 343.6 | 338.1 |
| $Pd^\circ$ | 340.7 | 335.5 |

$(Ph_3P)_3RhCl$ and $(Ph_3P)_3RhBr$ and polymer-bound versions of these.[51] From phosphine/halogen ratios, it was possible to conclude whether the bulk coordination involved monomeric or dimeric metal species. With low-cross-linked polystyrene (2% DVB) both halides existed as dimers. In a polymer with 20% cross-linking, the bromide was present as the monomer (the chloride was not tested), demonstrating that increased cross-linking did lead to isolation of metal sites.

Also determined by this method are some interatomic distances (see Table 3). Metal–halide and metal–phosphorus distances found by EXAFS for $(Ph_3P)_3RhCl$ are in reasonable agreement with those determined crystallographically.

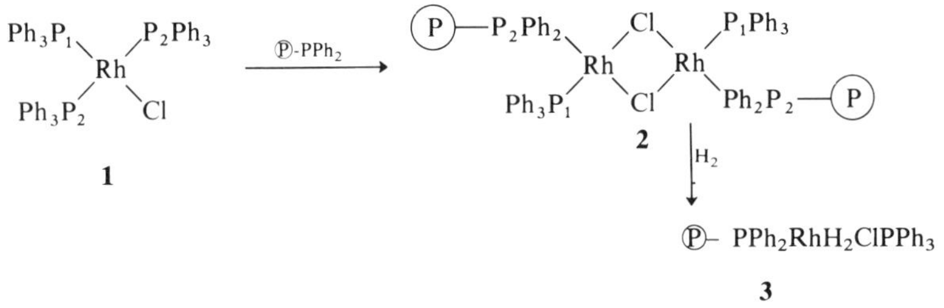

$$\underset{1}{\underset{Ph_3P_2}{Ph_3P_1}\diagdown\underset{Cl}{\overset{P_2Ph_3}{Rh}}} \xrightarrow{\text{Ⓟ-PPh}_2} \underset{2}{\underset{Ph_3P_1}{Ⓟ—P_2Ph_2}\diagdown\underset{Cl}{\overset{Cl}{Rh}}\diagdown\underset{Ph_2P_2—Ⓟ}{\overset{P_1Ph_3}{Rh}}} \xrightarrow{H_2} \underset{3}{Ⓟ-PPh_2RhH_2ClPPh_3}$$

Table 3. Summary of Interatomic Distances (in Å)

| | Wilkinson's 1 | | Polymer-bound Wilkinson's 2 | Hydrogenated polymer-bound Wilkinson's 3 |
|---|---|---|---|---|
| | X ray | EXAFS | | |
| Rh-Cl | 2.376 | 2.35 | 2.33 | 2.29 |
| Rh-$P_1$ | 2.214 | 2.23 | 2.23 | 2.20 |
| Rh-$P_2$ | 2.326 | 2.35 | 2.16 | 2.38 |

*Electron Spin Resonance (ESR).* ESR has received but limited use to establish metal oxidation states.[7,52-56] It was shown, for example, that norbornadiene complexes of Rh(I) on a phosphinated polystyrene-DVB formed low levels of Rh(II) during the course of the hydrogenation of ketones and olefins. Supported $RhCl_3$ was assigned a structure ⓟ–$PPh_2Rh(II)Cl_2$ following ESR and ESCA characterization.[47]

*X-Ray Microprobe Analysis.* This technique aids in determining the distribution of metal and ligand in a polymer-bound catalyst.[57,58] In highly cross-linked beads, it was observed that while chloromethylation proceeded homogeneously throughout the polymer beads, phosphine and metal attachment depended upon polymer morphology. Polymers with large pores (1300 Å) allowed phosphination and complex formation to proceed throughout the polymer beads. In polymers with small pores (<50 Å), penetration was limited to the outer bead areas.[58] In attaching the metal to the support, it was found that the metal was distributed in the outer portions of the bead if a deficiency of the metal was used. With excess metal the distribution was similar to that of the phosphine groups.[57]

## 4. INFLUENCE OF THE SUPPORT

In designing a catalyst it is certainly intended that the support will be inert at the conditions of the desired reaction. Normally this has been observed. This does not mean, however, that the support does not play an integral role in establishing the activity of the metal center. To state the obvious, the support is part of the ligand structure and exerts significant influence on performance at the phosphine–metal centers. This is not at

Table 4. Influence of the Support in the Hydrogenation of $\alpha$-Olefins.[87] Rh(I)-phosphines

| Phosphinated support | Relative rate |
|---|---|
| None[a] | 1.0–2.3 |
| Amberlite XAD-2 | 0.14 |
| Biobeads SX-2 | 0.05 |
| Dowex 50 | 0.20 |
| PVC | 0.15 |

[a] Homogeneous catalyst.

Table 5. Influence of the Support in the Hydrogenation of Cyclopentene.[4] Rh(I)-phosphines

| Phosphinated support | Relative activity |
|---|---|
| None[a] | 1.0 |
| Soluble polystyrene[b] | 0.91 |
| Insoluble macroreticular Polystyrene | 0.74 |

[a] Homogeneous catalyst.
[b] $M = 250,000$.

all surprising since we know from homogeneous catalysis that even seemingly minor changes in either electronic or steric factors can impact dramatically on reaction rates and/or product distributions. It would be anticipated, therefore, that the polymer matrix in polymer-bound catalysts would have a significant role in determining its properties (see Tables 4 and 5).

## 4.1. Changes in Selectivity

### 4.1.1. Regioselectivity

If a reagent must diffuse into a polymer to react with metal sites, it is to be expected that the framework of the catalyst might exert some influence upon the orientation in which the reagent approaches the catalytic site. This concept was tested with steroids[59] having two potential sites of reactivity:

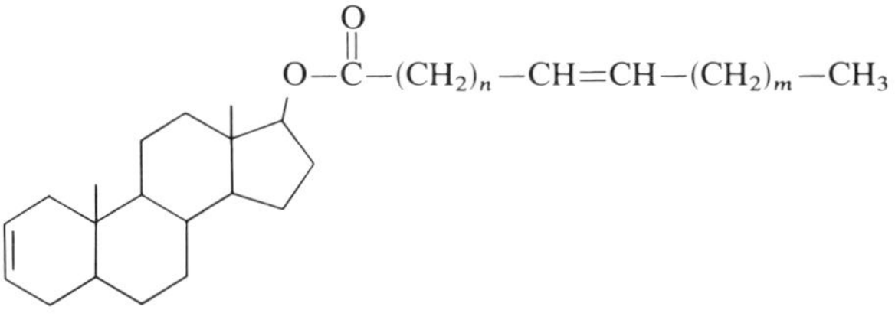

$$O-\overset{O}{\underset{\|}{C}}-(CH_2)_n-CH=CH-(CH_2)_m-CH_3$$

Limiting the reaction to the addition of one mole of $H_2$ showed that the heterogenized catalyst was 2–4 times more selective toward side-chain hydrogenation [polystyrene 2% DVB, anchored $RhCl(PPh_3)_3$] in benzene than the homogeneous rhodium complex. In the solvent benzene:ethanol (1:1), a poorer swelling solvent, the selectivity was enhanced.

### 4.1.2. Size

The steric influence exerted by a polymer matrix is also manifested in the hydrogenation of olefins of varying sizes.[9] Cyclohexene was reduced 32 times faster than the bulky olefin $\Delta^2$-cholestene with polymer-bound $RhCl(PPh_3)_3$, whereas in the homogeneous reaction the rate difference was less than 2. The support in this case was 200–400 mesh polystyrene beads (1.8% DVB) with 10% of the aromatic rings chloromethylated.

This size selectivity may not be entirely attributable to macroscopic polymer properties, e.g., small pore channels. It may also be a matter of the bulkiness of the ligand itself, the ligand in this case being represented not just by the immediate environment around the phosphine, but also the more distant polymeric backbone. Evidence for a steric influence of distant groups comes from studies on phosphines of the type[10]:

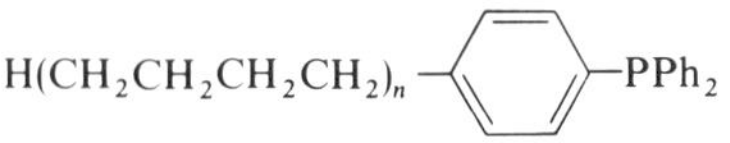

$$H(CH_2CH_2CH_2CH_2)_n-\langle\bigcirc\rangle-PPh_2 \qquad (n = 0, 1, 2, 3)$$

In this case, the tail simulates the polymer backbone. When the Wilkinson's catalyst analog of this ligand was prepared, it was found that the rate of hydrogenation of cyclohexene increased when $n$ was increased from $n = 0$ to $n = 3$. This increase was ascribed to the electron-donating effect of the para-alkyl group. Use of the bulkier olefin camphene instead of cyclohexene shows a slowing of the reaction with increasing length of the tail, lending support to the notion that quite distant groups may exert a measurable steric influence. Application of this finding to polymeric materials should be disciplined by the understanding that polymers are conformationally less flexible than the examples just discussed.

### 4.1.3. Chemical Selectivity

The selectivity of catalysts with respect to various functional groups may change upon attachment of a metal complex to a support. With soluble $PdCl_2(PPh_3)_2$, conjugated dienes are more readily hydrogenated to monoenes than isolated dienes. The reverse is observed for the supported catalyst[10,61]; the basis for this reversal is not yet clear.

### 4.1.4. Polarity

A nonpolar reagent in a mixture of a polar solvent and a hydrophobic catalyst will not be distributed uniformly throughout the mixture. Because of the polarity difference between catalyst support and the solvent, the nonpolar reagent will be concentrated to some extent within and near the polymer. A polar reagent, on the other hand, will be dispersed in the solvent. Capitalizing upon the inequality in distribution and the proximity of the nonpolar reagent to the catalytic sites, it has been shown that the relative rate of cyclohexene hydrogenation was enhanced over that of allyl alcohol[62]; another study found the same trend for cyclopentene and acrylonitrile.[4]

## 4.2. Asymmetric Induction

While the choice of ligand for inducing high-optical yields has been something of a matter of chance,[63] success is known to depend heavily upon steric factors.[64] One intuitively would expect that a polymer-bound catalyst would be more sterically hindered than a homogeneous analog, thus providing higher optical yields than are available from soluble catalysts. Optical yields often comparable to those achieved with soluble catalysts were observed in the hydrogenation of $\alpha$-amino acid precursors[54,37,65,66,67,68,69] and the hydrosilylation of ketones.[67,78] Much poorer optical yields were observed in the hydrogenation of $\alpha$-substituted styrene,[67] unsaturated carboxylic acids,[71a] and ketones,[71b] and during hydroformylation.[3a,46,72]

The simplest explanation for the generally lesser optical yields from polymer-bound catalysts lies in two considerations. The polymer backbone would increase the steric hindrance at the posterior of a ligand, but posterior variations have little effect upon optical yields.[73] What is more important are the steric influences in the immediate vicinity of the metal–phosphine coordination. The motion of the DIOP phosphines is known to be limited considerably in a polymer, thus reducing the volume swept per second.[46] This, in turn, reduces the effective bulkiness of the ligand, a factor of importance in inducing asymmetry.

## 4.3. Matrix Isolation

One facet of structure that has aroused considerable attention is the matter of the flexibility of the polymeric chains and the pendant phosphines. The possibility of interaction between two or more sites along the same or different chains is partially dependent upon this. Intuitively, one would think that as the degree of cross-linking increased the flexibility of the chains would decrease, effecting a restriction in interactions between sites. This has important consequences both in synthesis and catalysis. In catalysis, site–site interactions can lead to, for example, dimeric metal species that may have much reduced activities in comparison to those of monomeric complexes. One obviously would seek to design a matrix that would avoid site–site interaction for such catalysts. Quite a number of experiments have been designed that test the conditions under which site–site interactions occur.

The growing interest in the synthesis of macrocyclic ring systems has resulted in novel synthetic methods. A severe restriction in macrocyclic ring formation is that precursors to large rings often react intermolecularly instead of intramolecularly. Polymers were recognized to provide the opportunity of isolating such precursors within a matrix, thus forcing the ring closures to proceed intramolecularly. However, when such experiments were carried out the enhancement in intramolecular ring closures was often disappointing and the main product was still derived from intermolecular (site–site) reactions.[74]

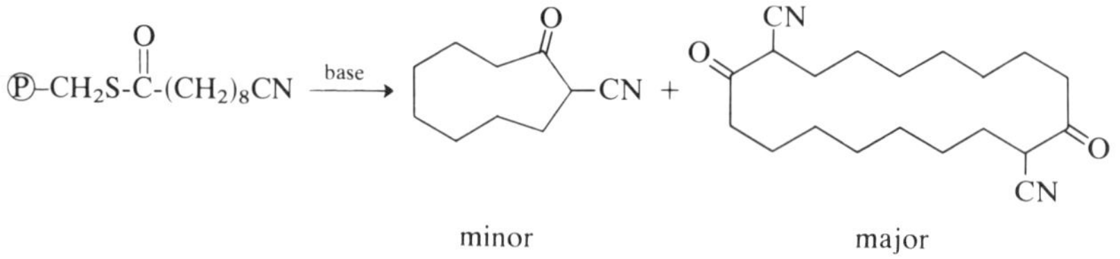

$$Ⓟ-CH_2S-\overset{O}{\overset{\|}{C}}-(CH_2)_8CN \xrightarrow{base}$$ minor + major

The example shown above is by no means atypical. Treatment of a polycarboxylic acid with dicyclohexylcarbodiimide (DCC) under forcing conditions effected dehydration even with a 20% DVB-polystyrene polymer.[75]

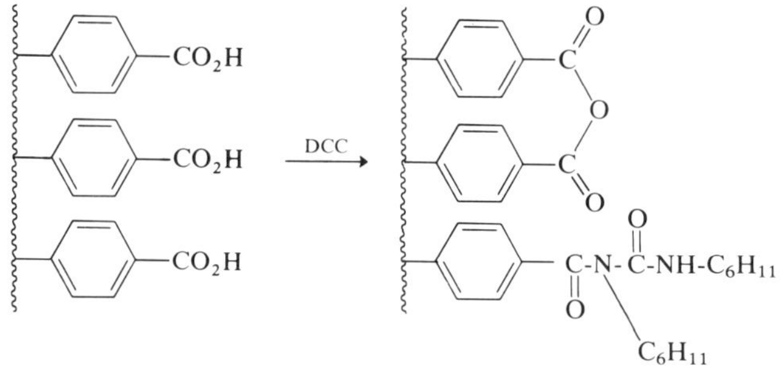

There is other evidence, however, which demonstrates that matrix isolation does reduce or even prevent site–site interaction. One very impressive experiment dealt with the generation of benzyne on a polystyrene support having but 1% cross-linking.[76]

When generated in solution the coupling of benzyne is diffusion controlled, but when benzyne was generated on a polymer, the lifetime was extended to over one minute. Moreover, there was no evidence for benzyne–benzyne coupling.

Another demonstration of the isolation of sites, in this case involving metal centers, came from the EXAFS experiments by Reed.[51] On a 2%

cross-linked polymer, supported $Rh(PPh_3)_3Br$ was present in the dimeric state, whereas on a polymer with 20% cross-linking the monomer prevailed.

The seeming contradiction between these various experiments may be resolved if site isolation is considered to be partially a kinetic phenomenon. For slow reversible reactions, such as anhydride formation or condensation reactions, the polymer chains have sufficient time to exercise several degrees of freedom, permitting site–site interaction. For reactions that occur over shorter periods, such as the decomposition of benzynes, the probability of close contact between reactive centers is greatly reduced. Metal dimer formation was obviated simply because there was no important thermodynamic driving force.

The principle of matrix isolation is important in catalysis if a catalyst can decompose *via* site–site interactions. In the dimerization–methoxylation of butadiene, catalyst decomposition occurs through metal agglomeration[77]:

$$\diagup\!\!\!\diagup\diagdown\!\!\!\diagdown + CH_3OH \xrightarrow[\text{\textcircled{P}-(PPh}_2)_x Pd(PPh_3)_y]{Pd(PPh_3)_4 \text{ or}} \diagup\!\!\!\diagup\diagdown\!\!\!\diagdown\diagup\!\!\!\diagup\diagdown\!\!\!\diagdown OCH_3$$

$$+$$

isomers

A polystyrene with just 1% cross-linking was reasonably effective in reducing catalyst decomposition in this reaction.

### 4.4. Coordinative Unsaturation

It is desirable at times to create coordinative unsaturation in metal complexes to either permit the coordination of species that are very poor ligands and thus cannot readily displace ones already in place, or for reasons of rate enhancement. Examples of the former effect are seen spectacularly in the works of Crabtree[78] and Bergman,[79] in which alkanes are coordinated by transiently unsaturated metal centers. (There are no comparable studies of heterogenized systems.)

The case for coordinative unsaturation on polymers is rather weak and comes principally from the enhanced rates of hydrogenation observed for polymeric catalysts. Kaneda *et al.* supported $PdCl_2$ on a linear phosphinated polystyrene, and the hydrogenation of nonconjugated dienes was examined.[49] The supported catalyst was active even under conditions where $PdCl_2(PPh_3)_2$ was inert. ESCA studies established a 2+ oxidation state for the metal. The increase in hydrogenation activity was considered to possibly have its origin in unsaturation at palladium.

A second observation of hydrogenation activation superior to that of the homogeneous situation was found for $Ir(CO)Cl(PPh_3)_2$ on phosphinated polystyrene-1% DVB with P/Ir ratios varying between 3 and 22.[80,81] At high ratios the homogeneous and supported complexes displayed similar activities, but at low ratios, i.e., 3 or 4, the supported catalyst displayed rates much enhanced over those of the soluble catalyst. This difference was attributed to the lesser availability of phosphine from the polymer, permitting a shift in the equilibrium between (4) and (5) toward the unsaturated (5). Such a shift would indeed enhance the rate, since it is (5) that adds hydrogen in the rate-determining step:

$$\text{\textcircled{P}-PPh}_2)_2\text{IrCl(CO)(olefin)} \rightleftarrows \text{\textcircled{P}-PPh}_2 + \text{\textcircled{P}-PPh}_2\text{IrCl(CO)(olefin)}$$
$$\qquad\qquad 4 \qquad\qquad\qquad\qquad\qquad\qquad 5$$

A third experiment in which the rates of reaction for polymer-supported catalyst were more rapid than those of the soluble analog was one already mentioned, the dimerization–methoxylation of butadiene.[77] Rates were particularly fast with catalysts having low-ligand loadings (on a polystyrene 1% DVB) and low P/Pd ratios.

While coordinative unsaturation is not widely documented for polymer-bound catalysts, it could be argued that few authors have looked for it. Few investigations have systematically varied P/metal ratios, for example, so the phenomenon may be more widespread than suspected. Spectroscopic examinations would be of value. It is a bit curious that the instances in which coordinative unsaturation was argued employed polymers either lacking or having very low levels of cross-linking! One certainly would expect the phosphines in these polymers to be highly mobile. The fact that published accounts of coordinative unsaturation employed non-cross-linked or lightly cross-linked supports may not reflect a requirement for the observation of coordinative unsaturation, i.e., the use of low-cross-linked polymers, but rather may simply reflect the fact that most studies have employed low-cross-linked polymers. Based upon the better documented concept of matrix isolation, one would anticipate that more highly cross-linked polymers would more likely display coordinative unsaturation.

## 5. REACTIONS OF OLEFINS AND DIENES

### 5.1. Isomerization

Numerous catalysts that are used for hydrogenation or hydroformylation are also active isomerization catalysts. These catalysts require the presence of hydrogen to be activated, probably to form a metal hydride. Carbonyls of cobalt,[13] iron,[8] and osmium[82] are activated in this manner and effect hydrogen shifts.

Irradiation of $Fe(CO)_5$ and $Fe_3(CO)_{12}$ supported on phosphinated styrene-DVB at 25°C created isolated, coordinatively unsaturated metal atoms that are catalytically active for olefin isomerization and hydrosilylation.[8] Additional observations include: 1) the extent of phosphine substitutions affects both the initial catalytic activity and the isomeric ratio of the products; and 2) the photocatalysis required continuous irradiation.

Several potentially interesting reactions apparently have not received attention. Several unsaturated steroids are obtained *via* homogeneous isomerization with $RhCl_3$.[83] The interest in examining a similar isomerization with polymer-bound catalysts lies in the potential of an altered isomeric ratio.

The greatest potential for industrial applications may lie with nickel hydride catalysts. Homogeneous catalysts may be prepared from $Ni[P(OR_3)_3]_4$ or by reduction of $NiS_2(PR_3)_2$ with alkylaluminum compounds.[84,85] Carbon skeleton rearrangements are known with these complexes. For example, *cis*-1,4-hexadiene is converted to *trans*-2-methyl-1,3-pentadiene and 2,4-hexadienes at room temperature[84]:

$$\diagup\!\!\!\diagdown\!\!\diagup\!=\!\diagdown \;\longrightarrow\; \diagup\!\!\!\diagdown\!=\!\diagdown \;+\; \diagdown\!\!=\!\!\diagup\!\!\diagdown$$

Apparently there are no similar studies with polymer-bound catalysts.

## 5.2. Hydrogenation

Mechanistically, hydrogenation places but limited demands on metals and ligands, so it is not at all surprising that a great variety of complexes have demonstrated activity. Before considering some specific examples of this chemistry, a few more general characteristics are listed.

   a. Polymer-bound catalysts are fashioned after homogeneous analogs. Those homogeneous catalysts having the highest activities also display high activities when bound to a polymer.
   b. Polymer-bound catalysts are generally less active than homogeneous analogs. The lesser activity may be related to reduced catalyst accessibility and/or diffusion control.
   c. Hydrogenation studies have focused on simple hydrocarbon unsaturation (alkenes, dienes, trienes). Ketone and arene hydrogenation are known.

Interest in the hydrogenation reaction is not, however, limited to a documentation of the organic transformations that are possible. This reaction has also been used more than any other to probe the characteristics of polymer-bound catalysts to determine the details of the effects of catalyst attachment.

## 5.2.1. Rhodium Catalysts

The most commonly supported catalyst is the heterogenized version of Wilkinson's catalyst: $RhCl(PPh_3)_3$. The mechanistic details of this highly active homogeneous catalyst are so well established that the influences supporting the catalyst are readily accessible. Many of these effects have been discussed already.

The activity and selectivity of the bound Wilkinson catalyst parallels in many ways the homogeneous system. Terminal olefins are hydrogenated more rapidly than internal olefins, *cis*-olefins react faster than *trans*-olefins, and olefins are reduced faster than acetylenes.[86] There were some differences with the sterically hindered 1(7)-*p*-menthene: the polymeric catalyst, for example, produced appreciably more isomerized material. Furthermore, the attached complex was more selective in the hydrogenation of 1,3-cyclooctadiene.

The activity of a catalyst depends very much on the type of support employed. This was documented for several types of phosphinated polymers to which $RhCl(PPh_3)_3$ was bound. All the polymers listed in Table 4 possess small pore sizes: The macroreticular XAD-2 has 90 Å pores, the others lack permanent pore structures. For these catalysts, activity is an order of magnitude lower than for the homogeneous catalyst. When a polymer with large pores (1000 Å) was employed as a support, the rate was almost up to that of the homogeneous complex. It was concluded from these observations that the lesser activity of the catalysts possessing low cross-linkings was attributable to diffusional effects. Alternative explanations are possible. It could be argued that in low-cross-linked polymers the form of the metal complex was different from that in more rigid materials, that in the former the metal was present to some extent as dimer. Justification for this argument comes from the EXAFS study by Reed *et al.* In homogeneous reactions, the dimer of Wilkinson's catalyst displays a lower activity than the monomer. One could also argue that in the macroreticular support there is a degree of coordinative unsaturation caused by the rigidity of the polymeric structure. But the simplest explanation, that the lesser activity is based upon diffusional factors, would seem to be the most reasonable choice. This is suggested by the very similar rates for three very different catalysts: homogeneous $RhCl(PPh_3)_3$, a noncross-linked polystyrene, and a macroreticular polystyrene with 1000 Å pores. In these cases the polymers would permit ready access to catalytic sites. (It should be added that macroreticular supports with smaller pores are more prone to diffusional limitations.)

High activities can be obtained for supported catalysts having a low degree of cross-linking if the particle size is kept quite small[12]; use of small particles minimizes diffusional effects.

One of the important factors that contributes to such things as catalyst activity and metal loss is the ratio of phosphine to metal. In a study by Bernard *et al.* it was shown with soluble polystyrene that upon reaching a P/Rh ratio of 4/1 the catalyst was deactivated.[4] The behavior resembles that of the homogeneous catalyst in its sensitivity to excess phosphine and suggests that the support was sufficiently flexible for the metal to become coordinated to three phosphines; the presence of the fourth phosphine simply inhibits the dissociation of the complex so that hydrogen might add. On the other hand, there was little change in activity for a macroreticular support having P/Rh between 2/1 and 10/1.

Metal loss from the polymers may occur through various means including *via* the oxidation of phosphine to its oxide. This is a facile process and is the reason that feedstocks must be purified so carefully of oxygen. Since the resulting phosphine oxide is a poor ligand, the metal is released from the oxidized binding site. If the polymer contains a high P/metal ratio, the metal is more likely to be retained at another site within the polymer. For macroreticular polymers, which tend to concentrate the metal near the outer portion of the support as opposed to being evenly dispersed throughout the polymer, metal loss is more likely.

As we have just seen, the activities of polymer-bound catalysts are generally somewhat less than those of corresponding homogeneous complexes, but there are exceptions beyond those already discussed. Two reports of enhanced activities are claimed using phosphinated silica. By one report, attaching $[Rh(C_2H_4)_2Cl]_2$ resulted in a catalyst displaying hydrogenation activities of up to fifty times those for homogeneous $RhCl(PPh_3)_3$.[88] The behavior of catalysts prepared from supporting cobalt, nickel, rhodium, or palladium salts on a silica phosphinated with $Ph_2P(CH_2)_3Si(OMe)_3$ was reported by Kochloefl *et al.*[89] Hydrogenation rates for cyclohexene were $10^2$–$10^4$ times faster than for the unattached salts. As a test for the presence of metal the CO uptake was measured and found to be very low.

The origin of these enhanced activities is not clear. Some investigators argue that all cases of enhanced activity are based upon the presence of small amounts of zero-valent metal. This point is extremely difficult to prove either way. Given the number of reports of enhanced activity in a rather broad variety of reactions, using both phosphine and other ligands, it would appear that such an argument is inadequate.

There is mixed evidence with regard to the question of whether supported catalysts are more immune to poisoning than are the soluble complexes. After supporting $[RhCl(COD)]_2$ on a phosphinated silica, the activity for 1-hexene hydrogenation was lowered slightly in the presence of thiophene, but more markedly by *n*-butyl mercaptan.[20] A further effect of the mercaptan was the stabilization of the rhodium to prevent metal formation. By contrast, the activity of RhCl(PPh$_3$)$_3$ was greatly reduced by the presence of mercaptans. In another study by this same group at British Petroleum, rhodium(I) carboxylates were anchored to phosphinated polymers. This catalyst displayed as much sensitivity to sulfur as the homogeneous catalyst.[90]

Just how important the support can be in establishing activity is illustrated by the example of phosphite polymers of polymethallylalcohol!

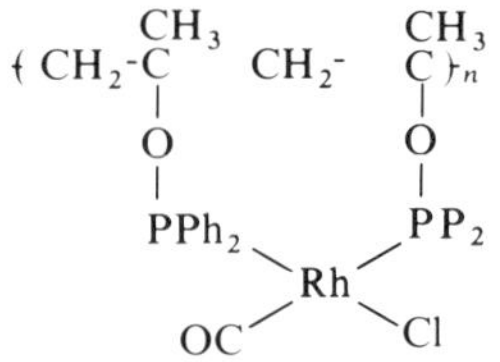

$$\begin{array}{c} CH_3 \quad\quad\quad CH_3 \\ \{CH_2\text{-}C\quad CH_2\text{-}\ C\}_n \\ | \quad\quad\quad\quad\quad | \\ O \quad\quad\quad\quad\ O \\ | \quad\quad\quad\quad\quad | \\ PPh_2\quad\quad\ PP_2 \\ \diagdown\ \ \diagup \\ Rh \\ \diagup\ \ \diagdown \\ OC \quad\quad Cl \end{array}$$

While catalysts derived from the atactic polymer were active, those from the syndiotactic material were not.[91] Homogeneous rhodium–phosphite complexes are not active hydrogenation catalysts.

More polar functional groups such as ketones are hydrogenated effectively with homogeneous cationic rhodium[92] and ruthenium[93] complexes. Polymer catalysts active for both olefin and ketone hydrogenation activity were synthesized from ionic precursors such as Rh(norbornadiene)(*acac*) + HClO$_4$ or Rh(norbornadiene)(PE$t_3$)$_2$ + ClO$_4^-$. Two types of coordination were observed, depending on the method of attachment[53-55]:

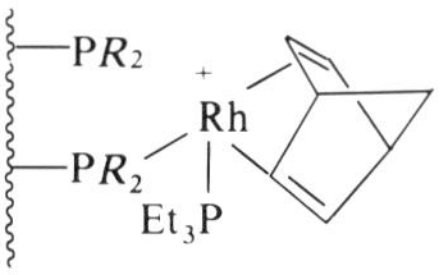

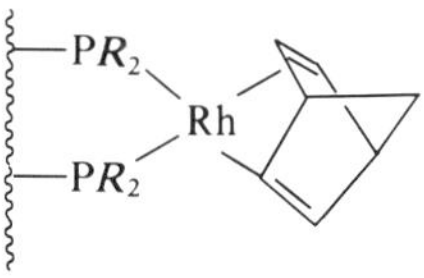

Modification of the phosphine to compare activities of Ⓟ–PPh$_2$ with Ⓟ–PPhMe demonstrates that the hydrogenation of ketones is faster with the latter, more basic, phosphine. Alkene hydrogenation is slower with these more basic ligands. Incorporation of a very bulky alkyl group, P–PPh(menthyl), results in a dramatic increase in the loss of rhodium. Similar losses upon the incorporation of a bulky ligand into the phosphine do not occur for olefin hydrogenation; the losses during the ketone reduction are partially due to the high-coordinating ability of the product alcohol. Hydrogenation of cyclohexene is highly solvent dependent and is rapid in THF, slow in CH$_2$Cl$_2$. Some Rh(II) is formed during the reaction, but this was considered to be inactive. During olefin hydrogenation there was an induction period, and this was ascribed to the hydrogenation of the initially coordinated diene.

Rhodium trichloride on a noncross-linked phosphinated polystyrene was characterized by esr and ESCA techniques as having a ⓟ–PPh$_2$Rh(II)Cl$_2$ structure.[47] While a variety of monoenes and conjugated and nonconjugated dienes were reduced, there was but slight activity with alkynes, cyano, keto, and nitro groups. Catalyst activities were highly solvent dependent; the highest rates were in alcohols, the least activity was in hydrocarbon or chlorinated hydrocarbon solvents.

### 5.2.2. Iridium

Vaska's complex IrCl(CO)(PPh$_3$)$_2$ was attached to microporous polystyrene (1% DVB) with different phosphine loadings and P/Ir ratios.[80,81,94] 1,5-Cyclooctadiene was hydrogenated at the rather high temperature of 170°C. The results showed that when P/Ir was $<\frac{5}{1}$ the rates were appreciably faster with the supported catalyst; when the ratio reached or exceeded $\frac{12}{1}$ the homogeneous catalyst was more active. When the rate of hydrogenation at 170° was compared with that at 80°, it was observed for some loadings that the rate was actually lower at the higher temperature. At higher temperatures, it was considered that the effects of matrix isolation were overcome and more phosphine became available for coordination, reducing the concentration of the 4-coordinate intermediate involved in the catalytic cycle.

### 5.2.3. Palladium and Platinum

Use of rhodium catalysts to effect the hydrogenation of polyenes results in high conversions to saturated products. On the other hand, if palladium or platinum is employed, the hydrogenation can be stopped at the monoene stage. This has important implications, particularly for the partial hydrogenation of vegetable oils. The linoleic and linolenic residues in vegetable oils are prone to oxidation, so from the standpoint of stability it is desirable to hydrogenate them. But to avoid the build-up of low-density cholesterols in humans, it is important to retain *cis*-monoene units. It is not surprising that much of the research on palladium and platinum complexes has been in the context of vegetable oil hydrogenation.[10,61] The complexes most commonly supported are PdCl$_2$(PhCN)$_2$ and PtCl$_2$(PhCN)$_2$. Hydrogenation of soybean oil methyl ester proceeds fastest in alcohol solvents, but the highest selectivity to monoenes occurs in dichloromethane.

It is very clear that differences exist between the attached and the homogeneous model of the attached complex, Pd(PPh$_3$)$_2$Cl$_2$. Ir studies show that the polymer-attached *bis*(phosphine)dichloropalladium is in a *trans* configuration, whereas the homogeneous complex exists as both *cis* and *trans* isomers.[10] Furthermore, the rate of hydrogenation by the homogeneous complex is not accelerated by alcoholic solvents. Also, whereas stannous chloride activates the homogeneous complex by producing coordinative unsaturation, the polymeric version is deactivated by $SnCl_2$. Analogous (supported) platinum complexes are activated by stannous chloride. Homogeneous $Pd(PPh_3)_2Cl_2$ hydrogenates nonconjugated dienes faster than conjugated ones, while the supported version displays the opposite selectivity.

Generally, the activities of polymer-bound palladium catalysts are less than those of unsupported ones, but here too there are exceptions. Kaneda et al. reported that $PdCl_2$, on phosphinated polystyrene was more active than the homogeneous situation.[49] Rates were very solvent dependent—hydrogenation of styrene was slow in dimethyl sulfoxide; optimum activity was obtained in solvents of moderate coordinating ability (see Table 6).

### 5.3. Dimerization, Oligomerization, and Polymerization

#### 5.3.1. Dimerization

Investigations with polymer-bound catalysts are very limited and are patterned generally after homogeneous analogs. Thus, the linear dimerization of butadiene was effected by the homogeneous catalyst $NiBr_2(PPh_3)_2/NaBH_4$.[40] The polymer-bound version gave essentially the same results, though its activity ceased after about 1500 turnovers.

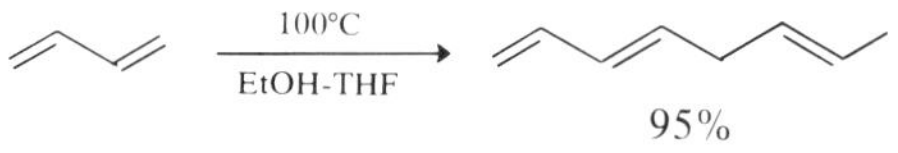

$$\diagup\!\diagdown\!\diagup \xrightarrow[\text{EtOH-THF}]{100°C} \diagup\!\diagdown\!\diagup\!\diagdown\!\diagup\!\diagdown$$
$$95\%$$

Nickel-based Ziegler catalysts are very effective for propylene dimerization.[116] The unsupported catalyst is prepared by mixing $NiCl_2$, $Et_3Al$, and butadiene in chlorobenzene to yield a $C_{12}$ $\pi$-allyl complex of nickel. A phosphine is then added, followed by liquid propylene at 15 atm pressure. A mixture of n-hexenes, 2-methylpentenes, and 2,3-dimethylbutenes in 85–90% yield are produced rapidly at 30–40°C. Apparently this reaction is not reported for a polymer-bound version. This would be of interest since it is observed that the product distribution is quite sensitive to the nature of the phosphine.

#### 5.3.2. Cyclodimerization and Cyclotrimerization

The nickel carbonyl catalyst, $Ni(CO)_2(PPh_3)_2$, is excellent for the cyclooligomerization of butadiene, providing a mixture of 4-vinyl-

Table 6. Hydrogenation with Phosphine-Bound Catalysts

| Support | Ligand | Metal complex supported | Illustrative reactant | Product | Reference |
|---|---|---|---|---|---|
| Soluble polystyrene | ⓟ–$PR_2$ | $CoCl_2$ | — | — | 30 |
|  | ⓟ–$PR_2$ | $PdCl_2$ | 1,5-COD | COD | 49 |
|  | ⓟ–$PR_2$ | $[RhCl(C_2H_4)_2]_2$ | Cyclopentene | Cyclopentene | 4 |
|  | ⓟ–$PR_2$ | $[RhCl(C_8H_{14})_2]_2$ | Cyclohexene |  | 95 |
|  | ⓟ–$PR_2$ | $[RhCl(CO)_2]_2$ |  |  |  |
|  | ⓟ–$PR_2$ | $Rh(acac)(CO)_2$ | Cyclopentene |  | 3 |
|  | ⓟ–$PR_2$ | $RhCl(PPh_3)_3$ |  |  |  |
| Styrene/DVB | ⓟ–$PR_2$ | $NiCl_2/NaMH_4$ |  |  | 96 |
|  | ⓟ–$PR_2$ | $H_2PtCl_6$ | 1-Heptene |  | 16 |
|  | ⓟ–$PR_2$ | $PdCl_2(PhCN)_2$ | Soybean oil methyl ester | Selective to monoene | 10, 97 |
|  | ⓟ–$PR_2$ | $RhCl_3$ | Olefin |  | 82, 98 |
|  | ⓟ–$PR_2$ | $RhCl(PPh_3)_3$ | Cyclolefin trienes, dienes |  | 14, 12, 9, 62, 100, 99 |
|  | ⓟ–$PR_2$ | $[Rh(diene)(PR_3)_2]^+$ | Olefins, ketones |  | 4, 43, 53, 54, 55, 101 |
|  | ⓟ–$PR_2$ | $[RhCl(olefin)_2]_2$ |  |  |  |
|  | ⓟ–$PR_2$ | $Rh(acac)(olefin)_2$ | Alkenes |  | 43, 53, 102 |
|  | ⓟ–$PR_2$ | $Rh(acac)(CO)_2$ |  |  | 102 |
|  | ⓟ–$PR_2$ | $RuCl_2(CO)_2(PPh_3)_2$ | 1,5-COD | Cyclooctene | 103 |
|  | ⓟ–$PR_2$ | $Ir(COD)(acac)$ | Cyclohexene |  | 53 |
|  | ⓟ–$PR_2$ | $IrCl(CO)(PPh_3)_2$ | Diene |  | 80, 81, 94 |

Table 6. (Continued)

| Support | Ligand | Metal complex supported | Illustrative reactant | Product | Reference |
|---|---|---|---|---|---|
| Styrene/DVB | Ⓟ-$PR_2$ | $Rh_4(CO)_{12}$ | Arenes | | 104 |
| | Ⓟ-$PR_2$ | $Rh_6(CO)_{16}$ | Arenes, olefins | | 104–106 |
| | Ⓟ-$PR_2$ | $H_4Ru_4(CO)_{12}$ | Ethylene | | 11, 107 |
| | Ⓟ-$PR_2$ | $Ir_4(CO)_{12}$ | | | 108, 109 |
| | Ⓟ-$PR_2$ | $Fe_2Pt(CO)_{12}$ | Ethylene | | 11 |
| | Ⓟ-$PR_2$ | $RuPt_2(CO)_8$ | Ethylene | | 110 |
| | Ⓟ-$PR_2$ | $Pd(PPh_3)_4$, $Pd(OAc)_2$ | Unsat. esters | | 14, 111 |
| | Ⓟ-$PR_2$ | $Ni(CO)_2(PPh_3)_2$ + $RhCl(PPh_3)_2$ | Dienes, trienes | | |
| Pdy(imino-methylene) | Ⓟ-$PR_2$ | $[RhCl(olefin)_2]_2$ | Cyclohexene | | 112 |
| Poly-1, 6-bis (p-toluene sulfonate)-2, 4-hexadiyne | Ⓟ-$PR_2$ | $[RhCl(olefin)_2]_2$ | 1, 3-COD, 1-octene, benzene | | 35 |

## POLYMER-BOUND PHOSPHINE CATALYSTS

| | | | | |
|---|---|---|---|---|
| Polyphenylene-isophthalamide | ⓟ–$PR_2$ | $RhCl(PPh_3)_3$ | Olefin | 17 |
| Soluble poly-siloxanes | ⓟ–$PR_2$ | $[RhCl(olefin)_2]_2$ | 1-Octene | 18 |
| Butadiene/maleic anhydride | ⓟ–$PR_2$ | $PdCl_2$<br>$H_2PtCl_6$ | | 113 |
| Chiral methacrylate | ⓟ–DIOP | $[RhCl(C)D)]_2$ | $\alpha$-Acetamidoacrylic acid | 36, 37, 65 |
| | ⓟ–DIOP | $[RhCl(COD)]_2$ | $\alpha$-Acetamidoacrylic acid | 65, 38, 66, 68, 114 |
| Polymethallyl alcohol | ⓟ–$OPR_2$ | $[RhCl(CO)_2]_2$ | Terminal olefins | 115 |
| $SiO_2$ | ⓢⓘ–$PPh_2$ | $[RhCl(COD)]_2$ | 1-Hexene | 20, 22 |
| | ⓢⓘ–$PPh_2$ | | Alkenes | 19 |
| | ⓢⓘ–$PPh_2$ | | 1-Hexene | 23 |
| | ⓢⓘ–$PPh_2$ | $RhCl(PPh_3)_3$ | Alkenes | 86 |
| | ⓢⓘ–$PPh_2$ | | Cyclohexene | 26 |

cyclohexene, cycloocta-1,5-diene, and cyclodeca-1,5,9-triene. For the supported complex, prepared through phosphine exchange, the product distribution was similar to that of the soluble catalyst but the rate was only one-third that of the soluble complex.[12,13,94] Both types of catalysts responded similarly to changes in CO pressure. After 1100–1200 catalytic cycles, the polymeric catalyst was no longer active—an infrared spectrum revealed no CO absorptions. There was a slow leaching of metal, but at a rate slower than that of catalyst deactivation.

A similar Ni(O) species is derived from *bis*-(cyclooctadiene) nickel and a phosphinated polystyrene.[43] The catalyst has little intrinsic activity in butadiene cyclodimerization to cyclooctadiene and vinylcyclohexene, but this was enhanced to a level of about 60–100 g-product/g-Ni/hr by the addition of AlEt$_2$(OEt). Cyclododecatriene was not produced, indicating coordination of a phosphine to the nickel throughout the process.

Polymer-bound Ni(CO)$_2$(PPh$_3$)$_2$ is active in the cyclotrimerization of acetylenes in benzene or THF. The lifetime of this catalyst is rather short, and there are apparently changes of product distribution during the course of the reaction.[12]

The Shell Higher Olefins, a large-scale process, is a nickel-based oligomerization of ethylene to linear $\alpha$-olefins.[117] The catalyst, a homogeneous one, is a nickel hydride generated by reduction of a nickel salt in the presence of a chelating ligand such as diphenylphosphinoacetic acid. A polymer-bound catalyst for this reaction would be attractive if the olefins in the C$_{10}$–C$_{18}$ range could be increased above the 40% produced by the homogeneous version. The homogeneous reaction is performed in ethylene glycol and, for efficient operation with a polymer-bound catalyst, it might be necessary to use a hydrophilic polymer.

### 5.3.3. Polymerization

Work in this area has focused on nonphosphine supports.[2] Polymerization of ethylene was reported with TiCl$_4$ on a homopolymer or copolymer of vinylphenylphosphine; AlR$_3$ was added to activate the catalyst. With a nickel salt on a phosphinated polystyrene, addition of NaBH$_4$ resulted in a catalyst active in the oligomerization and polymerization of acetylenic monomers[118] (see Table 7).

### 5.4. Addition

Polymer-attached catalysts facilitate addition to olefins:

$$RCH = CH_2 + HY \rightarrow RCH_2CH_2Y + RCH - CH_3$$
$$\phantom{RCH = CH_2 + HY \rightarrow RCH_2CH_2Y + RCH - CH_3xxx}|$$
$$\phantom{RCH = CH_2 + HY \rightarrow RCH_2CH_2Y + RCH - CH_3xxx}Y$$

$$HY = HSiR'_3, HSi(OR)_3, HCN$$

Table 7. Dimerization, Oligomerization and Polymerization

| Support | Ligand | Metal complex supported | Reactant | Product | Reference |
|---|---|---|---|---|---|
| Styrene/DVB | Ⓟ–$PR_2$ | $NiCl_2/NaBH_4$ | Butadiene | Linear dimerization | 118 |
| | Ⓟ–$PR_2$ | $NiBr_2(PPh_3)_2/NaBH_4$ | | | 40 |
| | Ⓟ–$PR_2$ | $Pd(PPh_3)_4$ | 1,3-Butadiene + MeOH | Dimerization-methylation | 77 |
| | Ⓟ–$PR_2$ | $PdCl_2$ | Styrene, alkenes | Cotrimers | 119, 120 |
| | Ⓟ–$PR_2$ | $Pd(OAc)_2$ | Butadiene | Octadienyl acetates | 121 |
| | Ⓟ–$PR_2$ | $Pd(OAc)_2, PdCl_2(PhCN)_2$ $Pd(PPh_3)_4$ | 1,3-Butadiene | Linear dimers | 110 |
| | Ⓟ–$PR_2$ | $[PdCl(\eta^2\text{-—pinenyl})]_2$ | | | 122 |
| | Ⓟ–$PR_2$ | $Ni(COD)_2/AlEt_2OEt$ | 1,3-Butadiene | Cyclodimers | 43 |
| | Ⓟ–$PR_2$ | $Ni(COD)_2/Al_2Et_3Cl_3$ | | | 123 |
| | Ⓟ–$PR_2$ | $Ni(CO)_2(PPh_3)_2$ | Butadiene | Cyclooligomers | 12, 13, 94 |
| Poly(vinylchloride) | Ⓟ–$PR_2$ | $CoCl_2/NaBH_4$ | | | 124 |
| Polyglycolphosphate | Ⓟ–$OP(OR)_2$ | $Ni(CO)_4$ | | | 125 |

The additions illustrated in the equation are of commercial interest. Hydrosilylation is used for the preparation of silicone polymers. Silicone rubbers are "cured" through addition of silanes, a process that converts the rubber to a hard material, suitable, for example, as dental cement. The usual homogeneous catalyst is chloroplatinic acid.[126] For supported $RhCl_3$, conversions for $HSiEt_3$ addition were very poor when polystyrene was the support, but improved when the support was a phosphinated allyl chloride-DVB.[16] Addition of $HSi(OR)_3$ to 1-hexene with this catalyst was efficient.

In a study with dimethylaminomethyl polystyrene/DVB polymer, a close correlation was observed between pore size and catalytic activity.[127] Polymers with low cross-linking, i.e., those with virtually no large pores, displayed no catalytic activity. A catalyst with 60% DVB, having a high percentage of pores over 300 Å, was most active. Catalysts with intermediate cross-linkings displayed intermediate activities. Metal loss was particularly high for the high-cross-linked support.

The synthesis of adiponitrile commercialized by DuPont requires the high regioselective addition of two moles of HCN to butadiene:

$$\diagup\!\!\!\diagdown\!\!\diagup \rightarrow \diagup\!\!\diagdown\!\!\diagup\!\!_{CN} \rightleftarrows \diagup\!\!\diagdown\!\!\diagup\!\!_{CN} \rightarrow NC\text{-}(CH_2)_4\text{-}CN$$

These reactions are effected by $Ni[P(OAr)_3]_4$.[128] Apparently no studies of this reaction with polymer-bound catalysts are reported (see Table 8).

## 5.5. Reactions with CO

### 5.5.1. Hydroformylation

$$CH_2=CH_2 \xrightarrow{M\text{-}H} \underset{\underset{M}{|}\phantom{XX}\underset{H}{|}}{CH_2-CH_2} \xrightarrow{CO} \underset{\underset{O=C\diagdown_{M}}{|}\phantom{XX}\underset{H}{|}}{CH_2-CH_2} \xrightarrow{H_2} \underset{\underset{CHO}{|}}{CH_2-CH_3}$$

The essential mechanistic steps in hydroformylation are addition of a metal hydride to an olefin, CO insertion into the alkyl-metal bond, and reaction with hydrogen. Hydroformylation of olefins with homogeneous catalysts is a major industrial process:

$$CH_3CH=CH_2 \xrightarrow[H_2]{CO} CH_3CH_2CH_2CHO \xrightarrow{H_2} CH_3CH_2CH_2CH_2OH$$

Butyraldehyde, for example, is produced from propylene and synthesis gas on a scale of about three million tons per year worldwide. Also commercially

Table 8. Olefin Hydrosilylation

| Support | Ligand | Metal complex supported | Addition | Reference |
|---|---|---|---|---|
| Styrene/DVB | ⓟ–PPh$_2$ | RhCl$_3$ | 1-Hexene + triethoxysilane | 16, 129 |
| Styrene/DVB | ⓟ–PPh$_2$ | PtCl$_2$, H$_2$PtCl$_6$ | 1-Hexene + HSiEt$_3$ | 16 |
| Styrene/DVB | ⓟ–PPh$_2$ | Fe(CO)$_5$, Fe$_3$(CO)$_{12}$ | 1-Pentene + triethylsilane | 8, 130 |
| Polymethyl-acrylate allyl-chloride/DVB | ⓟ–PPh$_2$ | RhCl$_3$, H$_2$PtCl$_6$ | 1-Hexene + HSi(OEt)$_3$ | 16 |
| Styrene/DVB | ⓟ–DIOP | [RhCl(C$_2$H$_4$)$_2$]$_2$ | Acetophenone + HSi(OEt)$_3$ or H$_2$SiPh$_2$ | 67, 131 |

important is the production of fatty alcohols via hydroformylation for use in biodegradable detergents:

$$\text{CH}_2=\text{CHCH}_2\text{CH}_3 \xrightarrow[\text{H}_2]{\text{CO}} \underset{\text{normal}}{\text{CH}_3\text{CH}_2\text{CH}_2\text{CH}_2\text{CHO}} + \underset{\text{branched}}{(\text{CH}_3)_2\text{CHCH}_2\text{CHO}}$$

Two types of catalysts are of greatest commercial interest: those based on cobalt and those based on rhodium. Commercial operations are based typically upon $Co_2(CO)_8$, even though both higher temperatures and pressures are required than for rhodium catalysts. Furthermore, cobalt catalysts produce greater percentages of the usually less desirable branched products than do rhodium-catalyzed reactions.

Yet in spite of these negative aspects, the cobalt catalysts are predominant in industrial applications. The reason is very simple: the cost of rhodium. Given its cost, it is imperative that the catalyst be recovered "completely." This consideration was partially responsible for the very early and intense interest in hydroformylation with polymer-supported catalysts.

From this broad experience several generalizations are apparent.

a. The activities of polymer-bound complexes are generally less than activities displayed by soluble catalysts, though the difference narrows at higher temperatures.
b. As the P:metal ratio increases the $n/b$ ratio increases. This trend parallels that observed in homogeneous catalysis and is explained in terms of an equilibrium between tetra- and pentacoordinate species **6** and **7**. Both electronic and steric factors favor more selective addition of **7** to form terminal metal intermediates. Within a polymer matrix, *bis*-phosphine species analogous to 7 are favored by high P/metal ratios (see Scheme 2).

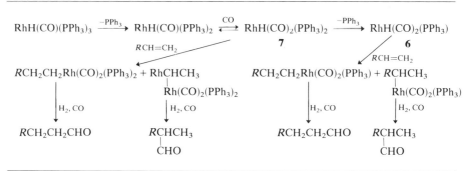

Scheme 2

c. To restrict the loss of metal from the support: 1) Keep feeds free of oxygen. Oxygen converts phosphines to phosphine oxides, thereby reducing the phosphine content in the support. Phosphine oxides are poor ligands for rhodium. 2) Use resins with high P:Rh ratios. This favors *bis*-phosphine complexes of the attached complexes. 3) Keep olefin conversions at modest levels. Aldehyde is able to coordinate rhodium relatively well, so at high conversions metal losses increase. 4) Avoid the use of polar, coordinating solvents. 5) Operate at high temperatures. Coordination of the metal with phosphine is thermodynamically more favorable, with respect to CO coordination, at higher temperatures. 6) Use a polymer with low cross-linking. In macroreticular beads the metal is concentrated at the surface. 7) Use a chelating ligand.
d. Bidentate phosphines give low $n/b$ ratios.
e. Functionalized silicas have given slightly lower $n/b$ ratios than those derived from polystyrene.

Pittman has found unusual selectivities in phosphinated polystyrenes with attached $RhH(CO)(PPh_3)_3$. At low phosphine loadings (8% phenyl rings phosphinated) at low (3.5) P/Rh ratios, the $n/b$ values at 400 psi were 3 or 2 at 40° and 110°. These values are comparable to those of the homogeneous catalyst at these conditions. With a very high P/Rh ratio[19] and a very high phosphine loading (40% phenyl rings phosphinated) the $n/b$ ratio increases to 6 at 100 psi and even greater (up to 15) upon increasing the pressure to 400 psi. The homogeneous catalyst does not display a similar trend. These very high selectivities can be understood to be a consequence of species 7 being favored in the polymer because of the proximity of phosphine.

### 5.5.2. Carboxylation

In the hydroformylation reaction, the acyl metal intermediate reacts with hydrogen to form aldehyde. If, however, a nucleophile is present, the course of the reaction may be altered, producing carboxylic acids or derivatives:

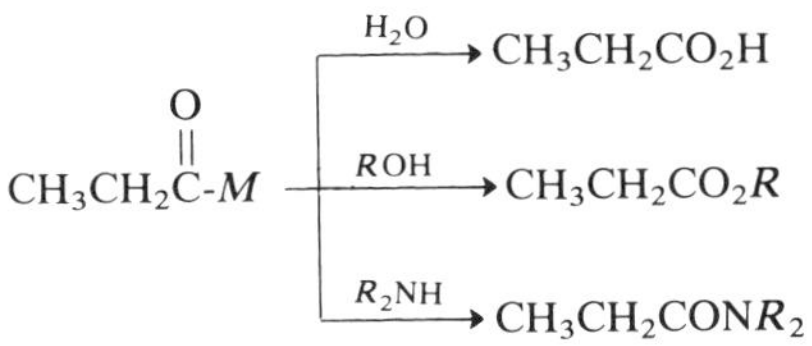

$$CH_3CH_2\overset{O}{\overset{\|}{C}}\text{-}M \begin{cases} \xrightarrow{H_2O} CH_3CH_2CO_2H \\ \xrightarrow{ROH} CH_3CH_2CO_2R \\ \xrightarrow{R_2NH} CH_3CH_2CONR_2 \end{cases}$$

Investigations in this area with polymer-bound catalysts are rather limited. This modest interest is due, in part, to the fact that excellent industrial routes to large-volume carboxylic acids already exist. One aspect that would be of interest would be the effect of a polymeric catalyst on the $n/b$ ratio. In the ethoxycarbonylation of 1-pentene:

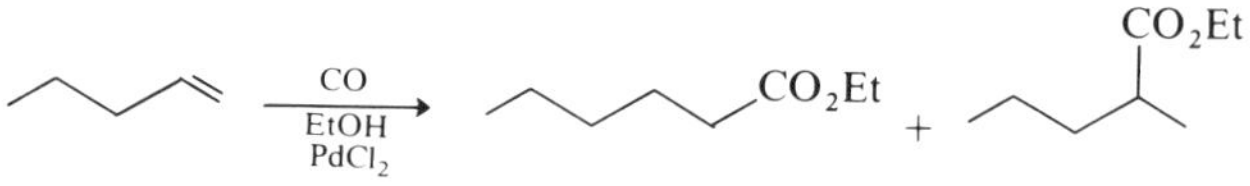

$$CH_2=CHCH_2CH_2CH_3 \xrightarrow[\text{PdCl}_2]{\text{CO, EtOH}} CH_3CH_2CH_2CH_2CH_2CO_2Et + CH_3CH_2CH_2CH(CO_2Et)CH_3$$

it was found that for the same P:Pd ratios, the bound catalyst gave higher $n/b$ ratios than the soluble one.[139]

The Monsanto process for acetic acid production from methanol has been a very successful application of homogeneous catalysis with rhodium.[132]

$$CH_3OH \xrightarrow{CO} CH_3CO_2H$$

When the polymer-bound version of the catalyst was investigated,[44] it was discovered that the supported catalyst was not very active and, furthermore, that the metal was rapidly lost from the support. Reactions of CO in a polar medium, in this case methanol, should be expected to lead to loss of metal (see Table 9).

### 5.5.3. Decarboxylation

Insertion of CO into an alkyl–metal bond is a reversible process. Through abstraction of CO from an aldehyde, alkanes may be synthesized:

$$RCHO \rightarrow RH + CO$$

This is a well-known reaction for homogeneous catalysts, but apparently there are no examples using polymer-bound catalysts.

### 5.6. Metathesis

$$2R'CH=CHR^2 \rightleftharpoons R'CH=CHR' + R^2CH=CHR^2$$

During the late 1960s and the 1970s, olefin metathesis was one of the most intensively investigated organometallic reactions. This reaction was intriguing mechanistically and presented the possibility of selectively producing attractive olefins and polymers. Ultimately, the mechanism was proven to be a carbene process. There has been but modest activity using phosphine-bound catalysts.

Basset investigated the catalytic properties of $Mo(CO)_6$ attached to a phosphinated polystyrene-DVB resin.[148] Upon treatment with ethylaluminum dichloride and oxygen, this slightly active system gave a conversion of cis-2-pentene of 3.4% in 20 min at room temperature.

Table 9. Hydroformylation

| Support | Ligand | Metal complex supported | P/M | Illustrative conditions ||||| Reference |
| --- | --- | --- | --- | --- | --- | --- | --- | --- | --- |
| | | | | Temp | Pressure | Reactant | n/b | | |
| Styrene/DVB | ⓟ-$PR_2$ | $Co_2(CO)_8$ | — | 150 | 1000 psi | 1-Pentene | 61/24 | | 13 |
| Styrene/DVB | ⓟ-$PR_2$ | $RhCl_3/NaBH_4$ | — | — | — | — | — | | 133 |
| Styrene/DVB | ⓟ-$PR_2$ | $[RhCl(CO)_2]_2$ | — | 100 | 1 atm | Ethylene | — | | 43, 134 |
| Styrene/DVB | ⓟ-$PR_2$ | $RhCl(CO)(PPh_3)_2$ | 3.7% Rh 3.3% P | 90 | 20 atm | Methanol | — | | 44 |
| Styrene/DVB | ⓟ-$PR_2$ | $RhH(CO)(PPh_3)_3$ | 4.4 | 62 | 1000 psi | 1-Pentene | 4.4 | | 42, 12, 14, 135–138 |
| Styrene/DVB | ⓟ-$PR_2$ | $Ru(CO)_3(PPh_3)_2$ | 3.1 | 140 | 1000 psi | 1-Pentene | 3.7 | | 103 |
| Styrene/DVB | ⓟ-$PR_2$ | $PdCl_2/EtOH$ | — | — | — | — | — | | 139 |
| Styrene/DVB | ⓟ-$PR_2$ | $[RhCl(CO)_2]_2$ | — | 130 | 15 atm | Ethylene | — | | 140 |
| Styrene/DVB | ⓟ-$PPhCH_2CH_2PPh_2$ | $RhH(CO)(PPh_3)_3$ | 14.3 | 80 | 800 psi | Styrene | 1/6.9 | | 137 |
| Styrene/DVB | ⓟ-DIOP | $RhH(CO)(PPh_3)_3$ | 4.0 | 40 | 400 psi | Styrene | 1/4.8 | | 141, 142 |
| Styrene/DVB | ⓟ-DIOP | $Rh(CO)(PPh_3)(acac)$ | — | 22 | 80 atm | 1-Pentene | — | | 3b |
| Poly(vinyl chloride) | ⓟ-$PR_2$ | $[Rh(CO)_2(acac)]_2$ | 4.8% P 7.9% Rh | 80–90 | 42 atm | 1-Hexene | 1.5–2.5 | | 21 |
| Poly(butadiene) | ⓟ-$PR_2$ | $Co_2(CO)_8$ | — | — | — | — | — | | 145 |
| Poly(dimethyl-siloxane) | | $RhCl(CO)(PPh_3)_2$ | 0.50% P 0.79% Rh | 100 | 1000 psi | 1-Hexene | 0.9 | | 144 |
| Polystyrenes | Chelating phosphines | $RhH(CO)(PPh_3)_3$ | 1.9 | 20 | 1 atm | 1-Hexene | 6.4 | | 145 |
| Styrene resins | ⓟ-$P(OMe)_2$ | $[RhCl(CO)_2]_2$ | — | — | — | — | — | | 146 |
| $SiO_2$ | ⓢ-$PPh_2$ | $[RhCl(CO)_2]_2$ | — | — | — | — | 0.7 | | 147 |

*Table 10. Metathesis*

| Support | Ligand | Metal complex supported | Reactant | Reference |
|---|---|---|---|---|
| Styrene/DVB | Ⓟ–$PR_2$ | $Mo(PPh_3)_2(NO)_2Cl_2$ | 1,7-Octadiene | 15 |
| Styrene/DVB | Ⓟ–$PR_2$ | $Mo(CO)_6/AlEt_2Cl + O_2$ | *cis*-2-Pentene | |
| Styrene/DVB | Ⓟ–$PR_2$ | $W(CO)_6/iBu_3Al + O_2$ | *trans*-3-Heptene | 149 |

A similar catalyst based upon $W(CO)_6$ was studied by Warwel and Buschmeyer.[149] Adding isobutylaluminum and oxygen to the catalyst effected a 50% conversion of *trans*-3-heptene. Upon recycling the catalyst, the activity decreased dramatically. The tungsten was displaced from the polymer by the alkylaluminum compound, and the tungsten complex then performed as a homogeneous catalyst (see Table 10).

## 6. TRENDS

### 6.1. Catalyst Characterization

One of the fruitful endeavors that might be continued in the future is that of catalyst characterization. Surface properties such as pore size and distribution are inadequately considered when catalytic activity is interpreted. There is still but a poorly formed concept of long-range interactions between the support and the reaction center. We know, for example, that the structures of conventional heterogeneous catalysts are often crucial in establishing their activity. This is seen perhaps most dramatically in the zeolite catalysts, where their electrostatic environments are largely responsible for their chemistry. It is clear that polymeric supports affect reactions, but the precise influences of cavity size, edge effects (surface *vs.* interior), solvation, and polymer flexibility are unknown. It appears that coordination patterns closely parallel those seen in homogeneous catalysis. There is no direct spectroscopic evidence yet that polymers may force unusual bond angles at coordinated metal centers. Are polymers, even highly cross-linked ones, so flexible that strained metal coordinations are not formed?

Upon examination of the literature of homogeneous catalysis one is struck by the remarkable paucity of kinetic studies. The same comment could be made about polymer-bound catalysis. It is clear that in any relatively new field the normal initial thrust is to define the possible reaction chemistry. This search for interesting reactions certainly has not become passé, but there is a need to mature the field by learning more of the details of the catalytic cycles.

It is also of note that a virtual void exists in evaluating the influence of the polymer on stereoselectivity. Furthermore, isotope studies are largely lacking. We might also expect additional supports, ones mechanically and thermally more stable than polystyrene, to gain increased use in the future. Phosphines have been attached to electrical insulators; would the use of semiconducting or conducting supports provide access to catalysis more akin to conventional metal heterogeneous catalysts? Could one usefully study classical heterogeneous catalysts by attaching small aggregates to polymer supports? A rigid support might provide the matrix isolation necessary to examine the chemistry of small units.

## 6.2. Reactions

Since the researcher normally looks to the chemistry of soluble complexes in designing polymer-bound catalysts, it is notable that some areas that have proven fruitful in homogeneous catalysis have been omitted from investigations using polymer-bound catalysts. One of these areas concerns the reactions of arenes. Benzene, for example, may be hydrogenated with homogeneous cobalt phosphite and ruthenium phosphine complexes, but the corresponding supported versions are not reported. Aryl halides may be carboxylated in the presence of a soluble palladium catalyst:

$$\text{Ph-}X \xrightarrow[H_2O]{CO} \text{Ph-}CO_2H$$

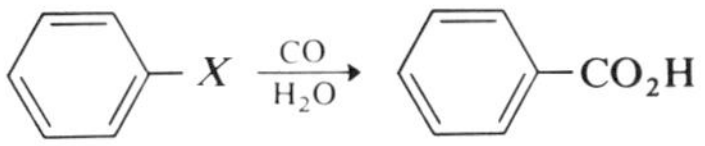

Similarly, the coupling of aryl halides and olefins with soluble palladium catalysts has but one polymer-bound analog.[150]

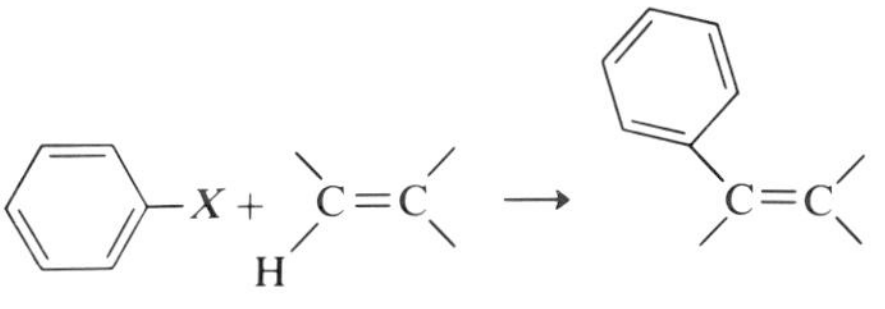

$$\text{Ph-}X + \underset{H}{\overset{}{>}}C=C\overset{}{<} \rightarrow \underset{Ph}{\overset{}{>}}C=C\overset{}{<}$$

Soluble catalysts are also employed for coupling reactions between aryl halides and metal alkyls. The most effective catalysts are those derived from $NiCl_2(PPh_3)_2$.

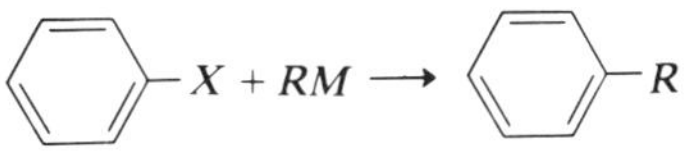

$$\text{Ph-}X + RM \rightarrow \text{Ph-}R$$

Rearrangement of the alkyl moiety may occur, and the product distribution is highly dependent upon the nature of the ligand.

When we wish to reduce a compound catalytically, we typically think of using hydrogen. The same transformation can sometimes be effected using hydrogen already found in organic compounds. Hydrogen transfer from alcohols (e.g., isopropyl alcohol), using soluble catalysts, is highly effective for the reduction of ketones, olefins, and even nitro compounds. Apparently there are no published examples of hydrogen transfer using polymer-bound catalysts.

For all the examples discussed thus far, it is assumed that the metal complexes remain attached to the polymer during the course of the reaction. This is not the only type of catalyst, however, that would be attractive. Equally attractive is the situation in which the metal complex decoordinates from the polymer under certain conditions, e.g., high temperature, and functions as a normal homogeneous catalyst, then is recoordinated when the conditions are changed. The basic feature of catalyst recovery would not be lost if a catalyst were to operate in this manner. There is an account of this concept, using pyridine as the ligand. Apparently there are no examples in phosphine chemistry.

## REFERENCES

1. C. U. Pittman, Jr., *Polymer-Supported Reactions in Organic Synthesis*, edited by P. Hodge and D. C. Sherrington (John Wiley and Sons, New York, 1980), p. 249.
2. Y. Chauvin, D. Commereuc, and F. Dawans, *Prog. Polym. Sci.* **5**, 95–226 (1977).
3. a. E. Bayer and V. Schurig. *Chem. Technol.* **1976**, 212. b. E. Bayer and V. Schurig, *Angew. Chem. Intl. Edn.* **14**, 493–494 (1975).
4. G. Bernard, Y. Chauvin, and D. Commereuc, *Bull. Soc. Chim.* **1163**, 1168–72 (1976).
5. W. Heitz, *Adv. Pdym. Sci.* **23**, 1 (1977).
6. M. J. Farrall and J. M. Frechet, *J. Org. Chem.* **43**, 2618 (1978).
7. K. A. Abdula, N. P. Allen, A. H. Badran, J. Dwyer, C. A. McAuliffe, and N. D. A. Toma, *Chem. Ind.* **273** (1976).
8. R. D. Sanner, R. G. Austin, M. S. Wrighton, W. D. Honnick, and C. U. Pittman, Jr., "Interfacial Photoprocesses: Energy Conversion and Synthesis," *ACS Adv. in Chem. Ser.* **184**, 1980, 13; C. U. Pittman, Jr., W. D. Honnick, M. S. Wrighton, R. D. Sanner, R. G. Austin, *Fundamental Research in Homogeneous Catalysis*, edited by M. Tsutsui (Plenum Publishing, 1979), p. 603.
9. R. H. Grubbs and L. C. Kroll, *J. Am. Chem. Soc.* **93**, 3062 (1971).
10. H. S. Bruner and J. C. Bailar, *Inorg. Chem.* **12**, 1465 (1973).
11. R. Pierantozzi, K. J. McQuade, B. C. Gates, M. Wolf, H. Knözinger, and W. Ruhmann, *J. Am. Chem. Soc.* **101**, 5436 (1979).
12. C. U. Pittman, Jr., L. R. Smith, and R. M. Hanes, *J. Am. Chem. Soc.* **97**, 1742 (1975).
13. G. O. Evans, C. U. Pittman, Jr., R. McMillan, R. T. Beach, and R. Jones, *J. Organometal. Chem.* **67**, 295 (1974).
14. C. U. Pittman, Jr. and L. R. Smith, *J. Am. Chem. Soc.* **97**, 1749 (1975).
15. R. H. Grubbs, S. Swetnick, and S. C. H. Su, *J. Mol. Catal.* **3**, 11–15 (1977–1978).

16. M. Capka, P. Svoboda, M. Kraus, and J. Hetflejs, *Chem. Ind.*, 650–651 (1972).
17. T. H. Kim and H. F. Rase, *Ind. Eng. Chem. Prod. Res. Dev.* **15**, 249–254 (1976).
18. M. Czakova and M. Capka, *J. Mol. Catal.* **11**, 313–322 (1981).
19. K. G. Allum, R. D. Hancock, S. McKenzie, and R. C. Pitkethly, *Proceedings of the Fifth International Congress on Catalysis 1972*, edited by J. W. Hightower (North Holland, Amsterdam, 1973), Vol. 1, p. 447.
20. R. D. Hancock, I. V. Howell, R. C. Pitkethly, and P. J. Robinson, *Catalysis Heterogeneous and Homogeneous*, edited by B. Delmon and G. Jannes (Elsevier, Amsterdam, 1975), pp. 349–359.
21. K. G. Allum, R. D. Hancock, I. V. Howell, R. C. Pitkethly, and P. T. Robinson, *J. Catalysis* **43**, 322–330 (1976).
22. K. G. Allum, R. D. Hancock, I. V. Howell, T. E. Lester, S. McKenzie, R. C. Pitkethly, and P. J. Robinson, *J. Catalysis* **43**, 331 (1976).
23. K. G. Allum, R. D. Hancock, I. V. Howell, T. E. Lester, S. McKenzie, R. C. Pitkethly, and P. J. Robinson, *J. Organometallic Chem.* **107**, 393 (1976).
24. K. G. Allum, R. D. Hancock, I. V. Howell, S. McKenzie, R. C. Pitkethly, and P. J. Robinson, *J. Organomet. Chem.* **87**, 203 (1975).
25. J. M. Moreto, J. Albaiges, and F. Camps, *Catalysis Heterogeneous and Homogeneous*, edited by B. Delmon and G. Jannes (Elsevier, Amsterdam, 1975), pp. 339–347.
26. K. Kochloefl, W. Liebelt, and H. Knozinger, *J. Chem. Soc., Chem. Comm.* **1977**, 510.
27. M. Capka and J. Hetflejs, *Coll. Czech. Chem. Comm.* **39**, 154 (1974).
28. Badische Anilir und Soda-Fabrik, A. G., French Patent No. 2 053 300.
29. K. G. Allum, R. D. Hancock, British Patent No. 1 287 566.
30. J. Manassen, *Israel J. Chem.* **8**, 5p (1970).
31. S. V. McKinley and J. W. Rakshys, United States Patent No. 3 708 462 (to Dow Chemical Company).
32. Y. Nonaka, S. Takahashi, and N. Hagihara, *Mem. Inst. Sci. Ind. Res., Osaka Univ.* **31**, 23 (1974).
33. A. J. Naaktgeboren, R. J. M. Nolte, and W. Drenth, *Recl. Trav. Chim. Pays-Bas* **97**, 112 (1978).
34. A. J. Naaktgeboren, R. J. M. Nolte, and W. Drenth, *J. Amer. Chem. Soc.* **102**, 3350 (1980).
35. J. Kiji, S. Kadoi, and J. Furukawa, *Angew. Makromol. Chem.* **46**, 163 (1975).
36. N. Takaishi, H. Imai, C. A. Bertelo, and J. K. Stille, *J. Am. Chem. Soc.* **98**, 5400 (1976).
37. N. Takaishi, H. Imai, C. A. Bertelo, and J. K. Stille, *J. Am. Chem. Soc.* **100**, 264 (1978).
38. G. L. Baker, S. J. Fritschel, and J. K. Stille, *J. Org. Chem.* **46**, 2960–2965 (1981).
39. I. Dietzmann, D. Tomanova, and J. Hetflejs, *Coll. Czech. Chem. Comm.* **39**, 135 (1974).
40. C. U. Pittman, Jr. and L. R. Smith, *J. Amer. Chem. Soc.* **97**, 341 (1975).
41. C. U. Pittman, Jr. and R. F. Felis, *J. Organometallic Chem.* **72**, 399 (1974).
42. C. U. Pittman, Jr. and R. M. Hanes, *J. Amer. Chem. Soc.* **98**, 5402 (1976).
43. K. G. Allum, R. D. Hancock, I. V. Howell, R. C. Pitkethly, and P. J. Robinson, *J. Organometallic Chem.* **87**, 189 (1975).
44. M. S. Jarrell and B. C. Gates, *J. Catalysis* **40**, 255 (1975).
45. D. Doskocilova, B. Schneider, and J. Jakes, *J. Magn. Reson.* **29**, 79 (1978).
46. S. J. Fritschel, J. J. H. Ackerman, T. Keyser, and J. K. Stille, *J. Org. Chem.* **44**, 3152 (1979).
47. T. Imanaka, K. Kaneda, S. Teranishi, and M. Terasawa, *Proceedings of the Sixth International Congress on Catalysis, 1976*, edited by G. C. Bond, P. B. Wells, and F. C. Tompkins (The Chemical Society, London, 1977), Vol. 1, p. 509.
48. N. Takahashi, I. Okura, and T. Keij, *J. Amer. Chem. Soc.* **97**, 7489 (1975).
49. K. Kaneda, M. Terasawa, T. Imamaka, and S. Teranishi, *Chem. Lett.* 1005–1008 (1975).

50. N. L. Holy, *Fundamental Research in Homogeneous Catalysis 3*, edited by M. Tsutsui (Plenum Publishing, 1979), pp. 691–706.
51. J. Reed, P. Eisenberger, B.-K. Teo, and B. M. Kincaid, *J. Amer. Chem. Soc.* **99**, 5217–5218 (1977); **100**, 2375–2378 (1978).
52. L. D. Rollmann, *J. Amer. Chem. Soc.* **97**, 2132 (1975).
53. F. Pinna, M. Bonivento, G. Strukul, M. Graziani, E. Cernia, and N. Palladino, *J. Mol. Catalysis* **1**, 309 (1975–1976).
54. G. Strukul, M. Bonivento, M. Graziani, E. Cernia, and N. Palladino, *Inorg. Chim. Acta* **12**, 15 (1975).
55. M. Graziani, G. Strukul, M. Bonivento, F. Pinna, E. Cernia, and N. Pallidino, *Catalysis Heterogeneous and Homogeneous*, edited by B. Delmon and G. Jannes (Elsevier, Amsterdam, 1975), pp. 331–338.
56. G. Strukul, P. Dolimpio, M. Bonivento, F. Pinna and M. Graziani, *J. Mol. Catalysis* **2**, 179 (1977).
57. D. Tatarsky, D. H. Kohn, and M. Cais, *J. Polym. Sci., Polym. Chem. Ed.* **18**, 1387–1397 (1980).
58. R. H. Grubbs and E. M. Sweet, *Macromolecules* **8**, 241–242 (1975).
59. R. H. Grubbs, C. Gibbons, L. C. Kroll, W. D. Bonds, and C. H. Brubaker, Jr., *J. Amer. Chem. Soc.* **95**, 2373 (1973).
60. J. Manassen, Reference 20, 293–306.
61. M. S. Bruner and J. C. Bailor, *J. Amer. Oil Chem. Soc.* **49**, 533 (1972).
62. R. H. Grubbs, L. C. Kroll, and E. M. Sweet, *J. Macromol. Sci., Chem. A* **7**, 1047 (1973).
63. J. D. Morrison, W. F. Masler, and M. F. Neuberg, *Adv. Catal.* **25**, 81 (1976).
64. B. D. Vineyard, W. S. Knowles, M. J. Sabacky, G. L. Bachman, and D. J. Weinkauff, *J. Am. Chem. Soc.* **99**, 5946–5952 (1977).
65. T. Masada and J. K. Stille, *J. Am. Chem. Soc.* **100**, 268 (1978).
66. G. C. Baker, S. J. Fritschel, J. R. Stille, and J. K. Stille, *J. Org. Chem.* **46**, 2954–2960 (1981).
67. W. Dumont, J. C. Poulin, T. P. Dang, and H. B. Kagan, *J. Am. Chem. Soc.* **95**, 8295–8299 (1973).
68. K. Achiwa, *Chem. Lett.* 905 (1978).
69. M. E. Wilson and G. M. Whitesides, *J. Am. Chem. Soc.* **100**, 306 (1977).
70. F. Ciardelli, E. Chiellini, C. Carlini, and R. Nocci, *Polym. Preprints* **17**, 188 (1976).
71. a. K. Ohkubo, K. Fujimori, and K. Yoshinaga, *Inorg. Nucl. Chem. Lett.* **15**, 231–234 (1979).
    b. K. Ohkubo, M. Setoguchi, and K. Yoshinaga, *Inorg. Nucl. Chem. Lett.* **15**, 235–238 (1979).
72. C. U. Pittman, Jr., A. Hirao, J. J. Yang, Q. Ng, R. Hanes, and C. C. Lin, *Preprints Div. Petrol. Chem. (ACS)* **22**, 1196 (1977).
73. C. F. Hobbs and W. S. Knowles, *J. Org. Chem.* **46**, 4422–4427 (1981).
74. J. I. Crowley and H. Rapoport, *Acc. Chem. Res.* **9**, 135–144 (1976).
75. L. T. Scott, J. Rebek, L. Oysyanko, and C. L. Sims, *J. Am. Chem. Soc.* **99**, 625–626 (1977).
76. P. Jayalekshmy and S. Mazur, *J. Am. Chem. Soc.* **98**, 6710–6711 (1976).
77. C. U. Pittman, Jr. and Q. Ng, *J. Organometal. Chem.* **153**, 85 (1978).
78. R. H. Crabtree, M. F. Mellea, J. M. Mihelcic, and J. M. Quirk, *J. Am. Chem. Soc.* **104**, 107–113 (1982).
79. A. H. Janowicz and R. G. Bergman, *J. Am. Chem. Soc.* **104**, 352 (1982).
80. S. Jacobson, W. Clements, H. Hiramoto, and C. U. Pittman, Jr., *J. Mol. Catal.* **1**, 73 (1975).
81. C. U. Pittman, Jr., S. E. Jacobson, and H. Hiramoto, *J. Am. Chem. Soc.* **97**, 4774 (1975).
82. B. C. Gates and J. Lieto, *Chem. Technol.* 248 (1980).

83. J. Andrieux, D. H. R. Barton, and H. Patin, *J. Chem. Soc. Perkin* **I**, 359–000 (1977).
84. C. A. Tolman, *J. Am. Chem. Soc.* **94**, 2994–2999 (1972).
85. R. G. Miller, P. A. Pinke, R. D. Stauffer, H. J. Golden, and D. J. Baker, *J. Am. Chem. Soc.* **96**, 4211–4220 (1974); P. A. Pinke and R. G. Miller, op. cit., 4221–4229; P. A. Pinke, R. D. Stauffer, and R. G. Miller, *op. cit.* 4229–4234.
86. J. M. Moreto, J. Albaiges, and F. Camps, Ref. 20, 339–348.
87. P. A. Gosselain, Reference 20, 107–132.
88. J. Conan, M. Bartholin, and A. Guyot, *J. Mol. Catalysis* **1**, 375 (1975–1976).
89. K. Kochloefl, W. Liebelt, and H. Krozinger, *J.C.S. Chem. Commun.* 510 (1977).
90. I. V. Howell and R. D. Hancock, British Patent No. 1 408 013.
91. W. R. Cullen, D. J. Patmore, A. J. Chapman, and A. D. Jenkins, *J. Organometal. Chem.* **102**, C12 (1975).
92. R. R. Schrock and J. Osborn, *J.C.S., Chem. Commun.* 567 (1970).
93. G. Pez, R. A. Grey, and J. Corsi, *J. Am. Chem. Soc.* **103**, 7528 (1981).
94. S. E. Jacobson and C. U. Pittman, Jr., *J. Chem. Soc. Chem. Comm.* 187 (1975).
95. M. H. J. M. DeCroon and J. W. E. Coenen, *J. Mol. Catal.* **11**, 301–311 (1981).
96. S. Lecolier, French Patent No. 2 270 238.
97. M. Terasawa, K. Kaneda, T. Imanaka, and S. Teranishi, *J. Catal.* **51**, 406 (1978).
98. K. Kaneda, M. Terasawa, T. Imanaka, and S. Teranishi, *Chem. Lett. Japan* 995 (1976).
99. R. H. Grubbs, L. C. Kroll, and E. M. Sweet, *Polymer Preprints* **13**, 828 (1972).
100. R. H. Grubbs, E. M. Sweet, and S. Phisanbut, *Catalysis In Organic Synthesis*, edited by P. Rylander and H. Greenfield (Academic Press, New York, 1975), p. 153.
101. A. Guyot, C. Graillat, and M. Bartholin, *J. Mol. Catal.* **3**, 39 (1977–1978).
102. K. G. Allum, R. D. Hancock, and R. C. Pitkethly, British Patent No. 1 295 675.
103. C. U. Pittman, Jr. and G. Wilemon, *Ann. N.Y. Acad. Sci.* **333**, 67 (1980).
104. J. P. Collman, L. S. Hegedus, M. P. Cooke, J. R. Norton, G. Golcetti, and D. N. Marquart, *J. Am. Chem. Soc.* **94**, 1789 (1972).
105. M. S. Jarrell, B. C. Gates, and E. D. Nicholson, *J. Am. Chem. Soc.* **100**, 5727 (1978).
106. M. S. Jarrel and B. C. Gates, *J. Catal.* **54**, 81 (1978).
107. Z. Otero-Schipper, Z. Lieto, and B. C. Gates, *J. Catal.* **63**, 175 (1980).
108. J. Lieto, J. J. Rafalko, and B. C. Gates, *J. Catal.* **62**, 149 (1980).
109. J. J. Rafalko, J. Lieto, B. C. Gates, and G. L. Schrader, Jr., *J. Chem. Soc., Chem. Commun.* 540 (1978).
110. C. U. Pittman, Jr., S. K. Wuu, and S. E. Jacobson, *J. Catal.* **44**, 87 (1976).
111. C. U. Pittman, Jr., L. R. Smith, and S. E. Jacobson, *Catalysis: Heterogeneous and Homogeneous*, edited by B. Delmon and G. Jannes (Elsevier, Amsterdam, 1975), p. 393.
112. A. J. Naaktgeboren, R. J. M. Notte, and W. Drenth, *J. Mol. Catal.* **11**, 343–351 (1981).
113. R. G. Muratova, R. Z. Khairullina, S. V. Shulyndin, B. E. Ivanov, and R. I. Izmailov, *Kin. and Catal.* **15**, 115 (1974).
114. G. L. Baker, S. J. Fritschel, and J. K. Stille, *Polymer Preprints* **22**(1), 155 (1981).
115. W. R. Cullen, D. J. Patmore, A. J. Champan, and A. D. Jenkins, *J. Organomet. Chem.* **102**, C12 (1975).
116. B. Bogdanovic, H. Biserka, H. G. Karmann, H. G. Nussel, D. Walter, and G. Wilke, *Ind. Eng. Chem.* **62**, 34–38 (1970).
117. R. S. Bauer, H. Chung, P. W. Glockner, and W. Keim, United States Patent No. 3 644 563 (1972); R. F. Mason, United States Patent No. 3 737 475 (1973).
118. Badische Anilin und Soda Fabrik A.G., French Patent No. 2 053 300.
119. K. Kaneda, M. Terasawa, T. Imanaka, and S. Teranishi, *Tetrahedron Lett.* 2957 (1977).
120. K. Kaneda, T. Uchiyama, M. Terasawa, T. Imanaka, and S. Teranishi, *Chem. Lett. Japan* 449 (1976).
121. C. U. Pittman, Jr. and S. E. Jacobson, *J. Mol. Catal.* **3**, 293–297 (1977–1978).
122. F. Hobjabri, *Polymer* **17**, 58 (1976).

123. H. Pracejus, German Patent No. 2 230 739.
124. K. G. Allum and R. D. Hancock, *Fifth International Congress of Catalysis, Preprint 31* (Miami, 1972).
125. R. F. Clark, C. D. Storrs, and G. B. Barnes, United States Patent No. 3 364 273.
126. J. L. Speier, "Homogeneous Catalysis by Transition Metals," *Adv. Organomet. Chem.* **17,** 407 (1979).
127. I. Dietzmann, D. Tomanova, and J. Hetflejs, *Coll. Czechoslov. Chem. Comm.* **39,** 123 (1974).
128. V. D. Luedeke, *A dipronitrile*, edited by J. J. McKetta and W. A. Cunningham, Encyclopedia of Chemical Processing and Design (Marcel Dekker, 1977), Vol. 2, p. 146.
129. M. Capka, P. Svoboda, M. Corny, and J. Hetflejs, *Tetrahedron Lett.* 4787 (1971).
130. R. D. Sanner, R. G. Austin, M. S. Wrighton, W. D. Honnick, and C. U. Pittman, Jr., *Inorg. Chem.* **18,** 928 (1979).
131. J. C. Poulin, W. Dumont, T. P. Dang, and H. B. Kagan, *Compt. Rend. Ser.* C **277,** 41 (1973).
132. J. F. Roth, J. H. Craddock, A. Hershman, and F. E. Paulik, *Chem. Tech.* **1,** 600 (1971); H. D. Grove, *Hydrocarbon Proc.*, 76 (1972); F. E. Paulih, United States Patent No. 3 769 329 (1973).
133. I. V. Howell and R. D. Hancock, British Patent No. 1 408 013.
134. H. Arai, T. Kanedo, and T. Kunugi, *Chem. Lett. Japan*, 265 (1975).
135. C. U. Pittman, Jr. and W. D. Honnick, *J. Org. Chem.* **45,** 2132 (1980).
136. C. U. Pittman, Jr., A. Hirao, C. Jones, R. M. Hanes, and Q. Ng, *Ann. N.Y. Acad. Sci.* **295,** 15 (1977).
137. C. U. Pittman, Jr. and C. C. Lin., *J. Org. Chem.* **43,** 4928 (1978).
138. C. U. Pittman, Jr., W. D. Honnick, and J. J. Yang, *J. Org. Chem.* **45,** 684 (1980).
139. C. U. Pittman, Jr. and Q. Y. Ng, United States Patent No. 4 258 206 (1981).
140. H. Arai, *J. Catal.* **51,** 135–142 (1978).
141. C. U. Pittman, Jr., A. Hirao, J. J. Yang, Q. Ng, R. Hanes, and C. C. Lin, *Preprints Div. Petrol. Chem. (ACS)* **22,** 1196 (1977).
142. S. J. Fritschel, J. J. H. Ackerman, T. Keyser, and J. K. Stille, *J. Org. Chem.* **44,** 3152 (1979).
143. Badische Anilin und Soda-Fabrik A.G., French Patent No. 2 053 300.
144. M. O. Farrell and C. H. VanDyke, *J. Organometal. Chem.* **172,** 367–376 (1979).
145. A. R. Sanger and L. R. Schallig, *J. Mol. Catal.* **3,** 101 (1977–1978).
146. W. O. Haag and D. D. Whitehurst, "Catalysis," edited by J. W. Hightower, paper 29, Vol. 1, p. 465 (1973).
147. R. D. Hancock, I. V. Howell, R. C. Pitkethly, and P. J. Robinson, Reference 20, 361–371.
148. J. M. Basset, R. Mutin, G. Descotes, and D. Sinou, *Compt. Rend. Ser. C* **280,** 1181 (1975).
149. S. Warwel and P. Buschmeyer, Angew. Chem., *Int. Ed. Engl.* **17,** 131–000 (1978).
150. M. Terasawa, K. Kaneda, T. Smanaka, and S. Teranishi, *J. Organometal. Chem.* **162**(3), 403 (1978).

# Index

Acetylene trimerization, 207
Acrylic acid esters from alcohols, acetylene, and CO, 5
Active catalyst, defined, 14
Acyl carbonyl complexes, structures, 124t
Acyl complexes, structures, 125t
A-frame, 175
A-frame complexes, hydrogenation catalytic activity, 219t
Aldehydes
  adduct with $[Rh(dppp)_2]^+$, 364
  decarbonylation, 313, 343
  hydrogenation, 318–319
  oxidative addition of, 352
Alkane activation, 310
Alkane dehydrogenation, 311
Alkenylphosphine ligands, 251
Alkyl carbonyl complexes, structures, 123t
Alkyldiazenido complexes, 418
Alkyl olefin complexes, structures, 127t
Alkyne to alkene hydrogenation, 221
Amines from $N_2$, 6
Aminobis (difluorophosphines), 170
Aminoketones, hydrogenation of, 333–335
Anchimeric assistance, in rate of oxidative addition, 241
Aryl halides, coupling reactions, 479
Asymmetric bisphosphines, 138f
Asymmetric hydrogenation, 9, 138
  deuterium isotope effects, 160
  energetics, 161f
  of ketones, 328t
  kinetic studies, 151
  mechanism and stereoselectivity, 154
  optical yields, 140t
  pressure effects, 149–150t
  reaction pathway, 159–160
  substrate complexes, 155

Asymmetric induction, 41
  with supported catalysts, 457
$Au_2[\mu\text{-}(CH_2)_2PR_2]_2$, 199

BINAP, 138
Binuclear diphosphine bridging geometries, 175f
Binuclear elimination reactions, 226
Bis(diphenylarsino)methane bridged complexes, structural parameters, 181t
Bis(2-diphenylphosphinoethyl)amine ligands, 244
1,5-Bis(diphenylphosphino)-3-oxopentane, complexes of, 242–243f
Bonding modes in diphosphine complexes, 217
$BPh_4^-$ as a ligand, 299
Bridging carbonyl ligands, 196t
Bridging methyl groups, 33
Bulky phosphines, 258
Butadiene, Ni catalyzed cyclooligomerization, 64–80
Butene isomerization, 49
  mechanism using $Ni[P(OEt)_3]_4$, 50f

Carbon monoxide bridged complexes, 223t
Catalyst deactivation product, 15
Catalyst precursor, definition, 14
Cationic bisphosphine rhodium diene complexes, 148
Cationic complexes, 303, 305, 307
Cationic complexes, synthetic methods, 303–304
Celanese hydroformylation studies, 280f
Charge effects on catalysts, 297–303
Chelate ring size, effect on catalytic rates, 225
Chelating bisphosphine rhodium complexes, structures, 152f

Chiral ferrocenyldiphosphine ligand, 253f
Chiral shift reagents, 40
CHIRAPHOS, 42, 138
Cluster hydrides, 309
C–N bond formation from coordinated $N_2$, 414
Co-catalysts, 14
Coordinative unsaturation in supported catalysts, 459–460
Co-oxidation reactions
  of alkenes and hydrogen, 392
  of alkenes and tertiary phosphines, 390–391f
  of isocyanides and CO, 392f
  of triphenylphosphines and CO, 393
$Co_2[\mu\text{-PNP}]_3(CO)_2$, 172f
Counter ion effects, 302
$CS_2$ coupling reaction, 203
CTSRDS, 18
Cyclometalation reactions, 310
Cyclooligomerization of butadiene
  ligand concentration control maps, 69
  ligand effects, 65
  mechanism, 76f
  Ni catalyzed, 64–80
Cyclopropanes, 126–129
CYPHOS, 140

Deactivation of catalysts, 15, 308
Dead catalyst, 15
Decarbonylation
  acid chlorides, 345, 347–352, 372, 348t
  aldehydes, 352–355
  aldehyde substituent effects, 363
  catalytic, 355–372, 357t, 360t
  diphosphine complexes, 346, 359–372
  Ir catalysts, 370t
  kinetic studies, 354–355, 363–368
  ligand steric effects, 360t, 368–369
  oxidative addition intermediates, 352
  P–N complexes of Rh, 370
  $RhCl(PPh_3)_3$, 345, 347–355
  tetraphenylporphyrin complex, 346
Dehydrogenation of alkanes, 311
Deuterium isotope effects, 160
Diastereomers (NMR), 34, 37
Diastereoselection in enamide complexes, 162f
Diazenido complexes of Mo and W, 414–415
Dinitrogen complexes, 410–412f
  reactivity with electrophiles, nucleophiles, and radicals, 433–434

Dinitrogen to ammonia, 406
(−)-DIOP, 9, 137
DIOP complexes, conformational analysis, 142f
DIOXOP, 143
Dioxygen complexes, reactions with electrophiles, 380–382
  ketones, 381f
DIPAMP, 138, 139
2(Diphenylphosphino)pyridine, 182
Diphosphine bridge binuclear complexes, catalytic reactions, 206–208
Diphosphines
  attached to organic polymers, 288–289
  bridging ligands, 170
  dppb, 360, 368
  dppe, 360, 361
  dppm, 360, 368–369
  dppp, 359–363
Diphosphite ligands, 170
Double A-frame complexes, 175
Dppm bridged complexes, structural parameters, 179t

E-$\alpha$-benzamidocinnamic acid, asymmetric hydrogenation of, 144–145
Enantiomer excesses in hydrogenation of $\alpha$-benzamidocinnamic acids, 145t
Enantiomeric ratio determination by polarity, 38
Enantioselective hydrogenation of ketones, 324t
Enantioselective reduction of C=N double bond, 337t
Equilibrium isotope effects, 30
ESCA of polymer bound catalysts, 452–453t
Ethane decomposition, 21
EXAFS, 154
  of polymer bound catalysts, 453

Face-to-face binuclear complex, 175
$Fe_2Cp_2(CO)_4$, 169
Ferrocene-based $P_2$ ligands, 281f
Fischer-Tropsch reaction, 30

Hemilabile chelating agents, 241
High dilution synthesis of phosphorus macrocycles, 269–271
Hydrazido complexes of Mo and W, 414–415
Hydride attack on ligands, 297
Hydrido alkene complexes, structures, 114t

# INDEX

Hydrido-carbonyl complexes, structures, 119t
Hydroformylation
   catalysis with chelating diphosphine ligands, 279–285
   effect of diphosphine, 227–229f
   effect of triphosphine, 230f
   Rh-catalyzed, 81–99
     ligand steric and electronic effects, 95
     mechanism, 86f
     olefin structure effects, 94t
     product distribution, 97t
     rates for various olefins, 88t
     spectroscopic studies, 82
Hydroformylation catalysts, Co, Rh, and Ir comparison, 99–102
Hydrogenation
   in aqueous solution, 288
   catalysis with chelating triphosphine ligands, 275–279
   catalytic activity of A-frame complexes, 219t
   with cationic Rh and Ir complexes, 304–310
   of olefins, influence of solid support, 454t
   using $Pd_2(dppm)_3$, 207
   of phenylacetylene, 221f
   with phosphine bound catalysts, 467–469t
   rates as a function of diphosphine chelate ring size, 225
Hydrogen atom abstraction from toluene, 252
Hydrosilation of ketones, 329

Imines, asymmetric reduction of, 336
Iminophosphines, 250
Infrared at high temperature and pressure, 81, 82, 100
Insertion into metal-metal bonds, 189
Ion cyclotron resonance, 312
Ionic catalysts, 303
$IrCl(CO)(PPh_3)_2$, 344, 345
$[Ir(diphosphine)_2]^+$, 369–372
$IrL_2(H)_2^+$, 311
$Ir_2(CO)_2(\mu\text{-dppm})_2(\mu\text{-S})$, 199
Isocyanide bridged dimers, 198
Isotopic labeling
   cross-over experiments, 23
   decarbonylation, 353, 363
   effects, 27–30
   IR, 22
   kinetic isotope effects, 27
   NMR, 22

Keto acids and esters, hydrogenation, 331, 332t
Ketones,
   asymmetric hydrogenation, 328t
   hydrogenation, 319–324, 320f
Kinetic isotope effect, decarbonylation, 350, 355

L-DOPA, catalytic synthesis, 138
Ligand properties important in catalysis, 6
Long chain diphosphines, 258

Matrix isolation in supported catalysis, 457–459
Mechanism of catalyst system, general, 15
Metallacycle complexes, structures, 130–131t
Metal–metal bonds, insertion, 189
Methylaminobis(difluorophosphine) bridged complexes, structural parameters, 181t
Microwave spectroscopy, 50
Mixed donor ligands, PS, PAS, PN, 267–269
Mixed metal PN bridge dimers, 250f
M–M separations in binuclear complexes, 192–193
$Mn_2(\mu\text{-dppm})_2(CO)_6$, 202
$MoBr_2(NNH_2)(triphos)$, 425
$Mo(N_2)_2(dppe)_2$, 409
$Mo_2[\mu\text{-CH}_3N(PF_2)_2]_4Cl_2$, 171f

$Ni(CO)_2(PPh_3)_2$, 5
$Ni[P(OR)_3]_4$, 49
Nitrile group activation, 298
Nitrite complexes
   oxygen atom transfer, 396
   oxygen transfer to alkenes, 398
   reaction with CO, 396–397
Nitrogenase, 406
NMR
   determination of stereochemistry, 34
   ($^1$H and $^{31}$P) of $\mu$-dppm binuclear complexes, 183
   saturation transfer studies, 155
   shift reagents, 40
NOPAPHOS, 140

$o$-diphenylphosphinoanisole, 241
$o$-diphenylphosphinobenxoylpinacolone, Pt and Cu complexes, 248–249f

Olefin carbonyl complexes, structures, 120–121t
o-diphenphosphinobenzoic acid, 247
Olefin hydrogenation
  kinetics and mechanism with RhCl(PPh$_3$)$_3$, 53–61
  rate and equilibrium constants, 58t
  scheme, 57f
  using RhClL$_3$
    ligand dependence on rate, 62t
    olefin structure dependence, 63t
Olefin isomerization, 307
Olefin metathesis, 476
Optical activity, aldehydes—mechanistic probe, 353, 361
Optical purity by optical rotation, 39
OsBr$_2$(PPh$_3$)$_3$, 344
Os$_2$(CO)$_8$R$_2$, 23
Oxidation reactions
  metal catalyzed, 377–378
    of aldehydes and ketones, 386–387
    of alkenes, 383–385
    of carbon monoxide, 207, 385–386, 394–395
    of cumene, 387
    of isocyanides, 385
    of nitric oxide, 394–395
    of phosphines, 387–389
Oxidative addition in diphosphine bridged complexes, 197–200
Oxygen atom transfer reactions, 393–394
Oxygen bridged binuclear complexes, 192
Oxygen complexes, 378–380

Paramagnetic species, 19
Pd(PEt$_3$)$_3$, 39
PdO$_2$(PPh$_3$)$_2$, 395
Pd$_2$Cl($\mu$-dppm)$_2$(SnCl$_3$), 177f
Pd$_2$Cl$_2$($\mu$-dppm)$_2$($\mu$-SO$_2$), 178f
Peracyls, 396
P–H addition to vinyl compounds, 266–267
PHELLANPHOS, 140
PHEPHOS, 140
Phoran, Ph(H)P(OCH$_2$CH$_2$)$_2$N, 245
Phosphide anions, R$_2$P$^-$, 171
Phosphinocyclopentadiene ligands, 254f
Phosphoranes as ligands, 245
Phosphorus macrocycles, 270
Phosphorus–nitrogen donor ligands, 371
$^{31}$P NMR coupling constants in ($\mu$-dppm)$_2$M$_2$ complexes, 186–187t

$^{31}$P NMR for polymer-bound catalysts, 451–452
$^{31}$P NMR of RhCl(PPh$_3$)$_3$ and H$_2$ system, 59f
Polydentate ligands
  phosphines, 259f
  synthesis, 261–275
Polyetherphosphonite ligands, 243
Polyethylene, 8
Polymer bound
  optically-active diphosphine ligands, 263
  phosphine catalysts, 444
  rhodium catalysts in ketone reduction, 321
Polynuclear complexes of N$_2$, 428
Polypropylene, 8
Polystyrene as a solid support, 444–448
Prochiral $\alpha\beta$-unsaturated carboxylic acids, 147
PROPHOS, 139
Protein-bound rhodium hydrogenation catalyst, 245f
Protonation of NiL$_4$ complexes, equilibrium constants, 53t
$^{31}$P spin-lattice relaxation times of polymer-bound catalysts, 451
Pt(C$_2$H$_4$)$_2$L$_2$, 17
  oxidative addition of MeI to, 17
PtO$_2$(PCy$_3$)$_2$, 382
Pt$_2$(CH$_3$)$_3$($\mu$-dppm)$_2$$^+$, 174f
Pt$_2$(CH$_3$)$_4$($\mu$-dppm)$_2$, 174f
Pt$_2$Cl$_2$($\mu$-dppm)$_2$, reaction chemistry, 190f
Pt$_3$($\mu$-CO)$_3$(PR$_3$)$_3$, 24
$^{195}$Pt NMR, 25

Radical
  CIDNP, 29, 46
  esr spectra of, 44
  pairs and stereochemistry, 48
  pathway in hydroformylation reactions, 101
  reaction pathways, 29, 43
  reactions, 21
  spin traps, 45
Rate laws, 16–17, 18, 20
Reduction of C=O and C=N functions, 310, 317–318
RhCl(diop), 337
RhCl(PPh$_3$)$_3$, 8, 112, 345
RhCl(CO)(PPh$_3$)$_2$, 81
RhCl$_2$(CO)(R)(PPh$_3$)$_2$, 351
RhCl$_2$(COR)(PPh$_3$)$_2$, 349
Rh(Cl)$_2$(dppp)(PhCO), 372
Rh(CO)(dppp)$_2$$^+$, 363

INDEX

Rh(CO)(PPh$_3$)$_2$$^+$, 359
Rh(diene)(P$_2$)$^+$, 322
Rh(diphosphine)$_2$$^+$, 359–372
Rh(dppe)$_2$$^+$, 225
[Rh(dppp)]BF$_4$, 360
RhHCl(CO)(PPh$_3$)$_3$, 343
RhH(CO)(PPh$_3$)$_3$, 87, 228, 280, 475
RhH[(−)-diop)]$_2$, 225
Rh(H)$_2$(dppp)$_2$$^+$, 225
RhH(PPh$_3$)$_3$, 112
Rh$_2$($\mu$-Cl)($\mu$-CO)(CO)$_2$($\mu$-dppm)$_2$$^+$, 178f
Rh$_2$Cl(CO)$_2$(N$_3$)($\mu$-dppm)$_2$, 224
Rh$_2$Cl$_2$(CO)$_2$($\mu$-dppm)$_2$, 176f
Rh$_2$Cl$_2$(CO)$_2$[$\mu$-Ph$_2$P(CH$_2$)$_2$O(CH$_2$)$_2$PPh$_2$], 175f
Rh$_2$(CNR)$_4$($\mu$-dppm)$_2$$^{2+}$, 195, 197
Rh$_2$(CO)$_2$Cp$_2$($\mu$-dppb), 182f
Rh$_2$(CO)$_4$(PPh$_3$)$_4$, 386
Rh$_2$(CO)$_2$X$_2$($\mu$-dppm)$_2$, 218
RhPd(Cl)$_3$(CO)($\mu$-Ph$_2$P$_{py}$)$_2$, 204–206f
$^{103}$Rh NMR, 23
Rhodium alkyl hydride complex in asymmetric hydrogenation, 158
Rhodium-phosphine catalyzed hydroformylation, mechanisms, 283f, 284f
[Rh(P–N)$_2$]$^+$, 371
1,2-R,R-bis(anisylphenylphosphino)-ethane ligand, synthesis, 240
[Ru(bipy)$_2$Cl]$_2$($\mu$-dppm)$^{2+}$, 182
RuCl$_2$(PPh$_3$)$_3$, 9, 112
[Ru$_2$Cl$_3$(PEt$_2$Ph)$_6$]Cl, 344
Ru$_2$(CO)$_4$Cp$_2$, 169
RuH($\eta^2$-BH$_4$)(ttp), 277
RuHCl(PPh$_3$)$_3$, 9
Ru(PPh$_3$)$_2$(tetraphenylporphyrin), 346
Ru(RCN)$_3$L$_3$$^{2+}$, 298

Saturation labeling in NMR, 25
Saturation transfer, 25, 26
Selectivity, 9
  in aldehyde decarbonylation, 361t, 367–368
  in polymer-bound catalysts, 455–456
Shift reagents, 40
Side-by-side binuclear complex, 175
Sixteen- and eighteen-electron rule, 19
SKEWPHOS, 141

Solid supports
  polymethycrylate allyl chloride, 446
  polyphenylene-isophthalamide, 447
  polysiloxanes, 446
  polystyrene, 444–448
  polyvinyl chloride, 446
Solvent effects, 299, 303, 308, 311
Solvent stabilized species, 19
Spin-state labeling, 23
Steric and electronic effects, 16, 77–78, 95
Steroid reductions, 309
Sulfur bridged binuclear complexes, 193
Supported catalysts
  hydroformylation, 472–475, 477t
  hydrogenation, 461
  hydrosilylation, 473t
  iridium complexes, 465
  olefin isomerization, 460–461
  palladium, platinum complexes, 465–466
  rhodium complexes, 462–465
Supported phosphine catalysts, 285–289
Surface bound catalysts, 10

Template synthesis of polyphosphine ligands, 271–274
Transition state, composition, 18
Trialkylphosphines, preparations, 2
Trimethylphosphine, 1
Tri-$n$-butylphosphine, 7
Triphosphine attached to glass beads, 287
Triphosphine ligands, 260
Tripod ligands, 261, 290, 291
  HC(PPh$_2$)$_3$, 208
Tris(diphenylphosphino)methane, 262
Tritium labeling, 147
Turnover number, 15

Wacker oxidation of ethylene to acetaldehyde, 21
Water gas shift reaction, 313–314
Water soluble phosphine complexes, 288

Z-dehydroamino acids, asymmetric hydrogenation of, 140f
Ziegler-Natta polymerization, 7